# PERIODIC TABLE OF THE CHEMICAL ELEMENTS

| 1 | 2 | 3 | 4 | 5 | 6 | 7 | 8 | 9 | 10 | 11 | 12 | 13 | 14 | 15 | 16 | 17 | 18 |
|---|---|---|---|---|---|---|---|---|---|---|---|---|---|---|---|---|---|
| $H^1$ $1s$ | | | | | | | | | | | | | | | | | $He^2$ $1s^2$ |
| $Li^3$ $2s$ | $Be^4$ $2s^2$ | | | | | | | | | | | $B^5$ $2s^2 2p$ | $C^6$ $2s^2 2p^2$ | $N^7$ $2s^2 2p^3$ | $O^8$ $2s^2 2p^4$ | $F^9$ $2s^2 2p^5$ | $Ne^{10}$ $2s^2 2p^6$ |
| $Na^{11}$ $3s$ | $Mg^{12}$ $3s^2$ | | | | | | | | | | | $Al^{13}$ $3s^2 3p$ | $Si^{14}$ $3s^2 3p^2$ | $P^{15}$ $3s^2 3p^3$ | $S^{16}$ $3s^2 3p^4$ | $Cl^{17}$ $3s^2 3p^5$ | $Ar^{18}$ $3s^2 3p^6$ |
| $K^{19}$ $4s$ | $Ca^{20}$ $4s^2$ | $Sc^{21}$ $3d$ $4s^2$ | $Ti^{22}$ $3d^2$ $4s^2$ | $V^{23}$ $3d^3$ $4s^2$ | $Cr^{24}$ $3d^5$ $4s$ | $Mn^{25}$ $3d^5$ $4s^2$ | $Fe^{26}$ $3d^6$ $4s^2$ | $Co^{27}$ $3d^7$ $4s^2$ | $Ni^{28}$ $3d^8$ $4s^2$ | $Cu^{29}$ $3d^{10}$ $4s$ | $Zn^{30}$ $3d^{10}$ $4s^2$ | $Ga^{31}$ $4s^2 4p$ | $Ge^{32}$ $4s^2 4p^2$ | $As^{33}$ $4s^2 4p^3$ | $Se^{34}$ $4s^2 4p^4$ | $Br^{35}$ $4s^2 4p^5$ | $Kr^{36}$ $4s^2 4p^6$ |
| $Rb^{37}$ $5s$ | $Sr^{38}$ $5s^2$ | $Y^{39}$ $4d$ $5s^2$ | $Zr^{40}$ $4d^2$ $5s^2$ | $Nb^{41}$ $4d^4$ $5s$ | $Mo^{42}$ $4d^5$ $5s$ | $Tc^{43}$ $4d^6$ $5s$ | $Ru^{44}$ $4d^7$ $5s$ | $Rh^{45}$ $4d^8$ $5s$ | $Pd^{46}$ $4d^{10}$ $-$ | $Ag^{47}$ $4d^{10}$ $5s$ | $Cd^{48}$ $4d^{10}$ $5s^2$ | $In^{49}$ $5s^2 5p$ | $Sn^{50}$ $5s^2 5p^2$ | $Sb^{51}$ $5s^2 5p^3$ | $Te^{52}$ $5s^2 5p^4$ | $I^{53}$ $5s^2 5p^5$ | $Xe^{54}$ $5s^2 5p^6$ |
| $Cs^{55}$ $6s$ | $Ba^{56}$ $6s^2$ | $La^{57}$ $5d$ $6s^2$ | $Hf^{72}$ $4f^{14}$ $5d^2$ $6s^2$ | $Ta^{73}$ $5d^3$ $6s^2$ | $W^{74}$ $5d^4$ $6s^2$ | $Re^{75}$ $5d^5$ $6s^2$ | $Os^{76}$ $5d^6$ $6s^2$ | $Ir^{77}$ $5d^9$ $-$ | $Pt^{78}$ $5d^9$ $6s$ | $Au^{79}$ $5d^{10}$ $6s$ | $Hg^{80}$ $5d^{10}$ $6s^2$ | $Tl^{81}$ $6s^2 6p$ | $Pb^{82}$ $6s^2 6p^2$ | $Bi^{83}$ $6s^2 6p^3$ | $Po^{84}$ $6s^2 6p^4$ | $At^{85}$ $6s^2 6p^5$ | $Rn^{86}$ $6s^2 6p^6$ |
| $Fr^{87}$ $7s$ | $Ra^{88}$ $7s^2$ | $Ac^{89}$ $6d$ $7s^2$ | | | | | | | | | | | | | | | |

| $Ce^{58}$ $4f^2$ $6s^2$ | $Pr^{59}$ $4f^3$ $6s^2$ | $Nd^{60}$ $4f^4$ $6s^2$ | $Pm^{61}$ $4f^5$ $6s^2$ | $Sm^{62}$ $4f^6$ $6s^2$ | $Eu^{63}$ $4f^7$ $6s^2$ | $Gd^{64}$ $4f^7$ $5d$ $6s^2$ | $Tb^{65}$ $4f^8$ $5d$ $6s^2$ | $Dy^{66}$ $4f^{10}$ $6s^2$ | $Ho^{67}$ $4f^{11}$ $6s^2$ | $Er^{68}$ $4f^{12}$ $6s^2$ | $Tm^{69}$ $4f^{13}$ $6s^2$ | $Yb^{70}$ $4f^{14}$ $6s^2$ | $Lu^{71}$ $4f^{14}$ $5d$ $6s^2$ |
|---|---|---|---|---|---|---|---|---|---|---|---|---|---|
| $Th^{90}$ $-$ $6d^2$ $7s^2$ | $Pa^{91}$ $5f^2$ $6d$ $7s^2$ | $U^{92}$ $5f^3$ $6d$ $7s^2$ | $Np^{93}$ $5f^5$ $7s^2$ | $Pu^{94}$ $5f^6$ $7s^2$ | $Am^{95}$ $5f^7$ $7s^2$ | $Cm^{96}$ $5f^7$ $6d$ $7s^2$ | $Bk^{97}$ | $Cf^{98}$ | $Es^{99}$ | $Fm^{100}$ | $Md^{101}$ | $No^{102}$ | $Lr^{103}$ |

# FUNDAMENTALS OF
# SOLID STATE PHYSICS

## J. RICHARD CHRISTMAN
U.S. Coast Guard Academy

**JOHN WILEY & SONS**
New York   Chichester   Brisbane   Toronto   Singapore

Copyright © 1988, by John Wiley & Sons, Inc.

All rights reserved. Published simultaneously in Canada.

Reproduction or translation of any part of
this work beyond that permitted by Sections
107 and 108 of the 1976 United States Copyright
Act without the permission of the copyright
owner is unlawful. Requests for permission
or further information should be addressed to
the Permissions Department, John Wiley & Sons.

*Library of Congress Cataloging-in-Publication Data:*
Christman, J. Richard.
    Fundamentals of solid state physics.

    1. Solid state physics.   I. Title.
QC176.C47   1988       530.4'1      87-21651
ISBN 0-471-81095-9
Printed in the United States of America
10  9  8  7  6  5  4  3  2  1

# PREFACE

This is a text for a first course in solid state physics at the advanced undergraduate level. Emphasis is placed on understanding fundamental ideas in the hope that students will gain a firm foundation for further study. Electron and phonon states are studied early, then used to discuss basic thermodynamic, electric, magnetic, and optical phenomena.

An important goal of the book is to provide strong links between solid state phenomena and the basic laws of quantum mechanics, electromagnetism, and thermodynamics. An attempt is made to describe each phenomenon clearly and to show in a straightforward way how it follows from the basic laws of physics. One dividend that accrues from a study of the solid state is a deeper understanding of fundamental laws, and this aspect is not neglected. For many students the course is the first in which they must bring to bear knowledge from a diversity of lower level courses. The text is written with an eye to helping in this difficult task.

A second goal is to show the experimental roots of the subject. Descriptions of many important experiments have been included. They have been selected to illuminate important ideas and to show how fundamental quantities can be measured.

To cover the fundamentals well means that some technological applications normally included in a first solids course are mentioned only briefly or not at all. Some are relegated to problems at the ends of chapters. Extensive reading lists appended at the ends of chapters often include references to works that discuss applications and are highly recommended to students who are interested in technological aspects of the subject.

Chapter 14, on the physics of semiconductor devices, has been included as an aid to the many students who will enter the semiconductor industry directly from their undergraduate schooling. Although the emphasis is on fundamental concepts, particularly recombination and diffusion, several important devices are discussed. The instructor has the option of covering the material at any point after Chapter 9. If it is studied before Chapter 10, however, a short explanation of optical absorption may be necessary.

Numerical calculations are emphasized for the purpose of determining the sizes of various important quantities. Students should be aware of the orders of magnitude of phonon frequencies, electron velocities, binding energies, energy gaps, Fermi energies, absorption frequencies, Debye temperatures, and a host of other parameters that characterize solids. With this in mind, many worked out examples are included throughout and roughly one-half the problems involve calculations of this type.

Other problems are more involved and can be used to test for an understanding of the material in the text. The large number of problems and wide range of difficulty should provide the instructor with ample options. Assignments can be tailored to the backgrounds of the students. Some problems extend the material of the text and can be used as the basis for short lectures or to inspire extra reading.

Except for angstroms and electron volts, SI units are used exclusively. Having completed introductory and intermediate courses in which these units are emphasized, most students are comfortable with them. On the other hand, the use of units that are new to them might tend to draw attention away from fundamental ideas. Since much work is reported in the literature in cgs-gaussian units, a conversion table is provided on the inside front cover and Maxwell's equations are given in Appendix A.

Students should bring to the course some knowledge of quantum mechanics, electromagnetic theory, and thermodynamics at the intermediate undergraduate level. They should be familiar with solutions to the Schrödinger equation for one-dimensional square well and harmonic oscillator potentials. They should also be acquainted with some aspects of three-dimensional quantum mechanics, particularly kinetic energy and momentum operators. Angular momentum operators and allowed values of the angular momentum are used in the chapters on magnetism. Some knowledge of atomic orbitals, at the level of the usual course in modern physics, helps in understanding parts of the chapters on binding and electron states.

Maxwell's equations in both integral and differential form are used freely. The ideas of electric polarization and magnetization are discussed but some previous experience is helpful. The chapter on superconductivity makes use of the vector potential. A few boundary value problems are included at the ends of some chapters, but the emphasis here is on the physics that can be learned from the solutions, not on techniques for finding solutions.

The ideas of energy, temperature, and entropy are used freely without defining them. The canonical ensemble is also used with only a short statement to describe it. Fermion and boson distribution functions, however, are thoroughly discussed when they are introduced.

Mathematical preparation should include some rudimentary knowledge of the Laplacian, gradient, divergence, and curl operators and their use in quantum mechanics and electromagnetic theory. Students should be familiar with Gauss's and Stokes' theorems. Complex notation is used for wave functions, and students should be able to find the complex conjugate and magnitude of a complex number. Fourier series are also used, and students should be familiar with the idea that a nicely behaved but otherwise arbitrary function can be expanded as such a series. A review is contained in an appendix.

Many people helped with the book. I am especially grateful to Frank J. Blatt (Michigan State University), Louis Buchholtz (California State University at Chico), Cheuk-Kin Chau (California State University at Chico), D. R. Chopra (East Texas State University), Robert J. DeWitt (Southern Arkansas University), Eric Dietz (California State University at Chico), L. Edward Millet (California State University at Chico), Robert L. Paulson (California State University at Chico), and Lawrence Slifkin (University of North Carolina). They carefully read the manuscript and made many valuable suggestions. Their efforts have greatly improved the book, to the benefit of the students.

While writing of the book I spent an exhilarating year at Rensselaer Polytechnic Institute and gained much from association with members of the Physics Department there. Special thanks go to Robert Resnick, who supported my writing efforts and who buoyed my spirits when the road seemed long. Gerhard Salinger used a draft in his course and passed along comments from students. Thomas Furtak (now at Colorado School of Mines) provided information about surface phonon states, Gwo-Chiang Wang introduced me to LEED experiments, and John Schroeder helped with amorphous materials. I thank all of them.

The continuing support of Saul Krasner, Hugh Costello, Edward Wilds, Joseph Pancotti, and Robert Fuller, all members of the Physics Section at the U.S. Coast Guard Academy, is greatly appreciated. Mary McKenzie, of the Academy library, was extremely helpful in tracking down references. I also appreciate the support of Captain David A. Sandell, Dean of Academics at the Academy.

Robert A. McConnin, formerly Physics Editor at Wiley, and Blanca Ferreris, his assistant, eased the task of writing and publishing in innumerable ways. Thanks also go to Deborah Herbert, who supervised the editing, to Virginia Dunn, who edited the manuscript, to Ann Renzi, who designed the book, and to Pam Pelton, who supervised production. Illustrations were drawn by Sigmund Malinowski, Alfred Corring, Richard Schaffer, and John Bukofsky. It is a great pleasure to be associated with this group of hard working, highly professional people.

Thanks are also due the technical support personnel at Image Processing Systems of Madison, Wisconsin, who helped set up a fine word processing system for technical writing, and at The Computer Establishment of Old Saybrook, Connecticut, who helped with the hardware.

I express my gratitude to my wife, Mary Ellen, who helped with the manuscript and who exhibited enormous patience and good spirit during the writing of the book. Her support is greatly appreciated.

This book is dedicated to Hillard B. Huntington, who introduced me to the marvels of the solid state. He continues to pursue the study of solids with a zeal that is contagious. As a teacher, he has the enviable ability to communicate both the subject matter and his love for it.

*J. Richard Christman*
*New London, CT 06320*

# CONTENTS

# Chapter 1

## A SURVEY OF SOLID STATE PHYSICS

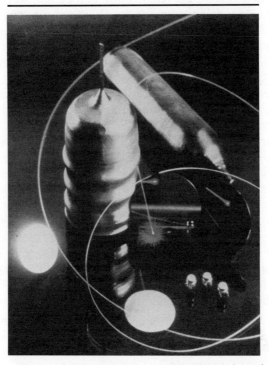

Boules of gallium arsenide. A crystal is grown by slowly pulling it from the melt by the long needle at the upper end.

- 1.1 SOLID STATE THEORY
- 1.2 ATOMIC BACKGROUND
- 1.3 PROPERTIES OF SOLIDS
- 1.4 SOLID STATE EXPERIMENTS

Solid state physicists are concerned primarily with understanding the properties of materials in terms of their constituent particles: electrons and nuclei. Why are some materials good conductors of electricity while others are not? Why are some materials good thermal insulators? Why are some materials opaque and others transparent? At a more sophisticated level, we might ask why a particular solid absorbs light within a certain narrow band of frequencies but transmits light with neighboring frequencies, or why the magnetization of a particular solid depends on the temperature in a certain way. To answer these and similar questions, detailed knowledge about the motions of electrons and nuclei in solids and about the interactions of these particles with externally applied fields is required. Most properties of solids are produced by the simultaneous interaction of the particles with each other and with the applied field.

Once underlying physical mechanisms are understood, they can be exploited to produce materials and devices that are tailor-made for particular applications. The technological consequences are far-reaching: Witness the materials and devices that have been developed for electronic applications.

But the electronics industry is certainly not the only one to benefit from burgeoning knowledge of the solid state. For example, many modern advances in both linear and nonlinear optics can be attributed to the development of new optical materials and solid state lasers. The study of superconductivity has led to the development of high-field magnets and powerful dc motors which do not suffer resistance losses. It has also led to the design of sensitive magnetometers for the measurement of magnetic fields. The communications industry has benefited from new fibers for the transmission of light and from the development of microwave amplifiers. High-energy and nuclear physics now make use of semiconductor-based particle detectors.

Solid state physics has also aided in more mundane but equally important ways. Materials with special thermal, mechanical, or dielectric properties owe their existence to our fundamental understanding of solids. In addition, techniques developed for the study of solids are now being applied successfully to the study of biological materials.

## 1.1 SOLID STATE THEORY

The laws of physics used to understand properties of solids are the same as those used to understand atoms and molecules: Maxwell's equations for electromagnetic fields and the Schrödinger equation for particle wave functions. To these must be added the laws of thermodynamics and statistical mechanics.

Electrons and nuclei are charged and their interactions with each other are predominantly electrostatic. At a fundamental level, a solid is a collection of nuclei and electrons, interacting with each other via Coulomb's law. Greater accuracy can be obtained in some calculations by including magnetic interactions and using relativistically correct expressions, but these produce relatively minor corrections in most instances.

Particle wave functions are found using the Schrödinger equation and these contain information about dynamical quantities, such as energy, associated with the particles. In the next step, dynamical quantities are used to calculate the contribution of an electron in any state to various material properties. We will follow this sequence many times in this book.

Not all states are occupied. We will be concerned chiefly with solids that are in thermodynamic equilibrium or only slightly removed from equilibrium by external forces. Although particles continually make transitions between states, the thermodynamic probability that any given state is occupied can be calculated using the ideas of statistical mechanics. This probability gives the average number of particles occupying a given state when the solid is in thermodynamic equilibrium at a given temperature. To evaluate a physical property, the contribution of each state is weighted by the average occupation number and the results for all states are summed.

An external force changes particle wave functions and values of various dynamical quantities. To investigate the influence of an applied force we carry out the program outlined above but with terms describing the force added to the Schrödinger equation.

*The Structure of Solids.*    Atoms of a solid are not stationary; rather, each atom vibrates with small amplitude about a fixed equilibrium position. The immobility of atomic equilibrium positions gives a solid a fixed structure and distinguishes it from a liquid or gas. In liquids and gases, atoms move over large distances and the structure is not fixed.

The distribution of atomic equilibrium positions defines the structure of a solid. There are three major classes: crystalline, amorphous, and polycrystalline. In crystals, atomic equilibrium positions form a geometric pattern that is exactly repeated throughout the solid without change in composition, dimension, or orientation. Atomic equilibrium positions in an amorphous solid do not form such a repeating pattern. A polycrystalline solid is made up of a large number of small crystals, called crystallites. The atoms form a pattern just as in a crystal but the orientation of the pattern changes abruptly at crystallite boundaries. Any given material may be either crystalline, polycrystalline, or amorphous, depending on how it is prepared.

Because equilibrium positions in crystals form a repeating pattern, these solids have been more extensively studied than amorphous or polycrystalline solids. The structure of a crystal tremendously reduces the labor involved in many calculations. The Schrödinger equation, for example, need be solved only for points within a single pattern rather than for points throughout the whole crystal. For many experiments, data for crystals are much easier to interpret than data for amorphous solids. In addition, the influence of periodicity on crystalline properties is interesting for its own sake. Many of the concepts developed in the study of crystals, however, are valid for amorphous materials and properties of these solids are also discussed in this book.

*Electrons.*    Electrons obey the Pauli exclusion principle: no more than one electron is in a given state of a system. Two electrons may have the same spatial wave function but only if one has its spin up and the other has its spin down.

This has important consequences. An electron system composed of $N$ particles has the lowest possible total energy when the $N$ states lowest in energy are each occupied by a single electron and all other states are unoccupied. Occupation of the lowest states forces other electrons to occupy higher states.

Because they obey the Pauli exclusion principle, all electrons do not play equally important roles in the determination of material properties. Electrons that are tightly bound to atoms, and therefore have low energy, are influenced only slightly by neighboring atoms or by applied forces. They contribute little to most properties of solids because there are no empty states nearby in energy to which they can be excited. These electrons are called core electrons and the combination of a nucleus with its complement of core electrons is called an ion core.

Electrons outside the core contribute most to material properties. These states are closely related to atomic $s$ and $p$ states in partially filled shells: the $4s$ states of potassium and calcium and the $4s$ and $4p$ states of gallium and germanium, for example. In the shell just below, $s$ and $p$ subshells are completely filled and these states, along with those of lower energy, are core states.

Electrons outside the core produce forces of attraction that hold atoms together and, in conjunction with ion cores, produce restoring forces that are responsible for the vibratory motion of the atoms. These electrons may flow from place to place in an electric field or a temperature gradient and are responsible for the electrical conductivity and most of the thermal conductivity of metals. They absorb light and so contribute to optical properties. In fact, outer electrons participate directly or indirectly in all solid state phenomena.

Core electrons also have an important role to play. They are responsible for short-range forces of repulsion between atoms and so prevent the collapse of solids. In conjunction with forces of attraction due to outer electrons, forces of repulsion determine atomic equilibrium separations. The potential energy associated with the interaction of two atoms might look like the function shown in Fig. 1-1. The equilibrium separation $R_0$ is at the minimum of the curve: for separations greater than $R_0$ the atoms attract each other, but for smaller separations they repel each other. Core electrons resist the compression of a solid.

The description given above of core and noncore states leaves out the partially filled $d$ and $f$ subshells of atoms in transition series. The energies of these states are roughly equivalent to those of $s$ and $p$ states in the next higher shell, but the wave functions do not extend as far from the nucleus as $s$ and $p$ functions. Electrons in these states contribute to some properties but not to others and must be dealt with on a case-by-case basis.

*Atomic Vibrations.* Potential energy functions with minima lead to oscillations of the atoms about their equilibrium positions. If the amplitude of the motion is small, the restoring force is proportional to the displacement and the motion of any atom can be described as a linear combination of simple harmonic motions.

When one atom moves, it forces its neighbors from their equilibrium positions and they, in turn, exert forces on their neighbors. As a consequence, atomic

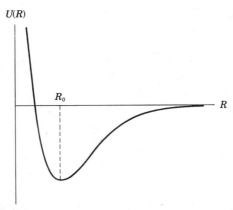

**FIGURE 1-1** A typical potential energy function for the interaction of two atoms, separated by $R$. For $R > R_0$ the atoms attract each other, a phenomenon associated with electrons from partially filled atomic shells. For $R < R_0$ the atoms repel each other, the force of repulsion becoming strong when core electron wave functions overlap.

oscillations are described as normal modes of vibration, collective oscillations for which all atoms have the same frequency. A given system can oscillate only with certain frequencies and, because the allowed frequencies are determined by the equilibrium positions of the atoms and by the strength of the forces between them, they differ from solid to solid. Its normal mode spectrum is an important characteristic of a solid and a great deal of both theoretical and experimental effort goes into the investigation of these spectra.

Normal mode energies are quantized. The energy associated with the normal mode of angular frequency $\omega$ can be changed, but only in multiples of $\hbar\omega$, where $\hbar$ is the Planck constant divided by $2\pi$ ($\hbar = 1.05 \times 10^{-34}$ J $\cdot$ s). The quantum of energy $\hbar\omega$ is called a phonon, a name similar to that for a quantum of electromagnetic energy, a photon. Phonons contribute to many properties of solids and, along with electrons, may be considered a basic ingredient of a solid.

Except in unusual circumstances, as when electron-positron annihilation or $\beta$ decay occurs, the number of electrons in a solid remains constant. The number of phonons does not: it is temperature dependent and routinely changes in interactions involving ion cores.

*Defects.* Defects are anomalies in the atomic structure, regions where the structure deviates from that which prevails throughout most of the solid. Impurity atoms, for example, are defects. For crystals, any deviations from regularity in the pattern of atomic positions are also defects. Some of these may be vacancies: places where atoms should be according to the crystal pattern but are not. Others may be interstitials: places where atoms should not be but are. In addition, crystalline compounds may be disordered, with some atoms of one type occupying sites that should be occupied by atoms of another type.

All crystals have impurities, interstitials, vacancies, and other defects. Some impurity atoms are present in the melt from which the crystal is grown and others enter during the growing process or diffuse in from the surface. Vacancies and interstitials are thermally generated and their concentrations increase with temperature.

The absence of a pattern is not considered a defect for an amorphous structure, but such structures may have other types of defects. Impurities are one type. For covalently bonded amorphous solids, another type is called a dangling bond. Such a defect occurs if an atom normally forms a certain number of bonds with neighboring atoms but one of the atoms forms fewer.

Defects influence nearly all properties of materials to some extent. The scattering of electrons from defects and from vibrating host atoms causes the electrical conductivities of metals and semiconductors to be finite. Impurities added to semiconductors significantly alter their electrical properties by greatly changing the number of electrons that contribute to the electrical current. Vacancies in some solids act like charged particles and attract or repel electrons. They may change the electromagnetic absorption spectrum of the material and thereby change the color or cause a transparent host solid to become opaque.

## 1.2 ATOMIC BACKGROUND

The periodic table of chemical elements, showing ground state electron configurations, is displayed on the inside front cover. The elements are arranged, left to right, in order of increasing atomic number. Elements in the same column have similar chemical properties and solids formed by them usually have similar physical properties.

*Inert Gas Elements.* Atoms of inert gas elements, in the last column to the right, interact only weakly with other atoms. This behavior is a result of their electron energy level structures and their ground state configurations. Except for helium, the highest occupied shell for any of these atoms has completely filled $s$ and $p$ subshells, with all higher energy states empty. For an inert gas atom in row $n$, the $ns$ and $np$ subshells are the highest filled subshells and the lowest empty states are the $s$ states of the $n + 1$ shell. In each case, there is a wide energy gap between the highest filled state and the lowest empty state and any change in the electron configuration requires considerable energy.

Inert gas atoms bind together to form solids at temperatures below 100 K. Binding is accomplished chiefly through van der Waals interactions. Slight deformations in electron wave functions lead to the formation of an electric dipole moment and the associated electric field induces dipole moments on nearby atoms. Dipoles on neighboring atoms attract each other weakly and the atoms crowd as closely together as core-core repulsion permits. Each atom in a crystal has 12 nearest neighbors.

*Columns IA and IIA.* The ground state of any atom from another column of the table consists of electrons in an inert gas configuration and one or more electrons in higher energy states. The first electrons are in the core and the core configuration for an atom from any given row is that of the inert gas atom of the previous row.

Except for hydrogen, which has a single electron, atoms from column IA of the table each have a single *s* electron outside an inert gas core, while atoms from column IIA each have two *s* electrons outside their cores. In each case, these *s* electrons are weakly bound to the atoms and their wave functions extend well beyond the core. For example, a 3*s* electron of sodium has an energy of about $-5.2$ eV and the outer peak of its probability density is about 1.3 Å from the nucleus. This is to be compared with a 2*p* core electron, which has an energy of about $-36$ eV and an outer peak that is 0.26 Å from the nucleus.

IA and IIA elements solidify at temperatures in the range from 300 to 1600 K. Upon formation of a solid, the outer *s* electrons are easily pulled from their original atoms and become free to move throughout the material. Energies of these electrons are reduced and the atoms are bound together. These materials are metals. Their large free electron concentrations make them excellent electrical and thermal conductors and give them high optical reflectivities. Except for hydrogen, column IA elements are known as alkali metals and column IIA elements are known as alkaline earth metals.

*The Right Side of the Table.*   Atoms from columns III, IV, V, VI, and VII have completely filled *s* subshells and partially filled *p* subshells outside their cores. These elements have a large variety of physical characteristics. Nitrogen, oxygen, fluorine, and chlorine are gases at room temperature but become solid at temperatures under 100 K. Bromine is a liquid at room temperature and the rest are solids. Gallium melts at about room temperature, silicon at about 1670 K, and carbon in the form of diamond at over 3700 K.

Aluminum, tin, and lead are metals. Arsenic, antimony, and bismuth have many of the characteristics of metals but they are not typical metals. Some forms of carbon and tellurium are moderately good conductors of electricity but these elements are not metallic. Silicon and germanium are semiconductors. At low temperatures the pure materials are insulators but they become moderately good conductors at high temperatures or if certain impurities are added. Other elements from this side of the periodic table are insulators.

Except for the metals, atoms from the right side of the table bond covalently to each other to form solids. In a covalent bond electrons occupy the region along the line joining two atoms and the energy is lowered, chiefly because such electrons are attracted by two ion cores rather than just one. Such bonds are extremely strong.

An atom can bond covalently to four other atoms at most. In germanium, silicon, and diamond each atom tends to have only four nearest neighbors, arranged at the vertices of a regular tetrahedron, and the solid structure displays a packing that is much more loose than that of a typical metal or van der Waals solid. Electrons in covalent bonds do not readily move through the solid and these materials are much poorer electrical conductors than metals.

Not all covalently bonded atoms form tetrahedral structures. In graphite (another form of carbon), for example, covalent bonds join each atom to three others, all in a plane, but planes of atoms attract each other via van der Waals forces. Nitrogen atoms form molecules, consisting of two atoms each, by means of covalent bonds, but solid nitrogen is formed by van der Waals attraction of the molecules for each other. Solid oxygen is formed in a similar manner.

*Transition Elements.* The wide central sections of the fourth, fifth, sixth, and seventh rows of the periodic table contain the transition elements. In these atoms, $d$ and $f$ states are filled in order of increasing energy. In the fourth row, for example, $4s$ states are filled in the first two columns, then $3d$ states are filled in the central columns. A similar order is followed in the fifth row. In the sixth row, lanthanum has a single $5d$ electron, but $4f$ states are then filled before the filling of $5d$ states resumes with hafnium. The seventh row displays a similar pattern.

Although $d$ states in the transition series have energies that are comparable to those of the outer $s$ states, $d$ wave functions have peaks that are well inside the outer peaks of the $s$ wave functions and are generally quite close to the core. For scandium, in the $3d$ series, the outer peak of the $4s$ probability density is at $r = 1.6 \,\text{Å}$, while the peak of the $3d$ probability density is at $r = 0.53 \,\text{Å}$.

Since one or more $s$ electrons per atom become free in the solid, transition elements form metals. On the other hand, $d$ states form covalent-like bonds with neighbors and so bring about extremely tight bonding of atoms. Of all the metallic elements, tungsten, in the $6d$ series, forms the most tightly bound solid. As we will see, $d$ electrons in partially filled subshells are important for the magnetic properties of transition metal solids. Iron, the prototype ferromagnetic material, is a $3d$ transition metal.

Although energies of $f$ electrons in atoms of the sixth and seventh rows are comparable to $d$ state energies, these electrons are buried deep in the cores. In the lanthanide series, for example, the outer peaks of the $4f$ functions are positioned at distances ranging from 0.21 to 0.37 Å from the nucleus. The $5p$ core functions, on the other hand peak between 0.63 and 0.85 Å from the nucleus. As a consequence, $4f$ electrons do not interact significantly with neighboring atoms. In some solids, however, one of the $f$ electrons may make a transition to a $d$ state and thereby increase its interaction with neighboring atoms. Electrons in partially filled $f$ states are important for the magnetic properties of sixth- and seventh-row elements.

*Columns IB and IIB.* The three transition series end with nickel, palladium, and platinum. Atoms of the next column, IB, have completely filled $d$ subshells and each has one electron in its outer $s$ subshell. These are the noble metals and they are similar in many respects to the alkali metals. Atoms of the zinc family, in column IIB, have both outer $d$ and $s$ subshells completely filled. They are somewhat similar to the alkaline earth metals. For both sets of atoms, the $d$ states are well below the outer $s$ states in energy. They are important for some, but not all, properties.

## 1.3  PROPERTIES OF SOLIDS*

The number of properties that have been studied is quite large. A short list is given below, and details are discussed at appropriate places in the remainder of this book.

*Mechanical Properties.*   Mass densities of most common solids are between $1 \times 10^3$ and $25 \times 10^3$ kg/m$^3$. In the final analysis, the density of a solid is determined by the masses of the atoms and the bonding forces between them. Bonding forces dictate the atomic equilibrium positions and hence the volume occupied. For most solids the concentration of atoms is within an order of magnitude of $10^{28}$ atoms/m$^3$ and the average distance between atoms is a few angstroms (1 Å $= 10^{-10}$ m).

Bonding forces also dictate the cohesive energy of a solid. This is the energy per atom required to separate the solid into neutral atoms, at rest and far removed from each other. Cohesive energies range from about 0.02 eV/atom to over 10 eV/atom. They are small for van der Waals bonding and large for covalent bonding, with values for metallic bonding falling between these extremes.

Externally applied mechanical forces deform solids. The deformation, or strain as it is called, may take the form of a compression, a shear, or some combination of these. In any case, if the applied forces are sufficiently weak, most solids behave like ideal springs, with the strain proportional to the stress (the applied force per unit area). The constants of proportionality are called the elastic constants and their values are important parameters of a solid. The bulk modulus, which gives the pressure required to produce a given fractional change in the sample volume, is related to these constants. Of the chemical elements that are solid at room temperature, cesium is the easiest to compress. A pressure of $5 \times 10^8$ Pa reduces the volume of a cesium sample by about 16%. Most other elements are compressed 5% or less by the same pressure.

Elastic waves propagate in solids. Wave speeds, called the speeds of sound, are determined by atomic masses and interatomic forces. Shear forces are typically much weaker than compressional forces so the speeds of transverse or nearly transverse waves are much less than the speeds of longitudinal or nearly longitudinal waves. In aluminum, for example, longitudinal waves travel at about 6000 m/s while shear waves travel at about 3000 m/s.

*Thermal Properties.*   The heat capacity is the energy per kelvin that must be supplied to a solid to raise its temperature. Because other externally controllable characteristics of a solid might change during the heating process, there are several different heat capacities. Of these, the heat capacity at constant volume is most simply related to particle motions and is the one we will consider in detail.

---

*Data in the following section were taken from the *American Institute of Physics Handbook*, 3rd ed. (New York: McGraw-Hill, 1972).

When a solid is heated at all but the lowest temperatures, the primary recipients of the energy are the vibrating atoms. As a solid is heated, the energy of vibration increases or, put another way, the number of phonons increases. Classical statistical mechanics predicts that the heat capacity at constant volume should be $3Nk_B$, where $k_B$ is the Boltzmann constant and $N$ is the number of atoms in the sample. This result is found experimentally at high temperatures, but at low temperatures the phonon contribution to the heat capacity is proportional to the cube of the absolute temperature. The low-temperature behavior of the heat capacity is evidence for the quantization of vibrational energy.

Electrons also receive some of the energy absorbed on heating. At moderate and high temperatures this is a negligible fraction of the whole, but it is significant for metals at low temperatures. The electronic contribution to the heat capacity increases in proportion to the temperature, thereby providing evidence that electrons do not share the added energy equally.

The thermal conductivity is a measure of the rate at which energy is transferred from one region of a solid to another when the two regions are held at different temperatures. For most materials the energy flux $Q$ is proportional to the temperature gradient ($dT/dx$ for flow in the $x$ direction).* Specifically, $Q = -\kappa(dT/dx)$, where $\kappa$ is the thermal conductivity.

Both electrons and phonons are instrumental in transferring energy from place to place in a solid. Electrons are the primary carriers in metals and these materials generally have large thermal conductivities. For example, at room temperature the thermal conductivity of aluminum is about 235 W/m·K and that of copper is about 400 W/m·K. Most nonmetals, for which phonons are the primary energy carriers, have small thermal conductivities. That of sodium chloride is about 6 W/m·K, for example. The thermal conductivities of diamond and silicon, however, are over 100 W/m·K.

Nearly all properties of materials depend on the temperature. Most materials expand with increasing temperature. Some become better electrical and thermal conductors; others become worse. Magnetized iron loses its magnetization above a certain critical temperature unless an external field is applied. Much effort, both theoretical and experimental, goes into understanding the temperature dependence of material properties and a large part of this book deals with the results.

*Electrical Properties.* For materials that obey Ohm's law, the relationship between an applied electric field $\mathcal{E}$ and the current density $\mathbf{J}$ it produces is $\mathbf{J} = \sigma\mathcal{E}$, where $\sigma$ is the conductivity.† The electrical conductivity of most materials depends on the concentration of mobile electrons and on the velocities they attain in an applied electric field. Since electron velocities are limited by scattering from vibrating atoms and defects in the atomic structure, the conductivity de-

---

*To find the energy flux we consider an infinitesimal area perpendicular to the direction of energy flow. The flux is the energy that crosses this area per unit time per unit area.

†Current density is flux of charge: the charge per unit area per unit time which crosses an infinitesimal area perpendicular to the flow.

pends on the temperature and on the concentrations of impurities, vacancies, and other defects.

The value of its electrical conductivity indicates the quality of a solid as an electrical conductor. Good conductors, such as pure copper, silver, and gold, have conductivities of roughly $5 \times 10^7 \, (\Omega \cdot m)^{-1}$ at room temperature. By way of contrast, the conductivity of sodium chloride at room temperature is on the order of $10^{-11} \, (\Omega \cdot m)^{-1}$. Note the tremendous range represented by these values. Semiconductors have electrical conductivities that are intermediate between those of conductors and insulators. At low temperatures pure semiconductors are insulators but their conductivities increase rapidly with increasing temperature and above room temperature are usually only a few orders of magnitude less than those of metals. At room temperature the conductivity of pure germanium is about $2 \, (\Omega \cdot m)^{-1}$. Semiconductor conductivities can be increased dramatically, by a factor of several hundred or more, when certain impurity atoms are added. This behavior, which occurs in spite of increased impurity scattering, is extremely important for most electronic uses of semiconducting materials.

Superconductors have zero electrical resistance at low temperatures. When current exists in a superconducting loop, Joule heating does not occur and the current continues unabated, even if no source of electromotive force is present. Above a critical temperature the sample becomes a normal conductor with nonzero electrical resistance. A sufficiently high magnetic field also switches a superconductor to the normal state. Lead, for example, is a superconductor below 7.23 K, but it is normal above this temperature. Near 0 K a field of about $8 \times 10^{-2}$ T turns a lead sample normal. Superconducting compounds have been developed with critical temperatures above 90 K and with critical fields estimated to be over 100 T.

The electrical quantity of primary interest in the study of insulators is the dielectric constant. Bound electrons and ions are displaced in different directions by an applied electric field and charge separation, or polarization, occurs. The center of the electron distribution for each atom moves slightly away from the nucleus, so the atoms possess electric dipole moments. The field tends to align the dipole moments and, since the moments are sources of an electric field, the total field in a polarized solid may be significantly different from the applied field.

The polarization **P** at a point in a solid is the dipole moment per unit volume there and for most materials it is proportional to the total electric field at the point. If $\mathcal{E}$ is the local electric field then $\mathbf{P} = \epsilon_0 (K - 1)\mathcal{E}$, where $K$ is the dielectric constant of the material $\epsilon_0$ is the permittivity of empty space. A solid with a large dielectric constant polarizes easily and thereby alters the electric field significantly. When a material with dielectric constant $K$ is placed between the plates of a capacitor, the capacitance increases by the factor $K$. Gaps between the plates of capacitors used to store energy are often filled with a high dielectric constant material. For a given potential difference across the plates, the energy stored is proportional to the dielectric constant of the gap material.

Most alkali halides have dielectric constants of around 5. Dielectric constants of elements from column IV of the periodic table are larger: 13 for carbon in the form of diamond, 12 for silicon, and 16 for germanium. Some compounds, particularly compounds of lead, have extremely large dielectric constants: about 280 for lead selenide and about 400 for lead telluride. Some solids, such as barium titanate, are polarized even in the absence of an applied field.

*Magnetic Properties.* A magnetic field changes electron orbits and spin directions and, because of this, atoms in a field often have nonvanishing magnetic dipole moments. If they do, the field is altered by their presence. The magnetization **M** of a solid is the dipole moment per unit volume and, for many materials, it is proportional at any point to the local magnetic field. This situation is analogous to the electric polarization of a dielectric but the mathematical description is somewhat different. The magnetic field **H** at any point is defined by $\mathbf{H} = (1/\mu_0)\mathbf{B} - \mathbf{M}$, where **B** is the magnetic induction field, and the magnetic susceptibility $\chi_m$ is given by $\mathbf{M} = \chi_m\mathbf{H}$ for many materials.

A solid with a positive magnetic susceptibility is said to be paramagnetic. If such a solid is present the total field is greater in magnitude than the applied field. A solid with a negative magnetic susceptibility is said to be diamagnetic and, for this type material, the total field is smaller in magnitude than the applied field. Aluminum, for example, is paramagnetic with a room temperature magnetic susceptibility of $+2.1 \times 10^{-5}$. Bismuth is diamagnetic with a room temperature magnetic susceptibility of $-1.64 \times 10^{-5}$.

Below a certain temperature, called the Curie point, magnetization is spontaneous in a ferromagnet. That is, the sample is magnetized even in the absence of an applied field. At temperatures above its Curie point, a ferromagnetic sample is paramagnetic. The Curie point of pure iron, for example, is about 1043 K.

*Optical Properties.* When light shines on the surface of a solid, some is reflected and some enters the material. Furthermore, light that enters the material generally travels in a different direction and with a different phase velocity than the incident light. Some is absorbed. The reflectance, the index of refraction, and the absorption coefficient, respectively, are used to describe these phenomena.

The index of refraction is the ratio of the velocity of light in vacuum to the phase velocity in the material. It can be used, via Snell's law, to find the direction of propagation in the material, given the direction of incidence. Reflectance is the ratio of the reflected intensity to the incident intensity. For the same angle of incidence this ratio is greater for good reflectors than for poor reflectors. A distance $x$ into the material, the intensity $I$ is given by $I = I_0e^{-\alpha x}$, where $\alpha$ is the absorption coefficient and $I_0$ is the intensity just inside the surface, at $x = 0$. Transparent materials have small absorption coefficients; opaque materials have large absorption coefficients.

The index of refraction, the reflectance, and the absorption coefficient are all dependent on the frequency of the incident light. A solid might be transparent at the blue end of the optical spectrum and opaque at the red end, for example. Some solids are used as filters to produce light with a narrow band of wavelengths. The frequency dependence of the index of refraction is responsible for the rainbow of colored light that emerges from a prism on which white light is incident.

Fused silica is transparent for wavelengths from about 100 to about 4500 nm, or from well into the ultraviolet region to well into the infrared. Absorption is strong at both ends of the region of transparency and over the region the index of refraction varies from just above 1.5 to just below 1.4. By way of comparison, germanium is transparent only in the infrared, from 1800 to 23,000 nm and its index of refraction throughout this region is about 4. Metals are highly reflective and have large absorption coefficients. For yellow light, platinum has a reflectance of about 0.7 and an absorption coefficient of about $9 \times 10^7 \, \mathrm{m}^{-1}$.

There are other important optical phenomena. Many materials emit light, or luminesce, under various conditions. Materials that luminesce when struck by an electron beam are used to coat television receiver tubes. Light emission from other solids can be stimulated by an electric field and these are used to fabricate light-emitting diodes, commonly used for displays. The light-producing properties of certain semiconductors are utilized to make lasers.

## 1.4  SOLID STATE EXPERIMENTS

Experimentally determined values for properties of materials have obvious practical use. Data are collected in handbooks and used by design engineers, for example. Of greater interest for our purposes, however, is their use in verification of theoretical calculations. Solid state physics has profited enormously from the interplay between theory and experiment. The desire to explain experimental data has led to detailed calculations and a deeper understanding of interactions between particles in solids. The desire to verify theory has led to the design of experiments that probe at the atomic level to obtain information about atomic structure, electron states, normal mode frequencies, and defects.

X rays, electrons, and neutrons scattered from solids are used for a variety of measurements. Incident x rays and electrons interact primarily with electrons of the material while neutrons interact, via the strong nuclear force, with nuclei. After scattering, electromagnetic, electron, or neutron waves produce a diffraction pattern that depends on the distribution of scattering particles in the sample. In one type of experiment, the scattered intensity as a function of angle is used to infer the atomic structure.

Electrons with useful energies do not penetrate the solid more than about 50 Å, so the use of these particles is confined primarily to the study of surfaces. X rays and neutrons, on the other hand, penetrate much more deeply and are used to study the structure of bulk materials. Neutron scattering is influenced by the spins of the scattering particles, so neutrons are often used to determine the positions of magnetic dipoles in magnetic materials. Neutron diffraction patterns are also influenced by atomic vibrations and one of their most important uses is the determination of normal mode spectra. X rays with sufficient energy to excite deep-lying electrons are used to study the distribution of core electrons.

An extremely important experiment, especially for the study of semiconductors, is based on the Hall effect. An electric field drives an electric current through a sample and a magnetic field is applied perpendicular to the current. The magnetic field causes charge accumulation along the sides of the sample and this charge produces an additional electric field, which is transverse to both the current and magnetic field. The transverse potential difference can be used to calculate the concentration of mobile electrons in some materials. It can also be used in combination with conductivity data to find the average electron velocity in an applied electric field.

Optical absorption and reflection properties yield a wealth of information about both electron energies and normal mode frequencies. When an electron absorbs light it makes a transition to a higher energy level. In a plot of the absorption coefficient as a function of frequency, the peaks reveal differences in the energies of electron states. Light is also absorbed by vibrating atoms, so frequencies for some peaks correspond to normal mode frequencies.

Magnetic resonance techniques are extremely useful for investigating interactions between electrons and ion cores in magnetic materials. Electron energy levels are split if a magnetic field is present and measurements of the splitting provide information about the levels. The extent of splitting can be inferred from the absorption of electromagnetic radiation, tuned so that it causes transitions between the levels in a magnetic field.

Many properties of metals, such as the conductivity and magnetic susceptibility, display periodic behavior when they are plotted as functions of the reciprocal of the magnetic field. The periodic behavior of the magnetic susceptibility, called the de Haas-van Alphen effect, has played an important role in the study of electronic contributions to magnetic properties. The effect is perhaps more important, however, because it can be used to obtain information about electron states in metals. For a given metal at low temperatures, the period of the susceptibility is related to the momenta of the most energetic electrons. In the simplest case, that of free electrons, it can be used to find the value of the maximum electron momentum.

## 1.5 REFERENCES

The books listed below are good general references for solid state physics. Each covers all the major topics. They should be kept in mind and consulted in conjunction with later chapters of this book.

**Introductory Texts**

These can be read with profit for qualitative descriptions of solid state phenomena and experimental techniques as well as for mathematical analyses at an introductory level.

J. S. Blakemore, *Solid State Physics,* 2nd ed. (Philadelphia: Saunders, 1974).
H. E. Hall, *Solid State Physics* (New York: Wiley, 1974).
M. A. Omar, *Elementary Solid State Physics* (Reading, MA: Addison-Wesley, 1975).
M. N. Rudden and J. Wilson, *Elements of Solid State Physics* (New York: Wiley, 1980).

### Intermediate Texts

These contain more quantitative analysis than texts listed above. A good background in modern physics and quantum mechanics, at the undergraduate level or above, is required for reading them.

A. O. E. Animalu, *Intermediate Quantum Theory of Crystalline Solids* (Englewood Cliffs, NJ: Prentice-Hall, 1977).

N. W. Ashcroft and N. D. Mermin, *Solid State Physics* (New York: Holt, Rinehart & Winston, 1976).

C. Kittel, *Introduction to Solid State Physics*, 6th ed. (New York: Wiley, 1986).

J. P. McKelvey, *Solid State and Semiconductor Physics* (New York: Harper & Row, 1966).

R. A. Smith, *Wave Mechanics of Crystalline Solids* (London: Chapman & Hall, 1961).

J. M. Ziman, *Principles of the Theory of Solids*, 2nd ed. (London: Cambridge Univ. Press, 1972).

### Advanced Texts

These should be consulted for more rigorous discussions of solid state phenomena. They are usually used for second courses in solid state physics, at the graduate level.

J. Callaway, *Quantum Theory of the Solid State* (New York: Academic, 1976).

W. Jones and N. H. March, *Theoretical Solid State Physics* (New York: Wiley-Interscience, 1973).

C. Kittel, *Quantum Theory of Solids* (New York: Wiley, 1963).

### Others

Every student of the solid state should become familiar with the series of books *Solid State Physics*, published twice yearly by Academic Press. Early volumes were edited by F. Seitz and D. Turnbull; H. Ehrenreich was added as an editor for later volumes. These books contain monographs dealing with nearly all important solid state topics. They are a good place to look for both introductory and advanced expositions, for experimental data, and for bibliographies.

Many introductory articles on solid state topics, written by leading researchers, appear in *Scientific American, Physics Today,* and *Spectrum.* These magazines should be read regularly.

Two interesting books that describe some of the history of solid state physics are:

S. Millman (ed.), *A History of Engineering and Science in the Bell System, Physical Sciences (1925–1980)* (Murray Hill, NJ: AT&T Bell Laboratories, 1983).

N. Mott (ed.), *The Beginnings of Solid State Physics* (London: The Royal Society, 1980).

# Chapter 2

## CRYSTAL LATTICES

Naturally occurring crystals of quartz. Because a crystal cleaves most easily along planes with high densities of lattice points, only a small number of different angles occur between adjacent facets.

■ 2.1 CRYSTALLINE PERIODICITY

■ 2.2 CRYSTAL SYMMETRY

■ 2.3 BRAVAIS LATTICES

■ 2.4 POSITIONS, DIRECTIONS, AND PLANES IN CRYSTALS

This chapter is about the geometry of crystals. Because equilibrium positions of atoms in a crystal are arranged so that the same pattern is repeated throughout, these positions can be described easily. Mathematical techniques are explained and used to define precisely what is meant by crystalline periodicity. For many crystals, the atomic structure is characterized by a high degree of symmetry and, in fact, structures are often classified by the symmetries they display. The description and use of crystalline symmetry are discussed. Notation in common use for specifying directions and planes in crystals is also given. The concepts and language introduced here are used in nearly every aspect of the study of crystalline solids and we will need them at many places throughout this book.

## 2.1 CRYSTALLINE PERIODICITY

Figure 2-1 illustrates the idea of crystalline periodicity. It shows atomic equilibrium positions on a portion of a plane through a crystal. Two atoms, indicated by ○ and ●, form an atomic grouping, called a basis, which is repeated throughout the crystal. Along the upper line, the basis is duplicated many times, always with the same separation between replicas. This line of basis replicas is then duplicated along other parallel lines, each separated from neighboring lines by the same displacement, with replicas shifted by the same amount relative to replicas of the preceding line. To complete the description of a three-dimensional crystal, one must imagine other planes, identical to the one shown and parallel to it, above and below the plane of the page. Neighboring planes are equidistant from each other and perhaps shifted parallel to each other. Each plane is shifted by the same amount and in the same direction relative to the plane below it. All replicas of the basis have the same orientation.

To describe the structure of a crystal, atomic equilibrium positions are given. Since the crystal consists of exact duplicates of the basis, spaced at regular intervals, the task divides neatly into two parts: a description of the relative positions of the atoms within the basis and a description of the positions of basis replicas in the crystal.

Positions of basis replicas are described in terms of a lattice: a three-dimensional periodic array of points, unbounded in all directions. For the crystal depicted in Fig. 2-1, centers of ○ atoms form a lattice. Centers of ● atoms form an identical lattice, shifted by a pure translation from the first lattice. For pur-

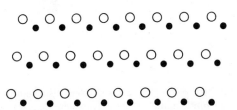

**FIGURE 2-1** Atomic configuration of a two-dimensional crystalline solid. The basis, consisting of two atoms, ○ and ●, is repeated periodically throughout the solid.

poses of describing crystal periodicity, these two lattices and any others related to them by a pure translation may be considered identical. To generate the lattice for a given crystal, we may pick any one of the basis atoms, then replace the crystal by an array of points, with one point at each of the equilibrium positions formerly occupied by the chosen atom and its duplicates. Figure 2-2 shows a plane of lattice points for the crystal of Fig. 2-1.

*Lattice and Basis Vectors.* Two displacement vectors **a** and **b** have been drawn on the diagram, each from lattice point $A$ to a neighboring lattice point. These are called fundamental lattice vectors and they are chosen so the position relative to $A$ of any lattice point in the plane is given by $n_1\mathbf{a} + n_2\mathbf{b}$, where $n_1$ and $n_2$ are integers (positive, negative, or zero). For example, the position vector for point $B$ is $3\mathbf{a}$, that for point $C$ is $2\mathbf{b}$, and that for point $D$ is $3\mathbf{a} + 2\mathbf{b}$. If the points were not arranged in a periodic array, their position vectors could not be written as linear combinations of the same two fundamental vectors with integer coefficients.

Three fundamental vectors **a**, **b**, and **c** are required to describe a lattice in three dimensions. The third vector gives the position of a lattice point neighboring $A$ in the plane above or below the one shown. In terms of these vectors, the position of any lattice point, relative to $A$, is given by $n_1\mathbf{a} + n_2\mathbf{b} + n_3\mathbf{c}$, where $n_1$, $n_2$, and $n_3$ are integers. Any vector of this form is called a lattice vector. The fundamental vectors need not be mutually orthogonal.

The point $A$ chosen for the origin is not special. Any lattice point can be picked, with the same result. A lattice has what is known as translational symmetry: if every point is translated by any one of the lattice vectors $n_1\mathbf{a} + n_2\mathbf{b} + n_3\mathbf{c}$, the lattice looks exactly as it did before the translation. That is, lattice points are at exactly the same places before and after the translation.

Strictly speaking, crystals can have translational symmetry only if they are unbounded. For discussions of bulk properties, it is mathematically convenient to consider idealized crystals, without boundaries, and to assume crystals and

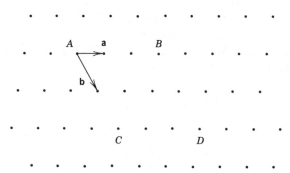

**FIGURE 2-2** A lattice for the crystal of Fig. 2-1. Lattice points give the relative positions of the basis replicas. **a** and **b** are fundamental lattice vectors: the displacement of any lattice point from $A$ can be written $n_1\mathbf{a} + n_2\mathbf{b}$, where $n_1$ and $n_2$ are integers.

their lattices extend infinitely far in all directions. Boundary effects must then be discussed separately.

One method sometimes used to select fundamental lattice vectors starts with finding the shortest displacement vector joining two lattice points and labeling it **a**. If there are several such vectors, any one will do. The second vector **b** is taken to be one of the shortest lattice vectors not parallel to **a** and the third vector **c** is taken to be one of the shortest lattice vectors not in the plane of **a** and **b**. Other choices for the fundamental vectors are often used when the lattice has a high degree of symmetry. Examples will be given later.

One replica of the basis is associated with each lattice point and the positions of its atoms, relative to the lattice point, are given by what are called basis vectors. There is one basis vector for each atom in the basis and they are labeled $\mathbf{p}_1$, $\mathbf{p}_2$, ..., as illustrated in Fig. 2-3. The crystal can be re-created by placing a duplicate of each basis atom near each lattice point, its displacement from the point being specified by its basis vector. Atomic positions in the crystal are then given by vectors of the form $n_1\mathbf{a} + n_2\mathbf{b} + n_3\mathbf{c} + \mathbf{p}_i$, where $i$ labels the basis atom being considered. The first three terms locate a lattice point and the last locates the atom relative to that point.

Even though we generated a lattice by replacing one basis atom and its duplicates by lattice points, a lattice need not be placed·with its points at atomic positions. It can be translated as a whole by any amount relative to the atoms. A translation can be accomplished by adding the same arbitrary vector to all basis vectors. Both the old and new basis vectors, along with the lattice vectors, describe the same crystal structure.

Neither the basis nor the fundamental lattice vectors are unique for a given crystal. The basis, for example, can be enlarged so that it contains twice the original number of atoms. Every finite portion of the crystal then contains half the original lattice points and new fundamental vectors must be chosen. Figure 2-4 shows a possible choice of fundamental vectors for a basis that is twice the size of the one we have been considering.

If the basis is as small as possible, the basis and the associated lattice are said to be primitive. The fundamental vectors shown in Fig. 2-2 are primitive

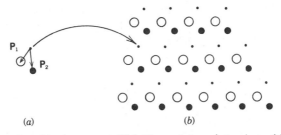

(a)                              (b)

**FIGURE 2-3** (a) Basis and basis vectors. (b) Lattice points and atomic positions. An atom, ○, is associated with each lattice point, displaced from it by $\mathbf{p}_1$. Similarly, an atom, ●, is associated with each lattice point, displaced from it by $\mathbf{p}_2$.

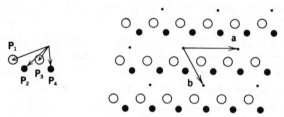

**FIGURE 2-4** Another basis and set of fundamental lattice vectors for the crystal of Fig. 2-1. The basis is twice the sizse of the basis in Fig. 2-1.

while those shown in Fig. 2-4 are not. Even if the lattice is primitive, there are several possible choices for the set of fundamental vectors; see Fig. 2-5 for some examples. In principle, any combination of basis and fundamental lattice vectors can be used as long as together they faithfully describe the atomic positions. In practice, some choices are more convenient than others.

**EXAMPLE 2-1** In a cesium chloride crystal, illustrated in Fig. 3-6a, planes of cesium atoms alternate with planes of chlorine atoms. On every plane, the atoms lie at the corners of squares, positioned like the squares of a checkerboard. Atoms on one plane are directly above the centers of the squares formed by atoms on the plane below. The distance between neighboring atoms on a plane is $a = 4.11$ Å and the separation of adjacent planes is $\frac{1}{2}a$. Find a set of primitive lattice vectors and associated basis vectors for this crystal.

**SOLUTION** There are just as many cesium atoms as chlorine atoms, so we try a basis consisting of one atom of each type. If the chlorine atoms form a lattice that is identical to the one formed by the cesium atoms, our choice of basis is satisfactory. Consider a plane of cesium atoms and place the origin of a Cartesian coordinate system at one of them, with its $x$ and $y$ axes parallel to edges of the squares described above and its $z$ axis perpendicular to the plane of the squares. Clearly, $a\hat{x}$ and $a\hat{y}$ are two fundamental lattice vectors since the position of every cesium atom on the plane is given by a vector of the form $n_1 a\hat{x} + n_2 a\hat{y}$. Planes of cesium atoms are separated by $a$ and, relative to each atom on a plane, there are cesium atoms at $n_3 a\hat{z}$ on other planes. Vectors of the form $n_1 a\hat{x} + n_2 a\hat{y} + n_3 a\hat{z}$ give the positions of

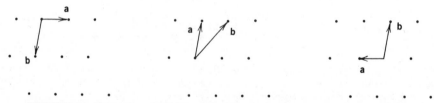

**FIGURE 2-5** Alternative sets of primitive lattice vectors for the crystal of Fig. 2-1.

every cesium atom, so $a\hat{x}$, $a\hat{y}$, and $a\hat{z}$ may be taken as fundamental lattice vectors for the lattice of cesium atoms. If the origin is now placed at a chlorine atom and the position vectors of all chlorine atoms are found, the result is the same: their positions form a lattice with fundamental vectors $a\hat{x}$, $a\hat{y}$, and $a\hat{z}$. These are therefore fundamental lattice vectors for the crystal and the basis consists of one cesium atom and one chlorine atom. Since the displacement vector from a cesium to a chlorine atom is $(a/2)(\hat{x} + \hat{y} + \hat{z})$, possible basis vectors are $\mathbf{p}_{Cs} = 0$ and $\mathbf{p}_{Cl} = (a/2)(\hat{x} + \hat{y} + \hat{z})$. Clearly, a smaller basis does not exist, so the basis and fundamental vectors we have found are primitive. ∎

*Unit Cells.* Crystalline periodicity can also be described in terms of three-dimensional geometric figures called unit cells. An example is shown in Fig. 2-6. A crystal may be thought of as a collection of unit cells, all identical and all containing the same distribution of atoms. To form the crystal, they are placed side by side and stacked on top of each other in such a way that they do not overlap and there is no space between any of them. Any one of the cells can be made to coincide with any other if it is translated by a lattice vector. Atoms associated with a unit cell form a basis for the crystal.

A unit cell can be constructed as a parallelepiped with the three fundamental lattice vectors as edges. Each side of the cell is a parallelogram with two of the fundamental lattice vectors as edges. For the cell shown in Fig. 2-6, **a** and **b** are used to form the top and bottom, **a** and **c** are used to form the left and right sides, and **b** and **c** are used to form the front and back. The cell is positioned so that its corners lie at lattice points, but this is not mandatory.

Just as there are various choices for the basis and lattice vectors used to describe a given crystal, so there are various choices for the unit cell. If the cell is the smallest in volume of all possible unit cells, it is said to be primitive. Such a cell contains a primitive basis and can be constructed using primitive lattice

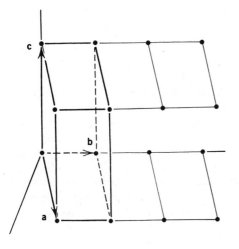

**FIGURE 2-6** A parallelepiped unit cell. The fundamental lattice vectors a, b, and c are along the edges. The crystal can be constructed by placing unit cells side by side and on top of each other. The atomic configuration is the same in each cell and every cell is displaced from any other by a lattice vector.

vectors as edges. Several different primitive cells are possible for a given crystal. They all have the same volume regardless of their shapes.

According to the criteria for a unit cell, there are exactly one lattice point and one set of basis atoms for each unit cell in the crystal. For the cell of Fig. 2-6, a lattice point is at each of the eight corners, but each of these is shared by eight cells. If the cell is moved slightly without moving the lattice points, one and only one point will be within it. Nonprimitive unit cells are often used to describe the periodicity of crystals. Clearly, such a cell contains more than one primitive lattice point.

If atoms are at the corners of a parallelepiped unit cell, they are separated by lattice vectors and only one of them may be included in the basis. The others belong to replicas associated with neighboring cells. Similarly, if an atom is on a cell edge, three other identical atoms must be on other edges and only one of these may be included in the basis. An atom on a cell face is separated by a lattice vector from an identical atom on the opposite face and only one of these is part of the basis. When we count the number of atoms per unit cell, we may include all atoms at corners, at edges, and on faces, but count them, respectively, as one-eighth, one-fourth, and one-half an atom each. Atoms in the interior of the cell, of course, are counted as one atom each.

The volume of a unit cell can be found in terms of fundamental lattice vectors. For the cell of Fig. 2-6 the area of the base, a parallelogram with edges $\mathbf{a}$ and $\mathbf{b}$, is $|\mathbf{a} \times \mathbf{b}| = ab \sin \gamma$, where $\gamma$ is the angle between $\mathbf{a}$ and $\mathbf{b}$. To find the volume of the cell, the base area is multiplied by the component of $\mathbf{c}$ along an axis perpendicular to the base. Since the vector product $\mathbf{a} \times \mathbf{b}$ is perpendicular to the base, the cell volume $\tau$ is given by

$$\tau = |\mathbf{c} \cdot (\mathbf{a} \times \mathbf{b})|. \tag{2-1}$$

Since all unit cells contain exactly the same distribution of atoms, the mass densities of the crystal and cell are the same and are given by $\rho = M/\tau$, where $M$ is the total mass in the cell.

**EXAMPLE 2-2**  For cesium chloride, take the fundamental lattice vectors to be $\mathbf{a} = a\hat{x}$, $\mathbf{b} = a\hat{y}$, and $\mathbf{c} = a(\hat{x} + \hat{y} + \hat{z})$. Describe the parallelepiped unit cell and find the cell volume.

**SOLUTION**  The base of the cell, formed by $\mathbf{a}$ and $\mathbf{b}$, is a square. Since $\mathbf{c} = a(\hat{x} + \hat{y} + \hat{z})$, the cell sides are oblique parallelograms, not perpendicular to the base. The cell volume is $\tau = a(\hat{x} + \hat{y} + \hat{z}) \cdot (a\hat{x} \times a\hat{y}) = a(\hat{x} + \hat{y} + \hat{z}) \cdot (a^2\hat{z}) = a^3$. This is exactly the same as the volume of the cell described in Example 2-1, a cube with edge $a$. For CsCl, with $a = 4.11$ Å, $\tau = 6.94 \times 10^{-29}$ m³. ∎

**EXAMPLE 2-3**  A unit cell for zinc has a base that is a rhombus with edge $a = 2.66$ Å and internal angle $\gamma = 60°$. The sides are rectangles perpendicular to the base and have length $c = 4.95$ Å. There are two zinc atoms per unit cell. Find the cell volume and the density of zinc.

**SOLUTION** The cell volume is $\tau = |\mathbf{c} \cdot (\mathbf{a} \times \mathbf{b})| = ca^2 \sin \gamma = 4.95 \times 10^{-10} \times (2.66 \times 10^{-10})^2 \sin 60° = 3.03 \times 10^{-29}$ m³. Zinc has atomic weight 65.38 and to find the mass of an average zinc atom in grams we divide by Avogadro's number: $65.38/6.022 \times 10^{23} = 1.086 \times 10^{-22}$ g $= 1.086 \times 10^{-25}$ kg. Since there are two zinc atoms per unit cell $\rho = 2 \times 1.086 \times 10^{-25}/3.03 \times 10^{-29} = 7.13 \times 10^3$ kg/m³. ■

Another type unit cell, called a Wigner-Seitz cell, is often used as an alternative to a parallelepiped unit cell, particularly if the crystal has only a single atom in its primitive basis. A Wigner-Seitz cell is designed so that a lattice point is at its center and every point in the cell is closer to the center than to any other lattice point. To construct such a cell, a central lattice point is chosen and lines from it to all nearby lattice points are drawn. Finally, each line is bisected by a plane perpendicular to it. The geometric figure with the smallest volume, bounded by these planes and centered at the chosen lattice point, is the Wigner-Seitz unit cell. If the parallelepiped cell is a cube, then so is the Wigner-Seitz cell. In other cases, the Wigner-Seitz cell has more faces than the parallelepiped cell and so appears more complicated.

## 2.2 CRYSTAL SYMMETRY

Most crystals display a high degree of symmetry. Certain symmetries are inherent in the periodic lattice of a crystal and lattices are categorized according to their symmetries. Symmetry is described by giving some physical operation that changes the positions of the lattice points, but in such a way that there are lattice points at exactly the same places after the operation as before. We have already considered one type, translational symmetry. Although there are many possible symmetry operations, we will be concerned chiefly with three types: rotations, mirror reflections, and inversions.

*Rotational Symmetries.* The rotation of a lattice through the angle $\alpha$ is diagrammed in Fig. 2-7. The axis of rotation is perpendicular to the plane of the page and is marked with a X. A single lattice point is shown in ($a$), at $A$ before the rotation and at $A'$ after the rotation. If the lattice is invariant, there must have been a lattice point at $A'$ before the rotation. In ($b$) a lattice that is invariant to a rotation by $\pi/2$ is illustrated. Each lattice point rotates to a position previously occupied by another lattice point.

If a lattice remains unchanged after a certain rotation, each point must return to its original position after some integer number of like rotations. So the angle of rotation must be one of the angles $2\pi/n$, where $n$ is an integer. If a lattice remains unchanged by a rotation of $2\pi/n$, it is said to have an $n$-fold axis of symmetry.

The translational symmetry of a lattice limits the values of $n$ to 1, 2, 3, 4, or 6. That is, the angle of rotation must be a multiple of $2\pi$, $\pi$, $2\pi/3$, $\pi/2$, or $\pi/3$ and it cannot, for example, be $2\pi/5$ or $2\pi/7$. To see how this comes about,

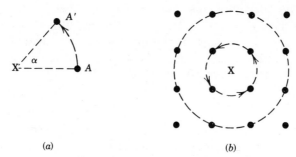

(a)                                                    (b)

**FIGURE 2-7** Rotation of a lattice through the angle $\alpha$. (a) Lattice point A goes to A'. (b) A two-dimensional lattice with fourfold symmetry. Arrows show the result of a rotation by $\frac{1}{2}\pi$ radians.

consider Fig. 2-8, which shows part of a plane of lattice points with a symmetry axis through $X$, perpendicular to the page. It is immaterial whether or not there is a lattice point at $X$.

Suppose the lattice is unchanged by a rotation through $\alpha$ and that lattice point $A$ is rotated to $B$ by such a rotation. Further, suppose there are no lattice points closer to $X$ than $A$ and $B$. If the separation $AB$ is $a$, then lattice points on the line $AB$ and on any line parallel to it must have separation $a$. Two points, $A'$ and $B'$, can be generated on a line parallel to $AB$ if the lattice is first rotated by $\alpha$ in the clockwise direction, then by the same angle in the counterclockwise direction. Other lattice points may be between $A'$ and $B'$, so their separation may not be $a$. It must, however, be a multiple of $a$.

Both $AB$ and $A'B'$ are chords of a circle and, as illustrated in $(b)$, a chord has length $2r\sin(\frac{1}{2}\theta)$, where $r$ is the radius of the circle and $\theta$ is the angle subtended by the chord. So the length of $AB$ is $2r\sin(\frac{1}{2}\alpha)$ and the length of $A'B'$ is $2r\sin(\frac{3}{2}\alpha)$. The first quantity must equal $a$ and the second must equal a multiple of $a$, so $2r\sin(\frac{1}{2}\alpha) = a$ and $2r\sin(\frac{3}{2}\alpha) = sa$, where $s$ is an integer. Division yields $[\sin(\frac{3}{2}\alpha)]/[\sin(\frac{1}{2}\alpha)] = s$ and use of the trigonometric identity $\sin(\frac{3}{2}\alpha) =$

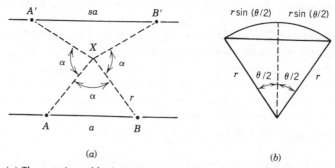

(a)                                                    (b)

**FIGURE 2-8** (a) The rotation of lattice points around $X$ to produce parallel lines of points. If the separation of $A$ and $B$ is $a$, then the separation of $A'$ and $B'$ must be a multiple of $a$. (b) The chord of a circle is $2r\sin(\frac{1}{2}\theta)$, where $\theta$ is the angle subtended by the chord.

$3 \sin(\frac{1}{2}\alpha) - 4 \sin^3(\frac{1}{2}\alpha)$ gives

$$\sin^2(\tfrac{1}{2}\alpha) = \frac{3 - s}{4}. \qquad (2\text{-}2)$$

The right side of this equation must be positive and less than or equal to 1. The only values of $s$ that satisfy these conditions are $s = -1$ ($\alpha = \pm\pi$), $s = 0$ ($\alpha = \pm 2\pi/3$), $s = 1$ ($\alpha = \pm\pi/2$), $s = 2$ ($\alpha = \pm\pi/3$), and $s = 3$ ($\alpha = 0$ or $2\pi$). Thus, only two-, three-, four-, and sixfold symmetry axes can occur in crystals.

*Mirror Symmetry.* Figure 2-9 shows an example of a mirror plane. It is perpendicular to the page and its intersection with a plane of lattice points is shown by the heavy line. Each lattice point can be paired with another, on the opposite side of the mirror plane, so that the two points are the same distance from the plane and the line joining them is perpendicular to the plane. That is, the points are mirror images of each other. If all lattice points can be paired in this way, the lattice is said to be invariant on reflection in the plane. For a two-dimensional lattice, the corresponding symmetry element is a mirror line.

*Inversion Symmetry.* If a crystal has a center of inversion at the origin, then the crystal is exactly the same at $\mathbf{r}$ and $-\mathbf{r}$ for every point $\mathbf{r}$. The distribution of atoms, for example, around $\mathbf{r}$ is the same as the distribution around $-\mathbf{r}$.

For a lattice alone, without a basis, every lattice point is a center of inversion. If there is a lattice point at $\mathbf{r}$ then there is one at $-\mathbf{r}$ since the negative of any lattice vector is also a lattice vector. There may be other inversion centers which do not coincide with lattice points.

*Two-Dimensional Lattices.* Lattice types are distinguished from each other by the symmetry they display. There are five distinct types in two dimensions, shown in Fig. 2-10 with their symmetry elements identified. A twofold axis (diad) is labeled $\bigcirc$, a threefold axis (triad) is labeled $\triangle$, a fourfold axis (tetrad) is labeled $\square$, and a sixfold axis (hexad) is labeled $\bigcirc$. Mirror lines are shown as heavy lines. All symmetry axes are, of course, perpendicular to the plane of the page.

An oblique lattice, with only twofold axes of symmetry and no mirror lines,

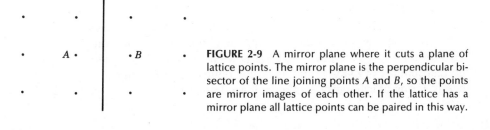

**FIGURE 2-9** A mirror plane where it cuts a plane of lattice points. The mirror plane is the perpendicular bisector of the line joining points $A$ and $B$, so the points are mirror images of each other. If the lattice has a mirror plane all lattice points can be paired in this way.

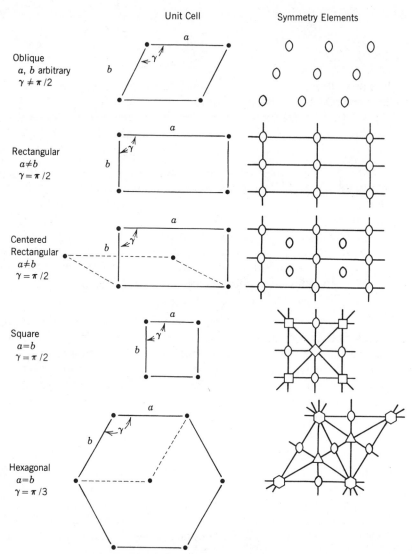

**FIGURE 2-10** Two-dimensional lattices: unit cells and symmetry elements.

is the least symmetric of the two-dimensional lattice types. The axes pass through lattice points and through cell centers, as shown. If $a \neq b$, the angle $\gamma$ between two adjacent cell sides can have any value except $\pi/2$. If $a = b$, $\gamma$ cannot be $\pi/3$, $\pi/2$, or $2\pi/3$. For these special situations other symmetry elements automatically appear and the lattice has higher symmetry than an oblique lattice.

Mirror lines force the lattice to have at least rectangular symmetry. Note that the distribution of diads is the same for oblique and rectangular lattices and that, in terms of symmetry elements, these lattices differ only in the number of mirror lines. Another lattice has only diads and mirror lines; the centered rectangular lattice. It is similar to the rectangular lattice but additional lattice points are at rectangle centers. Two primitive lattice points are associated with each rectangular unit cell so the rectangle cannot be primitive. A primitive cell is shown by dotted lines in the diagram.

A square lattice results when there is a fourfold axis of symmetry. Tetrads pass through lattice points and the cell center. In addition, four mirror lines pass through the cell center, 45° apart. A centered square lattice, similar to a centered rectangular lattice, can be created but it is again a square lattice and so cannot be listed as a distinct lattice type; See Fig. 2-11.

Finally, a sixfold axis produces a hexagonal lattice. The primitive unit cell is not shaped like a hexagon but is the figure shown by the dotted lines: a rhombus with an internal angle of 60°. The primitive cell may be constructed from two identical equilateral triangles, inverted with respect to each other. Threefold axes pass through the centers of the triangles and twofold axes pass through their edge centers. Mirror lines join each cell corner to the midpoint of the opposite edge.

If a symmetry element passes through a lattice point, parallel elements of the same kind must pass through all lattice points and, if a symmetry element passes through the interior of a unit cell, parallel elements must pass through corresponding points of all unit cells of the crystal. The set of all symmetry elements that pass through a lattice point is called the point group of the lattice. For example, the point group of a square lattice consists of a fourfold axis and four mirror lines, while the point group of a hexagonal lattice consists of a sixfold axis and six mirror lines.

A crystal has lower symmetry than its lattice if the basis is not invariant under one or more of the lattice symmetry operations. Figure 2-12a shows a lattice and basis which both have fourfold symmetry. The crystal therefore also has fourfold symmetry. On the other hand, (b) shows a crystal for which the lattice has fourfold symmetry but the basis has only twofold symmetry. As demonstrated in (c), the crystal is different after a rotation by $\pi/2$ about the symmetry axis. Although every lattice point is a center of inversion for a lattice, a crystal does not have centers of inversion unless the basis is also invariant under inversion.

Three noncolinear lattice points define a lattice plane in a three-dimensional crystal. Clearly in any crystal, a great many lattice planes exist, at many different orientations. No matter which is examined, lattice points on the plane must form one of the two-dimensional lattices described above.

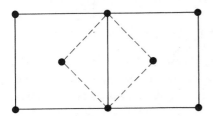

**FIGURE 2-11**  A centered square lattice. Solid lines outline squares with lattice points at corners and centers. The unit cell indicated by dotted lines is also a square, but it has lattice points only at its corners.

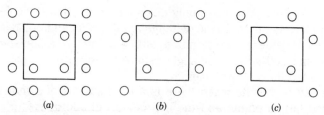

**FIGURE 2-12** (a) Both lattice and basis have fourfold symmetry. (b) The lattice has fourfold symmetry but the basis has only twofold symmetry. (c) The crystal of (b) after a rotation by $\frac{1}{2}\pi$. It is not the same as (b).

## 2.3 BRAVAIS LATTICES

In three dimensions, there are 14 different lattice types, called Bravais lattices; see Fig. 2-13. They are grouped into seven lattice systems, with the lattices of any one system having the same point group of symmetry elements. In these diagrams, **a** and **b** form the edges of the unit cell base and **c** points upward from the base. $\alpha$ is the angle between **c** and **a**, $\beta$ is the angle between **c** and **b**, and $\gamma$ is the angle between **a** and **b**. These angles are identified on the unit cell drawn at the top of the diagram. Also shown there is a coordinate system used to describe lattice vectors in the following discussion.

If lattice points are only at cell corners, the cell is primitive and labeled P. Cells with primitive lattice points at both corners and body centers are labeled I (from the German innenzentrierte) and cells with primitive lattice points at both corners and face centers are labeled F. In the diagram, there is one cell with primitive lattice points at its corners and at the centers of its top and bottom faces only. It is labeled C since the **c** axis is usually taken to be perpendicular to these faces.

*Cubic Lattices.* Lattices of the cubic system can all be constructed using unit cells in the shape of cubes. These lattices have the greatest number of symmetry elements of all lattice types: three tetrads, four triads, six diads, and nine mirror planes through the center of each cube. Symmetry elements are shown in Fig. 2-14. Each tetrad runs through the centers of two opposing faces, the triads are along body diagonals, and each diad crosses the cube from the center of one edge to the center of the diametrically opposite edge. Three of the mirror planes are parallel to the cube faces and pass through the body center, while each of the other six cut across the cube through diametrically opposite face diagonals.

A simple or primitive cubic lattice has lattice points only at cube corners and the cube is a primitive unit cell. If the cube has edge $a$, then $\mathbf{a} = a\hat{\mathbf{x}}$, $\mathbf{b} = a\hat{\mathbf{y}}$, and $\mathbf{c} = a\hat{\mathbf{z}}$ are three primitive lattice vectors.

For a body-centered cubic lattice, there are two primitive lattice points per cube, one at a corner and one at the cube center. Each cube contains two replicas of the primitive basis. Fundamental lattice vectors are usually taken to be vectors from a cube center to three neighboring cube corners: $\mathbf{a} = \frac{1}{2}a(\hat{\mathbf{x}} + \hat{\mathbf{y}} - \hat{\mathbf{z}})$, $\mathbf{b} = \frac{1}{2}a(-\hat{\mathbf{x}} + \hat{\mathbf{y}} + \hat{\mathbf{z}})$, and $\mathbf{c} = \frac{1}{2}(\hat{\mathbf{x}} - \hat{\mathbf{y}} + \hat{\mathbf{z}})$. These vectors are shown

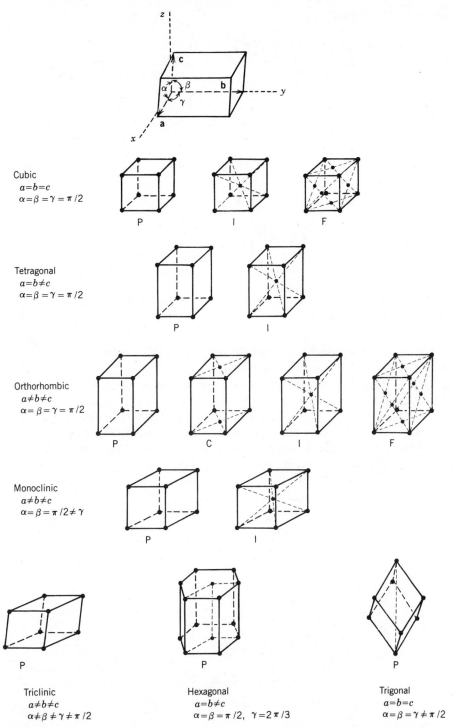

**FIGURE 2-13** The 7 three-dimensional lattice systems and 14 Bravais lattices. Additional conditions on edge lengths and angles are described in the text.

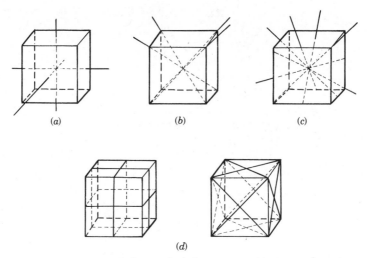

**FIGURE 2-14** Symmetry axes and planes of a cube: (a) tetrads; (b) triads; (c) diads; and (d) mirror planes.

in Fig. 2-15a, along with the parallelepiped unit cell generated by them. Note that $\mathbf{a} + \mathbf{c} = a\hat{\mathbf{x}}$, $\mathbf{a} + \mathbf{b} = a\hat{\mathbf{y}}$, and $\mathbf{b} + \mathbf{c} = a\hat{\mathbf{z}}$, so positions of cube corners, as well as positions of cube centers, can be written as linear combinations of the fundamental vectors, with integer coefficients.

For a face-centered cubic lattice, one corner point and three face center points are associated with each cube, so there are four primitive lattice points per cube. Each cube contains four replicas of the primitive basis. The fundamental lattice vectors usually used are vectors from a cube corner to three neighboring face centers: $\mathbf{a} = \frac{1}{2}a(\hat{\mathbf{x}} + \hat{\mathbf{y}})$, $\mathbf{b} = \frac{1}{2}a(\hat{\mathbf{y}} + \hat{\mathbf{z}})$, and $\mathbf{c} = \frac{1}{2}a(\hat{\mathbf{x}} + \hat{\mathbf{z}})$. These vectors and a parallelepiped unit cell are shown in Fig. 2-15b.

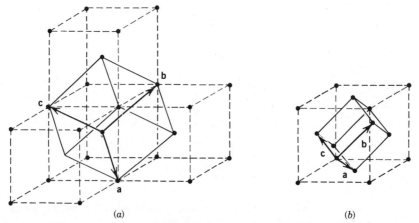

**FIGURE 2-15** Primitive translation vectors and parallelepiped unit cells for (a) a body-centered cubic lattice and (b) a face-centered cubic lattice.

Cubic lattices can be formed by stacking two-dimensional square lattices. To form a simple cubic lattice, the two-dimensional lattices are placed directly over each other and are separated by a distance equal to the cube edge $a$. To form a body-centered cubic lattice, each plane is placed with the corners of its squares over square centers on the plane below, as shown in Fig. 2-16$a$. The separation of the planes is $\frac{1}{2}a$. To form a face-centered cubic lattice, the square edges are taken to be half a cube face diagonal or $a/\sqrt{2}$. Planes are separated by $\frac{1}{2}a$ and again the corners of squares on one plane are over centers of squares on the plane below. This arrangement is shown in Fig. 2-16$b$.

For all three cubic lattices, lattice points on planes perpendicular to a cube body diagonal, a threefold axis, form two-dimensional hexagonal lattices. Figure 2-17 shows a hexagonal plane and a two-dimensional hexagonal unit cell for a simple cubic lattice. Identical parallel planes pass through all lattice points. As one of the cube triads passes from plane to plane, it goes through one of the points of threefold symmetry in a hexagonal cell, then the other point of threefold symmetry, and finally through a point of sixfold symmetry. This pattern is then repeated. Two hexagonal planes cut a cube body diagonal in the cube interior, dividing it into thirds. Since the length of a body diagonal is $\sqrt{3}a$, the separation of the planes is $a/\sqrt{3}$. Cube face diagonals form edges of the hexagonal cell, so the length of a hexagon edge is $\sqrt{2}a$.

To construct a face-centered cubic lattice, the same planes are used with the same separation, $a/\sqrt{3}$. Now, however, there is a lattice point at the center of each face diagonal, so the hexagon edge is half as long, or $a/\sqrt{2}$. To construct a body-centered cubic lattice, twice as many planes at half the separation must be used. Three new planes, through cube body centers, cut the body diagonal

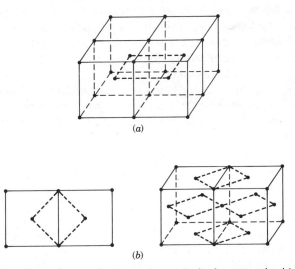

(a)

(b)

**FIGURE 2-16**  The stacking of square lattices to form (a) a body-centered cubic lattice and (b) a face-centered cubic lattice. For a face-centered cubic lattice, cube faces do not form two-dimensional primitive unit cells. The square drawn on the left with dotted lines, however, is a two-dimensional primitive cell.

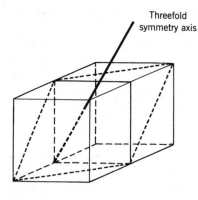

Threefold
symmetry axis

**FIGURE 2-17** A hexagonal plane in a simple cubic lattice. Dotted lines outline a two-dimensional hexagonal unit cell in the plane. A threefold symmetry axis, normal to the plane and along a cube body diagonal, is also shown. A parallel plane passes through each lattice point.

and divide it into sixths. So the plane separation is now $a/(2\sqrt{3})$. The hexagon edge is $\sqrt{2}a$.

**EXAMPLE 2-4**  Find the position vectors of the eight cube corners and six cube face centers in terms of the primitive lattice vectors $\mathbf{a} = \frac{1}{2}a(\hat{\mathbf{x}} + \hat{\mathbf{y}})$, $\mathbf{b} = \frac{1}{2}a(\hat{\mathbf{y}} + \hat{\mathbf{z}})$, and $\mathbf{c} = \frac{1}{2}a(\hat{\mathbf{x}} + \hat{\mathbf{z}})$ for a face-centered cubic lattice.

**SOLUTION**  Relative to the coordinate system shown in Fig. 2-13, the eight cube corners have position vectors $0$, $a\hat{\mathbf{x}}$, $a\hat{\mathbf{y}}$, $a\hat{\mathbf{z}}$, $a(\hat{\mathbf{x}} + \hat{\mathbf{y}})$, $a(\hat{\mathbf{x}} + \hat{\mathbf{z}})$, $a(\hat{\mathbf{y}} + \hat{\mathbf{z}})$, and $a(\hat{\mathbf{x}} + \hat{\mathbf{y}} + \hat{\mathbf{z}})$, respectively. Since $\mathbf{a} - \mathbf{b} + \mathbf{c} = a\hat{\mathbf{x}}$, $\mathbf{a} + \mathbf{b} - \mathbf{c} = a\hat{\mathbf{y}}$, and $-\mathbf{a} + \mathbf{b} + \mathbf{c} = a\hat{\mathbf{z}}$, these can be written $0$, $\mathbf{a} - \mathbf{b} + \mathbf{c}$, $\mathbf{a} + \mathbf{b} - \mathbf{c}$, $-\mathbf{a} + \mathbf{b} + \mathbf{c}$, $2\mathbf{a}$, $2\mathbf{c}$, $2\mathbf{b}$, and $\mathbf{a} + \mathbf{b} + \mathbf{c}$. The six face centers are at $\frac{1}{2}a(\hat{\mathbf{x}} + \hat{\mathbf{z}})$, $\frac{1}{2}a(\hat{\mathbf{y}} + \hat{\mathbf{z}})$, $\frac{1}{2}a(\hat{\mathbf{x}} + \hat{\mathbf{y}})$, $\frac{1}{2}a(\hat{\mathbf{x}} + 2\hat{\mathbf{y}} + \hat{\mathbf{z}})$, $\frac{1}{2}a(2\hat{\mathbf{x}} + \hat{\mathbf{y}} + \hat{\mathbf{z}})$, and $\frac{1}{2}a(\hat{\mathbf{x}} + \hat{\mathbf{y}} + 2\hat{\mathbf{z}})$, respectively. And these can be written $\mathbf{c}$, $\mathbf{b}$, $\mathbf{a}$, $\mathbf{a} + \mathbf{b}$, $\mathbf{a} + \mathbf{c}$, and $\mathbf{b} + \mathbf{c}$. Note that all these position vectors can be written as linear combinations of the primitive lattice vectors with integer coefficients. ∎

**EXAMPLE 2-5**  Consider a body-centered cubic lattice and find three primitive lattice vectors such that two of them form the sides of a two-dimensional hexagonal unit cell. Use the third vector to find the separation of the hexagonal planes.

**SOLUTION**  Use the plane shown in Fig. 2-17 and the coordinate system shown in Fig. 2-13. Two lattice vectors in the plane are $\mathbf{a} = a(-\hat{\mathbf{x}} + \hat{\mathbf{z}})$, across the left face of the cube, and $\mathbf{b} = a(-\hat{\mathbf{x}} + \hat{\mathbf{y}})$, across the bottom face. The angle $\theta$ between these vectors is given by $\cos\theta = (\mathbf{a} \cdot \mathbf{b})/(|\mathbf{a}||\mathbf{b}|) = a^2/(2a^2) = \frac{1}{2}$, so $\theta = 60°$. They have the same length, $\sqrt{2}a$, so they form the sides of a hexagonal unit cell. The third vector $\mathbf{c}$ joins two lattice points on neighboring planes. Since a neighboring plane passes through the cube body center, we may take $\mathbf{c}$ to be the vector from the lower left front cube corner to the body center, or $\mathbf{c} = \frac{1}{2}a(-\hat{\mathbf{x}} + \hat{\mathbf{y}} + \hat{\mathbf{z}})$. The plane separation is the component of $\mathbf{c}$ along the body diagonal normal to the hexagonal planes. Since

$(1/\sqrt{3})(\hat{\mathbf{x}} + \hat{\mathbf{y}} + \hat{\mathbf{z}})$ is a unit vector along this body diagonal, the separation of planes is $d = \frac{1}{2}a(-\hat{\mathbf{x}} + \hat{\mathbf{y}} + \hat{\mathbf{z}}) \cdot (1/\sqrt{3})(\hat{\mathbf{x}} + \hat{\mathbf{y}} + \hat{\mathbf{z}}) = a/(2\sqrt{3})$. The three vectors **a**, **b**, and **c** are primitive since there are lattice points at the corners of the parallelepiped cell they generate and none in the interior.  ■

*Tetragonal Lattices.*  A primitive tetragonal unit cell, with lattice points only at corners, has a square base and rectangular sides perpendicular to the base. It can be generated by stacking planes containing square lattices, with **c** perpendicular to the plane of **a** and **b** and with a plane separation that is different from the length of a square edge. Fourfold axes of the squared lattices are retained and are parallel to **c**. Possible primitive lattice vectors are $\mathbf{a} = a\hat{\mathbf{x}}$, $\mathbf{b} = a\hat{\mathbf{y}}$, and $\mathbf{c} = c\hat{\mathbf{z}}$.

In all, the cell has one tetrad, four diads, and five mirror planes through its center, as shown in Fig. 2-18a. Two diads pass through centers of opposing rectangles and two pass through centers of diametrically opposed rectangle edges. Four mirror planes are perpendicular to the square base, two of them

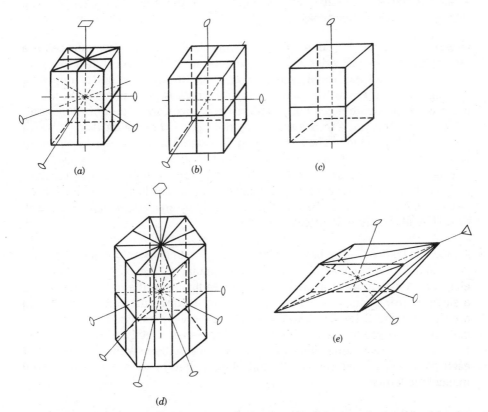

**FIGURE 2-18** Symmetry elements of unit cells: (a) tetrahedral; (b) orthorhombic; (c) monoclinic; (d) hexagonal; (e) trigonal.

cutting the square along diagonals and two cutting it from edge centers to edge centers. The fifth mirror plane is parallel to the base and cuts the rectangular cell sides at their midpoints. These symmetry elements can all be seen in the diagrams of the square and rectangular lattices shown in Fig. 2-10.

A tetragonal lattice results when a cubic lattice is distorted by compressing or elongating it along one of its fourfold axes. The other tetrads of the cube become diads and the body diagonals are no longer symmetry axes. Four axes that were diads also lose their symmetry character.

There is another Bravais lattice in the tetragonal system: a body-centered lattice with lattice points at the tetragonal cell corners and an additional point at the center. Primitive lattice vectors are $\frac{1}{2}a(\hat{x} + \hat{y}) - \frac{1}{2}c\hat{z}, \frac{1}{2}a(-\hat{x} + \hat{y}) + \frac{1}{2}c\hat{z}$, and $\frac{1}{2}a(\hat{x} - \hat{y}) + \frac{1}{2}c\hat{z}$, where $a$ is the square edge and $c$ is the height of the tetragonal unit cell.

The body-centered cell shown in Fig. 2-13 may be replaced by a face-centered cell. A new square base is formed by base diagonals of four adjoining body centered cells, in the fashion shown in Fig. 2-16 for the face-centered cubic cell. There are now lattice points at the new cell corners and face centers. We conclude that a face-centered tetragonal cell describes exactly the same lattice as a body-centered tetragonal cell and only one of these need be included in a listing of Bravais lattice types.

*Orthorhombic Lattices.* All six faces of a primitive orthorhombic unit cell are rectangles and the cell sides are perpendicular to the base. As shown in Fig. 2-18b, the unit cell has three diads, which pass through centers of opposing rectangles, and three mirror planes, each of which is perpendicular to a diad and passes through the cell body center. The lattice can be produced by compressing or elongating a tetragonal lattice along one or both of the diads through rectangular face centers. The base is then distorted from a square to a rectangle and what was a fourfold axis becomes a twofold axis.

There are four lattices in the orthorhombic system: primitive, base centered, body centered, and face centered. The orthorhombic cell is not primitive for the last three. All of them can be constructed easily by stacking either rectangular or rectangular centered lattices.

*Monoclinic Lattices.* The base of a monoclinic unit cell is an oblique parallelogram and its rectangular sides are perpendicular to the base. Its symmetry elements consist only of a single diad, perpendicular to the oblique base, and a single mirror plane, parallel to the base and through the cell center. These are shown in Fig. 2-18c. The twofold axes of the oblique unit cell that forms the base are preserved by stacking oblique lattice planes directly above each other to form a primitive monoclinic lattice or by stacking them with cell corners on each plane above cell centers on the plane below to form a body-centered monoclinic lattice.

*Triclinic Lattices.* A triclinic lattice has the least symmetry of all three-dimensional lattice types. It has no symmetry axes and no mirror planes. It can be formed by stacking two-dimensional oblique lattices, with the planes shifted so that no twofold axes are aligned.

*Hexagonal and Trigonal Lattices.* A three-dimensional hexagonal lattice can be formed by stacking two-dimensional hexagonal lattice planes so lattice points are directly above each other. The base of a primitive unit cell is a 60° rhombus. The sides are rectangles perpendicular to the base. Two primitive lattice vectors are parallel to rhombus edges and a third is perpendicular to the base. The third is usually designated **c**.

A unit cell with a hexagon base is used in Fig. 2-18d to show symmetry elements. A hexad runs parallel to the sides through the cell center. There are six diads, each of which pass through the cell center and either a face center or an edge center. A mirror plane is perpendicular to each diad and, in addition, there is a mirror plane parallel to the base, halfway between the base and top of the cell.

A trigonal cell can be constructed by stacking planes of two-dimensional hexagonal lattices in groups of three so that one sixfold and two threefold axes, associated with two-dimensional unit cells on different planes, are aligned. This is the same as the scheme used to construct cubic lattices, except the separation of the planes is not one of the special values required by those lattices. A cubic unit cell can be deformed to a trigonal cell by elongating or compressing it along a body diagonal so that the cube edges move toward or away from the diagonal, like ribs on a folding umbrella.

Symmetry elements are shown in Fig. 2-18e. In addition to the triad, there are three diads, each of which crosses the unit cell from the center of an edge to the center of the opposite edge, and three mirror planes, each of which cuts across a face along a diagonal and passes through the cell center. The triad is the common line of intersection of the three mirror planes.

*Unique Symmetry Elements.* A small set of symmetry elements uniquely identifies each lattice system. For example, if the lattice has sixfold axes, it must be hexagonal. The existence of hexads guarantees the appearance of all other symmetry elements of a hexagonal lattice. If the lattice has fourfold axes, it is either tetragonal or cubic: tetragonal if the axes are all parallel to each other and cubic if they are not. If the lattice has threefold axes but no sixfold axes, it is either trigonal or cubic, depending on whether or not the axes are all parallel to each other. If the axes of highest symmetry are twofold, the lattice is either monoclinic or orthorhombic, again depending on whether or not the axes are all parallel to each other. Finally, if no symmetry axes are present, the lattice is triclinic. Unique symmetry elements are often used in connection with x-ray scattering data to help identify the lattice associated with a particular sample.

## 2.4  POSITIONS, DIRECTIONS, AND PLANES IN CRYSTALS

Special notation is used to designate points within a unit cell, directions along lines through lattice points, and planes through lattice points. Position vectors of points and vectors parallel to lines are written as linear combinations of fundamental lattice vectors and the coefficients are used to identify the point or line. The orientation of a plane is specified by giving a vector perpendicular to it.

*Positions in a Cell.*   The position vector **r** of a point within a cell, relative to a cell corner, can be written

$$\mathbf{r} = u\mathbf{a} + v\mathbf{b} + w\mathbf{c}, \tag{2-3}$$

where $u$, $v$, and $w$ are numbers with magnitudes less than or equal to 1. The point is identified by giving these three numbers in the form $uvw$, not separated by commas. Since the unit of length is associated with the fundamental vectors, the coefficients $u$, $v$, and $w$ are unitless.

This notation is used to specify positions of atoms and other points of interest in unit cells. For example, the center of any parallelepiped unit cell has the position vector $\frac{1}{2}\mathbf{a} + \frac{1}{2}\mathbf{b} + \frac{1}{2}\mathbf{c}$ so the indices of that point are $\frac{1}{2}\frac{1}{2}\frac{1}{2}$. Points on the body diagonal from the origin to the diametrically opposite cell corner have indices $uuu$, so that, for example, the point one-fourth of the body diagonal from the origin is labeled $\frac{1}{4}\frac{1}{4}\frac{1}{4}$.

**EXAMPLE 2-6**  Find the indices of the left and right face centers of the general unit cell shown at the top of Fig. 2-13.

**SOLUTION**  The left face center is at $\frac{1}{2}\mathbf{a} + \frac{1}{2}\mathbf{c}$, so its indices are $\frac{1}{2}0\frac{1}{2}$. The right face center is at $\frac{1}{2}\mathbf{a} + \mathbf{b} + \frac{1}{2}\mathbf{c}$, so its indices are $\frac{1}{2}1\frac{1}{2}$.  ∎

**EXAMPLE 2-7**  For a primitive hexagonal unit cell, find the indices of the points where threefold axes intersect the base.

**SOLUTION**  Figure 2-19 shows the cell base and the fundamental vectors **a** and **b**. The lower symmetry point is at the intersection of the lines $y = a - \sqrt{3}x$ and $y = 0$, so its coordinates are $x = a/\sqrt{3}$, $y = 0$. Substitute $\mathbf{a} = (\sqrt{3}/2)a\hat{x} - (1/2)a\hat{y}$ and $\mathbf{b} = a\hat{y}$ into $\mathbf{r} = u\mathbf{a} + v\mathbf{b}$ to find $\mathbf{r} = (\sqrt{3}/2)ua\hat{x} - (1/2)(u - 2v)a\hat{y}$. The $x$ component must be $a/\sqrt{3}$ and the $y$ component must be 0 so $u = \frac{2}{3}$ and $v = \frac{1}{3}$. The indices are $\frac{2}{3}\frac{1}{3}0$. The upper symmetry point is at the intersection of the lines $y = a - \sqrt{3}x$ and $y = \frac{1}{2}a$. The same analysis leads to $u = \frac{1}{3}$ and $v = \frac{2}{3}$ so the indices are $\frac{1}{3}\frac{2}{3}0$.  ∎

*Directions in Crystals.*   To describe a direction, the indices $u$, $v$, and $w$ are chosen so $u\mathbf{a} + v\mathbf{b} + w\mathbf{c}$ is in the desired direction. They are given in the form $[uvw]$, enclosed in brackets and not separated by commas. A negative index is indicated by a bar above its magnitude: $[\bar{u}\bar{v}\bar{w}]$ is opposite in direction to $[uvw]$. Directions usually considered are those of lines through lattice points and since these lines are parallel to lattice vectors their directions can be specified by integer indices.

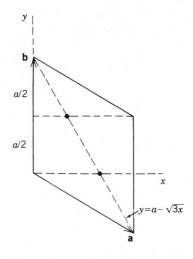

**FIGURE 2-19** A two-dimensional hexagonal cell with edge $a$, showing points where the two threefold symmetry axes intersect the plane of the cell. The lower point is at the intersection of $y = a - \sqrt{3}x$ and $y = 0$ while the upper point is at the intersection of $y = a - \sqrt{3}x$ and $y = \frac{1}{2}a$.

Furthermore, the integers are always chosen to be the smallest in magnitude of those that give the correct direction. That is, any common divisors of $u$, $v$, and $w$ are omitted.

For example, a [100] line is in the direction of **a**, a [010] line is in the direction of **b**, and a [001] line is in the direction of **c**. For any parallelepiped unit cell, the [111], $[11\bar{1}]$, $[1\bar{1}1]$, and $[\bar{1}11]$ directions are parallel to the four body diagonals, respectively.

For any given lattice, directions can be grouped so that members of each group are related to each other by symmetry operations that leave the lattice unchanged. Directions in the same group are said to be equivalent. In a cubic lattice [100], [010], and [001] directions and directions opposite to them are all equivalent. A rotation by $\pi/2$ about the [100] direction, for example, changes a [010] line into either a [001] line or a $[00\bar{1}]$ line, depending on the sense of the rotation. A set of equivalent directions is denoted by $\langle uvw \rangle$, where the indices of one of them are placed in angular brackets. This notation is useful when we do not need to distinguish between equivalent directions.

*Lattice Planes.*  A lattice plane is one that passes through lattice points; any particular plane can be identified by giving three lattice points on it. Of chief interest, however, is the orientation of the plane and this is described by giving a vector normal to it.

The vector product of any two nonparallel vectors is normal to the plane containing the vectors. For example, **b** × **c** is normal to the plane of **b** and **c**. As we will show, the three vectors **b** × **c**, **c** × **a**, and **a** × **b** form a fundamental set such that normals to all lattice planes can be written as linear combinations of them with integer coefficients.

Figure 2-20 shows the fundamental vectors of a lattice drawn from the origin $O$, at a lattice point. A lattice plane is also drawn and it intersects the **a** axis at $x$**a**, the **b** axis at $y$**b**, and the **c** axis at $z$**c**. The points of intersection are not necessarily lattice points. The two vectors $\mathbf{r}_1 = y\mathbf{b} - x\mathbf{a}$ and $\mathbf{r}_2 = z\mathbf{c} - x\mathbf{a}$ lie

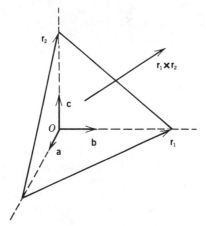

**FIGURE 2-20** A crystal plane and its intercepts on the crystal axes. The plane intersects the **a** axis at $x$**a**, the **b** axis at $y$**b**, and the **c** axis at $z$**c**. $\mathbf{r}_1 = y\mathbf{b} - x\mathbf{a}$ and $\mathbf{r}_2 = z\mathbf{c} - x\mathbf{a}$ are vectors in the plane and $\mathbf{r}_1 \times \mathbf{r}_2$ is a vector normal to the plane. A parallel plane passes through the origin.

in the plane so

$$\mathbf{r}_1 \times \mathbf{r}_2 = xyz \left[ \frac{\mathbf{b} \times \mathbf{c}}{x} + \frac{\mathbf{c} \times \mathbf{a}}{y} + \frac{\mathbf{a} \times \mathbf{b}}{z} \right] \qquad (2\text{-}4)$$

is a vector normal to the plane.

Because the plane is a lattice plane, the ratios $y/x$, $z/x$, and $y/z$ are rational numbers. To demonstrate, we consider the two-dimensional example diagrammed in Fig. 2-21. Lattice points in the plane of **a** and **b** are shown, along

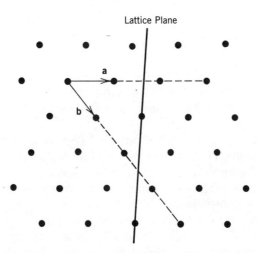

**FIGURE 2-21** The heavy line is the intersection of a lattice plane with the plane of **a** and **b**. The ratio of its **a** and **b** intercepts, in units of $a$ and $b$, respectively, is a rational number.

with the intersection of a second lattice plane with the plane of $\mathbf{a}$ and $\mathbf{b}$. If $\mathbf{r}_1 = n_a\mathbf{a} + n_b\mathbf{b}$ and $\mathbf{r}_2 = m_a\mathbf{a} + m_b\mathbf{b}$ are two lattice points on the line of intersection, then a general point on that line is given by

$$\mathbf{r} = \mathbf{r}_2 + \gamma(\mathbf{r}_1 - \mathbf{r}_2)$$
$$= [m_a + \gamma(n_a - m_a)]\mathbf{a} + [m_b + \gamma(n_b - m_b)]\mathbf{b}, \qquad (2\text{-}5)$$

where $\gamma$ is a number that is different for different points. To find the $\mathbf{a}$ intercept, take $\gamma = -m_b/(n_b - m_b)$ and, to find the $\mathbf{b}$ intercept, take $\gamma = -m_a/(n_a - m_a)$. In each case, substitute the value for $\gamma$ into Eq. 2-5. The $\mathbf{a}$ intercept is given by $\mathbf{r} = x\mathbf{a}$, with

$$x = \frac{m_a(n_b - m_b) - m_b(n_a - m_a)}{n_b - m_b} \qquad (2\text{-}6)$$

and the $\mathbf{b}$ intercept is given by $\mathbf{r} = y\mathbf{b}$, with

$$y = \frac{m_b(n_a - m_a) - m_a(n_b - m_b)}{n_a - m_a}. \qquad (2\text{-}7)$$

The ratio $y/x$ is $-(n_b - m_b)/(n_a - m_a)$, a ratio of integers. The proof can be extended to three dimensions, although the algebra is somewhat more complicated.

Since the intercepts of a lattice plane in units of the lattice constants are in the ratio of integers, Eq. 2-4 can be written

$$\mathbf{r}_1 \times \mathbf{r}_2 = A[\hbar\mathbf{b} \times \mathbf{c} + \hbar\mathbf{c} \times \mathbf{a} + \ell\mathbf{a} \times \mathbf{b}], \qquad (2\text{-}8)$$

where $\hbar$, $\hbar$, and $\ell$ are integers and $A$ is a number whose magnitude is immaterial since it does not affect the direction of the vector. Usually $\hbar$, $\hbar$, and $\ell$ are taken to be the three smallest integers that, when substituted into Eq. 2-8, give a vector in the correct direction. Any common integer factors are included in the number $A$ outside the brackets.

The three integers $\hbar$, $\hbar$, and $\ell$ are called the Miller indices of the plane and the plane is designated by the symbol $(\hbar\hbar\ell)$. If an index is negative a bar is placed above its magnitude. Miller indices specify a vector normal to a plane, and not a specific plane: all parallel planes have the same indices.

Equation 2-4 provides a technique for finding the Miller indices of a plane. Place the origin at a lattice point, then find the intercepts of the plane on the crystal axes. Express each intercept in units of the lattice spacing along the corresponding axis, then evaluate its reciprocal. Finally, multiply the three resulting numbers by a common factor to obtain the three smallest integers with the same ratios. These are $\hbar$, $\hbar$, and $\ell$. If the plane does not intersect one of the crystal axes, that intercept is taken to be infinitely far from the origin and the corresponding index is 0.

**EXAMPLE 2-8** What are the Miller indices of the plane that contains the three lattice points $\mathbf{r}_1 = \mathbf{a} - \mathbf{b}$, $\mathbf{r}_2 = 2\mathbf{a} + \mathbf{c}$, and $\mathbf{r}_3 = 3\mathbf{b} + \mathbf{c}$?

**SOLUTION** The vectors $r_1 - r_3$ and $r_2 - r_3$ lie in the plane so all points $r$ in the plane satisfy $r = r_3 + \alpha(r_1 - r_3) + \beta(r_2 - r_3) = (\alpha + 2\beta)a + (3 - 4\alpha - 3\beta)b + (1 - \alpha)c$, where $\alpha$ and $\beta$ are numbers. This has the form $r = xa + yb + zc$. At the $a$ intercept, $3 - 4\alpha - 3\beta = 0$ and $1 - \alpha = 0$ so $\alpha = 1$, $\beta = -\frac{1}{3}$, and $x = \alpha + 2\beta = \frac{1}{3}$. At the $b$ intercept, $\alpha + 2\beta = 0$ and $1 - \alpha = 0$ so $\alpha = 1$, $\beta = -\frac{1}{2}$, and $y = 3 - 4\alpha - 3\beta = \frac{1}{2}$. At the $c$ intercept $\alpha + 2\beta = 0$ and $3 - 4\alpha - 3\beta = 0$ so $\alpha = \frac{6}{5}$, $\beta = -\frac{3}{5}$, and $z = 1 - \alpha = -\frac{1}{5}$. The reciprocals are 3, 2, and $-5$, respectively, so the plane is a $(32\bar{5})$ plane. ∎

In most cases, positions, directions, and planes are indexed in terms of conventional rather than primitive lattice vectors. That is, primitive cubic, tetragonal, orthorhombic, or monoclinic lattices are used to find the indices even if the primitive lattice is face centered, body centered, or base centered. Except where explicitly noted otherwise, we will use that convention.

Some examples of planes in cubic lattices are shown in Fig. 2-22. (100), (010), and (001) planes are normal to the $a$, $b$, and $c$ axes, respectively. (110), (101), and $(011)$ planes cut through opposing face diagonals, as do $(1\bar{1}0)$, $(10\bar{1})$, and $(01\bar{1})$ planes. (111), $(\bar{1}11)$, $(1\bar{1}1)$, and $(11\bar{1})$ planes are normal to body diagonals.

Since the normal to a plane can be in either of two directions, $(hkl)$ and $(\bar{h}\bar{k}\bar{l})$ are alternative labels for the same set of planes. If, however, two parallel faces of a unit cell must be distinguished, then the outward normal is used. For the cell shown in Fig. 2-22, the left face is designated $(0\bar{1}0)$ with its normal in the negative $b$ direction and the right face is designated (010) with its normal in the positive $b$ direction.

The set of all planes that are equivalent by virtue of the symmetry operations of the lattice point group is denoted by $\{hkl\}$, where $h$, $k$, and $l$ are the indices for one of the planes. In this notation, for example, {100} represents all the faces of a cube.

*Interplanar Spacing.* The intercepts on crystal axes of all lattice planes with a given a set of Miller indices can easily be found. According to the discussion following Eq. 2-8, the intercepts of a plane on the $a$, $b$, and $c$ axes are

$$x = \frac{n}{h}, \qquad (2\text{-}9)$$

$$y = \frac{n}{k}, \qquad (2\text{-}10)$$

and

$$z = \frac{n}{l}, \qquad (2\text{-}11)$$

respectively. As $n$ takes on integer values these expressions give the intercepts of parallel lattice planes. We must be careful, however, if $h$, $k$, and $l$ have a

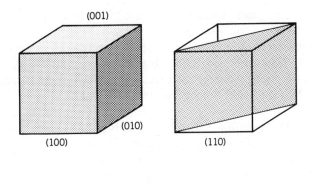

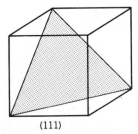

**FIGURE 2-22** High symmetry planes in cubic crystals: (a) (100), (010), and (001) planes; (b) a (110) plane; (c) a (111) plane.

common integer divisor other than 1. Then these expressions also give intercepts of planes that do not pass through lattice points. According to Eqs. 2-9, 2-10, and 2-11, a (200) plane in a simple cubic lattice intersects the **a** axis at $\frac{1}{2}$**a** but this is not a lattice plane since it does not contain any lattice points. If we use primitive lattice vectors to index planes, we must use indices with no common divisor other than 1. Then Eqs. 2-9, 2-10, and 2-11 correctly give the intercepts of all lattice planes and no others.

On the other hand, if we use nonprimitive lattice vectors to index planes and we wish Eqs. 2-9, 2-10, and 2-11 to give the intercepts of all planes, we must accept values for $h$, $k$, and $\ell$ which have a common factor. For example, suppose we use simple cubic lattice vectors to index planes in a body-centered cubic lattice. In addition to lattice planes through points at cube corners, there are planes through body centers. Planes parallel to (100) cube faces have intercepts at $\frac{1}{2}n$**a**, where $n$ is an integer, and we must use $h = 2$, $k = 0$, and $\ell = 0$ in Eqs. 2-9, 2-10, and 2-11.

The distance $d$ between adjacent planes with the same Miller indices is given by

$$d = \frac{\tau}{|\mathbf{g}|},\qquad(2\text{-}12)$$

where $\tau$ is the volume of the primitive unit cell and $\mathbf{g} = h\mathbf{b} \times \mathbf{c} + k\mathbf{c} \times \mathbf{a} + \ell\mathbf{a} \times \mathbf{b}$. Here **a**, **b**, and **c** are primitive and the indices have no common divisor except 1. Since the plane shown in Fig. 2-20 intersects the **a** axis at $x$**a**, the separation between it and the plane parallel to it through the origin is the projection of $x$**a** on the normal. Since $\mathbf{g}/|\mathbf{g}|$ is a unit vector normal to the plane, $d = |x\mathbf{a} \cdot \mathbf{g}|/|\mathbf{g}|$. Now $|x\mathbf{a} \cdot \mathbf{g}| = |x\hbar\mathbf{a} \cdot (\mathbf{b} \times \mathbf{c})| = |x\hbar\tau|$, so $d = |x\hbar\tau|/|\mathbf{g}| = |n|\tau/|\mathbf{g}|$. To obtain the interplanar spacing the plane shown must be adjacent to the one through the origin and $n$ must be 1; Eq. 2-12 follows.

**EXAMPLE 2-9**  For a simple cubic lattice with cube edge $a$, place the origin of a coordinate system at a lattice point and take the fundamental lattice vectors to be along cube edges. Find the Miller indices for a plane that intersects the **a** axis at 4**a**, the **b** axis at 3**b**, and the **c** axis at 2**c**. Then find the separation of adjacent planes parallel to the given plane. Finally, find the intercepts on the crystal axes of the parallel plane closest to the origin, exclusive of the one through the origin.

**SOLUTION**  For the plane described, $x = 4$, $y = 3$, and $z = 2$. The reciprocals are $\frac{1}{4}$, $\frac{1}{3}$, and $\frac{1}{2}$, respectively. To obtain the smallest integers with the same ratios, multiply each reciprocal by 12 and obtain $\hbar = 3$, $\ell = 4$, and $\ell = 6$. The plane is a (346) plane. If $\mathbf{a} = a\hat{\mathbf{x}}$, $\mathbf{b} = a\hat{\mathbf{y}}$, and $\mathbf{c} = a\hat{\mathbf{z}}$, then $\mathbf{g} = 3a^2\hat{\mathbf{x}} + 4a^2\hat{\mathbf{y}} + 6a^2\hat{\mathbf{z}}$ and $|\mathbf{g}| = \sqrt{61}a^2$. The primitive cell volume is $a^3$ so $d = \tau/|\mathbf{g}| = a^3/(\sqrt{61}a^2) = a/\sqrt{61} = 0.128a$. For the plane closest to the origin $x = 1/\hbar = \frac{1}{3}$, $y = 1/\hbar = \frac{1}{4}$, and $z = 1/\ell = \frac{1}{6}$ so its intercepts are $\frac{1}{3}a\hat{\mathbf{x}}$, $\frac{1}{4}a\hat{\mathbf{y}}$, and $\frac{1}{6}a\hat{\mathbf{z}}$. ∎

For any lattice the number of lattice points per unit area on any plane is given by $d/\tau$, where $\tau$ is the volume of the primitive unit cell. Imagine forming a primitive cell using two primitive lattice vectors in the plane and a third that extends from a lattice point in the plane to another in a neighboring plane. The number of lattice points per unit area of the plane is $1/A$, where $A$ is the area of the parallelepiped cell base in the plane. The unit cell volume is $Ad$, so $1/A = d/\tau$. For a face-centered cubic lattice, the largest concentration of lattice points occurs on (111) planes, while for body-centered cubic lattices, the largest concentration occurs on (110) planes.

**EXAMPLE 2-10**  Find the concentration of lattice points on a (111) plane of a face-centered cubic lattice.

**SOLUTION**  We take the primitive lattice vectors to be $\mathbf{a} = \frac{1}{2}a(\hat{\mathbf{x}} + \hat{\mathbf{y}})$, $\mathbf{b} = \frac{1}{2}a(\hat{\mathbf{y}} + \hat{\mathbf{z}})$, and $\mathbf{c} = \frac{1}{2}a(\hat{\mathbf{x}} + \hat{\mathbf{z}})$. The volume of a primitive unit cell is $\tau = |\mathbf{a} \cdot (\mathbf{b} \times \mathbf{c})| = \frac{1}{4}a^3$, as might be expected since there are four primitive lattice points per cube. The vector $\mathbf{g} = \mathbf{a} \times \mathbf{b} + \mathbf{b} \times \mathbf{c} + \mathbf{c} \times \mathbf{a} = \frac{1}{4}a^2(\hat{\mathbf{x}} + \hat{\mathbf{y}} + \hat{\mathbf{z}})$ is normal to (111) planes, so the plane separation is $d = \tau/|\mathbf{g}| = a/\sqrt{3}$ and the concentration of lattice points is $d/\tau = (a/\sqrt{3})/(a^3/4) = 4/(\sqrt{3}a^2) = 2.331/a^2$. ∎

## 2.5  REFERENCES

M. J. Buerger, *Contemporary Crystallography* (New York: McGraw-Hill, 1970).

M. J. Buerger, *Introduction to Crystal Geometry* (New York: McGraw-Hill, 1971).

D. McKie and C. McKie, *Crystalline Solids* (New York: Wiley, 1974).

H. D. Megaw, *Crystal Structures* (Philadelphia: Saunders, 1973).

D. E. Sands, *Introduction to Crystallography* (New York: Benjamin, 1969).

A. R. Verma and O. N. Srivastava, *Crystallography for Solid State Physics* (New York: Wiley, 1982).

E. J. W. Whittaker, *Crystallography* (Elmsford, NY: Pergamon, 1981).

## PROBLEMS

1. There are no two-dimensional lattices with triads as the axes of highest symmetry. Consider a threefold axis through a lattice point and show that, when translational symmetry is taken into account, it automatically becomes a sixfold axis.

2. Face-centered and body-centered tetragonal lattices were shown in the text to be identical Bravais types. Why are body-centered and face-centered orthorhombic lattices different Bravais types? To answer, consider four conventional body-centered unit cells with an edge in common and redraw the unit cell so it has lattice points at its corners and face centers.

3. Use the method outlined in the previous problem to show that both lattices in the following pairs are the same Bravais type: (a) face-centered and body-centered monoclinic; (b) base-centered and primitive monoclinic; (c) face-centered and primitive triclinic.

4. $SrTiO_3$ has what is known as the ideal perovskite structure. Strontium atoms are the corners of cubes, titanium atoms are at cube body centers, and oxygen atoms are at cube face centers. Take the cube edge length to be $a$. (a) What is the Bravais lattice type? (b) Verify that there are 3 oxygen atoms, 1 titanium atom, and 1 strontium atom for each primitive unit cell in the crystal. (c) Write down a set of primitive lattice vectors and associated basis vectors for this structure.

5. For each of the following sets of primitive lattice vectors, identify the Bravais lattice type and give the dimensions of the conventional unit cell in terms of $a$, $b$, and $c$:

   (a) $(a/2)\hat{x} + (a/2)\hat{y}, a\hat{y}, (a/\sqrt{2})\hat{z}$;

   (b) $(a/2)\hat{x} + (a/2)\hat{y}, a\hat{y}, a\hat{z}$;

   (c) $a\hat{x} + 2b\hat{y}, b\hat{y}, c\hat{z}$;

   (d) $\frac{1}{2}a\hat{x} + \frac{1}{2}b\hat{y}, b\hat{y}, c\hat{z}$.

6. Position vectors for lattice points in two different lattices are given by

   (a) $\mathbf{r} = (10n_1 + 9n_2 + 19n_3)(a/10)\hat{x} + 6(n_2 + n_3)(a/5)\hat{y} + 2n_3a\hat{z}$ and

   (b) $\mathbf{r} = \frac{1}{2}(2n_1 + n_2)a\hat{x} + \frac{1}{2}\sqrt{3}n_2a\hat{y} + 2n_3a\hat{z}$.

Here $n_1$, $n_2$, and $n_3$ are integers and $a$ is a length. In each case, find a set of primitive lattice vectors and identify the Bravais lattice type.

7. As $n_1$, $n_2$, and $n_3$ take on integer values, the vectors $\mathbf{r}_1$ and $\mathbf{r}_2$ given below describe the positions of atoms in a crystal. First assume $\mathbf{r}_1$ and $\mathbf{r}_2$ are associated with two different types of atoms and find a set of primitive lattice vectors and associated basis vectors for the crystal. Identify the Bravais lattice type. Then assume the atoms associated with $\mathbf{r}_1$ are identical to those associated with $\mathbf{r}_2$ and again find a set of primitive lattice vectors and associated basis vectors. Identify the Bravais lattice type.

(a) Take $\mathbf{r}_1 = (n_1 + n_3)a\hat{\mathbf{x}} + (n_2 + n_3)a\hat{\mathbf{y}} + n_3 a\hat{\mathbf{z}}$ and
$\mathbf{r}_2 = (n_1 + n_3 + \frac{1}{2})a\hat{\mathbf{x}} + (n_2 + n_3 + \frac{1}{2})a\hat{\mathbf{y}} + (n_3 + \frac{1}{2})a\hat{\mathbf{z}}$.

(b) Take $\mathbf{r}_1 = \frac{1}{5}(5n_1 + 3n_2)a\hat{\mathbf{x}} + \frac{4}{5}n_2 a\hat{\mathbf{y}} + 2n_3 a\hat{\mathbf{z}}$ and
$\mathbf{r}_2 = \frac{1}{5}(5n_1 + 3n_2 + 4)a\hat{\mathbf{x}} + \frac{2}{5}(2n_2 + 1)a\hat{\mathbf{y}} + (2n_3 + 1)a\hat{\mathbf{z}}$.

8. Take the primitive lattice vectors for a body-centered cubic lattice to be $\mathbf{a} = \frac{1}{2}a(\hat{\mathbf{x}} + \hat{\mathbf{y}} - \hat{\mathbf{z}})$, $\mathbf{b} = \frac{1}{2}a(-\hat{\mathbf{x}} + \hat{\mathbf{y}} + \hat{\mathbf{z}})$, and $\mathbf{c} = \frac{1}{2}a(\hat{\mathbf{x}} - \hat{\mathbf{y}} + \hat{\mathbf{z}})$, with the coordinate system as shown in Fig. 2-13. Express as linear combinations of these vectors: (a) position vectors for the eight cube corners; (b) position vectors for the eight points within the cubic unit cell that are one-fourth a body diagonal from cube corners; and (c) the lattice vector from $a\hat{\mathbf{y}}$ to $a\hat{\mathbf{x}}$, diagonally across the cell base.

9. Given a point with coordinates $x$, $y$ in the $xy$ plane, find its coordinates after each of the following operations: (a) a rotation by $\pi/2$ about the $z$ axis; (b) a rotation by $2\pi/3$ about the $z$ axis; (c) a rotation by $\pi/2$ about the $y$ axis; and (d) a reflection in the $yz$ plane.

10. Consider a simple cubic unit cell with edge $a$ and place a Cartesian coordinate system with its origin at a lattice point and its $z$ axis along a cube body diagonal. Choose any convenient orthogonal directions for the $x$ and $y$ axes. (a) Find the coordinates of all cube corners. (b) Find the coordinates of each corner after the cube is rotated by $2\pi/3$ about the $z$ axis. (c) Show that each corner rotates to a position previously occupied by another corner.

11. Primitive lattice vectors for a body-centered tetragonal lattice are $\mathbf{a} = \frac{1}{2}a(\hat{\mathbf{x}} + \hat{\mathbf{y}}) - \frac{1}{2}c\hat{\mathbf{z}}$, $\mathbf{b} = \frac{1}{2}a(-\hat{\mathbf{x}} + \hat{\mathbf{y}}) + \frac{1}{2}c\hat{\mathbf{z}}$, and $\mathbf{c} = \frac{1}{2}a(\hat{\mathbf{x}} - \hat{\mathbf{y}}) + \frac{1}{2}c\hat{\mathbf{z}}$, where $a$ is a side of the square base and $c$ is the height of the conventional unit cell. Initially $c > a$. The crystal is now compressed along the $z$ axis. (a) For what value of $c$ does the lattice become body-centered cubic? (b) For what value of $c$ does the lattice become face-centered cubic? Give your answers in terms of $a$.

12. Consider a trigonal unit cell and place a Cartesian coordinate system with its origin at a lattice point and its $z$ axis along the threefold symmetry axis of the cell. Three primitive lattice vectors have equal length $a$ and make the same angle $\theta$ with the symmetry axis. The projection of $\mathbf{a}$ on the $xy$

plane is along the positive $x$ axis, the projection of $\mathbf{b}$ makes an angle of 120° with the positive $x$ axis, and the projection of $\mathbf{c}$ makes an angle of 240° with the positive $x$ axis. (a) Write the primitive lattice vectors in terms of their Cartesian components. (b) For what value of $\theta$ is $\mathbf{a}$ perpendicular to $\mathbf{b}$? (c) Show that $\mathbf{c}$ is perpendicular to both $\mathbf{a}$ and $\mathbf{b}$ for the value of $\theta$ found in part b. (d) Show that the value of $\theta$ found in part b is the angle between an edge and body diagonal of a cube, through the same corner. (e) Find a general expression in terms of $a$ and $\theta$ for the length of the body diagonal along the symmetry axis, then show that when $\theta$ has the value found in part b the body diagonal has the length appropriate for a cube, $\sqrt{3}a$.

13. At 1190 K, iron has a face-centered cubic lattice with a cube edge of 3.647 Å, while at 1670 K, it has a body-centered cubic lattice with a cube edge of 2.932 Å. In each case the primitive basis contains one atom. Calculate the mass density of iron at these two temperatures. Naturally occurring iron has an average atomic weight of 55.85.

14. Consider a simple cubic unit cell with cube edge $a$. (a) Use the notation $uvw$ to write the positions of the eight points in the cell which are one-fourth a body diagonal from a cube corner. (b) Find the distance from one of these points to one of the nearest cube face centers.

15. For a simple cubic lattice with cube edge $a$, find the separation of lattice points along lines in the following directions: (a) [110]; (b) [111]; (c) [320]; and (d) [321].

16. Find Miller indices for the following lattice planes: (a) a plane parallel to both $\mathbf{a}$ and $\mathbf{c}$, in any lattice; (b) a plane parallel to both $3\mathbf{a} + \mathbf{c}$ and $\mathbf{b}$, in any lattice; (c) the plane containing the points $3\mathbf{a}$, $2\mathbf{b}$, and $\frac{1}{2}(\mathbf{a} + \mathbf{b} + \mathbf{c})$, in any lattice; and (d) a plane that contains a cube edge and cuts two other cube edges of the same cube at their midpoints, in a simple cubic lattice.

17. For each of the following sets of parallel planes, compare the spacing in a simple cubic lattice to the spacing of planes with the same Miller indices in face-centered cubic and body-centered cubic lattices with the same cube edge: (a) (100); (b) (110); (c) (111); (d) (210). Indices are referred to primitive simple cubic lattice vectors.

18. Compare the concentration of lattice points on (111) planes of a face-centered cubic lattice with the concentration on (110) planes of a body-centered cubic lattice with the same cube edge.

19. The concentration of lattice points on a (111) plane of a certain face-centered cubic lattice is the same as the concentration on a (110) plane of a certain simple cubic lattice. What is the ratio of their cube edges?

20. For a certain simple tetragonal lattice the ratio of the separation of (111) planes to the separation of (110) planes is 0.905. What is the ratio of the height of the conventional cell to the edge of its square base?

21. An orthorhombic lattice has primitive lattice vectors $\mathbf{a} = a\hat{\mathbf{x}}$, $\mathbf{b} = b\hat{\mathbf{y}}$, and $\mathbf{c} = c\hat{\mathbf{z}}$. Find the interplanar spacing and the number of lattice points per unit area for each of the following sets of planes: (a) (100); (b) (010); (c) (110); and (d) (101).

22. Crystals tend to cleave along planes with high concentrations of lattice points and usually these have small Miller indices. Angles between planes with small Miller indices are possible angles between crystal facets when the crystal is cut. For a simple cubic lattice, find the angle between planes of the following pairs: (a) (100) and (110); (b) (110) and (111); and (c) (111) and $(\bar{1}1\bar{1})$.

# Chapter 3

## STRUCTURES OF SOLIDS

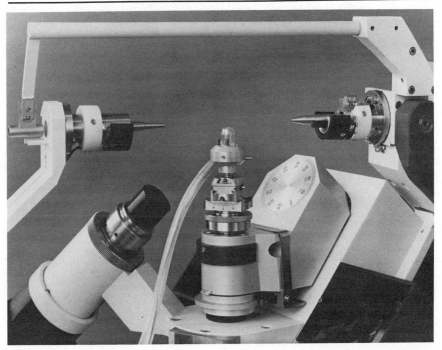

A goniometer used to orient crystals precisely for the determination of structure and orientation dependent properties. The crystal can be rotated about several axes and angles of rotation can be measured.

- 3.1 CRYSTAL STRUCTURES
- 3.2 STRUCTURAL DEFECTS
- 3.3 AMORPHOUS STRUCTURES
- 3.4 LIQUID CRYSTALS

This chapter contains detailed descriptions of many commonly occurring structures, both crystalline and amorphous. In most cases the structure of the solid is indicative of the bonding mechanism. Atoms in metals and inert gas solids, for example, are always packed tightly, with each atom having a large number of nearest neighbors. Atoms in covalent solids, on the other hand, have four or fewer nearest neighbors and are much more loosely packed. Other solids are structured so that the number of nearest neighbors is intermediate between that for a metal and that for a covalent solid. For this introduction to the subject, we consider simple structures for elemental solids, composed of a single type of atom, and binary compounds, composed of two types of atoms. References listed at the end of the chapter contain information about more complex structures. Because liquid crystals are technologically important a section dealing with their structures is also included.

## 3.1 CRYSTAL STRUCTURES

When a basis is replicated around each lattice point the result is called a crystal structure. Thousands of different crystal structures exist, but only a few are needed to describe crystals formed by most of the chemical elements and their simple compounds.

*Close-Packed Structures.* Most metals and all inert gas elements tend to crystalize in one or the other of the so-called close-packed structures, for which atoms are more tightly packed together than other structures allow. We might expect these materials to be close packed since the cohesive energy for metallic and van der Waals bonding increases with the number of nearest neighbors, up to the point where core overlap becomes significant.

Consider a crystal with one type atom and imagine the atoms to be spheres with radius $r$. The closest packing of spheres with centers on a plane is achieved if they are placed as shown in Fig. 3-1a. Sphere centers form a two-dimensional hexagonal lattice with primitive cell edge $a = 2r$. A sixfold symmetry axis passes through each sphere. A primitive hexagonal cell is shown on the diagram and points where threefold symmetry axes pierce the plane are marked with dots and X's: a dot marks an axis through the upper right portion of the cell shown and a X marks an axis through the lower left portion. These points are at the centers of regions between spheres and, to maintain close packing as other layers are added, spheres are placed over them so each sphere fits snugly into the well formed by three spheres below it.

Spheres with centers on the same plane do not fit over both sets of points. If the second layer of spheres is positioned so sphere centers are over dots, then no spheres on that plane can be centered over X's. If spheres of the second layer are over points marked with dots, one set of wells is directly over centers of spheres on the first layer and the second set is over points marked with X's. For purposes of this discussion, we use $A$ to designate a close-packed layer with spheres positioned as shown in the figure, $B$ to designate a layer with spheres centered on dots, and $C$ to designate a layer with spheres centered on X's.

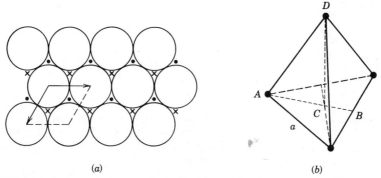

**FIGURE 3-1** (a) A plane of close-packed spheres. A side of the primitive hexagonal cell is $a = 2r$, where $r$ is the sphere radius. (b) A regular tetrahedron formed by four close-packed spheres. Each side has length $a$, $AB$ has length $\sqrt{1 - (1/2)^2}\, a = \sqrt{3}/2\, a$, and $AC$ is two-thirds of this or $a/\sqrt{3}$. $ACD$ is a right angle so $CD$ is $\sqrt{2/3}\, a$.

If the stacking sequence is $ABCABC\ldots$, then the atomic positions form a face-centered cubic (FCC) structure, with a face-centered cubic lattice and one atom in the primitive basis. The first layer is as shown in Fig. 3-1a, the second layer is shifted so cell corners lie above dots, and the third layer is shifted so cell corners lie above X's. Threefold axes pass through sphere centers on the first layer, then through threefold symmetry points on the left sides of second layer cells, and finally through threefold symmetry points on the right sides of third layer cells. Then the pattern repeats. This stacking scheme is clearly identical to the one described previously for cubic and trigonal lattices, using planes of two-dimensional hexagonal lattices; see Fig. 2-17.

To show that a face-centered cubic lattice and not a trigonal lattice is produced, we must find the ratio of the interplanar distance to the sphere separation. Recall that, for a face-centered cubic lattice with cube edge $a$, lattice points are $a/\sqrt{2}$ apart in a hexagonal plane and hexagonal planes are $a/\sqrt{3}$ apart. For the system of close-packed spheres, three adjacent spheres in a plane and a fourth sphere sitting in the well they form are at the vertices of a regular tetrahedron, as shown in Fig. 3-1b. The sides are equilateral triangles with edge length $2r$ and the altitude of the tetrahedron is $(2\sqrt{2/3})r$, so the interplanar spacing is $\sqrt{2/3}$ times the distance between spheres, the proper ratio for a face-centered cubic lattice. The translation vectors **a** and **b** shown in the diagram are along face diagonals of the cube and $2r$ is half the length of a face diagonal, so the cube edge is $2\sqrt{2}r$.

If the stacking sequence is $ABAB\ldots$, a hexagonal close-packed (HCP) structure results. Figure 3-2 shows the atomic positions for this structure. The primitive lattice is hexagonal and the primitive basis contains two atoms, one at a cell corner and one at $\frac{1}{3}\mathbf{a} + \frac{2}{3}\mathbf{b} + \frac{1}{2}\mathbf{c}$, for the cell shown. The second atom is on one of the threefold axes, halfway between the base and the top of the cell. Note that atoms are located along only one of the threefold axes; the other threefold axis marks an open channel, unoccupied by atoms. Also note that the displace-

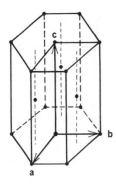

**FIGURE 3-2** Hexagonal close-packed structure. Dots represent atomic positions, at the corners of the primitive cell and on one of the interior threefold axes, halfway between the cell base and top. The other threefold axis is an open channel.

ment vector from one atom on the first layer to one on the second is not a lattice vector. There is no atom at twice this displacement, for example. For close-packed spheres of radius $r$, the length of a hexagon side is $a = 2r$ and the distance between layers is $(2\sqrt{2/3})r$, just as for FCC structures. Since the cell height is twice the interlayer distance, $c = (4\sqrt{2/3})r$ and the ratio $c/a$ is $2\sqrt{2/3} = 1.633$.

Many crystals have structures that are like an HCP structure in the shape of their unit cells and the indices of the atomic positions, but they differ from an HCP structure in their $c/a$ ratios. A comparison of the actual $c/a$ ratio with the ideal value of 1.633 can be used to determine the extent to which the structure resembles that formed by close-packed spheres. If $c/a$ is greater than the ideal value, we may think of spheres that are close packed on hexagonal planes but do not touch spheres on adjacent planes. Conversely, if $c/a$ is less than the ideal value, spheres touch others on adjacent hexagonal planes, but spheres on the same plane are pushed apart and, as a consequence, are not close packed.

The packing fraction is a number used to indicate how tightly atoms are packed in a given structure. Atoms are replaced by the largest spheres consistent with the cell size. The packing fraction is then the ratio of the total volume enclosed by these spheres to the total volume of the crystal. To find the packing fraction for a particular structure, first find the distance between closest atoms and take the sphere radius $r$ to be half that distance. Then the packing fraction $F$ is given by

$$F = N \frac{4\pi}{3} \frac{r^3}{\tau}, \qquad (3\text{-}1)$$

where $N$ is the number of atoms per unit cell and $\tau$ is the cell volume.

**EXAMPLE 3-1** (a) Find the packing fraction for a simple cubic structure, with a simple cubic lattice and one atom in the primitive basis. (b) Find the packing fraction for an FCC structure.

**SOLUTION** (a) Atoms at adjacent cube corners are nearest neighbors in a simple cubic structure. They are separated by the cube edge $a$ and the largest spheres consistent with the cube size have a radius of $\frac{1}{2}a$. The cell volume is $a^3$ and there is one sphere per cube, so the packing fraction if $F = (4\pi/3)(a/2)^3/a^3 = \pi/6 = 0.524$. (b) For an FCC structure, an atom at a cube corner and one at an adjacent face center are nearest neighbors, separated by half a face diagonal or $a/\sqrt{2}$. The appropriate sphere radius is $a/(2\sqrt{2})$ and there are four spheres per cube, so the packing fraction is $F = 4(4\pi/3)/(2\sqrt{2})^3 = \pi/(3\sqrt{2}) = 0.740$. Spheres arranged in a simple cubic structure occupy a little over half the available volume, while spheres arranged in an FCC structure occupy nearly three-fourths the available volume. ∎

An HCP structure has the same packing fraction as an FCC structure and this is the largest packing fraction of any crystal structure. Each atom in either structure is surrounded by 12 nearest neighbor atoms, the largest number of nearest neighbors of any crystal structure.

Of the chemical elements, 16 crystallize with FCC structures and 22 crystallize with HCP-like structures at room temperature. The noble metals (copper, silver, and gold) are FCC, while metals from columns IIA and IIB of the periodic table are HCP-like. Most of the transition elements are either FCC or HCP-like. In addition, the inert gases solidify at low temperatures to form FCC structures.

Magnesium has a $c/a$ ratio that is nearly ideal and many of the other HCP-like crystals have $c/a$ ratios that are only a few percent less than the ideal value. Structures for a few elements, such as zinc and cadmium, are HCP-like but with $c/a$ ratios larger than ideal. For zinc $c/a = 1.86$ and for cadmium $c/a = 1.89$.

FCC and HCP structures are quite similar and, for most solids mentioned above, the energy difference that decides the structure is extremely small. Cobalt, thallium, and cerium crystallize with both FCC- and HCP-like structures, depending on the temperature. Cubic forms are stable at high temperatures, hexagonal forms at low temperatures. Many cobalt crystals exhibit a stacking sequence such as *ABABCACA* .... Both of the sequences *ABAB* ... and *ACAC* ... are HCP-like but the hexagonal cells are shifted relative to each other. For the sequence given above, shifting is accomplished by means of the FCC-like sequence *ABCA*. Such shifts occur at irregular intervals throughout the solid. Small crystals of cobalt may exhibit random stacking, for which there is no repeated pattern to the order of the layers.

*Body-Centered Cubic Structures.*  At room temperature, 13 chemical elements form crystals with body-centered cubic (BCC) structures. These have body-centered cubic lattices and one atom in their primitive bases. While this structure is not close packed, it has a packing fraction of 0.680, only 8% less than that of the close-packed structures. Each atom has eight nearest neighbors, all of them half a cube body diagonal or $\sqrt{3}a/2$ away. This accounts for a packing that is less dense than that of FCC and HCP structures. However, there are six more atoms just a cube edge away, so each atom is surrounded by 14 others at nearly the same distance and the packing fraction remains high. All the alkali metals are BCC, as are other materials scattered throughout the left side of the periodic table and among the transition elements. Iron, for example, is BCC below 1179 K and above 1674 K. Between these two temperatures, it is FCC.

**EXAMPLE 3-2** Locate the nearest and next nearest neighbors of an atom in (a) an FCC structure and (b) a BCC structure.

**SOLUTION** (a) For an FCC structure, we concentrate on an atom at a cube corner. Twelve cube faces share this point and the nearest neighbors to the atom are at the centers of these faces. The nearest neighbor distance is half a face diagonal or $a/\sqrt{2}$, where $a$ is the cube edge. Six cube edges meet at the atom and one next nearest neighbor is along each of these, a cube edge away. All atoms of the FCC structure are equivalent so an atom at a face center also has 12 nearest neighbors, $a/\sqrt{2}$ away, and 6 next nearest neighbors, $a$ away. (b) For a BCC structure, we concentrate on an atom at a cube center. Its 8 nearest neighbors are at the corners of the cube enclosing it, each half a body diagonal or $\sqrt{3}a/2$ away. Its 6 next nearest neighbors are at the centers of the six neighboring cubes and are a cube edge away. An atom at a cube corner has the same configuration of atoms around it as one at a cube center. ∎

*Covalent Structures.* Because each atom is bonded to a maximum of only four neighbors, covalently bonded crystals are loosely packed. We examine three simple but important structures: diamond, zinc blende, and wurtzite.

The diamond structure has a face-centered cubic lattice and a primitive basis of two similar atoms. The positions taken by either of the basis atoms, when replicated, form a face-centered cubic lattice and the two lattices are displaced from each other by one-fourth a cube body diagonal. Figure 3-3a shows a perspective diagram and also a plan view of the structure. Atoms sit at cube corners and face centers. In addition, four atoms are in the interior of each cube, each displaced from one of the corners or face centers by one-fourth a body diagonal. For the structure shown, the displacements are in the direction from the lower left back corner of the cube toward the upper right front corner, but in other cases they may be along any of the other body diagonals. Each atom has four nearest neighbors, equally distant and at the vertices of a regular tetrahedron. The nearest neighbor distance is one-fourth a cube body diagonal or $\sqrt{3}a/4$, where $a$ is the cube edge.

This structure is very loosely packed. If the atoms are replaced by the largest spheres consistent with the cube size, their radius is half the nearest neighbor distance or $\sqrt{3}a/8$. There are eight atoms per cube so the packing fraction is $8(4\pi/3)(\sqrt{3}/8)^3 = 0.340$. Only about one-third of the volume is enclosed by spheres.

Silicon and germanium, two semiconducting materials, are the only chemical elements that crystallize with a diamond structure at room temperature. Below

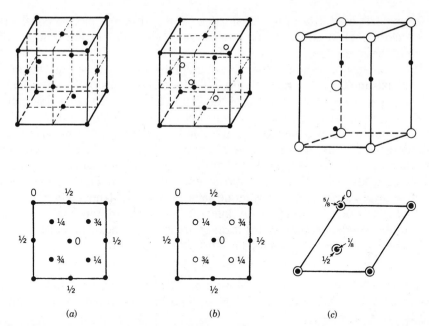

**FIGURE 3-3** Some crystal structures favored by covalent materials: (a) diamond, (b) zinc blende, and (c) wurtzite. In the plan views, elevations of atoms from the base plane are in units of cell height. In each structure the atoms are tetrahedrally bonded. In diamond all atoms are the same type while the other two structures contain equal numbers of two dissimilar atoms.

286.4 K, diamond is the stable structure for tin and, at high temperature and pressure, carbon solidifies with a diamond structure, hence the name.

The cubic zinc sulfide (ZnS) or zinc blende structure, shown in Fig. 3-3b, is similar to the diamond structure, but the two atoms of the primitive basis are of different types. The face-centered cubic lattice formed by one type atom is displaced from that formed by the other type by one-fourth a cube body diagonal and each atom sits at the center of a regular tetrahedron formed by four atoms of the other type. This is the structure of a great many binary semiconductors, such as CdS, InAs, InSb, AlP, and GaAs, as well as some noble metal halides, such as CuF, CuCl, and AgI.

In addition to crystallizing with a zinc blende structure, zinc sulfide has a hexagonal form called wurtzite, shown in Fig. 3-3c. The primitive basis consists of two atoms of each type and each type atom individually forms an HCP-like structure. The two HCP-like structures are displaced from each other along the c axis by five-eighths the cell height. Note that all atoms in the interior of the cell are along the same threefold axis. Each atom is at the center of a tetrahedron with four atoms of the other type at its vertices, but the tetrahedra are not quite regular, as they are for diamond and zinc blende. Other materials with this structure are ZnO, BeO, MgTe, SiC, and a second form of CdS. The $c/a$ ratios for these crystals are close to the ideal value, so the structure can be considered to be close-packed arrays of the larger atoms, with the smaller atoms filling some of the interstices.

The geometry of carbon in the form of graphite is not tetrahedral. One of the several existing forms is shown in Fig. 3-4, where elevations of the atoms above the base plane are given in units of the cell height. In each layer, the atomic positions are at the vertices of hexagons but these points do not form a lattice since no atoms occupy hexagon centers. Indeed, the primitive hexagonal cell has atoms at its corners and at one of the points where a threefold symmetry axis pierces the plane. The next layer has the same structure but with the interior atom at the other position of threefold symmetry in the cell. This pattern repeats as other layers are added. Atoms in the same hexagonal plane are held together by covalent bonds, while atoms in different layers are attracted to each other via van der Waals forces. For this form of graphite, the hexagon edge is about 1.4 Å and the distance between adjacent layers is about 3.35 Å.

**EXAMPLE 3-3**   For the (a) zinc blende and (b) wurtzite structures, as diagrammed in Fig. 3-3, locate the nearest neighbors to the atom at the lower left back cell corner.

**SOLUTION**   Figure 3-5 shows plan views of four cells for each structure. In each case the atom of interest has four nearest neighbors. (a) In zinc blende two are $\frac{1}{4}a$ above the plane of the page and are labeled $+\frac{1}{4}$. Two are $\frac{1}{4}a$ below the plane of the page and are labeled $-\frac{1}{4}$. (b) For wurtzite, three nearest neighbors are $\frac{1}{8}c$ above the plane of the page and are labeled $+\frac{1}{8}$ on the diagram. The other is $\frac{3}{8}c$ directly below the atom.   ■

*Other Cubic Structures.*   The cesium chloride (CsCl) structure, diagrammed in Fig. 3-6a, has a simple cubic lattice and a basis containing one atom of each type. Each atom is at the center of a cube, with atoms of the other type at the corners, all a distance $\sqrt{3}a/2$ away. Each atom also has six next nearest neighbors. These are of the same type and are each a distance $a$ away. If both atoms of the basis were the same, the structure would be BCC. As it is, a translation by half a cube body diagonal does not leave the crystal unchanged but, rather, interchanges the positions of the two types of atoms. CsCl structures are formed

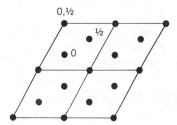

**FIGURE 3-4**   Plan view of a unit cell for a graphite structure. Elevations are in units of cell height.

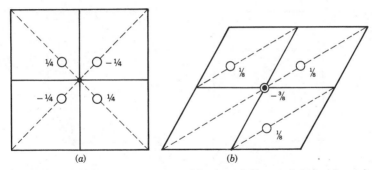

**FIGURE 3-5**   Plan views of unit cells for (a) zinc blende and (b) wurtzite, showing only a central atom (●) and it nearest nighbors (○). Elevations are in units of the cell height.

by a few halides, such as TlI and CsCl itself. A large number of intermetallic compounds, such as CuPd, CuZn (called β brass), AgMg, and AlNi, crystallize with this structure. This is not surprising considering the close resemblance of this structure to a BCC structure.

Other intermetallic compounds have the structure of $Cu_3Au$, shown in Fig. 3-6b. The lattice is simple cubic with gold atoms at cube corners and copper atoms at face centers. This structure would be FCC if all the atoms were identical. Each gold atom has 12 nearest neighbors, all $a/\sqrt{2}$ distant and all copper atoms. Each copper atom also has 12 nearest neighbors at the same distance, but 4 are gold and 8 are copper.

The sodium chloride (NaCl) structure, shown in Fig. 3-6c, has a face-centered cubic lattice and a basis consisting of one atom of each type. Note that atoms represented by dots are located at the corners and face centers of a cube and that atoms represented by open circles form an identical lattice, displaced from the first by half a cube edge. Each atom has six nearest neighbors, all of the other type and all half a cube edge distant. Some examples of materials that crystallize with NaCl structures are LiH, KCl, AgBr, MgO, and, of course, NaCl itself.

The lattice for the fluorite ($CaF_2$) structure, shown in Fig. 3-6d, is also face-centered cubic. The primitive basis consists of three atoms, two of one type and one of the other. In $CaF_2$ itself, the calcium atoms form a face-centered cubic lattice, while the fluorine atoms form two such lattices, each displaced from the calcium lattice by one-fourth a body diagonal, but along different body diagonals. This special combination of two face-centered cubic lattices produces a simple cubic lattice, with cube edge $\frac{1}{2}a$, imbedded in the calcium lattice, with cube edge $a$. There are four calcium atoms and eight fluorine atoms for each cubic unit cell. Each calcium atom has eight fluorine atoms for nearest neighbors, at the corners of a cube, while each fluorine atom sits at the center of a regular tetrahedron with calcium atoms at its vertices. The fluorite structure is favored by many fluorides, particularly those formed in partnership with column II elements. Some oxides, such as $ThO_2$, and some intermetallic compounds, such as $Mg_2Pb$, also crystallize with this structure.

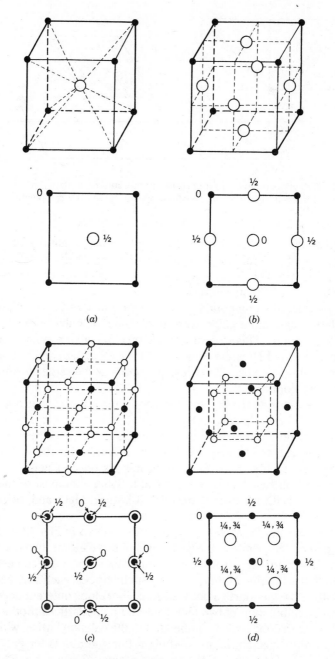

**FIGURE 3-6** Some crystal structures with cubic lattices: (a) CsCl, (b) Cu$_3$Au, (c) NaCl, and (d) CaF$_2$. Elevations from the base are in units of cell height. CsCl and Cu$_3$Au have simple cubic lattices while NaCl and CuF$_2$ have face-centered cubic lattices.

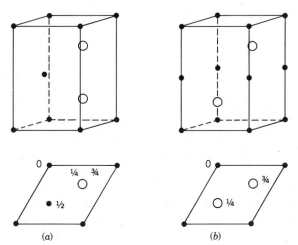

**FIGURE 3-7** Two crystal structures with hexagonal lattices: (a) NiAs and (b) CdI$_2$. Elevations are in units of cell height.

*Other Hexagonal Structures.* Figure 3-7 shows plan views of two important structures with hexagonal lattices. The nickel arsenide (NiAs) primitive basis consists of two atoms of each type, represented in the diagram by dots and open circles, respectively. Hexagonal planes of nickel atoms alternate with hexagonal planes of arsenic atoms, the layer separation being one-fourth the cell height. Using the same notation as was used to label hexagonal planes in FCC and HCP structures, the stacking sequence is $A(C)B(C)A(C)B(C)\ldots$, where labels for planes of nickel atoms are placed in parentheses. The arsenic atoms, taken alone, have an HCP-like structure but the $c/a$ ratio is usually very different from the ideal value. A great many sulfides, selenides, antimonides, and arsenides crystallize with this structure.

Figure 3-7b shows the cadmium iodine (CdI$_2$) structure, with two cadmium and four iodine atoms in its primitive basis. Taken by themselves, the iodine atoms (open circles) form an HCP-like structure and cadmium atoms occupy some of the interstices. Note that iodine atoms have other iodine atoms as well as cadmium atoms for nearest neighbors. Many halides, such as MgBr$_2$, CaI$_2$, and PbI$_2$, crystallize with this structure.

*Other Structures.* Although crystals of most chemical elements and their simple compounds have cubic or hexagonal lattices, some do not. For example, gallium, indium, and one form of maganese are tetragonal; iodine, oxygen, and one form of sulfur are orthorhombic; and arsenic, antimony, bismuth, mercury, and another form of sulfur are trigonal. For many of these solids, the basis is large and the structure is quite complex. Details can be found in the references listed at the end of this chapter.

## 3.2 STRUCTURAL DEFECTS

Deviations from an ideal structure, consisting of a basis replicated at lattice points throughout all space, are known as defects. Here we describe some of them; later we will see how they influence some properties of materials.

*Point Defects.* The three types of point defects are illustrated in Fig. 3-8. A vacancy is a site in the bulk material which would be occupied by an atom if the periodic pattern of the crystal were followed. Instead, the atom is missing. An interstitial is an atom that occupies a position between sites prescribed by the periodic pattern. An impurity is an atom of a type that is different from that of the host atoms. It may replace a host atom or occupy an interstitial position. In some cases, isotopes of the host must be considered impurities. Point defects may form groups: complexes of several thousand vacancies or more, called voids, have been observed, as have clusters of impurity atoms.

A closely related defect may occur in crystals of more than one type atom. As an example, consider a crystal with two different type atoms, A and B. If some A atoms occupy sites that, according to the periodic pattern, should be occupied by B atoms, while the displaced B atoms occupy A sites, then the crystal is said to be disordered.

In spite of the label, point defects disrupt the pattern of a crystal over several interatomic distances. Atoms neighboring a vacancy, for example, relax to positions that are closer to the vacant site than they would be if the site were occupied. Interstitials usually push neighboring atoms outward, while the solid may contract or expand in the vicinity of an impurity, depending on the size of the impurity atom.

Vacancies and interstitials are thermally generated. Equilibrium concentrations can be calculated by requiring the free energy $F$, defined by

$$F = E - TS, \tag{3-2}$$

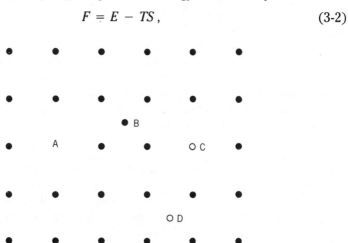

**FIGURE 3-8** Point defects in crystals. ●, Host atoms; ○, impurity atoms; A, a vacancy; B, an interstitial; C, a substitutional impurity; D, an interstitial impurity.

be a minimum. Here $E$ is the energy and $S$ is the entropy of the crystal. Both these quantitites depend on the number of defects. Although vacancies and interstitials increase the energy of a crystal, they also increase the entropy, so equilibrium concentrations do not vanish except when $T = 0$.

We calculate the concentration of isolated vacancies in a crystal with one atom in its primitive basis. The entropy is dominated by the configurational component, which is associated with the number of distinct arrangements of atoms. Suppose the crystal contains $N$ atoms and $N_v$ vacancies. These can be arranged $(N + N_v)!$ different ways on $N + N_v$ sites, but interchanges of atoms with other atoms or vacancies with other vacancies do not lead to different configurations. The number of distinct configurations is given by $(N + N_v)!/N!N_v!$ and the configurational entropy is

$$S = k_B \ln \left[ \frac{(N + N_v)!}{N!N_v!} \right]$$
$$\approx k_B[(N + N_v) \ln(N + N_v) - N \ln(N) - (N_v) \ln(N_v)] , \qquad (3\text{-}3)$$

where Stirling's approximation was used to obtain the last form: for large $x$, $\ln(x!) = x \ln(x)$. Other contributions to the entropy arise from changes in electron and phonon states, but they are small and we ignore them here.

Let $E_v$ be the energy required to create a vacancy. Since the number of atoms does not change when a vacancy is created, $E_v$ is essentially the energy required to move an atom from the interior to the surface of the sample. If we neglect multiple vacancy complexes, the energy of the crystal is increased by $N_v E_v$ when $N_v$ vacancies are created, so the free energy is

$$F = N_v E_v - k_B T[(N + N_v) \ln(N + N_v) - N \ln(N) - N_v \ln(N_v)] . \quad (3\text{-}4)$$

$F$ is a minimum when $N_v/(N + N_v) = e^{-\beta E_v}$, where $\beta = 1/k_B T$. Since $N_v \ll N$, the equilibrium number of vacancies is given by

$$N_v = Ne^{-\beta E_v} . \qquad (3\text{-}5)$$

According to Eq. 3-5, the number of vacancies is zero at $T = 0$ K and increases with temperature.

For most materials $E_v$ is several electron volts. If $E_v = 2$ eV, for example, then $N_v/N$ is about $3 \times 10^{-34}$ at room temperature and about $9 \times 10^{-11}$ at 1000 K. A graph of $\ln(N_v/N)$ as a function of $\beta$ is used to obtain a value for $E_v$ from experimental data. It is a straight line with slope $-E_v$.

The equilibrium interstitial concentration in a solid is calculated in much the same way. Only a small number of interstitial sites exist in a unit cell and the probability that any one of them is occupied depends on the energy required to bring an atom from the sample surface to the site. This energy is on the order of an electron volt for structures with small packing fractions, such as diamond, but is well over 10 eV for close-packed structures. Both an interstitial and a vacancy are created when an atom leaves its site and enters the region between atoms. The energy of interest for a Frenkel defect, as this combination is called, is clearly the sum of the energies required to create a vacancy and an interstitial individually.

Solids often have nonequilibrium vacancy and interstitial concentrations. Quenching, for example, produces this result. A large defect concentration is created by allowing a sample to come to equilibrium at a high temperature, then the sample is cooled quickly. Insufficient energy is available for atoms to move past each other, as they must to fill excess vacancies or get to the surface.

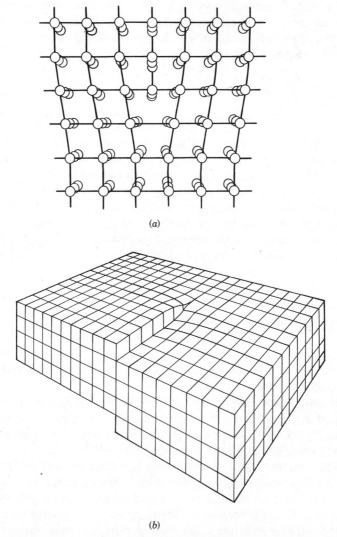

(a)

(b)

**FIGURE 3-9** Dislocations in crystals. (a) An edge dislocation. An extra plane of atoms starts in the center of the diagram and extends upward. (b) A screw dislocation. The plane of the dislocation extends part way into the crystal from the front. Atoms on one side are shifted downward by one unit cell relative to atoms on the other side. (From J. M. Hirth and J. Lothe, *Theory of Dislocations* (New York: Wiley, 1982). Used with permission.)

Bombardment of a sample with high-energy particles also produces excess defect concentrations. The study of radiation damage is an important part of modern material science.

*Dislocations.*    Figure 3-9 illustrates the two fundamental types of dislocations. In principle, an edge dislocation can be produced by slicing the crystal along a plane that extends part way through the sample, inserting an extra plane of atoms in the cut, and then allowing the atoms to relax to new equilibrium positions. The dislocation itself is along the line that forms the lower edge of the inserted plane. Far from this line atomic positions resume their usual pattern. After making the same type cut, a screw dislocation is produced by shifting atoms above and to one side of the cut one lattice constant parallel to the cut. There are more general types of dislocations but these can always be constructed as combinations of edge and screw dislocations.

Large numbers of dislocations are usually produced during crystal growth. If a dislocation line extends to the surface of a sample, the misalignment of crystal planes there results in a jagged edge where atoms from the melt are easily adsorbed. Dislocations are also produced by flexing a sample, a process called cold working.

*Two-Dimensional Defects.*    In a sense, the surface of a finite sample must be considered a defect. As we will see, most calculations of bulk properties are carried out with the assumption that the crystal displays translational periodicity and hence extends throughout all space. Surfaces are responsible for small corrections to the results and for phenomena in their own right. Electron states and atomic motions, unrelated to those of the interior, are associated with surfaces. We will discuss them at appropriate places in this book. Boundaries between crystallites in a polycrystalline sample may also be considered two-dimensional defects.

## 3.3  AMORPHOUS STRUCTURES

Structures of amorphous materials cannot be described with the same great detail as crystalline structures. Since atomic positions in amorphous materials do not display long-range periodicity, an accurate description of the structure requires a listing of all atomic positions. Even if that were possible, the enormous length of the list would render it virtually useless. Instead of such a listing, the average distribution of atoms relative to a central or reference atom is used to describe the structure. The average distribution is given by what is known as the radial distribution function.

*Radial Distribution Functions.*    To illustrate the physical significance of a radial distribution function, we consider an amorphous solid that consists of only one type atom and describe how the function might be evaluated if all atomic positions were known. First, with any one of the atoms chosen as a reference atom, we calculate the concentration $n(r)$ of atomic centers in a spherical shell extending from $r$ to $r + \Delta r$, measured from the center of the reference atom. That is, we divide the number of atoms with centers in the shell by the shell volume. The process is repeated using each atom in turn as the reference atom. In general, the result is different for different atoms since, unlike the situation for a crystal, the distribution of atoms is not the same around every atom. The average value of $n(r)$ for the system of atoms is computed and, finally, the limit as the shell thickness $\Delta r$ becomes small is evaluated. This is the average concentration $\bar{n}(r)$ of atoms a distance $r$ from an atom. The radial distribution function $\rho(r)$ is defined by

$$\rho(r) = 4\pi r^2 \bar{n}(r). \tag{3-6}$$

Of course, since the material contains a large number of atoms, the process just described cannot be carried out in practice. However, models of amorphous structures containing a few thousand atoms have been constructed and computers have been used to evaluate radial distribution functions for them. Radial distribution functions for actual solids are found experimentally using x-ray techniques.

Two limiting cases are of interest: a completely ordered crystal and a random distribution of atoms. We expect the radial distribution function for an amorphous solid to show features indicative of both these special cases.

First consider the radial distribution function for a crystal with an FCC structure. If the cube edge is $a$, then around each atom there are 12 other atoms at a distance of $a/\sqrt{2}$, 6 other atoms at a distance of $a$, 24 other atoms at a distance of $\sqrt{3/2}a$, etc. If we imagine an extremely narrow spherical shell centered on one of the atoms and count the number of atoms with centers in the shell, the result is zero unless the shell straddles one of the distances corresponding to an interatomic separation for the FCC structure. If it does straddle one of these distances, the concentration blows up. A finite number of atoms are within the shell and the shell volume becomes zero as $\Delta r$ decreases. For example, $n(r) = 0$ for $r$ between 0 and $a/\sqrt{2}$, then $n(r)$ is infinite for $r = a/\sqrt{2}$. As illustrated in Fig. 3-10, the radial distribution function for a crystal consists of a series of spikes, located at the various interatomic separations. It is zero between the spikes.

At the other extreme, if atomic centers are randomly distributed throughout the material, then the average concentration $\bar{n}$ does not depend on $r$. No matter what the value of $r$, pairs of atoms separated by that distance are located somewhere in the material and, in addition, the number of pairs with a given separation is the same as the number with any other separation. The average concentration is the same as the macroscopic concentration $n_0$ of atoms in the solid: the total number of atoms divided by the volume of the solid. For a random distribution

$$\rho(r) = 4\pi r^2 n_0. \tag{3-7}$$

This function is plotted in Fig. 3-11. Note that $\rho(r)$ for a random distribution is decidedly unrealistic for small $r$: atomic sites in a real solid are excluded from the near vicinity of other sites by core-core repulsion and $\rho(r) = 0$ for a range of separations near $r = 0$.

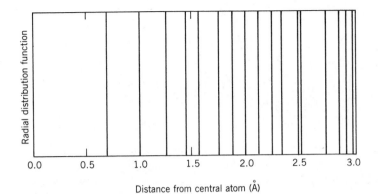

**FIGURE 3-10** Radial distribution function for a crystalline FCC structure. Distance from the central atom is given in units of the cube edge. The function consists of a series of infinite spikes, one for each of the various FCC interatomic separations.

A radial distribution function for an amorphous solid is shown in Fig. 3-12. The general trend of the curve follows the radial distribution function for a random distribution. At large $r$ the two curves nearly match. For small $r$, however, the function is significantly different from that for a random distribution. In fact, it has many features in common with the radial distribution function for a crystal. The function becomes zero for small but nonzero values of $r$, indicating there is a distance of closest approach for the atoms. At larger $r$ the function consists of a series of peaks, analogs of the spikes in the radial distribution function for a crystal. The peak that occurs at the smallest value of $r$, for example, corresponds to the separation of close neighbors. Atoms tend to congregate around each other and the concentration of close neighbors is much greater than the concentration for a random distribution. Unlike the spikes in a radial distribution

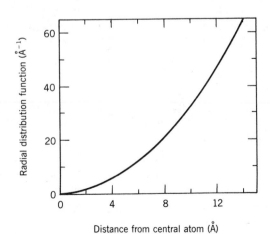

**FIGURE 3-11** Radial distribution function for a random distribution of atoms with macroscopic concentration $2.55 \times 10^{28}$ atoms/m³.

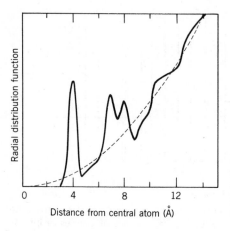

**FIGURE 3-12** Radial distribution function for an amorphous metal. The general trend follows the function for a random distribution, shown by the dotted curve. Deviations from the dotted curve indicate short-range order.

Radial distribution function

Distance from central atom (Å)

function for a crystal, however, the peaks have finite heights and nonzero widths. All close neighbors, for example, do not have exactly the same separation.

Just beyond the first peak, the radial distribution function dips to a value that is less than the value for a random distribution. Close neighbors, clustered around an atom, tend to exclude other atomic centers from their near neighborhood. As $r$ increases beyond the first few peaks, the peaks become broader and less pronounced. Amorphous materials display short-range order but not long-range order. Crystals, on the other hand, display both long- and short-range order.

A great deal of structural information is lost in the averaging process that leads to the radial distribution function. The function contains no information about the angular distribution of atoms about any atom, nor does it contain information about the separation of any particular pair of atoms. Nevertheless, details of the function are quite helpful in investigations of amorphous materials. Positions of peaks indicate dominant atomic separations. Widths of peaks are also of interest. A very narrow first peak, for example, indicates that nearly all close neighbors have about the same separation, while a wide first peak indicates a wide range of close-neighbor separations. The number of peaks is a measure of the range of order to the structure. If many peaks are discernible, the structure is ordered far from any atom but if only a few peaks can be seen, the structure is ordered only in the near neighborhood of an atom.

A radial distribution function can be used to find the average number of atoms in a given spherical shell. The average number of atoms with centers in the range from $r_1$ to $r_2$ is given by

$$\bar{N} = \int_{r_1}^{r_2} \bar{n}(r)\, d\tau, \tag{3-8}$$

where $d\tau$ is an infinitesimal volume element. Since $\bar{n}(r) = \rho(r)/(4\pi r^2)$ and $d\tau = 4\pi r^2\, dr$,

$$\bar{N} = \int_{r_1}^{r_2} \rho(r)\, dr. \tag{3-9}$$

If the integration is carried out over the first peak in the radial distribution function, for example, $\overline{N}$ represents the average number of close neighbors to an atom of the solid. We emphasize that this is an average over all atoms: some atoms have more close neighbors while others have fewer. $\overline{N}$ need not be an integer.

Since the radial distribution function gives the average separation of close neighbors, it can be used to estimate a packing fraction for an amorphous solid. First, the atoms are replaced by spheres of radius $R$, where $R$ is half the value of $r$ for the first peak of the function. If $n_0$ is the macroscopic concentration of atoms in the material, there are $N = n_0\tau$ atoms in the macroscopic volume $\tau$. The volume enclosed by spheres is $N(4\pi/3)R^3 = n_0\tau(4\pi/3)R^3$ and the packing fraction $F$ is the ratio of this volume to $\tau$, or

$$F = \frac{4\pi}{3} n_0 R^3 .\qquad(3\text{-}10)$$

*Amorphous Metals.* The function shown in Fig. 3-12 is typical of radial distribution functions for amorphous metals. Peaks in the function, which arise from the structure of the material, can be understood to a large extent in terms of a model using hard spheres to represent atoms. The model is quite similar to that used in the preceding section to develop FCC and HCP crystal structures.

A radial distribution function similar to that of Fig. 3-12 can be obtained by placing several thousand ball bearings in a container with irregular sides, shaking the container, and observing their positions when they have settled. Irregular sides are used to prevent the formation of a crystal structure. A distribution function can also be constructed entirely by computer. Coordinates of sphere centers are produced one at a time by a random number generator, subject to the conditions that no spheres overlap and each sphere touches at least three others. Once the positions are known, the computer is used again to evaluate the radial distribution function. Distributions such as these are known as random close-packed structures.

In many instances the closest neighbors of each sphere in a random close-packed structure form a ring with all of them in contact with the cental sphere. The configuration is similar to the distribution of nearest neighbors in an FCC or HCP crystal. A large number of spheres with close neighbor configurations like this give rise to the large first peak in the radial distribution function. It occurs at $r = 2R$, where $R$ is the sphere radius. Since no sphere centers can be closer than $2R$, the function vanishes for $r < 2R$. The first peak of a more realistic radial distribution function has a slight tail toward small $r$, indicating some core overlap occurs in an actual metal.

Not all spheres of the structure have the configuration described above. In some instances one or more close neighbors might not touch the reference sphere. Configurations with close neighbor distances slightly greater than $2R$ cause the first peak to have a finite width. The larger the sphere separation, the fewer the number of occurrences in the structure, so the function decreases beyond $r = 2R$.

There are other peaks, corresponding to other highly likely atomic configurations. For example, the dual peaks that are prominent on the graph are at distances of about $2\sqrt{3}R$ and $4R$, respectively. These correspond to atomic separations in close-packed crystal structures. Peaks at large $r$ are broad and the radial distribution function does not deviate much from that for a random distribution. Resemblance to a crystal structure is lost.

Both numbers of close neighbors and packing fractions are nearly the same for all random close-packed distributions. If the integral of the radial distribution function is evaluated for the range from $r = 0$ to the first minimum, the result is nearly always between 11.5 and 12. This result agrees quite well with the same quantity obtained from x-ray data for actual amorphous metals. In addition, the packing fraction for all random close-packed structures is close to 0.64, about 13% less than the packing fraction for FCC and HCP crystalline structures.

**EXAMPLE 3-4** Use the radial distribution function of Fig. 3-12 to estimate the sphere radius $R$ for an equivalent close-packed distribution of spheres. Then use the value obtained for $R$ to estimate the positions of the dual peaks and the packing fraction. The macroscopic concentration is 0.02 atoms/$\text{Å}^3$.

**SOLUTION** The first peak occurs at about 4 Å, so we may take the sphere radius to be 2 Å. Dual peaks should occur at about $r = 2\sqrt{3}R \approx 7$ Å and $r = 4R \approx 8$ Å. These values agree with the curve of Fig. 3-12. The packing fraction is $F = (4\pi/3)n_0R^3 = (4\pi/3) \times 0.02 \times 2^3 \approx 0.67$, somewhat greater than that for the usual random close-packed distribution but less than that for a close-packed crystal structure. ∎

*Amorphous Covalent Structures.* The radial distribution function for an amorphous covalent solid is shown in Fig. 3-13. The curve shows a fairly sharp first peak, after which it falls to zero before rising to the second peak. Thereafter it displays a few more peaks but generally follows the parabolic curve for a random distribution. Unlike the function for an amorphous metal, the second peak is not split.

The position of the first peak gives the close-neighbor distance and, for amorphous covalent solids, it is almost exactly the same as the nearest-neighbor distance for a crystal of the same material. Packing fractions of crystalline and amorphous covalent solids are the same, 0.34. The macroscopic concentration of atoms in an amorphous solid is also very nearly the same as the concentration in a crystal of the same material, so amorphous and crystalline forms of the same material have nearly the same density. The number of nearest neighbors to an atom can be found by integrating the radial distribution function over the first peak and, for covalent solids, the result is almost exactly 4 in all cases.

These characteristics suggest a structure that closely resembles that of a crystal, at least locally around each atom. In fact, the radial distribution function is duplicated quite accurately by a model in which each atom is bonded to four others and the bond length varies by only 1% or less from that in the crystal. The model seems reasonable on physical grounds since the atoms cannot form

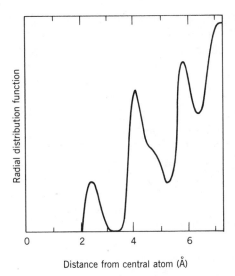

**FIGURE 3-13** Radial distribution function for an amorphous covalent solid. The function goes to zero after the first peak. Then, following a series of peaks, it approaches the function for a random distribution. The first peak represents almost exactly four nearest neighbors.

more than four bonds and the formation of fewer results in a considerable increase in energy. In addition, the bond length is determined to a large extent by the minimum in the energy of a single bond as a function of atomic separation, so we expect little difference in bond lengths.

Angles between adjacent bonds are also nearly the same in crystalline and amorphous forms of the same covalent material. In a diamond structure all bond angles are the same, about 109°. For most models of covalent amorphous structures all bond angles are within 20° of 109° and the standard deviation of the bond angle distribution is less than 10°.

The chief way in which amorphous and crystalline covalent structures differ is in the distribution of neighbors beyond the closest. In crystalline germanium, for example, second nearest neighbors to an atom at a cube corner are at face-centered sites. In amorphous germanium this pattern is destroyed. The bond system associated with a nearest-neighbor atom may have any angular position around the line joining the central atom and the nearest neighbor. In terms of the diamond structure, second nearest neighbors may lie anywhere on circles around the cube body diagonal and through the nearest face centers, not just at the face centers themselves. After several nearest-neighbor distances the distribution approaches a random distribution.

## 3.4 LIQUID CRYSTALS

Some aspects of liquid crystal structures display long-range crystal-like order while other aspects may display amorphous disorder. Liquid crystals are composed of organic molecules that are long and narrow, with either rod-like or plate-like shapes. Either their orientations or positions may be ordered. Furthermore, molecular positions may be ordered in one direction and disordered in another.

Those molecules that form liquid crystals by themselves are generally shaped like rods with lengths of about 25 Å and diameters of about 5 Å. Because the structures of most of these materials are temperature dependent, they are said to have thermotropic phases.

Other organic molecules form liquid crystals only when they are placed with high concentration in appropriate solvents. They are described by the term lyotropic, which means dispersion dependent. Synthetic polypeptide molecules, which are rodlike with lengths of about 300 Å and diameters of about 20 Å, are examples. Soaps, which have complex organic molecules, form liquid crystal structures at sufficiently high concentrations in water and other suitable solvents.

*Thermotropic Liquid Crystals.* Many different structures assumed by thermotropic liquid crystals have been identified. The most important of these are called nematic, smectic A, smectic B, and smectic C. The term nematic comes from the Greek word for thread while the term smectic comes from the Greek word for soap. These structures are diagrammed in Fig. 3-14.

Not all thermotropic liquid crystals form all the structures named above, but those that do are nematic at high temperatures. Molecular orientations are ordered, with the long axes of the molecules tending to be aligned. The tendency toward alignment is opposed by thermal motion and the number of aligned molecules increases as the temperature decreases. Molecular positions are not ordered and, strictly speaking, the material is liquid. It flows and its structure changes with time.

Below a transition temperature, the material has a smectic A phase, characterized by a layered structure. In any layer, the molecular arrangement is amorphous: it does not display long-range order. However, long-range order does occur in the direction perpendicular to the layers. More precisely, the molecular concentration is a periodic function of distance measured normally to the layers. Molecules of a smectic A structure are oriented with their long dimension perpendicular to the layers and the repeat distance is about the same as the length of a molecule.

At a still lower temperature the liquid crystal becomes smectic C. Just as for smectic A structures, this structure is layered and the material is a two-dimensional liquid with a molecular concentration that is periodic in the third direction. The molecules are still aligned. Now, however, the direction of alignment is not perpendicular to the layers. The layer separation decreases during the transition so that, afterward, it is $\ell \cos \theta$, where $\ell$ is the length of a molecule and $\theta$ is the angle between the long dimension of the molecule and the normal to the layer.

As the temperature is lowered past another transition temperature the material enters the smectic B phase, the last phase before complete crystallization. The molecules in each layer now form a regular array, so each layer is a two-dimensional crystal. Molecular positions on one layer are not related to molecular positions on other layers, however.

Thermotropic liquid crystals are technologically important. Molecular orientations can be controlled by both magnetic and electric fields. Since reflectivity

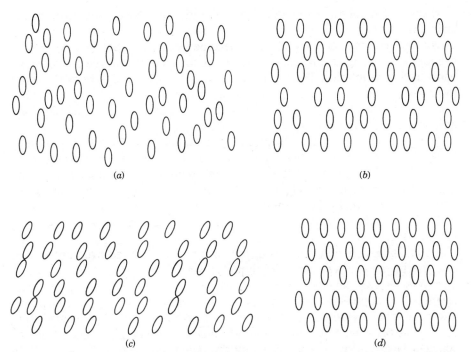

**FIGURE 3-14** Some liquid crystal structures. (a) Nematic. Molecular positions do not have long-range order but the molecules are aligned. (b) Smectic A. Molecules are concentrated in layers and are aligned perpendicular to the layers. (c) Smectic C. Molecules are concentrated in layers and are aligned, but their long dimension is not perpendicular to the layers. (d) Smectic B. Molecular positions form two-dimensional crystals. Molecular positions on different layers are not related.

depends on the molecular orientation, parts of a liquid crystal film with different molecular orientations can be distinguished optically. These two properties make liquid crystals extremely useful in display devices. The color of some liquid crystals depends on the molecular orientation. For some, the color is pressure dependent while, for others, it is temperature dependent. Liquid crystals are used as both pressure and temperature sensors.

*Lyotropic Liquid Crystals.* A typical soap molecule consists of a hydrocarbon chain bonded ionically to a positive ion, such as $Na^+$, $K^+$, or $NH_4^+$. The ionic end of the chain has an electric dipole moment and tends to attract water molecules. Water molecules tend to avoid the hydrocarbon end. In water, soap molecules collect in such a way that their dipolar ends shield their hydrocarbon ends from water molecules. At low concentrations this is accomplished by the formation of nearly spherical globules containing from 20 to 100 soap molecules, perhaps several hundred angstroms in diameter. The polar ends of the molecules form the surface of a globule, in contact with the water, while the hydrocarbon ends are inside.

As more soap is added, the globules become larger, elongate, and evolve into cylindrical rods. Again the polar ends are at the surface and the hydrocarbon ends are in the interior. The formation of globules and rods is responsible for the cleansing action of soap. The hydrocarbon ends have an affinity for hydrocarbon molecules in oil or grease and, if these are present, soap globules form around them, making the combination soluble.

Cylindrical rods of soap molecules are the building blocks of some lyotropic liquid crystal structures. When the concentration reaches a critical value and a sufficient number of rods are formed, the rods tend to become aligned and form a hexagonal arrangement. Figure 3-1 can be interpreted as a diagram of the structure, with the circles representing ends of rods. In reality, the rods do not quite touch each other and water molecules fill the interstitial space, bonding the rods together.

At still higher concentrations, soap molecules form a layered structure, similar to a smectic phase of a thermotropic liquid crystal. Molecules tend to concentrate on layers and are aligned perpendicularly to the layers. Chains on alternate layers point in opposite directions so the hydrocarbon end of a molecule points toward another hydrocarbon end from another layer. Water is excluded from the region between these layers. Polar ends face other polar ends from an adjacent layer and water molecules fill the region between these layers. Both the hexagonal arrangement of rods and the layered arrangement of molecules are liquid crystal structures.

## 3.5 REFERENCES

All of the references given at the end of Chapter 2 contain material on the structure of crystals. To these we add:

J. V. Smith, *Geometrical and Structural Crystallography* (New York: Wiley, 1982).
A. F. Wells, "The Structures of Crystals" in *Solid State Physics* (F. Seitz and D. Turnbull, Eds.), Vol. 7, p. 425 (New York: Academic, 1958).

Discussions of amorphous structures can be found in:

G. S. Cargill III, "Structure of Metallic Alloy Glasses" in *Solid State Physics* (H. Ehrenreich, F. Seitz, and D. Turnbull, Eds.), Vol. 30, p. 227 (New York: 1975).
J. C. Phillips, "The Physics of Glass." *Physics Today* **35**:27, Feb. 1982.
R. Zallen, *The Physics of Amorphous Solids* (New York: Wiley, 1983).

Discussions of liquid crystals can be found in:

P. G. deGennes, *The Physics of Liquid Crystals* (London: Oxford Univ. Press, 1975).
J. D. Litster and R. J. Birgeneau, "Phases and Phase Transitions." *Physics Today* **35**:26, May 1982.
P. S. Pershan, "Lyotropic Liquid Crystals." *Phys. Today* **35**:34, May 1982.

The following are chiefly compilations of crystal structure data:

C. S. Barrett and T. B. Massalski, *Structure of Metals: Crystallographic Methods, Principles, Data* (New York: McGraw-Hill, 1966).

J. D. H. Donnay and G. Donnay, *Crystal Data* (American Crystallographic Association, 1963). [Washington, D.C.]

International Union of Crystallography, *Structure Reports*, N. V. A. Oosthoek's Uitgevers MIJ (published annually).

W. B. Pearson; *Handbook of Lattice Spacings and Structures of Metals and Alloys* (Elmsford, NY: Pergamon, 1967).

A. Taylor and B. J. Kagle, *Crystallographic Data on Metal and Alloy Structures* (New York: Dover, 1963).

R. W. G. Wyckoff, *Crystal Structures* (New York: Wiley-Interscience, 1963).

Material dealing with defects can be found in:

L. A. Girifalco, *Statistical Physics of Materials* (New York: Wiley, 1973).

J. P. Hirth and J. Lothe, *Theory of Dislocations* (New York: Wiley, 1982).

F. A. Kroger and H. J. Vink, "Relations between the Concentrations of Imperfections in Crystalline Solids" in *Solid State Physics* (F. Seitz and D. Turnbull, Eds.), Vol. 3, p. 307 (New York: Academic, 1956).

## PROBLEMS

1. On each plane of a set of parallel planes, atomic positions form a two-dimensional hexagonal lattice with hexagon edge $a = 3.25$ Å. The planes are labeled according to the scheme discussed in Section 3.1 and all atoms are identical. For each of the following sequences of planes, identify the Bravais lattice and find the lengths of the conventional unit cell edges. (a) *ABAB* ..., with plane separation 4.88 Å; (b) *ABCABC* ..., with plane separation 2.66 Å; (c) *ABCABC* ..., with plane separation 2.15 Å; and (d) *ABCABC* ..., with plane separation 1.32 Å.

2. Show that the packing fraction for a body-centered tetragonal structure (a body-centered tetragonal lattice and a primitive basis of one atom) is $(\pi/3)(a/c)$ if $c > \sqrt{2}a$ and is $(\pi/24)(a/c)(2 + c^2/a^2)^{3/2}$ if $c < \sqrt{2}a$. Here $a$ is the edge of the square base and $c$ is the height of the conventional unit cell. Show that for $c = a$ the packing fraction for a BCC structure is obtained.

3. Consider an HCP-like structure with hexagonal edge $a$ and cell height $c$. (a) For $c/a > 2\sqrt{2/3}$, show that the nearest-neighbor distance is $a$ and that the packing fraction is $(2\pi/3\sqrt{3})(a/c)$. (b) For $c/a < 2\sqrt{2/3}$, show that the nearest-neighbor distance is $[(c^2/4) + (a^2/3)]^{1/2}$ and that the packing fraction is $(\pi/12\sqrt{3})[(c/a)^2 + (4/3)]^{3/2}(a/c)$. (c) For $c/a = 2\sqrt{2/3}$, show that the expressions found in parts a and b agree. In particular, show that the packing fraction is $\pi/(3\sqrt{2})$, the same as for an FCC structure.

4. The cubic unit cell of sodium chloride has an edge of 5.63 Å. (a) How many nearest neighbors does a sodium atom have? What type atom are they?

What is the nearest-neighbor distance? (b) How many next nearest neighbors does a sodium atom have? What type atom are they? What is the next-nearest-neighbor distance? (c) Consider (111) planes of atoms, indexed using simple cubic lattice vectors. Take the base plane to be a two-dimensional hexagonal lattice formed by sodium atoms and label it $A$. Continue to use the notation of Section 3.1 to label successive planes and tell what type atom is on each.

5. When a basis is added to a lattice, rotational and mirror symmetry may be reduced. For each of the structures listed below, tell what rotational and mirror symmetries of the hexagonal lattice are reduced or missing. Assume the atoms of the basis are spherically symmetric so they are unchanged by rotations or reflections. (a) Hexagonal close packed; (b) NiAs; and (c) wurtzite.

6. Cadmium iodide crystallizes with the structure shown in Fig. 3-7. Cell dimensions are $a = 4.24$ Å and $c = 6.84$ Å. (a) For the cell depicted in the diagram, find the position indices of all atoms. (b) What is the shortest Cd-Cd separation?; the shortest Cd-I separation? (c) The atomic weights of cadmium and iodine are 112.4 and 126.9, respectively. What is the density of $CdI_2$?

7. Sodium has atomic weight 23.0 and crystallizes with a BCC structure. At room temperature its density is $1.01 \times 10^3$ kg/m$^3$. Calculate (a) the length of a conventional unit cell edge; (b) the nearest-neighbor distance; and (c) the number of atoms per unit area on a (111) plane.

8. Zinc sulfide crystallizes with both zinc blende ($a = 5.41$ Å) and wurtzite ($a = 3.81$ Å, $c = 6.23$ Å) structures. Compare the nearest-neighbor distances and the densities for the two forms of zinc sulfide. Zinc has an atomic weight of 65.4 and sulfur has an atomic weight of 32.1.

9. Gallium arsenide crystallizes with a zinc blende structure. The Ga-As bond length is 2.45 Å. (a) What is the length of a cube edge? (b) What is the shortest Ga-Ga separation? (c) What is the density of GaAs? The atomic weights of gallium and arsenic are 69.7 and 75.0, respectively.

10. What is the $c/a$ ratio for a wurtzite structure in which each atom sits at the center of a *regular* tetrahedron with four atoms of the other type at the vertices? Find the nearest-neighbor distance in terms of $a$.

11. Suppose gallium arsenide crystallizes with a wurtzite structure instead of a zinc blende structure. Take the $c/a$ ratio to have the value found in Problem 10 and the bond length to have the value given in Problem 9, then calculate its density.

12. Suppose a crystal has $N$ normal sites and $N_1$ interstitial sites. (a) Show that the number of ways of arranging $N_f$ atoms on interstitial sites and $N - N_f$ atoms on normal sites is given by

$$W = \frac{N!}{N_f!(N - N_f)!} \frac{N_1!}{N_f!(N_1 - N_f)!}.$$

(b) Let $E_f$ be the energy required to move an atom from a normal site to an interstitial site and show that the free energy is a minimum when

$$\frac{(N - N_f)(N_1 - N_f)}{N_f^2} = e^{\beta E_f}.$$

(c) Show that the number of Frenkel defects is given by

$$N_f = (NN_1)^{1/2} e^{-\beta E_f/2}$$

if it is much smaller than $N$ and $N_1$.

13. An energy of 2.0 eV is required to form a Frenkel defect in a certain crystal with one atom in its primitive basis and eight interstitial sites in each primitive unit cell. Use the result of Problem 12 to find the equilibrium number of Frenkel defects per unit cell at 100, 300, and 1000 K.

14. List the chief similarities in radial distribution functions for amorphous metals and amorphous covalent solids. List the chief differences. Consider the position and width of the first peak, the depth of the well between the first and second peaks, the shape of the function near the second peak, and the behavior for large $r$.

15. A certain metal crystallizes with an FCC structure. It also forms an amorphous structure with a density 12% less than its crystalline density. Assume nearest-neighbor distances are the same for the two forms and estimate the packing fraction for the amorphous solid.

16. (a) Use the radial distribution function of Fig. 3-13 to estimate the overall concentration of atoms. Compare your result to the concentration of atoms in a crystal with a diamond structure and a cube edge of 5.66 Å. (b) Estimate the bond length for the amorphous solid and compare your result to the bond length in the crystal. (c) Estimate the number of atoms with centers between 3.5 and 4.5 Å from another atom in the amorphous solid. Compare your result to the analogous number in a random distribution with the same concentration.

# Chapter 4

## ELASTIC SCATTERING OF WAVES

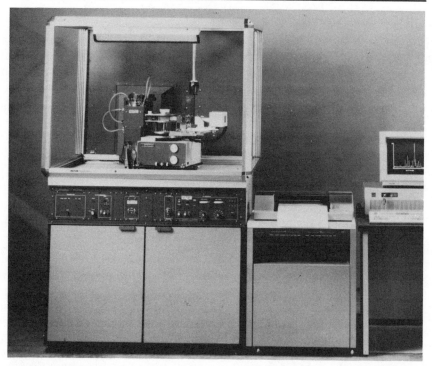

An x-ray diffraction apparatus. The camera and detector are mounted behind the glass window. Cabinets contain the high voltage power supply and a strip chart recorder to plot intensity as a function of scattering angle.

- 4.1 INTERFERENCE OF WAVES
- 4.2 ELASTIC SCATTERING BY CRYSTALS
- 4.3 EXPERIMENTAL TECHNIQUES
- 4.4 SCATTERING FROM SURFACES
- 4.5 ELASTIC SCATTERING BY AMORPHOUS SOLIDS

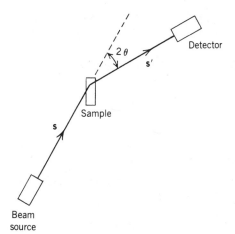

**FIGURE 4-1** Scattering from a sample. The beam may consist of x rays, electrons, or neutrons. Propagation vectors for the incident and scattered waves are **s** and **s'**, respectively, and the scattering angle is 2θ.

Figure 4-1 shows the geometry for a scattering experiment. A well collimated beam of x rays, electrons, or neutrons is directed onto the sample, where it is scattered. The intensity at the detector is measured as a function of the scattering angle 2θ and used to draw inferences about the atomic structure of the sample.

Electromagnetic and electron waves scatter from electrons in the sample, while neutron waves scatter from nuclei. In each case, waves scattered from particles of the solid interfere to produce a diffraction pattern. If the sample is crystalline, scattering angles corresponding to maxima in the pattern are used to determine the structure. If the sample is amorphous, the intensity as a function of scattering angle is used to obtain information about the radial distribution function. In this chapter the relationship between the intensity pattern and the sample structure is examined and experimental techniques used to obtain the intensity pattern are discussed.

## 4.1 INTERFERENCE OF WAVES

*Incident and Scattered Waves.* For most experiments the incident wave is nearly a monochromatic plane wave, characterized by a propagation vector **s** and an angular frequency ω. The propagation vector is in the direction the wave travels and has magnitude $2\pi/\lambda$, where λ is the wavelength. The wave function can be written

$$\xi(\mathbf{r}, t) = Ae^{i(\mathbf{s} \cdot \mathbf{r} - \omega t)}, \qquad (4\text{-}1)$$

where $A$ is the amplitude and $i$ is the unit imaginary number $\sqrt{-1}$.

A complex exponential function is used to describe the wave for two reasons. First, when waves scattered from particles of the system are summed, the sum can be evaluated more easily using an exponential function than a trigonometric function. Second, the same complex function can be used to describe electromagnetic, electron, and neutron waves. If the wave is electromagnetic, we take the real part of $\xi$ to represent one Cartesian component of the electric field, for example. For electrons and neutrons the complex function is a solution to the Schrödinger equation for a free particle. A solution cannot be represented by a real trigonometric function.

Scattering from a particle in the sample produces a spherical wave. Its wavefronts are spheres centered at the scattering particle and they move radially outward from the particle. If the scattering particle is at $\mathbf{r}'$ and the detector is at $\mathbf{r}$ then the scattered wave has the form

$$\xi_s(\mathbf{r}, t) = \frac{A'}{|\mathbf{r} - \mathbf{r}'|}\, e^{i(s|\mathbf{r} - \mathbf{r}'| - \omega t)} \tag{4-2}$$

at the detector. We consider elastic scattering, for which the frequency of the scattered wave is the same as the frequency of the incident wave and $s' = s$. A similar wave arrives at the detector from each scattering particle in the sample and the resultant wave is their sum.

Usually the sample size is small compared to the sample-detector distance and the wave scattered by any particle can be closely approximated by a plane wave at the detector. The particle-detector distance $|\mathbf{r} - \mathbf{r}'|$ is nearly the same for all particles in the sample, so we may treat it as a constant in the denominator of Eq. 4-2. In the exponent, however, small variations of $s|\mathbf{r} - \mathbf{r}'|$ with $\mathbf{r}'$ play an important role in determining the diffraction pattern and we must retain the $\mathbf{r}'$ dependence there.

Define a propagation vector $\mathbf{s}'$ that has magnitude $2\pi/\lambda$ and points from the scattering particle toward the detector. Since $\mathbf{r} - \mathbf{r}'$ is a vector in the same direction, $s|\mathbf{r} - \mathbf{r}'| = \mathbf{s}' \cdot (\mathbf{r} - \mathbf{r}')$ and the wave emanating from $\mathbf{r}'$ is given by

$$\xi_s(\mathbf{r}, t) = A_s e^{i[\mathbf{s}' \cdot (\mathbf{r} - \mathbf{r}') - \omega t]} \tag{4-3}$$

at the detector. This is a plane wave. Since the detector is far away, all waves that reach it travel along nearly parallel paths and we may take $\mathbf{s}'$ to be the same for all of them.

The amplitude $A_s$ of the scattered wave is proportional to the incident wave at the position of the scattering particle. That is, $A_s$ is proportional to $A e^{i\mathbf{s} \cdot \mathbf{r}'}$. For this discussion, the proportionality constant is not important and we may write

$$\begin{aligned}
\xi_s(\mathbf{r}, t) &= A e^{i(\mathbf{s} \cdot \mathbf{r}' - \omega t)} e^{i\mathbf{s}' \cdot (\mathbf{r} - \mathbf{r}')} \\
&= A e^{i(\mathbf{s}' \cdot \mathbf{r} - \omega t)} e^{-i(\mathbf{s}' \mathbf{s}) \cdot \mathbf{r}'} \\
&= A e^{i(\mathbf{s}' \cdot \mathbf{r} - \omega t)} e^{-i\Delta\mathbf{s} \cdot \mathbf{r}'},
\end{aligned} \tag{4-4}$$

where $\Delta\mathbf{s} = \mathbf{s}' - \mathbf{s}$ is the change on scattering of the propagation vector.

The magnitude of $\Delta\mathbf{s}$ can be written in terms of the scattering angle $2\theta$ and the wavelength $\lambda$. Since the propagation vectors $\mathbf{s}$ and $\mathbf{s}'$ have the same magnitude but differ in direction by $2\theta$, $\mathbf{s}' \cdot \mathbf{s} = s^2 \cos(2\theta)$ and

$$\begin{aligned}
|\Delta\mathbf{s}| = |\mathbf{s}' - \mathbf{s}| &= [2s^2 - 2\mathbf{s} \cdot \mathbf{s}']^{1/2} \\
&= \sqrt{2}s[1 - \cos(2\theta)]^{1/2}.
\end{aligned} \tag{4-5}$$

Use the identity $\cos(2\theta) = 1 - 2\sin^2(\theta)$ to write $|\Delta s| = 2s \sin \theta$ or, since $s = 2\pi/\lambda$,

$$|\Delta s| = \frac{4\pi}{\lambda} \sin \theta .$$

(4-6)

This relationship will be important later.

*Interference.* To understand how a diffraction pattern comes about, consider scattering from two particles, one at $r_1$ and the other at $r_2$. The resulutant wave at the detector is the sum of the two scattered waves or

$$\xi_s(\mathbf{r}, t) = A e^{i(\mathbf{s}' \cdot \mathbf{r} - \omega t)}[e^{-i\Delta s \cdot r_1} + e^{-i\Delta s \cdot r_2}]$$

$$= A e^{i(\mathbf{s}' \cdot \mathbf{r} - \omega t)} e^{-i\Delta s \cdot r_1}[1 + e^{i\Delta s \cdot (r_1 - r_2)}] .$$

(4-7)

In complex notation the wave intensity is proportional to the magnitude squared of $\xi$. This is the product of $\xi$ with its complex conjugate, formed from $\xi$ when $i$ is replaced by $-i$ and denoted by $\xi^*$. For the incident beam $\xi^*\xi = |A|^2$ since the exponential has magnitude 1. For the scattered wave

$$\xi_s^* \xi_s = |A|^2[2 + e^{i\Delta s \cdot (r_1 - r_2)} + e^{-i\Delta s \cdot (r_1 - r_2)}]$$

$$= |A|^2\{2 + 2 \cos[\Delta s \cdot (r_1 - r_2)]\}$$

$$= 4|A|^2 \cos^2[\tfrac{1}{2} \Delta s \cdot (r_1 - r_2)] ,$$

(4-8)

where the identities $e^{i\alpha} + e^{-i\alpha} = 2 \cos \alpha$ and $\cos \alpha = 2 \cos^2(\tfrac{1}{2}\alpha) - 1$, valid for any angle $\alpha$, were used. The detector reading is proportional to the magnitude squared of $\xi_s$.

Equation 4-8 indicates that if the particles are positioned so that $\Delta s \cdot (r_1 - r_2)$ is a multiple of $2\pi$, then the intensity is a maximum. If, on the other hand, $\Delta s \cdot (r_1 - r_2)$ is an odd multiple of $\pi$, then the intensity vanishes. The quantity $\Delta s \cdot (r_1 - r_2)$ is the difference in phase of the two waves. As Fig. 4-2 demonstrates,

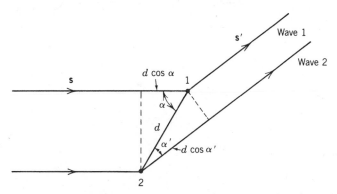

**FIGURE 4-2** 1 and 2 are scattering particles in the sample. Wave 2 goes ($d \cos \alpha' - d \cos \alpha$) further than wave 1, so the waves have a phase difference of $(2\pi/\lambda)(d \cos \alpha' - d \cos \alpha) = (s' - s) \cdot (r_1 - r_2)$.

it arises from the difference in the distances they travel. If $d = |\mathbf{r}_1 - \mathbf{r}_2|$ is the separation of the two particles, then wave 2 travels $d(\cos \alpha' - \cos \alpha)$ farther than wave 1 and their phase difference is $\Delta\phi = (2\pi/\lambda)(d \cos \alpha' - d \cos \alpha)$. Since $\mathbf{s}$ has magnitude $2\pi/\lambda$ and makes the angle $\alpha$ with $\mathbf{r}_1 - \mathbf{r}_2$ and $\mathbf{s}'$ has magnitude $2\pi/\lambda$ and makes the angle $\alpha'$ with $\mathbf{r}_1 - \mathbf{r}_2$, $\Delta\phi = (\mathbf{s}' - \mathbf{s}) \cdot (\mathbf{r}_1 - \mathbf{r}_2)$.

We suppose waves that reach the detector have been scattered only once. Multiple scattering is not important for x rays but is important for the scattering of electrons by electrons. To find the total scattered wave we must sum expressions like that of Eq. 4-4, one for each for scattering particle of the material. For concreteness, we consider the scattering of x rays by electrons and take $n(\mathbf{r}')$ to be the electron concentration. Then the sum over the scattered waves can be approximated by a volume integral with the integrand given by the product of Eq. 4-4 and $n(\mathbf{r}')$:

$$\xi_s(\mathbf{r}, t) = Ae^{i(\mathbf{s}' \cdot \mathbf{r} - \omega t)} \int e^{-i\Delta\mathbf{s} \cdot \mathbf{r}'} n(\mathbf{r}') \, d\tau' , \qquad (4\text{-}9)$$

where the primed coordinates are variables of integration and the integral is over the volume of the sample.

We carry out the integration atom by atom. Refer to Fig. 4-3. Suppose the nucleus of atom $j$ is located at $\mathbf{r}_j$ and the concentration of electrons around it is $n_j$. Let $\mathbf{r}''$ represent the displacement from the nucleus of an electron in the atom. Then $\mathbf{r}' = \mathbf{r}_j + \mathbf{r}''$ and

$$\xi_s(\mathbf{r}, t) = Ae^{i(\mathbf{s}' \cdot \mathbf{r} - \omega t)} \sum_j e^{-i\Delta\mathbf{s} \cdot \mathbf{r}_j} \int e^{-i\Delta\mathbf{s} \cdot \mathbf{r}''} n_j(\mathbf{r}'') \, d\tau'' , \qquad (4\text{-}10)$$

where the integral is over the electron distribution of atom $j$ and the sum is over all atoms of the solid. Except for a factor $A$, the integral

$$f_j(\Delta\mathbf{s}) = \int e^{-i\Delta\mathbf{s} \cdot \mathbf{r}''} n(\mathbf{r}'') \, d\tau'' \qquad (4\text{-}11)$$

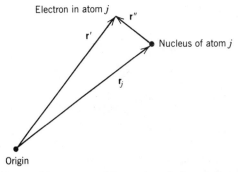

**FIGURE 4-3** The relative position vector $\mathbf{r}''$ is used to designate the position of an electron near nucleus $j$. It is at $\mathbf{r}' = \mathbf{r}_j + \mathbf{r}''$, where $\mathbf{r}_j$ is the position of the nucleus.

represents the sum of all waves scattered from electrons in atom $j$ and is called the atomic form factor for that atom. It is, in general, a complex number.

Many electrons in a solid cannot be assigned unambiguously to specific atoms but, fortunately, all assignments lead to the same results as long as the various atomic concentrations $n_j$, taken together, faithfully describe the total electron concentration. For our discussion of crystals, however, we assume the same atomic form factor for atoms that are equivalent by virtue of crystalline periodicity.

In terms of atomic form factors, the total scattered wave can be written

$$\xi_s(\mathbf{r}, t) = A e^{i(\mathbf{s}' \cdot \mathbf{r} - \omega t)} \sum_j f_j e^{-i\Delta \mathbf{s} \cdot \mathbf{r}_j} , \qquad (4\text{-}12)$$

where the sum is over all atoms of the sample. In this expression, waves from all atoms are summed, each weighted by the appropriate atomic form factor. This is as far as we can go without specifying atomic positions. In following sections we examine the intensity $|\xi_s|^2$ for both crystalline and amorphous structures. Before going on, however, we give an example of an atomic form factor.

**EXAMPLE 4-1**  As a rough model of an atom, assume the electron concentration has the uniform value $n$ inside a sphere of radius $R$ and is zero outside. (a) Find an expression for the atomic form factor as a function of $|\Delta s|R$. (b) Suppose the sphere has a radius of 2.5 Å and contains 10 electrons. Take the wavelength of the wave to be 2.0 Å and evaluate the atomic form factor for scattering angles of 5, 60, 90, and 120°.

**SOLUTION**  (a) If the $z$ axis is chosen to be along $\Delta s$ and spherical coordinates are used, the atomic form factor is given by

$$f = n \int_{r=0}^{R} \int_{\theta=0}^{\pi} \int_{\phi=0}^{2\pi} e^{-i|\Delta s|r\cos\theta} \, r^2 \sin\theta \, d\theta \, d\phi \, dr .$$

Here $\theta$ represents the usual spherical coordinate and is not half the scattering angle. The integral in $\phi$ yields $2\pi$. To evaluate the integral in $\theta$, set $\alpha = |\Delta s|r$ and use

$$\int_0^{\pi} e^{-i\alpha\cos\theta} \sin\theta \, d\theta = \left. \frac{e^{-i\alpha\cos\theta}}{i\alpha} \right|_0^{\pi}$$

$$= \frac{1}{i\alpha} (e^{i\alpha} - e^{-i\alpha}) = \frac{2}{\alpha} \sin\alpha ,$$

to obtain

$$f = 4\pi n \int_0^R \frac{\sin(|\Delta s|r)}{|\Delta s|} r \, dr .$$

Since $\int \sin(|\Delta s|r)r \, dr = (1/|\Delta s|^2)[\sin(|\Delta s|r) - |\Delta s| \cos(|\Delta s|r)$,

$$f = \frac{4\pi n}{|\Delta s|^3} [\sin(|\Delta s|R) - |\Delta s|R \cos(|\Delta s|R)] \ .$$

(b) If there are $N$ electrons in the sphere, the concentration is $n = 3N/4\pi R^3$ and, if $\beta = |\Delta s|R$, then

$$f = \frac{3N}{\beta^3} (\sin \beta - \beta \cos \beta) \ .$$

According to Eq. 4-6, $\beta = (4\pi R/\lambda) \sin \theta$, where $\theta$ is now half the scattering angle. For $2\theta = 5°$, $\beta = 0.685$ and $f = 9.54$; for $2\theta = 60°$, $\beta = 7.85$ and $f = 0.0619$; for $2\theta = 90°$, $\beta = 11.1$ and $f = -0.0488$; and for $2\theta = 180°$, $\beta = 15.7$ and $f = 0.122$. ∎

Figure 4-4 shows a plot of $|f|^2$ as a function of $\beta$ for the uniform sphere model of the example. For small scattering angles or small values of the ratio $R/\lambda$, $\beta$ is small and the atomic form factor is nearly equal to the number of electrons in the atom. It decreases quite rapidly with increasing $\beta$, indicating that the scattering intensity is small if the wavelength is much less than the atomic radius. As we will see, the wavelength cannot be much greater than the inter-atomic spacing or a diffraction pattern is not produced. Thus atomic form factors and the need to obtain a diffraction pattern place limits on the range of wavelengths used.

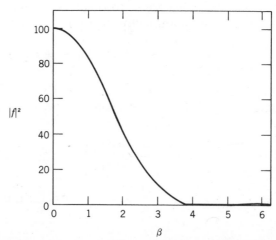

**FIGURE 4-4** The square $|f|^2$ of the atomic form factor for a sphere containing a uniform distribution of 10 electrons. $\beta = (4\pi R/\lambda)\sin \theta$, where $R$ is the sphere radius and $\theta$ is half the scattering angle. At small $\beta$, $|f|$ equals the number of electrons in the sphere, but it decreases rapidly with increasing $\beta$.

## 4.2 ELASTIC SCATTERING BY CRYSTALS

*Scattered Intensity.* If the atoms form a periodic array, Eq. 4-12 can be simplified considerably. Atomic positions are then given by vectors of the form $\mathbf{r}_j = n_1\mathbf{a} + n_2\mathbf{b} + n_3\mathbf{c} + \mathbf{p}_m$, where $\mathbf{a}$, $\mathbf{b}$, and $\mathbf{c}$ are fundamental lattice vectors and $\mathbf{p}_m$ is a basis vector. Equation 4-12 becomes

$$\xi_i = Ae^{i(\mathbf{s}'\cdot\mathbf{r}-\omega t)} \sum_{n_1}\sum_{n_2}\sum_{n_3}\sum_{m} f_m e^{-i\Delta\mathbf{s}\cdot(n_1\mathbf{a}+n_2\mathbf{b}+n_3\mathbf{c}+\mathbf{p}_m)}, \qquad (4\text{-}13)$$

where the first three sums on the right are over all lattice points and the fourth is over atoms of the basis. Different atoms of the basis may have different form factors but form factors for equivalent atoms in different unit cells are assumed to be identical. Since all unit cells of the crystal are exact duplicates of each other, the sum

$$F = \sum_m f_m e^{-i\Delta\mathbf{s}\cdot\mathbf{p}_m} \qquad (4\text{-}14)$$

is the same for all unit cells of the crystal and can be factored from each term of Eq. 4-13. $F$ is called the structure factor and gives the sum of the waves scattered from electrons in a single unit cell, apart from the amplitude factor $A$.

Once the structure factor is factored from all terms, Eq. 4-13 becomes

$$\xi_i = Ae^{i(\mathbf{s}'\cdot\mathbf{r}-\omega t)}F\left[\sum_{n_1} e^{-in_1\Delta\mathbf{s}\cdot\mathbf{a}}\right]\left[\sum_{n_2} e^{-in_2\Delta\mathbf{s}\cdot\mathbf{b}}\right]\left[\sum_{n_3} e^{-in_3\Delta\mathbf{s}\cdot\mathbf{c}}\right]. \qquad (4\text{-}15)$$

Because terms of each sum form a geometrical progression, the sums can be evaluated in closed form. We assume the crystal has $N$ unit cells along each of the three crystal axes, defined by $\mathbf{a}$, $\mathbf{b}$, and $\mathbf{c}$. Then $n_1$, for example, runs from 0 to $N - 1$ and the first sum is

$$\sum_{n_1=0}^{N-1} e^{-in_1\Delta\mathbf{s}\cdot\mathbf{a}} = \frac{e^{-iN\Delta\mathbf{s}\cdot\mathbf{a}} - 1}{e^{-i\Delta\mathbf{s}\cdot\mathbf{a}} - 1} = \frac{e^{-iN\Delta\mathbf{s}\cdot\mathbf{a}}}{e^{-i\Delta\mathbf{s}\cdot\mathbf{a}}}\frac{\sin(\frac{1}{2}N\,\Delta\mathbf{s}\cdot\mathbf{a})}{\sin(\frac{1}{2}\,\Delta\mathbf{s}\cdot\mathbf{a})}. \qquad (4\text{-}16)$$

We have factored $e^{-iN\Delta\mathbf{s}\cdot\mathbf{a}/2}$ from both terms in the numerator and $e^{-i\Delta\mathbf{s}\cdot\mathbf{a}/2}$ from both terms in the denominator, then used the identity $e^{i\alpha} - e^{-i\alpha} = 2i\sin\alpha$ in both numerator and denominator. When Eq. 4-16 and similar expressions for the other lattice sums are substituted into Eq. 4-15 and the magnitude of $\xi_s$ is squared to obtain the intensity at the detector, the result is

$$|\xi_s|^2 = |A|^2|F|^2 \left[\frac{\sin(\frac{1}{2}N\,\Delta\mathbf{s}\cdot\mathbf{a})}{\sin(\frac{1}{2}\,\Delta\mathbf{s}\cdot\mathbf{a})}\right]^2 \left[\frac{\sin(\frac{1}{2}N\,\Delta\mathbf{s}\cdot\mathbf{b})}{\sin(\frac{1}{2}\,\Delta\mathbf{s}\cdot\mathbf{b})}\right]^2 \left[\frac{\sin(\frac{1}{2}N\,\Delta\mathbf{s}\cdot\mathbf{c})}{\sin(\frac{1}{2}\,\Delta\mathbf{s}\cdot\mathbf{c})}\right]^2. \qquad (4\text{-}17)$$

Usually a crystal does not contain the same number of unit cells along each crystal axis. If it does not, the $N$'s that appear in different factors of this expression have different values.

We now examine the factor $\sin^2(\frac{1}{2}N\,\Delta\mathbf{s}\cdot\mathbf{a})/\sin^2(\frac{1}{2}\,\Delta\mathbf{s}\cdot\mathbf{a})$. For convenience,

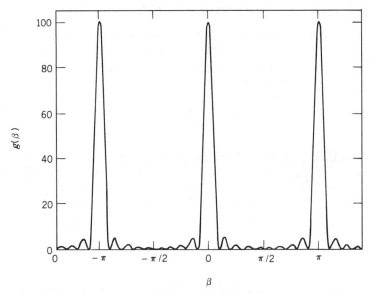

**FIGURE 4-5** The function $g(\beta) = \sin^2(N\beta)/\sin^2(\beta)$ for $N = 10$. It is periodic with period $\pi$ and has principal maxima at $\beta = \hbar\pi$, where $\hbar$ is an integer. $N - 2$ secondary maxima occur between adjacent principal maxima. As $N$ increases the principal maxima become higher and narrower and secondary maxima become less prominent.

let $\beta = \frac{1}{2}\,\Delta\mathbf{s} \cdot \mathbf{a}$ and $g(\beta) = \sin^2(N\beta)/\sin^2(\beta)$. Figure 4-5 shows a graph of $g(\beta)$ for $N = 10$. Except for a constant factor this function also gives the intensity of monochromatic light scattered from a 10 slit diffraction grating. It is periodic in $\beta$ with period $\pi$ and has principal maxima where $\beta$ is a multiple of $\pi$. At these maxima $\Delta\mathbf{s} \cdot \mathbf{a} = 2\hbar\pi$, where $\hbar$ is an integer, and $g(\beta) = N^2$. The last result follows when $2\hbar\pi$ is substituted for $\Delta\mathbf{s} \cdot \mathbf{a}$ on the left side of Eq. 4-16. Each term in the sum is 1.

The first zeros on either side of the $\beta = 0$ principal maximum occur at $N\beta = \pm\pi$, so the width of a principal peak may be taken to be $\Delta\beta = 2\pi/N$. As the sample size increases the intensity at a principal peak increases in proportion to $N^2$ and the width of the peak decreases in proportion to $1/N$. For macroscopic samples, $N$ is perhaps on the order of $10^8$ or more, so principal peaks are extremely intense and sharp.

$N - 2$ secondary maxima lie between each two consecutive principal maxima. They are positioned approximately where $\sin(N\beta) = \pm 1$ or $\beta = n\pi/2N$, with $n$ an odd integer. There are none, however, at $\beta = \hbar\pi \pm (\pi/2N)$, since these points are within principal maxima. For $N$ large, $g(\beta)$ is much smaller at secondary maxima than at principal maxima. For a crystal we need consider only principal maxima.

Similar results hold for the other lattice factors of Eq. 4-17. The most important situation for studies of crystalline structures arises when all three factors simultaneously have maximum values. Then an extremely intense scattered wave is produced. In terms of the change $\Delta\mathbf{s}$ in the wave propagation vector, the condition for an intense scattering peak is given by

$$\Delta \mathbf{s} \cdot \mathbf{a} = 2\hbar\pi, \tag{4-18}$$

$$\Delta \mathbf{s} \cdot \mathbf{b} = 2\hbar\pi, \tag{4-19}$$

and

$$\Delta \mathbf{s} \cdot \mathbf{c} = 2\ell\pi, \tag{4-20}$$

where $\hbar$, $\hbar$, and $\ell$ are integers. These equations are collectively known as the Laue condition. Many different peaks are possible, one for each set of $\hbar$, $\hbar$, and $\ell$ values. Peaks are identified by giving the indices $\hbar$, $\hbar$, and $\ell$ in the form $(\hbar\hbar\ell)$, just like the label for a lattice plane.

*Reciprocal Lattice Vectors.* Equations 4-18, 4-19, and 4-20 are easily solved for $\Delta \mathbf{s}$ in terms of three vectors **A**, **B**, and **C**, defined by

$$\mathbf{A} = 2\pi\,\frac{\mathbf{b} \times \mathbf{c}}{\mathbf{a} \cdot (\mathbf{b} \times \mathbf{c})}, \tag{4-21}$$

$$\mathbf{B} = 2\pi\,\frac{\mathbf{c} \times \mathbf{a}}{\mathbf{a} \cdot (\mathbf{b} \times \mathbf{c})}, \tag{4-22}$$

and

$$\mathbf{C} = 2\pi\,\frac{\mathbf{a} \times \mathbf{b}}{\mathbf{a} \cdot (\mathbf{b} \times \mathbf{c})}. \tag{4-23}$$

Each of these is proportional to one of the three vectors used in Chapter 2 to find normals to lattice planes and planar separations. **A** is perpendicular to both **b** and **c** and its projection on **a** is $2\pi/a$. This follows from $\mathbf{a} \cdot \mathbf{A} = 2\pi\mathbf{a} \cdot (\mathbf{b} \times \mathbf{c})/\mathbf{a} \cdot (\mathbf{b} \times \mathbf{c}) = 2\pi$. In a like manner, **B** is perpendicular to both **a** and **c** and its projection on **b** is $2\pi/b$; **C** is perpendicular to both **a** and **b** and its projection on **c** is $2\pi/c$. **A**, **B**, and **C** are not necessarily orthogonal to each other and none of them is necessarily parallel to any lattice vector. Figure 4-6 shows an example in which **A** is not parallel to **a** and **B** is not parallel to **b**.

The Laue condition for intense scattering is satisfied if

$$\Delta \mathbf{s} = \hbar\mathbf{A} + \hbar\mathbf{B} + \ell\mathbf{C}. \tag{4-24}$$

To show this, substitute Eq. 4-24 into the left side of Eq. 4-18 and obtain

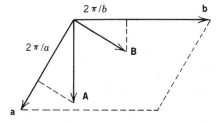

FIGURE 4-6   Reciprocal lattice vectors **A** and **B**. The fundamental lattice vector **c** and the reciprocal lattice vector **C** are perpendicular to the page. **A** is perpendicular to **b** and its projection on **a** is $2\pi/a$. **B** is perpendicular to **a** and its projection on **b** is $2\pi/b$.

$\Delta \mathbf{s} \cdot \mathbf{a} = \hbar \mathbf{A} \cdot \mathbf{a} + \ell \mathbf{B} \cdot \mathbf{a} + \ell \mathbf{C} \cdot \mathbf{a} = 2\pi \hbar$ since $\mathbf{B} \cdot \mathbf{a}$ and $\mathbf{C} \cdot \mathbf{a}$ both vanish and $\mathbf{A} \cdot \mathbf{a} = 2\pi$. The other two equations are satisfied in a similar manner.

Vectors of the form

$$\mathbf{G} = \hbar \mathbf{A} + \ell \mathbf{B} + \ell \mathbf{C}, \qquad (4\text{-}25)$$

where $\hbar$, $\ell$, and $\ell$ are integers, define the positions of points in a lattice. The fundamental vectors $\mathbf{A}$, $\mathbf{B}$, and $\mathbf{C}$ have units of reciprocal distance so the lattice is called a reciprocal lattice and vectors such as $\mathbf{G}$ are called reciprocal lattice vectors. To distinguish a reciprocal lattice from a lattice used to describe atomic positions, the latter is often called a direct lattice. Like a direct lattice, a reciprocal lattice is unbounded.

A reciprocal lattice is associated with every direct lattice. For a particular direct lattice, Eqs. 4-21, 4-22, and 4-23 are used to find fundamental reciprocal lattice vectors, then Eq. 4-25 is used to generate the reciprocal lattice.

**EXAMPLE 4-2**  Find fundamental reciprocal lattice vectors for a simple cubic lattice with cube edge $a$.

**SOLUTION**  Take $\mathbf{a} = a\hat{x}$, $\mathbf{b} = a\hat{y}$, and $\mathbf{c} = a\hat{z}$. Then $\mathbf{a} \cdot (\mathbf{b} \times \mathbf{c}) = a^3$ so $\mathbf{A} = (2\pi/a^3)a\hat{y} \times a\hat{z} = (2\pi/a)\hat{x}$, $\mathbf{B} = (2\pi/a^3)a\hat{z} \times a\hat{x} = (2\pi/a)\hat{y}$, and $\mathbf{C} = (2\pi/a^3)a\hat{x} \times a\hat{y} = (2\pi/a)\hat{z}$. Note that the array of points $\hbar \mathbf{A} + \ell \mathbf{B} + \ell \mathbf{C}$, with $\hbar$, $\ell$, and $\ell$ integers, is again a simple cubic lattice: the three fundamental vectors are mutually orthogonal and have the same magnitude. The cube edge of the reciprocal lattice unit cell is $2\pi/a$. ■

The reciprocal lattice corresponding to a face-centered cubic direct lattice is body-centered cubic while the reciprocal lattice corresponding to a body-centered cubic direct lattice is face-centered cubic. In each case, the cube edge is $4\pi/a$, where $a$ is the edge of the direct lattice cubic unit cell. The reciprocal lattice corresponding to a hexagonal direct lattice is again hexagonal, with hexagon edge $4\pi/(\sqrt{3}a)$ and cell height $2\pi/c$. Here $a$ and $c$ are the hexagon edge length and cell height, respectively, for the direct lattice cell.

In general, a direct lattice and its reciprocal belong to the same lattice system, although they may be different Bravais types. Since any operation of the point group of the lattice leaves the lattice unchanged it must also leave the reciprocal lattice unchanged. The volume of a reciprocal lattice unit cell is $\Omega = |\mathbf{A} \cdot (\mathbf{B} \times \mathbf{C})|$, and when Eqs. 4-21, 4-22, and 4-23 are used to substitute for $\mathbf{A}$, $\mathbf{B}$, and $\mathbf{C}$ respectively, the result is

$$\Omega = \frac{8\pi^3}{\tau}, \qquad (4\text{-}26)$$

where $\tau$ is the volume of the direct lattice unit cell. We call $\Omega$ a "volume" although it has units of reciprocal volume. Similarly, we speak of the "length" of a reciprocal lattice vector.

As Eq. 2-8 shows, the reciprocal lattice vector $\mathbf{G} = \hbar\mathbf{A} + \mathit{k}\mathbf{B} + \ell\mathbf{C}$ is normal to lattice planes with Miller indices $(\hbar\mathit{k}\ell)$. Equation 2-12, for the separation of parallel lattice planes, can be rewritten in terms of a reciprocal lattice vector normal to the planes. If $\hbar$, $\mathit{k}$, and $\ell$ have no common divisor other than 1, the separation $d$ is given by

$$d = \frac{2\pi}{|\mathbf{G}|}.\qquad(4\text{-}27)$$

This expression will prove useful later.

Finally we note that if fundamental reciprocal lattice vectors are known, then fundamental direct lattice vectors can be found. They are given by

$$\mathbf{a} = 2\pi\,\frac{\mathbf{B} \times \mathbf{C}}{\mathbf{A} \cdot (\mathbf{B} \times \mathbf{C})},\qquad(4\text{-}28)$$

$$\mathbf{b} = 2\pi\,\frac{\mathbf{C} \times \mathbf{A}}{\mathbf{A} \cdot (\mathbf{B} \times \mathbf{C})},\qquad(4\text{-}29)$$

and

$$\mathbf{c} = 2\pi\,\frac{\mathbf{A} \times \mathbf{B}}{\mathbf{A} \cdot (\mathbf{B} \times \mathbf{C})}.\qquad(4\text{-}30)$$

The proof can be carried out by substituting Eqs. 4-21, 4-22, and 4-23 into Eqs. 4-28, 4-29, and 4-30, then using vector identities found inside the back cover.

*Ewald Constructions.* To obtain an elastic diffraction peak, $\mathbf{s}' - \mathbf{s}$ must satisfy the Laue conditions and $\mathbf{s}'$ must have the same magnitude as $\mathbf{s}$. These conditions severely limit the number of elastic scattering peaks obtained for a fixed wavelength, incident direction, and crystal orientation. In fact, for arbitrary choices of these variables, elastic peaks are probably not produced. To produce elastic peaks, the crystal usually must be rotated or the wavelength changed.

An Ewald construction, as in Fig. 4-7, illustrates the idea. Reciprocal lattice points are shown, with positions given by $\hbar\mathbf{A} + \mathit{k}\mathbf{B} + \ell\mathbf{C}$, where $\hbar$, $\mathit{k}$, and $\ell$ are integers. The propagation vector for the incident wave is drawn with its head at any one of the points, then a sphere with radius $2\pi/\lambda$ is drawn with its center at the tail of the propagation vector. If the surface of the sphere passes through another reciprocal lattice point, then an elastic scattering peak is produced.

For the situation shown in Fig. 4-7a, the condition is met and the propagation vector for the scattered beam is drawn, from the center of the sphere to a reciprocal lattice point on its surface. The vector connecting the heads of the two propagation vectors is clearly a reciprocal lattice vector so the condition $\mathbf{s}' - \mathbf{s} = \mathbf{G}$ is satisfied. Furthermore, each of the propagation vectors clearly has magnitude $2\pi/\lambda$.

If the incident propagation vector is in a slightly different direction relative to the crystal, as in Fig. 4-7b, then the sphere center is at a slightly different place and its surface does not pass through a second reciprocal lattice point. The Laue condition is not met and no high-intensity scattering peak is obtained.

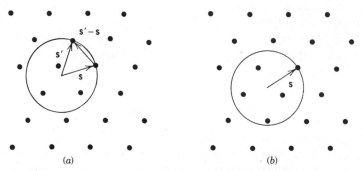

**FIGURE 4-7** An Ewald construction. The points are reciprocal lattice points, s is the propagation vector for the incident wave, and s' is the propagation vector for the scattered wave. The radius of the Ewald sphere is $2\pi/\lambda$. In (a) the Laue condition s' = s + G, where G is a reciprocal lattice vector, is met and an intense elastic scattering peak is produced. In (b) the condition is not met and no elastic peaks are produced.

*The Bragg Condition.* The Laue condition $\Delta\mathbf{s} = \hbar\mathbf{A} + \hbar\mathbf{B} + \ell\mathbf{C}$ for a scattering peak is identical to the condition for the constructive interference of waves reflected from $(\hbar\hbar\ell)$ planes. First note that when the $(\hbar\hbar\ell)$ peak is produced, the incident and scattered propagation vectors make equal angles with $(\hbar\hbar\ell)$ planes. Their magnitudes are equal and their difference is a reciprocal lattice vector normal to the plane. Thus the plane must bisect the angle between them, as shown in Fig. 4-8a.

Now consider two waves incident at the angle $\theta$ on two parallel planes with separation $d$, as shown in (b). The propagation vectors of the reflected waves also make the angle $\theta$ with the planes. Wave 1, reflected from the lower plane, travels $2d \sin \theta$ further than wave 2, reflected from the upper plane. The reflected

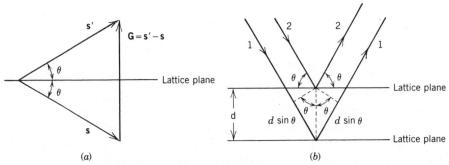

**FIGURE 4-8** Bragg scattering. (a) Incident and scattered propagation vectors each make the angle $\theta$ with a crystal plane. G = s' − s is a reciprocal lattice vector perpendicular to the plane. (b) Scattering from two parallel planes. Wave 1 travels $2d \sin \theta$ further than wave 2 and, for maximum constructive interference, the difference in the distances traveled is a multiple of the wavelength.

waves interfere constructively to produce a maximum in intensity if this distance is a multiple of the wavelength, so

$$2d \sin \theta = n\lambda \qquad (4\text{-}31)$$

for a scattering peak. Waves reflected from other planes of the set, not shown, also interfere constructively if this condition is met. Equation 4-31 gives the Bragg condition for a scattering peak.

The Bragg condition follows from the Laue condition. If the indices of the peak have a common divisor $n$, factor it and write $\Delta \mathbf{s} = n\mathbf{G}$, where $\mathbf{G} = \hbar\mathbf{A} + \hbar\mathbf{B} + \ell\mathbf{C}$ and $\hbar$, $\hbar$, and $\ell$ have no common divisor other than 1. Now equate the magnitudes of $\Delta \mathbf{s}$ and $n\mathbf{G}$, then use Eqs. 4-6 and 4-27 to obtain $(4\pi/\lambda) \sin \theta = 2\pi n/d$ or $2d \sin \theta = n\lambda$, the Bragg condition.

We can use the Bragg condition to understand the limitations on wavelengths used in structure analysis. Since $\sin \theta$ must be less than or equal to 1, no intensity peaks can occur if $\lambda$ is greater than twice the largest crystal plane separation. This condition is also evident in the Ewald construction. It is impossible for two reciprocal lattice points to be on the Ewald sphere unless the diameter of the sphere is greater than or equal to the magnitude of the shortest reciprocal lattice vector. The sphere diameter is $4\pi/\lambda$ and the shortest reciprocal lattice vector has magnitude $2\pi/d_{max}$, where $d_{max}$ is the plane separation for the set of most widely spaced lattice planes. Thus $4\pi/\lambda$ must be greater than or equal to $2\pi/d_{max}$ or $\lambda$ must be less than or equal to $2d_{max}$ for a peak to occur. If the sample has a simple cubic lattice with cube edge $a$, for example, the wavelength must be less than or equal to $2a$.

If the wavelength is nearly twice the spacing of a set of planes, $\theta$ is nearly $90°$ and the scattering angle is nearly $180°$ for the peak associated with those planes. If the planes are widely spaced, only a few peaks can occur and these are in the backscattering direction. As the wavelength is made smaller, the peaks move toward the forward scattering direction and other peaks, corresponding to more narrowly spaced planes, appear in the backscattering direction.

**EXAMPLE 4-3**   Consider a simple cubic crystal structure with cube edge $a = 3.50$ Å and suppose it is used to scatter 3.10-Å x rays. Find all sets of planes that satisfy the Bragg condition and, for each peak, find the Bragg angle $\theta$.

**SOLUTION**   According to the Bragg condition $\sin \theta = n\lambda/2d$. For a simple cubic lattice the separation of planes with indices $(\hbar\hbar\ell)$ is $d = a/(\hbar^2 + \hbar^2 + \ell^2)^{1/2}$, so $\sin \theta = (n\lambda/2a)(\hbar^2 + \hbar^2 + \ell^2)^{1/2} = n(3.10 \,/\, 7.00)(\hbar^2 + \hbar^2 + \ell^2)^{1/2} = 0.443n(\hbar^2 + \hbar^2 + \ell^2)^{1/2}$. This must be less than or equal to 1. We evaluate $0.443 \, (\hbar^2 + \hbar^2 + \ell^2)^{1/2}$ for various values of $\hbar$, $\hbar$, and $\ell$. If the result is 1 or less, the peak can be obtained by proper orientation of the crystal, provided the structure and atomic form factors are not too small. We check to see if multiplication by an integer $n$ also produces a value that is less than 1. If the peak is obtainable, we use $\sin \theta = 0.443n(\hbar^2 + \hbar^2 + \ell^2)^{1/2}$ to find $\theta$. The table below shows the results:

| $(\hbar\hbar\ell)$ | $n$ | $0.443n(\hbar^2 + \hbar^2 + \ell^2)^{1/2}$ | $\theta$ |
|---|---|---|---|
| (100) | 1 | 0.443 | 26.3° |
| | 2 | 0.886 | 62.3° |
| (110) | 1 | 0.626 | 38.8° |
| (111) | 1 | 0.767 | 50.1° |
| (210) | 1 | 0.990 | 82.0° |

For any of these planes, of course, the indices can be permuted and their signs can be changed to find other planes associated with peaks at the same scattering angle. For this example, these are the only planes for which the Bragg condition is satisfied. ∎

*Structure Factors.*   For $\Delta\mathbf{s} = \mathbf{G}$, Eq. 4-14 becomes

$$F = \sum_m f_m e^{-i\mathbf{G}\cdot\mathbf{p}_m}, \tag{4-32}$$

where the sum is over all atoms of the basis. Write $\mathbf{p}_m = u_m\mathbf{a} + v_m\mathbf{b} + w_m\mathbf{c}$ and $\mathbf{G} = \hbar\mathbf{A} + \hbar\mathbf{B} + \ell\mathbf{C}$. Then $\mathbf{G}\cdot\mathbf{p}_m = 2\pi(u_m\hbar + v_m\hbar + w_m\ell)$ and

$$F = \sum_m f_m e^{-2\pi i(u_m\hbar + v_m\hbar + w_m\ell)}. \tag{4-33}$$

For some values of $\hbar$, $\hbar$, and $\ell$, waves scattered from atoms of the basis tend to interfere constructively, while for other values they tend to interfere destructively. Consider, for example, a CsCl structure. Take the primitive direct lattice vectors to be $\mathbf{a} = a\hat{\mathbf{x}}$, $\mathbf{b} = a\hat{\mathbf{y}}$, and $\mathbf{c} = a\hat{\mathbf{z}}$. Possible basis vectors are then $\mathbf{p}_{Cs} = 0$ and $\mathbf{p}_{Cl} = \frac{1}{2}(\mathbf{a} + \mathbf{b} + \mathbf{c})$, so

$$F = f_{Cs} + f_{Cl}e^{-i\pi(\hbar+\hbar+\ell)}. \tag{4-34}$$

If $\hbar + \hbar + \ell$ is an odd integer, then $e^{-i\pi(\hbar+\hbar+\ell)} = -1$ and $F = f_{Cs} - f_{Cl}$. If $f_{Cs}$ and $f_{Cl}$ are real and have the same sign, these lines are extremely weak. If, on the other hand, $\hbar + \hbar + \ell$ is an even integer, then $e^{-i\pi(\hbar+\hbar+\ell)} = +1$ and $F = f_{Cs} + f_{Cl}$. These lines are strong.

X-ray peaks are often indexed using lattice vectors associated with a nonprimitive lattice. For example, peaks from FCC and BCC structures are usually indexed using simple cubic lattice vectors. Equation 4-17 still predicts the intensity of the peaks, provided all atoms of the nonprimitive basis are included in the structure factor. To find the structure factor for a BCC structure, indexed to a simple cubic lattice, we may use the results for the CsCl structure. Both atoms of the basis have the same atomic form factor $f$, so $F = 2f$ for $\hbar + \hbar + \ell$ even and $F = 0$ for $\hbar + \hbar + \ell$ odd. In the first case, waves from the two atoms interfere constructively, while in the second they interfere destructively. The following example gives results for an FCC structure.

**EXAMPLE 4-4**  Find the structure factor for an FCC crystal structure when scattering peaks are indexed to a simple cubic lattice. In particular, show that the structure factor vanishes unless $\hbar$, $\hbar$, and $\ell$ are either all even or all odd and find its value in these instances. Take the atomic form factor to be $f$.

**SOLUTION**  Take the fundamental direct lattice vectors to be $\mathbf{a} = a\hat{\mathbf{x}}$, $\mathbf{b} = a\hat{\mathbf{y}}$, and $\mathbf{c} = a\hat{\mathbf{z}}$ and the basis vectors to be $\mathbf{p}_1 = 0$, $\mathbf{p}_2 = \frac{1}{2}(\mathbf{a} + \mathbf{b})$, $\mathbf{p}_3 = \frac{1}{2}(\mathbf{a} + \mathbf{c})$, and $\mathbf{p}_4 = \frac{1}{2}(\mathbf{b} + \mathbf{c})$. Then

$$F = f[1 + e^{-i\pi(\hbar+\ell)} + e^{-i\pi(\hbar+\ell)} + e^{-i\pi(\ell+\ell)}] .$$

The sum of two even integers is even, the sum of two odd integers is even, and the sum of an even and odd integer is odd. If $\hbar$, $\hbar$, and $\ell$ are all even or all odd, each of the sums in the exponents is even, the exponentials are each 1, and $F = 4f$. If two of the indices are even and one is odd or if two are odd and one is even, two of the exponentials are $-1$ while the other is $+1$ so $F = 0$. There are no other possibilities.  ∎

For a BCC structure there are no (100), (111), or (210) peaks. For an FCC structure there are no (100), (110), or (210) peaks. Once a crystal is determined to be cubic, the sequence of scattering angles for which peaks occur can be used to classify it as SC, BCC, or FCC. Experimental values of the relative intensities at scattering peaks are helpful for the determination of atomic positions in a unit cell.

## 4.3  EXPERIMENTAL TECHNIQUES

*Beam Production and Detection.*  To obtain an easily observable diffraction pattern, the incident wavelength should be roughly the same as the interatomic separation. In most experiments, the wavelength is controlled by regulating the energy of the beam particles. For any particle, the wavelength $\lambda$ of its wave and the magnitude $p$ of its momentum are related by the de Broglie relation $\lambda = 2\pi\hbar/p$. For photons, $E = cp$, so $\lambda = 2\pi\hbar c/E$, while for nonrelativistic electrons or neutrons $E = p^2/2m$, so $\lambda = (2\pi^2\hbar^2/mE)^{1/2}$, where $m$ is the mass of the particle. Photons useful for structure analysis have energies on the order of $10^{-15}$ J (10 keV). Useful electrons have energies on the order of $10^{-17}$ J ($10^2$ eV) and useful neutrons have energies on the order of $10^{-20}$ J (0.10 eV).

To produce an x-ray beam, electrons from a heated filament, usually made of tungsten, are accelerated in an electric field and caused to strike a target, usually made of molydenum, copper, nickel, cobalt, or iron. The spectrum of x rays emitted from the target consists of two overlapping parts: a fairly broad continuous spectrum and a series of narrow, intense peaks at certain wavelengths. Both parts are used for structure analysis, in different types of experiments.

The continuous portion of the spectrum, called bremsstrahlung (braking radiation) or white radiation, is produced by the incident electron as it decelerates in the electric field of atoms in the target. Bremsstrahlung is independent of the target material but does depend on the potential difference used to accelerate the bombarding electrons. Typical curves for two different accelerating potentials are shown in Fig. 4-9a.

For each accelerating potential there is a minimum wavelength wave, produced when all the kinetic energy of an incident electron appears as a single photon. If the accelerating potential is $V$ the electron energy is eV, so the angular frequency of the x-ray wave is $\omega = eV/\hbar$ and the minimum wavelength is $\lambda = 2\pi c/\omega = 2\pi\hbar c/eV$. To obtain a minimum wavelength of 1 Å, an accelerating potential of 12.4 kV is required. Larger accelerating potentials produce spectra with shorter minimum wavelengths.

Most electrons undergo a series of photon-producing decelerations, losing only a fraction of their initial energy in each event. Each photon produced has less energy than the incident electron so the wavelengths of their waves are longer than the minimum wavelength. The maximum x-ray intensity in the brems-

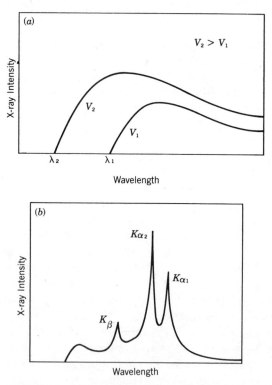

**FIGURE 4-9** (a) Bremsstrahlung spectrum for two accelerating potentials, with $V_2 > V_1$. The cutoff wavelengths are $\lambda_1$ and $\lambda_2$, respectively. The spectrum is independent of the target. (b) Bremsstrahlung and characteristic spectrum for a typical target in an x-ray tube. Different targets produce different characteristic lines.

strahlung spectrum occurs at about 1.5 $\lambda_{min}$ and thereafter the intensity slowly decreases toward zero. The upper limit to the useful portion of the spectrum is determined by absorption in air, the target, or other parts of the instrument. It is usually in the range from 2 to 5 Å.

In contrast to bremsstrahlung, peaks in the x-ray spectrum are characteristic of the target. If the incident electron energy is great enough, the electron may collide with and remove a bound electron from an atom in the target. An electron from a higher state falls into the empty state and the excess energy appears as a photon. Another state is then vacant and another photon results when an electron fills it. The net result is a cascade of photons with different wavelengths.

X-ray radiation with a wavelength appropriate for structure analysis is produced when an electron is removed from the innermost shell of a target atom with moderate atomic number. Characteristic lines for a typical target are shown in Fig. 4-9b. $K_\alpha$ radiation occurs when an electron from the second shell falls to the vacated state in the first shell, while $K_\beta$ radiation results when an electron falls from the third to the first shell. Since atomic shells contain electrons with different energies, several $K_\alpha$ and $K_\beta$ lines are produced. For a molydenum target, the average of the $K_\alpha$ lines is at 0.71 Å and the average of the $K_\beta$ lines is at 0.63 Å. For copper these averages are at 1.54 and 1.39 Å, respectively. A filter or monochromter must be used to obtain a monochromatic beam.

Electron beams for diffraction experiments are produced in much the same way as in an x-ray tube. A filament is heated to produce a flux of electrons, which are then accelerated in an electric field. The energy of the electrons, and hence the wavelength of their waves, is determined by the accelerating potential. A series of electromagnetic lenses is used to focus the beam and an aperture is used to reduce the beam diameter.

Neutrons are obtained from nuclear reactors and particle accelerators. As produced, they have a much higher energy and their waves have a much shorter wavelength than is useful for diffraction experiments. Energies are decreased by directing the neutrons into a moderator of heavy water or graphite and allowing them to reach thermal equilibrium. For a moderator temperature of about 600 K, the average wavelength of the neutron waves is about 1 Å, a useful wavelength for structure studies. There are, of course, waves with wavelengths both greater and less than the average, so a monochromter must be used.

Several types of x-ray detectors are employed. Photographic film has been used extensively in the past but most present day experiments make use of electronic counters. Scintillation counters, proportional counters, and solid state detectors are all currently in use. Individual x-ray photons ionize gas atoms or promote electrons to high-energy levels in the counter material, and the excitation is used to create an electrical pulse, which is then detected. In some cases the intensity of the electrical pulse is proportional to the photon energy, so the instrument can be adjusted to record photons associated with a narrow range of wavelengths. This insures that only elastically scattered waves are observed.

Film, phosphor screens, and Faraday cups are all used to record scattered electrons. Film with the proper coating can also be used to detect neutrons, but the most widely used neutron detectors make use of substances, such as boron trifluoride, that absorb neutrons and, as a result, emit particles. The $\alpha$'s are then detected by means of a charged particle detector.

When photographic film is used as a detector it is usually formed into a cylinder and placed around the sample with its axis perpendicular to the incident beam, as shown in Fig. 4-10. The film records the intensity pattern. If a counter is used, it is moved from place to place and the intensity is recorded at each position. In any event, directions of scattered propagation vectors for scattering peaks are found from the data.

There are three chief experimental methods: the Laue method, the rotating crystal method, and the powder method. In each case, the purpose is to assure that a reasonable number of peaks are obtained, either by using a wide spectrum of wavelengths or a wide variety of crystal orientations. In the following sections each method is discussed as it pertains to x-ray analysis.

*Laue Method.*    Laue patterns are obtained by illuminating the sample with bremsstrahlung from a suitable target. The incident beam contains a mixture of waves with wavelengths ranging from $\lambda_{min}$, determined by the accelerating potential, to $\lambda_{max}$. Typically, $\lambda_{min}$ is about 0.2 Å and $\lambda_{max}$ is about 3 Å. A Laue pattern for crystalline sodium chloride is reproduced in Fig. 4-11. It consists of a collection of spots, positioned where intense scattered waves struck the film. We wish to understand the formation of this and similar patterns for other crystals.

Figure 4-12 shows an Ewald construction for the Laue method. There are two incident propagation vectors shown, in the same direction and with their heads at the same reciprocal lattice point. The longer vector has magnitude $2\pi/\lambda_{min}$ and the shorter has magnitude $2\pi/\lambda_{max}$. For each, the appropriate Ewald sphere is drawn, with its center at the tail of the vector and with its radius equal to the magnitude of the vector. We may imagine a continuous distribution of Ewald spheres, one for each wavelength in the beam and all with centers lying between

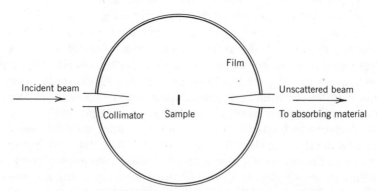

**FIGURE 4-10** Geometry of a cylindrical diffraction camera. The beam enters through a collimator and the unscattered portion exits to an absorber. Photographic film is formed into a cylinder around the sample and records the scattered intensity. If a counter is used instead of film, it is moved on an arc around the sample.

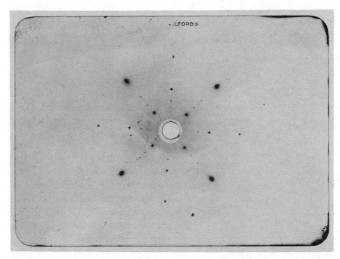

**FIGURE 4-11** A Laue pattern for crystalline sodium chloride. The pattern clearly displays fourfold symmetry. (From Walter Kiszenick, unpublished. Used with permission.)

the two in the diagram. A scattering peak occurs for each reciprocal lattice point in the region between the two limiting spheres, provided, of course, the structure factor is not too small.

Consider the peak associated with the reciprocal lattice vector **G**, as diagrammed in Fig. 4-13. The head of the incident propagation vector is at a reciprocal lattice point and its tail is on the perpendicular bisector of **G**. If $\hat{s}$ is a

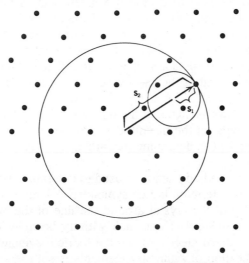

**FIGURE 4-12** An Ewald construction for the Laue method. The two incident propagation vectors $s_1$ and $s_2$ are associated with the longest and shortest wavelengths, respectively, in the beam. Barring a vanishing structure factor, a scattering peak is produced for each reciprocal lattice point between the two spheres and on their surfaces.

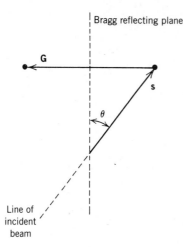

**FIGURE 4-13** To obtain the scattering peak associated with the reciprocal lattice vector **G**, the propagation vector for the incident wave must extend from a point on the perpendicular bisector of **G** (the reflecting plane) to the reciprocal lattice point at the tail of **G**.

unit vector in the direction of the incident beam, then $\hat{\mathbf{s}} \cdot \mathbf{G} = -|\mathbf{G}| \sin \theta$ and

$$\sin \theta = -\frac{\hat{\mathbf{s}} \cdot \mathbf{G}}{|\mathbf{G}|}. \tag{4-35}$$

From the diagram, $\sin \theta = \frac{1}{2}|\mathbf{G}|/|\mathbf{s}| = (\lambda/4\pi)|\mathbf{G}|$. This is substituted into Eq. 4-35 and the result is solved for $\lambda$:

$$\lambda = -4\pi \frac{\hat{\mathbf{s}} \cdot \mathbf{G}}{|\mathbf{G}|^2}. \tag{4-36}$$

Equation 4-36 can be used to decide if a given peak is produced. For each reciprocal lattice vector the wavelength is evaluated and if it falls between $\lambda_{min}$ and $\lambda_{max}$ and if the structure factor does not vanish, then that peak occurs.

In most cases, the Laue method cannot be used to map the reciprocal lattice. Equation 4-35 shows that spots associated with different but parallel reciprocal lattice vectors are formed at the same place on the film. The scattering angle can be used to find the direction of **G** but not its magnitude.

Since the range of wavelengths in the incident beam is limited, Eq. 4-36 can be used to place limits on the magnitudes of reciprocal lattice vectors. Occasionally this is sufficient to determine the lattice, but it is not a generally useful technique.

Although the Laue method cannot be used to determine details of the lattice, it is widely used to determine lattice symmetry. A Laue pattern displays any rotational symmetry of the crystal around the line of the incident beam. For example, the pattern of Fig. 4-11 was made with the beam incident along a $\langle 100 \rangle$ direction and the pattern clearly shows the fourfold symmetry of the crystal. Recall that identification of symmetry elements is sufficient to determine the lattice system.

Laue patterns are often used to orient crystals. If the lattice is cubic, for example, it can be rotated until the pattern shows fourfold symmetry about the direction of the incident beam. That direction is then one of the $\langle 100 \rangle$ directions. Similarly, threefold symmetry in the pattern indicates that the beam is incident along a $\langle 111 \rangle$ direction.

*Rotating Crystal Method.* In this method, a monochromatic beam is used and the crystal is rotated during exposure about an axis perpendicular to the beam. The geometry is like that shown in Fig. 4-10, with the axis of rotation perpendicular to the page. The reciprocal lattice rotates with the crystal. Refer to Fig. 4-7 and imagine the reciprocal lattice rotating about an axis perpendicular to the page, through the point at the head of the incident propagation vector. As it rotates, various reciprocal lattice points cross the surface of the Ewald sphere and, when a point is on the surface, the corresponding intensity peak is produced.

Since a reciprocal lattice point may cross the Ewald sphere at more than one place, more than one spot on the film may be associated with the same point. To reduce the number of redundant spots, the angle of rotation is limited. The crystal is made to oscillate so the same pattern is recorded many times.

Rotating crystal techniques are often used to determine the shape and size of the unit cell. The analysis is particularly simple if the lattice has a high degree of symmetry and the crystal is oriented with a symmetry axis along the axis of rotation. As an example, consider an orthorhombic crystal with the axis of rotation along one of the diads. Suppose the fundamental lattice vector **c** is along this axis, while **a** and **b** are perpendicular to each other and to **c**. They rotate with the crystal. A photographic film is formed into a cylinder and placed concentrically about the sample with its axis along the axis of rotation. When it is exposed and unwrapped, the scattering peaks recorded are as shown in Fig. 4-14a, with the horizontal rows of spots perpendicular to the axis of rotation. The distance from the central row to each of the others is measured and used to find the separation of lattice points along **c**.

The analysis proceeds as follows. A Cartesian coordinate system is oriented with its $z$ axis along the axis of rotation and its $x$ axis along the incident beam. Then $\mathbf{c} = c\hat{\mathbf{z}}$ and both **a** and **b** rotate in the $xy$ plane. One fundamental reciprocal lattice vector is $\mathbf{C} = (2\pi/c)\hat{\mathbf{z}}$. The other two, **A** and **B**, have magnitudes $2\pi/a$ and $2\pi/b$, respectively, and rotate in the $xy$ plane.

As illustrated in Fig. 4-14b, spots occur where the vector $\mathbf{s}' = \mathbf{s} + h\mathbf{A} + k\mathbf{B} + \ell\mathbf{C}$ intersects the film. Since **s**, **A**, and **B** all lie in the $xy$ plane, the $z$ component of $\mathbf{s}'$ is $2\pi\ell/c$. If $\mathbf{s}'$ makes the angle $\alpha$ with the $xy$ plane, then $s_z' = (2\pi/\lambda) \sin\alpha$ and

$$c = \frac{\ell\lambda}{\sin\alpha}. \tag{4-37}$$

The central row of spots corresponds to $\ell = 0$ and the value of $\ell$ for another row is simply the number of the row, counting from the $\ell = 0$ row. The angle $\alpha$ for a given row can be found from the pattern on the film: if $R$ is the radius formed by the film and $\Delta z_\ell$ is the distance on the film of row $\ell$ from the central row, then $\tan\alpha = \Delta z_\ell/R$. Once $\alpha$ is known for a row, Eq. 4-37 is used to compute

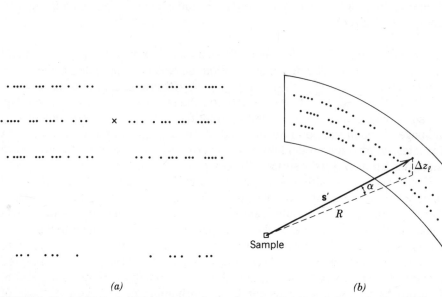

(a)                                                              (b)

**FIGURE 4-14** (a) Diagram of intensity peaks produced using the rotating crystal method. An orthorhombic crystal was rotated about its $c$ axis, which was perpendicular to the incident beam. Only peaks in the forward direction are shown. X marks the place where the unscattered portion of the beam struck the film. (b) The separation $\Delta z_\ell$ between a row of spots and the central row is used to calculate $\alpha$ and the lattice constant along the axis of rotation.

$c$, given the wavelength. A similar analysis, with the crystal reoriented, is used to find $a$ and $b$.

When the fundamental vectors are known, all scattering peaks can be indexed. We start with the $\ell = 0$ peaks, in the $xy$ plane. For them, $|\mathbf{s}' - \mathbf{s}|^2 = 4\pi^2[(h/a)^2 + (k/b)^2]$, and since $|\mathbf{s}' - \mathbf{s}|^2 = (4\pi/\lambda)^2 \sin^2 \theta$,

$$4 \sin^2 \theta = \lambda^2 \left[ \frac{h^2}{a^2} + \frac{k^2}{b^2} \right]. \tag{4-38}$$

For each peak, the half scattering angle $\theta$ is found from the position of the spot on the film, then values of $h$ and $k$ are selected so that Eq. 4-38 is satisfied. Finally, other rows of spots are considered, corresponding to other values of $\ell$.

*Powder Method.*   The sample is ground into a large number of small randomly oriented crystals and illuminated by a monochromatic beam. Since the powder contains a large number of crystals with essentially every orientation, all scattering peaks corresponding to reciprocal lattice vectors shorter than $4\pi/\lambda$ and nonvanishing structure factors are produced.

Imagine a crystal oriented so that an intense scattered beam occurs with scattering angle $\theta$. If the crystal is rotated about the direction of the incident beam, the scattered beam rotates around the surface of a cone with apex at the

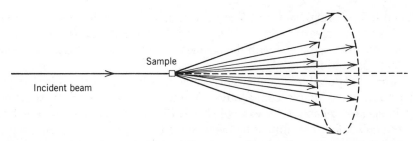

**FIGURE 4-15** A cone of scattered beams from a powder sample. All beams shown have the same scattering angle. They are produced by different groups of crystals, related to each other by rotations about the line of the incident beam. Other cones, not shown, correspond to other scattering angles.

crystal and with an apex angle equal to twice the scattering angle. Crystals are not rotated to obtain powder patterns, but the effect is much the same. For a powder, scattered beams corresponding to the various orientations are produced simultaneously. They form concentric cones, one for each possible scattering angle, with the incident beam along their common axis, as illustrated in Fig. 4-15.

As shown for tungsten in Fig. 4-16, the pattern on film is a series of concentric rings, one for each possible scattering angle. Before exposure the film was formed into a cylinder and placed around the sample, as described in connection with the rotating crystal method. This configuration produces two sets of rings, one formed by backscattered waves and the other formed by forward scattered waves. The two small circles at the centers of the ring systems are entrance and exit holes for the beam.

To show how powder patterns are indexed, we consider a cubic crystal and suppose the length of the cube edge is $a$. The value of $a$ is to be determined experimentally. If a simple cubic lattice is used as a basis for the indices, the separation of $(\hbar k \ell)$ planes is given by $d = a/(\hbar^2 + k^2 + \ell^2)^{1/2}$ and the Bragg condition becomes $[2a/(\hbar^2 + k^2 + \ell^2)^{1/2}] \sin \theta = n\lambda$, or

$$\sin^2 \theta = \frac{\lambda^2 N}{4a^2}, \qquad (4\text{-}39)$$

where $N = n^2(\hbar^2 + k^2 + \ell^2)$. $N$ is a different integer for each ring. The scattering angle is measured for each ring and $\sin^2 \theta$ is calculated, then a value of $N$ is

**FIGURE 4-16** Powder pattern for tungsten. Each ring corresponds to a different scattering angle. The pattern consists of two sets of rings, for beams scattered in the back (on the left) and forward (on the right) directions, respectively. (From Walter Kiszenick, unpublished. Used with permission.)

assigned to each ring. Values of $N$ are selected so their ratios are the same as the ratios of the corresponding values of $\sin^2 \theta$. If more than one set of integers has the same ratios, the one with the smallest values is usually selected. Finally, Eq. 4-39 is used to find $a$.

Knowledge of possible sequences of $N$ values is extremely helpful for assigning values to $N$. For a simple cubic lattice, $N$ must be the sum of the squares of three integers and cannot, for example, be 7, 15, 23, or 28. If the lattice is body-centered cubic, $h + k + \ell$ must be even. Since the square of an even integer is even and the square of an odd integer is odd, this means $N$ must be even. If the lattice is face-centered cubic, the indices must be all even or all odd so the sequence of $N$ values is 3, 4, 8, 11, 12, 16, 19, 20, .... .

Once $N$ is found for a diffraction ring, the indices are determined by finding three integers such that $h^2 + k^2 + \ell^2 = N$. In some cases more than one set of indices is associated with a given ring. For example, both the (300) and (221) peaks are associated with the $N = 9$ ring and both the (410) and (322) peaks are associated with the $N = 17$ ring.

## 4.4  SCATTERING FROM SURFACES

Low-energy electron diffraction (LEED) provides an important technique for the study of crystal surfaces. Electrons from a hot filament are accelerated in an electric field and formed into a collimated beam incident on the sample. The backscattered intensity is measured. Energies used are usually from several electron volts to several hundred electron volts. The wavelength, of course, must be roughly the same as the interatomic separation.

The geometry is quite similar to that of an x-ray scattering experiment. A phosphor screen is usually used for the detector and a wire grid is placed between the sample and the screen. If elastic scattering is studied, the grid is maintained at the proper electric potential relative to the sample so the only electrons to reach the screen are those with the same energy as the incident electrons. After elastically scattered electrons pass through the grid they are accelerated toward the screen and produce a bright spot where they hit. Usually the beam is focused so only a small region on the surface is probed. Typically, linear dimensions of the region studied are on the order of 100 Å or less.

If surfaces of pure crystals are studied, they must be clean, with a minimum of impurity atoms. Crystals are cleaved in a high vacuum and a high vacuum must be maintained in the scattering apparatus. In spite of these precautions, sample surfaces are often contaminated. On the other hand, the purpose of many LEED experiments is to study the disposition of impurity atoms adsorbed on crystal surfaces.

Atomic equilibrium positions on a clean crystal surface can be described in terms of a two-dimensional lattice and basis. For most metals the pattern is nearly the same as the pattern on a parallel plane in the interior. For example, the lattice corresponding to a (100) surface of a cubic metal is a two-dimensional square lattice, while that corresponding to a (111) plane is a two-dimensional hexagonal lattice. For tetrahedrally bonded crystals the surface pattern may be somewhat different from the pattern on a parallel plane in the interior. The absence of bonds pointing toward the exterior results in an increase in energy, which can be minimized if the pattern distorts.

To investigate the elastic scattering pattern, we write $\mathbf{s}_\perp + \mathbf{s}_\parallel$ for the propagation vector of the incident wave and $\mathbf{s}'_\perp + \mathbf{s}'_\parallel$ for the propagation vector of the reflected wave. Here the subscripts $\perp$ and $\parallel$ refer to components that are perpendicular and parallel, respectively, to the surface. For an intense scattered peak to occur, the parallel components must satisfy the condition

$$\mathbf{s}'_\parallel = \mathbf{s}_\parallel + \mathbf{G}, \qquad (4\text{-}40)$$

where $\mathbf{G}$ is a reciprocal lattice vector associated with the two-dimensional surface lattice, not the three-dimensional lattice of the interior. Specifically, if $\mathbf{a}$ and $\mathbf{b}$ are two primitive vectors of the surface lattice and $\mathbf{n}$ is a vector normal to the surface, then

$$\mathbf{A} = 2\pi \frac{\mathbf{b} \times \mathbf{n}}{\mathbf{a} \cdot (\mathbf{b} \times \mathbf{n})} \qquad (4\text{-}41)$$

and

$$\mathbf{B} = 2\pi \frac{\mathbf{a} \times \mathbf{n}}{\mathbf{a} \cdot (\mathbf{b} \times \mathbf{n})} \qquad (4\text{-}42)$$

are two fundamental vectors of the reciprocal lattice and

$$\mathbf{G} = \hbar\mathbf{A} + \ell\mathbf{B}, \qquad (4\text{-}43)$$

where $\hbar$ and $\ell$ are integers. $\mathbf{A}$, $\mathbf{B}$, and $\mathbf{G}$ are all in the plane of the surface. Scattering peaks are indexed using the two integers $\hbar$ and $\ell$ that appear in Eq. 4-43.

Equation 4-40 is the two-dimensional counterpart to Eq. 4-24. The atomic structure near the surface is not periodic along a line perpendicular to the surface, so a similar condition is not placed on perpendicular components of propagation vectors. The magnitude of $\mathbf{s}'_\perp$ is determined by the requirement that electron energy be conserved. Since the electron energy is given by $(\hbar^2/2m)s^2$, $|\mathbf{s}|^2 = |\mathbf{s}'|^2$ and

$$|\mathbf{s}'_\perp|^2 = |\mathbf{s}|^2 - |\mathbf{s}_\parallel|^2 - |\mathbf{G}|^2 - 2\mathbf{G} \cdot \mathbf{s}_\parallel, \qquad (4\text{-}44)$$

where Eq. 4-40 was used to eliminate $\mathbf{s}'_\parallel$. The first two terms on the right are known: $|\mathbf{s}|^2$ is determined by the energy of the incident electrons and ultimately by the accelerating potential. The parallel component of $\mathbf{s}$ is given by $\mathbf{s}_\parallel = s \cos\theta$, where $\theta$ is the angle between the incident beam and the surface.

We suppose the beam is incident normal to the surface. Then $\mathbf{s}_\parallel = 0$, so

$$\mathbf{s}'_\parallel = \mathbf{G} \qquad (4\text{-}45)$$

and

$$|\mathbf{s}'_\perp|^2 = |\mathbf{s}|^2 - |\mathbf{G}|^2. \qquad (4\text{-}46)$$

For extremely low energy, the right side of Eq. 4-46 is negative for every $\mathbf{G}$ except $\mathbf{G} = 0$ and the only reflected wave produced is the one with $\mathbf{s}'_{\parallel} = 0$ and $\mathbf{s}'_{\perp} = -\mathbf{s}_{\perp}$. This wave is said to be specularly reflected. As the energy is increased, it reaches values such that the right side of Eq. 4-46 is positive for one or more of the shortest reciprocal lattice vectors. For each such $\mathbf{G}$ there is then a scattered wave with $\mathbf{s}'_{\parallel}$ given by Eq. 4-45 and the magnitude of $\mathbf{s}'_{\perp}$ given by Eq. 4-46. When one of these waves first appears the perpendicular component of its propagation vector is nearly zero and its direction of travel is nearly parallel to the surface. With further increase in the energy, the angle between its propagation direction and the surface increases until $\mathbf{s}'_{\perp}$ reaches $\mathbf{s}_{\perp}$ as a limiting value. In the usual experiment the energy is sufficiently high that several scattering peaks are observed.

Figure 4-17$a$ is a LEED pattern produced by a beam of 68-eV electrons incident normally on a (110) surface of a copper crystal. The specularly reflected beam is blacked out. The pattern clearly shows the twofold symmetry expected of a rectangular surface lattice. Scattering angles are measured and the results are used to calculate primitive unit cell dimensions.

Because impurity atoms may be adsorbed from the environment or may be purposely placed on the surface, some atoms there may differ in type from atoms of the interior. If coverage is sufficiently great, adsorbed atoms form periodic arrays with sufficient extent to contribute to the LEED pattern. Since sites at which adsorption takes place are determined by the structure of the host surface, the lattice associated with impurities is related to the surface lattice of the host, but the impurity and host lattices need not be identical and their LEED patterns may be different. Figure 4-17$b$ shows the pattern for oxygen atoms adsorbed on a (110) copper surface. Compared with the pattern for a clean surface, the peaks have the same spacing horizontally but are closer together vertically. Figure 4-17$c$ illustrates one of many possible surface structures that could have produced the pattern. The impurities form a rectangular lattice, but with one unit cell side twice as long as the corresponding side of the host cell. For such a structure, one of the reciprocal lattice vectors is only half as large as for the host and diffraction peaks appear between the peaks produced by a clean surface.

The structure illustrated in Fig. 4-17$c$ is called a $(2 \times 1)$ structure, a notation which indicates that one primitive lattice vector is twice the length of the corresponding vector for the host while the other has the same length as the corresponding vector for the host. In general, a structure is labeled $(n \times m)$ if the unit vectors are $n\mathbf{a}$ and $m\mathbf{b}$, where $\mathbf{a}$ and $\mathbf{b}$ are host lattice vectors. In many cases, the same type impurity may form different structures on different parts of the same surface. The various regions are called domains. Typically, they are larger than the region probed by an electron beam, so the LEED pattern is usually indicative of a single domain.

Scattering angles for intense scattering peaks give information only about the size and shape of the unit cell. To obtain information about the type and position of basis atoms, scattered beam intensities must be compared with theoretical calculations. Both the theoretical prediction of intensities from known atomic patterns and the unraveling of experimental intensities to find actual patterns

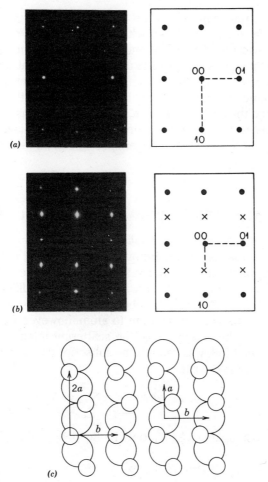

**FIGURE 4-17** LEED patterns for a (110) surface of copper. In (a) the surface is clean and the pattern clearly shows the twofold symmetry of the surface lattice. The reciprocal lattice for the surface is shown to the right, with indices given. In (b) the surface has a high concentration of oxygen impurities and additional peaks occur. (c) shows a surface structure which produces the pattern of b. Underlying copper atoms are indicated by larger circles while oxygen atoms, on the surface, are indicated by smaller circles. **a** and **b** are primitive lattice vectors for the copper surface lattice. One side of the unit cell formed by oxygen is twice as long as the corresponding copper unit cell and the associated reciprocal lattice vector is halved. Any displacement of the oxygen lattice parallel to the surface results in the same diffraction pattern. (From J. R. Noonan, Oak Ridge National Laboratory, unpublished. Used with permission.)

are far more difficult to carry out than the analogous calculations for x rays. The situation is complicated because incident electrons are strongly scattered by electrons of the material and many are scattered more than once before exiting the solid. Nevertheless, much progress has been made in the interpretation of LEED intensity data and this technique is responsible for much knowledge about the structure of surfaces.

Analysis of electrons emitted via the Auger mechanism is also important for the study of surfaces. In a collision with an atom on the surface, an incident electron may cause the removal of another electron from one of the atomic states. A higher energy electron then drops to the vacant level and, in most cases, photons are emitted. On occasion, however, the energy is transferred to still another atomic electron, which is then ejected from the solid. This is the Auger process and the energy spectrum of emitted electrons can be used to find the energy levels of surface atoms. Since energy level diagrams exist for all atoms, experimentally determined Auger spectra can be used to identify atoms on the surface.

## 4.5 ELASTIC SCATTERING BY AMORPHOUS SOLIDS

Equation 4-12 holds for amorphous solids as well as for crystals. We consider the simplest case, that of an amorphous solid consisting of only one type of atom, and we suppose that the atomic form factor $f_j$ is the same for all atoms. This is not strictly true since different atoms may have different distributions of neighbors and hence different distributions of electrons around them. Slight variations of the form factor from atom to atom, however, do not significantly influence dominant characteristics of the scattering intensity pattern and we neglect them. In what follows, we omit the subscript on the atomic form factor.

Start with Eq. 4-12 and write

$$\xi_s^* = A^* e^{-i(\mathbf{s}' \cdot \mathbf{r} - \omega t)} f^* \sum_k e^{i\Delta \mathbf{s} \cdot \mathbf{r}_k}, \tag{4-47}$$

for the complex conjugate of $\xi_s$. Multiply by Eq. 4-12 to find

$$|\xi_s|^2 = |A|^2 |f|^2 \sum_j \sum_k e^{i\Delta \mathbf{s} \cdot (\mathbf{r}_k - \mathbf{r}_j)}, \tag{4-48}$$

where each sum runs over all atoms in the material. The intensity at the detector is proportional to $|\xi_s|^2$.

The quantity

$$S_a = \sum_j \sum_k e^{i\Delta \mathbf{s} \cdot (\mathbf{r}_j - \mathbf{r}_k)} \tag{4-49}$$

is called the amorphous structure factor. It differs from the crystal structure factor in two ways. First, the sums run over all atoms in the solid. Second, it is a factor in the expression for the intensity, not in the expression for the wave function. In terms of the amorphous structure factor, the scattered intensity is

$$|\xi_s|^2 = |A|^2 |f|^2 S_a. \tag{4-50}$$

If there are $N$ atoms in the solid, the double sum in Eq. 4-49 contains $N$ terms for which $\mathbf{r}_k = \mathbf{r}_j$ and each of these has the value 1. They sum to $N$, so the amorphous structure factor can be written

$$S_a = N + \sum_j \sum_k{}' e^{i\Delta s \cdot (r_k - r_j)},$$   (4-51)

where the prime on the second summation symbol indicates that $k = j$ terms are omitted.

For any particular value of $j$ in the first sum, the second sum is over the distribution of atoms around atom $j$. Note that the displacement $\mathbf{r}_k - \mathbf{r}_j$ of atom $k$ from atom $j$ enters the sum. Since different atoms of an amorphous solid have different distributions of atoms around them, the sum over $k$ is different for different values of $j$. Unlike lattice sums for a crystal, the sums in Eq. 4-51 cannot be evaluated in closed form. In practice, however, the second sum is replaced by its average value.

To compute the average value of the second sum in Eq. 4-51, let $\mathbf{r} = \mathbf{r}_k - \mathbf{r}_j$ and suppose $\bar{n}(r)$ is the average number of atoms per unit volume with centers a distance $r$ from a reference atom. On average, the infinitesimal volume element $d\tau$ at $r$ contains $\bar{n}(r)\, d\tau$ other atoms and the average of the sum can be expressed as the volume integral $\int e^{i\Delta s \cdot r}\, \bar{n}(r)\, d\tau$. Since this is an average over all atoms of the solid, it is the same for all values of $j$ in the first sum. There are $N$ such terms, so the amorphous structure factor can be approximated by

$$S_a = N + N \int e^{i\Delta s \cdot r} \bar{n}(r)\, d\tau$$   (4-52)

and the scattered intensity by

$$|\xi_s|^2 = |A|^2 |f|^2 N \left[ 1 + \int e^{i\Delta s \cdot r} \bar{n}(r)\, d\tau \right],$$   (4-53)

where the integrals are over the volume of the sample.

If photographic film is wrapped in a cylinder around the sample and exposed to the scattered waves, the pattern consists of a series of rings, somewhat like a crystal powder pattern. The rings, however, are broad and diffuse, and only a few of them can be seen. As Eq. 4-53 shows, the scattering intensity depends on the magnitude of $\Delta s$ and not on its direction, so in practice the pattern is scanned only on a circular arc centered at the sample. For each scattering angle, $|\Delta s|$ is computed using $|\Delta s| = (4\pi/\lambda) \sin \theta$, where $\theta$ is half the scattering angle, then the intensity is plotted as a function of $|\Delta s|$. A typical result is shown in Fig. 4-18. Four rings are clearly discernible.

For a given change $\Delta s$ in the propagation vector we consider those atomic pairs with separation $\mathbf{r}$ such that $\Delta s \cdot \mathbf{r}$ is a multiple of $2\pi$. Waves from the atoms of these pairs interfere constructively at the detector. If there are many such pairs in the solid the scattering intensity is large and an intense ring is produced. Alternatively, relatively dark areas between rings are produced if $\Delta s \cdot \mathbf{r}$ is nearly an odd multiple of $\pi$ for a large fraction of atom pairs.

Rings, rather than spots, occur for the same reason they occur for a powder

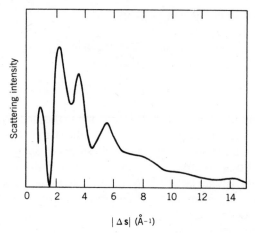

**FIGURE 4-18** The x-ray scattering intensity for amorphous germanium as a function of the change $|\Delta s|$ is the propagation vector. (From H. Krebs and R. Steffen, *Z. Anorg. Chem.* **327:** 224, 1964.)

pattern. If waves scattered from two atoms with separation **r** interfere constructively, then constructive interference also occurs for waves scattered from another pair with their relative displacement rotated from **r** by any angle about the direction of the incident beam.

Analysis of a pattern is complicated because all atom pairs, except those for which $\Delta \mathbf{s} \cdot \mathbf{r}$ is an odd multiple of $\pi$, contribute to the intensity for any $\Delta \mathbf{s}$. Suppose, for example, $\Delta \mathbf{s} \cdot \mathbf{r}$ has a value between 0 and $\pi$ for a great many pairs. Waves scattered from atoms of one pair interfere to produce an intensity that is less than the maximum possible, but there are so many pairs that the total intensity may be fairly high.

Nevertheless, scattering data can be used to determine the average concentration of atoms around a central atom and, through Eq. 3-2, the radial distribution function. The integrand in Eq. 4-52 has the form of a Fourier integral, the analog of a Fourier series for a nonperiodic function. If its value is known for all values of $\Delta \mathbf{s}$, techniques of Fourier analysis can be used to solve for $\bar{n}(r)$. The analysis is mathematically advanced and we will not pursue it here.

## 4.6 REFERENCES

All of the books referenced in Chapter 3 contain discussions of x-ray, electron, and neutron scattering from crystals. In addition, see:

B. D. Cullity, *Elements of X-ray Diffraction* (New York: Addison-Wesley, 1956).

R. W. James, "The Dynamical Theory of X-Ray Diffraction" in *Solid State Physics* (F. Seitz and D. Turnbull, Eds.), Vol. 15, p. 53 (New York: Academic, 1963).

C. G. Shull and E. O. Wollan, "Applications of Neutron Diffraction to Solid State Problems" in *Solid State Physics* (F. Seitz and D. Turnbull, Eds.), Vol. 2, p. 137 (New York: Academic, 1956).

Scattering by crystal surfaces is discussed in:

K. A. R. Mitchell, "Low-Energy Electron Diffraction." *Contemporary Physics* **14**:251, 1973.
G. A. Somorjai and H. H. Farrell, "Low-Energy Electron Diffraction" in *Advances in Chemical Physics* (I. Prigogine and S. A. Rice, Eds.) (New York: Wiley, 1971).
M. B. Webb and M. G. Lagally, "Elastic Scattering of Low-Energy Electrons from Surfaces" in *Solid State Physics* (H. Ehrenreich, F. Seitz, and D. Turnbull, Eds.), Vol. 28, p. 301 (New York: Academic, 1973).

For information on elastic scattering from amorphous solids, see the book by Zallen, referenced in Chapter 3, and:

R. Grigorovici, "The Structure of Amorphous Semiconductors" in *Electronic and Structural Properties of Amorphous Semiconductors* (P. C. LeComber and J. Mort, Eds.) (New York: Academic, 1972).

For x-ray diffraction data see the compilations listed in the previous chapter and *Crystal Data*, published jointly by the U.S. Department of Commerce, the National Bureau of Standards, and the Joint Committee on Powder Diffraction Standards.

## PROBLEMS

1. 1.54-Å x rays are incident along the $z$ axis on two identical atoms separated by 3.2 Å. Their relative displacement is in the $yz$ plane and makes the angle $\theta_r$ with the $z$ axis. A detector is moved in the $yz$ plane around the atoms. For (a) $\theta_r = 0$ and (b) $\theta_r = 45°$, find the positions of the detector where it records maxima of scattered intensity.

2. The wave function for an electron in the ground state of a hydrogen atom is $\psi(r) = (1/\sqrt{\pi})(1/a_0)^{3/2}e^{-r/a_0}$, where $a_0$ is the Bohr radius. (a) Take the electron concentration to be $n(r) = \psi^*\psi$ and evaluate the atomic form factor. (b) Find limiting values as the wavelength of the incident wave becomes large and as it becomes small. (c) With what atomic dimension should the wavelength be compared to decide if it is large or small?

3. The form factor you found in Problem 2 is independent of the scattering angle. (a) What characteristics should the electron concentration have for the form factor to depend on the scattering angle? (b) For what solids do you expect the angular dependence to be the strongest?

4. An orthorhombic lattice has primitive lattice vectors $\mathbf{a} = a\hat{\mathbf{x}}$, $\mathbf{b} = b\hat{\mathbf{y}}$, and $\mathbf{c} = c\hat{\mathbf{z}}$. Find a set of fundamental reciprocal lattice vectors and use them to derive an expression for the separation of adjacent parallel lattice planes. Take $a = 3.17$ Å, $b = 4.85$ Å, and $c = 2.13$ Å, then find the separation of planes in each of the following sets: (100), (110), (011), and (111).

5. For each of the following direct lattices, find a set of fundamental reciprocal lattice vectors and classify the reciprocal lattice according to Bravais lattice type: (a) hexagonal; (b) body-centered tetragonal; and (c) body-centered

orthorhombic. In each case, give the dimensions of the conventional reciprocal unit cell in terms of the dimensions of the conventional direct lattice cell.

6. Consider a trigonal direct lattice and use primitive lattice vectors with equal lengths and with the same angle $\theta$ between any pair of them. Find a set of fundamental reciprocal lattice vectors and show that the reciprocal lattice is trigonal. Also show that the angle $\theta'$ between any pair of them is given by $\cos \theta' = -(\cos \theta)/(1 + \cos \theta)$.

7. Use Eqs. 4-21, 4-22, and 4-23 to prove (a) Eq. 4-26 and (b) Eqs. 4-28, 4-29, and 4-30.

8. Suppose two atoms in the primitive basis of a zinc blende structure have atomic form factors $f_a$ and $f_b$, respectively. (a) Obtain an expression for the structure factor $F$ for the $(\hbar k \ell)$ scattering peak, indexed using a simple cubic lattice. Show that $F = 0$ unless $\hbar$, $k$, $\ell$ are all even or all odd. Then show that $F = 4(f_a + f_b)$ for $\hbar + k + \ell = 4n$, $F = 4(f_a - if_b)$ for $\hbar + k + \ell = 4n + 1$, $F = 4(f_a - f_b)$ for $\hbar + k + \ell = 4n + 2$, and $F = 4(f_a + if_b)$ for $\hbar + k + \ell = 4n + 3$, where $n$ is any integer. (b) Suppose both form factors are real and $f_b = \frac{1}{2}f_a$. Find the relative intensities of the four peaks associated with the same value of $n$. The two atoms in the primitive basis of the diamond structure usually have slightly different form factors because their bond systems are oriented differently so the results obtained in part a for zinc blende also hold for diamond.

9. (a) Find an expression for the structure factor $F$ associated with an ideal HCP structure. Use indices based on a primitive lattice. (b) Suppose the two atoms of the basis have the same atomic form factor and it is independent of $|\Delta \mathbf{s}|$. Rank the following scattering peaks in order of increasing intensity: (100), (110), (111), ($\bar{1}$11), (210), and (211).

10. Calculate the energy of (a) a photon, (b) an electron, and (c) a neutron if the wavelength of each of their waves is 1.00 Å.

11. Start with the Laue condition $\Delta \mathbf{s} = \mathbf{G}$ and show that an intensity peak for elastic scattering occurs if $\mathbf{s} \cdot \mathbf{G} = -\frac{1}{2}|\mathbf{G}|^2$.

12. A tetragonal lattice has fundamental lattice vectors $\mathbf{a} = a\hat{x}$, $\mathbf{b} = a\hat{y}$, and $\mathbf{c} = c\hat{z}$. The propagation vector for an incident beam of x rays is in the $xy$ plane and makes the angle $\alpha$ with the $x$ axis. (a) Take the wavelength to be $\lambda$ and show that the $(\hbar k \ell)$ elastic scattering peak occurs if

$$\hbar \cos \alpha + k \sin \alpha = -\frac{\lambda a}{2} \left[ \frac{\hbar^2 + k^2}{a^2} + \frac{\ell^2}{c^2} \right].$$

(b) Show that no peaks with indices of the form $(00\ell)$ are produced. (c) Suppose $\lambda = 1.54$ Å, $a = 4.73$ Å, and $c = 5.71$ Å. Find the angle $\alpha$ the beam must make with the $x$ axis to produce each of the following peaks: (100), (101), and (111). (d) Find the scattering angle for each of the peaks of part c.

13.  1.54-Å x rays are used to make a rotating crystal pattern for a simple cubic crystal with cube edge 4.51 Å. The film forms a cylinder with radius 57.3 mm about the sample and the sample rotates about a fourfold symmetry axis. (a) How many rows of spots are there? (b) What is the separation between rows on the film? (c) For the central row and the first row above it, find the positions of the two spots on the same side of the forward direction having the smallest scattering angle. (d) If the lattice is body-centered cubic, which of the spots found in part c are missing, if any?

14.  A rotating crystal pattern is made by scattering 1.39-Å x rays from an orthorhombic crystal, rotating about its c axis. The radius of the film cylinder is 57.3 mm and 5 rows of spots are obtained. (a) Show that, since there are only 2 rows of spots above the central row, the length of the unit cell edge along the c axis must be less than $3\lambda = 4.17$ Å. (b) The central row and the first row above it are separated by 12.1 mm. What is the length c of the cell edge? (c) What is the separation between the central row and the second row above it on the film? (d) The other edges of the conventional orthorhombic unit cell are determined to be $a = 4.34$ Å and $b = 3.23$ Å. Along the central row, the five spots closest to the unscattered beam are 9.20, 12.4, 15.5, 18.7, and 22.7 mm, respectively, from the point where the unscattered beam pierces the film. Index these spots.

15.  Consider a crystal with a simple cubic lattice and one atom in its primitive basis. The cube edge is 4.50 Å. (a) For each of the following peaks find the wavelength so that the peak occurs with a scattering angle of 40°: (100), (110), and (111). (b) If the wavelength is such that the scattering angle for the (110) peak is 40°, what are the scattering angles for (100) and (111) peaks? (c) If the wavelength is such that the scattering angle for the (110) peak is 40°, what is the scattering angle for the (220) peak?

16.  A tetragonal crystal is ground into a powder and 1.54-Å x rays are used to make a powder pattern. The conventional unit cell has a square base with an edge of 3.20 Å and a height of 4.63 Å. Find the scattering angle for each of the three smallest diameter rings in the forward scattering direction and for each of the three smallest diameter rings in the backscattering direction. (b) If the basis consists of an atom at a cell corner and a different type atom at the cell center, which of these rings are weak? Assume the atomic form factors are real and have the same sign.

17.  A cubic crystal is ground into a powder and 1.39-Å x rays are used to obtain a powder pattern. The smallest five rings in the forward scattering direction are associated with scattering angles of 24.4, 28.2, 40.3, 47.7, and 50.0°, while the smallest five rings in the backscattering direction are associated with scattering angles of 172.8, 154.6, 139.0, 131.7, and 123.1.° (a) Calculate $\sin^2 \theta$ for each of the rings in the forward scattering direction. Compare their ratios to the ratios you expect for simple cubic, face-centered cubic, and body-centered cubic lattices. Use the comparison to identify the lattice. (b) Assume the smallest diameter ring in the forward scat-

tering direction is the smallest possible for the lattice identified in part a and calculate the cube edge. (c) Index all the rings. (d) Use the smallest diameter ring in the backscattering direction to calculate the cube edge.

18. The surface lattice for a clean (001) surface of a copper sample is square and has an edge of 5.61 Å. In a LEED experiment, monoenergetic electrons are incident normally on the surface. (a) Calculate the minimum energy for which intense scattering, other than specular reflection, occurs. At this energy how many scattered beams are produced? What are their directions? (b) Suppose the energy is 4.50 times as great as that found in part a. Find the angle made with the surface by each intensely scattered beam. (c) For the energy of part b, sketch the LEED pattern as it might appear on a screen, originally placed around the sample as film is placed in an x-ray camera.

19. For the copper sample of Problem 18, suppose alternate rows of copper atoms, parallel to one of the square edges, are replaced by rows of impurity atoms. For a LEED experiment with a normal incident beam and an electron energy of 20.0 eV, identify all intense scattered beams that are obtained and find the angle each of them makes with the surface. Which of these beams would be missing if the surface did not contain impurities?

# Chapter 5

## BONDING

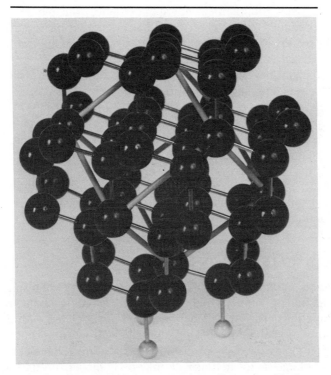

Stick model of the diamond structure. A cubic cell is outlined by heavy sticks.

109

Atoms bound together in a solid have a lower total energy than the same atoms when they are well separated. Energy reduction comes about in all cases because electrons in a solid interact with more than one ion core and, as a result, most occupy states with lower energy than they would in the ground state of an isolated atom. As atoms are brought together, electron wave functions for states that contribute to bonding spread into regions between atoms, where the potential energy is lower than when the atoms are far apart. The spreading of wave functions also means a decrease in electron kinetic energy, a result which follows easily from the Heisenberg uncertainty principle. The momentum of an electron confined to a small region of space generally has large fluctuations, so the average value of the momentum squared and hence the energy is greater than if the electron is less confined.

Bonding mechanisms can be distinguished qualitatively on the basis of the extent and form of the electron distribution. For covalent bonds the electron concentration is greater in the neighborhood of lines that join atoms than in regions away from these lines. In a metal, on the other hand, the electron distribution is much more uniform throughout interstitial regions. In the case of ionic solids, one or more electrons are transferred from the neighborhood of one atom to the neighborhood of another and the atoms can be treated as oppositely charged ions attracting each other electrostatically. We will examine some of the details of these and other bonding mechanisms in this chapter. A brief introduction to the calculation of electron energies is given first.

## 5.1 ENERGY CALCULATIONS

*The Schrödinger Equation.* Calculation of the total energy of a solid begins, in principle, with finding solutions to the Schrödinger equation for electron energies and wave functions. The wave function $\Psi(\mathbf{r}, t)$, associated with an electron, contains information about the behavior of the electron. For example, the quantity $dP = |\Psi(\mathbf{r}, t)|^2 \, d\tau$ gives the probability that, at time $t$, the electron is in the infinitesimal volume $d\tau$ located at $\mathbf{r}$. $\Psi$ may be complex and its squared magnitude is computed as the product of $\Psi$ with its complex conjugate, denoted by $\Psi^*$. $|\Psi|^2$ is called the probability density for the electron.

The wave function and hence the probability density are determined by the potential energy function $U(\mathbf{r})$ for the electron. More specifically, the wave function is a solution to the Schrödinger equation,

$$-\frac{\hbar^2}{2m} \nabla^2 \Psi(\mathbf{r}, t) + U(\mathbf{r})\Psi(\mathbf{r}, t) = i\hbar \frac{\partial \Psi(\mathbf{r}, t)}{\partial t}. \tag{5-1}$$

Here $m$ is the mass of the electron and $\nabla^2$ is the Laplacian differential operator, given by

$$\nabla^2 \Psi = \frac{\partial^2 \Psi}{\partial x^2} + \frac{\partial^2 \Psi}{\partial y^2} + \frac{\partial^2 \Psi}{\partial z^2} \tag{5-2}$$

in Cartesian coordinates.

For the approximation we will use, the potential energy function is independent of time and $\Psi(\mathbf{r}, t)$ can be written as the product of two functions, one depending only on spatial coordinates and the other depending only on the time: $\Psi(\mathbf{r}, t) = \psi(\mathbf{r})f(t)$. The time-dependent function has the form $e^{-i\omega t}$, where $\omega$ is an angular frequency. Both its real and imaginary parts oscillate sinusoidally with time but, since its magnitude is 1, it leads to a probability density that is independent of time: $|\Psi(\mathbf{r}, t)|^2 = |\psi(\mathbf{r})|^2$.

The angular frequency of the wave function is related to the energy of the particle by

$$E = \hbar\omega. \tag{5-3}$$

This relationship was used in Chapter 4 where we considered plane waves scattered by a sample. Wave functions considered here are significantly more complicated but Eq. 5-3 is still valid.

To obtain the differential equation for $\psi(\mathbf{r})$, $\Psi(\mathbf{r}, t) = \psi(\mathbf{r})e^{-i\omega t}$ is substituted into Eq. 5-1 and the exponential factor is canceled from each term. Once Eq. 5-3 is used, the result is the time-independent Schrödinger equation,

$$-\frac{\hbar^2}{2m}\nabla^2\psi(\mathbf{r}) + U(\mathbf{r})\psi(\mathbf{r}) = E\psi(\mathbf{r}). \tag{5-4}$$

A potential energy function is first constructed, then solutions to Eq. 5-4 obeying appropriate boundary conditions are sought. A great many states are possible for an electron bound in a solid, each with a wave function and energy. We use a subscript to distinguish the states from each other: $\psi_i$ and $E_i$ represent the wave function and energy, respectively, associated with state $i$.

In principle, the Schrödinger equation should also contain a term that depends on the spin angular momentum of the electron, but we can neglect this term in most discussions. Spin is important, however, when the occupation of electron states is considered. The $z$ component $S_z$ of the spin angular momentum must be either $+\frac{1}{2}\hbar$ or $-\frac{1}{2}\hbar$. Two electrons with the same spatial wave function $\psi(\mathbf{r})$ are in different states if $S_z$ is different for them. Furthermore, electrons obey the Pauli exclusion principle, so no more than one electron occupies any state and no more than two have the same spatial wave function.

*The Potential Energy Function.* Two terms dominate the potential energy function for an electron in a solid. The first results from electrostatic electron-nuclei interactions and is given by

$$U_{en}(\mathbf{r}) = -\frac{e^2}{4\pi\epsilon_0}\sum_i \frac{Z_i}{|\mathbf{r} - \mathbf{R}_i|}. \tag{5-5}$$

Here $\mathbf{R}_i$ is the position of nucleus $i$, assumed to have $Z_i$ protons, and the sum is over all nuclei in the solid. Because nuclei attract the electron, $U_{en}$ is negative relative to the potential energy of an electron far from all nuclei.

The second important term is due to electrostatic interactions between the electron under consideration and all other electrons. To calculate this term, the electrons are replaced by a continuous distribution with electron concentration $n(\mathbf{r}')$ proportional to the electron probability density. Specifically, the electron-electron contribution to the potential energy function is written

$$U_{ee}(\mathbf{r}) = \frac{e^2}{4\pi\epsilon_0} \int \frac{n(\mathbf{r}')}{|\mathbf{r} - \mathbf{r}'|} \, d\tau' , \qquad (5\text{-}6)$$

where

$$n(\mathbf{r}') = \sum_i |\Psi_i(\mathbf{r}', t)|^2 . \qquad (5\text{-}7)$$

The primed coordinates are the variables of integration in Eq. 5-6 and the sum in Eq. 5-7 is over all occupied electron states except the one for which the Schrödinger equation is being solved.

For the study of most material properties, we need to know electron wave functions and energy levels only for stationary nuclei, at their equilibrium sites. $U_{en}$ is then independent of time. Furthermore, we assume all electrons have wave functions of the form $\Psi(\mathbf{r}, t) = \psi(\mathbf{r})e^{-i\omega t}$, so $U_{ee}$ is also time independent. Since the total potential energy is independent of time, solutions to the Schrödinger equation do indeed have the assumed form.

Equation 5-6 is called the Hartree formulation of the electron-electron interaction. Note that electron wave functions are used to calculate the potential energy function and this function is used in turn to solve for the wave functions. A self-consistent method is employed. Wave functions are obtained using a trial potential energy function in the Schrödinger equation, then the wave functions are used to evaluate the potential energy function. If the result does not agree with the original trial function, that function is adjusted and the procedure is repeated.

More sophisticated formulations of the electron-electron interaction have been developed. The most important of these leads to the Hartree-Fock potential energy function, which takes into account the tendency of electrons with parallel spins to avoid each other, not because they are charged, but because they obey the Pauli exclusion principle. This function is more complicated than the Hartree function and we will not discuss it here.

The program outlined above is formidable and can be carried out accurately only if large-memory, high-speed computers are used. Nevertheless a great deal can be learned about bonding by studying simple situations involving only a small number of electrons. We begin with the hydrogen molecular ion.

## 5.2 THE HYDROGEN MOLECULAR ION

*Bonding States.* The system consists of two protons and an electron, as diagrammed in Fig. 5-1a. One of the protons, labeled $a$, is at the origin while the other, labeled $b$, is at $\mathbf{R}$. The displacement of the electron from proton $a$ is $\mathbf{r}$ and its displacement from proton $b$ is $\mathbf{r} - \mathbf{R}$, so the electron-proton potential

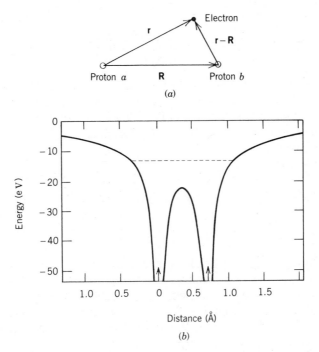

**FIGURE 5-1**  (a) Geometry of a hydrogen molecular ion. The protons are separated by **R**; the electron is at **r** relative to proton a and at **r** − **R** relative to proton b. (b) Electron potential energy along the line joining the protons. Atomic positions are indicated by arrows. The energy of the lowest electron state is indicated by a dotted line.

energy is given by

$$U(\mathbf{r}) = -\frac{e^2}{4\pi\epsilon_0}\frac{1}{r} - \frac{e^2}{4\pi\epsilon_0}\frac{1}{|\mathbf{r} - R|}. \tag{5-8}$$

A plot of this function along the line joining the protons is shown in Fig. 5-1b. Near each proton there is a nearly coulombic potential energy well and in the region between protons the potential energy is lower than the potential energy due to either proton alone.

Near each proton we expect the electron wave function to closely resemble a wave function for an electron in a hydrogen atom. The wave function tail, however, extends past the potential energy barrier into the neighboring well and joins smoothly to an atomic orbital centered on the proton there.

To illustrate, we suppose that near either proton the wave function is nearly a hydrogen $1s$ orbital and, as a first approximation to the wave function for the electron in the molecular ion, we write

$$\psi(\mathbf{r}) = N[\chi(r) + \chi(|\mathbf{r} - \mathbf{R}|)], \tag{5-9}$$

where

$$\chi(r) = \frac{1}{\sqrt{\pi}} \left[ \frac{1}{a_0} \right]^{3/2} e^{-r/a_0}.$$ (5-10)

Here $a_0$ is the Bohr radius ($a_0 = 4\pi\epsilon_0\hbar^2/me^2 = 0.529$ Å). The orbital is normalized: $\int |\chi(r)|^2 \, d\tau = 1$, where the integral is over all space.

The first function in the brackets of Eq. 5-9 is centered on proton a while the second is centered on proton b. $N$ is a normalization constant, chosen so $\int |\psi|^2 \, d\tau = 1$. Since $\chi$ is real

$$|\psi(\mathbf{r})|^2 = |N|^2 [\chi^2(r) + \chi^2(|\mathbf{r} - \mathbf{R}|) + 2\chi(r)\chi(|\mathbf{r} - \mathbf{R}|)].$$ (5-11)

If $\Delta = \int \chi(r)\chi(|\mathbf{r} - \mathbf{R}|) \, d\tau$, then we may take $N$ to be

$$N = \left[ \frac{1}{2(1 + \Delta)} \right]^{1/2}.$$ (5-12)

The wave function given by Eq. 5-9 and plotted in Fig. 5-2 describes the sharing of the electron by the two protons. Notice that atomic wave functions, not probability densities, are superposed. The probability density for the electron in the molecular ion, given by Eq. 5-11, includes the interference term $2\chi(r)\chi(|\mathbf{r} - \mathbf{R}|)$, which is large in regions where *both* $\chi(r)$ and $\chi(|\mathbf{r} - \mathbf{R}|)$ are significant. This is just the region where the potential energy for the electron in a molecular ion is lower than the potential energy for an electron in a hydrogen atom.

To estimate the energy of the state, we calculate its average value $\langle E \rangle$, given by

$$\langle E \rangle = \int \psi^*(\mathbf{r}) \left[ -\frac{\hbar^2}{2m} \nabla^2 \psi(\mathbf{r}) - \frac{e^2}{4\pi\epsilon_0 r} \psi(\mathbf{r}) - \frac{e^2}{4\pi\epsilon_0 |\mathbf{r} - \mathbf{R}|} \psi(\mathbf{r}) \right] d\tau.$$ (5-13)

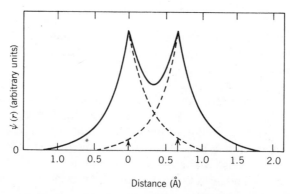

**FIGURE 5-2** Approximate wave function for the lowest energy state of an electron in a hydrogen molecular ion, along the line joining the protons. Dotted lines indicate the tails of 1s atomic orbitals, one centered on each proton.

Equation 5-9 is used to substitute for $\psi$, then the Schrödinger equations for $\chi(r)$ and $\chi(|\mathbf{r} - \mathbf{R}|)$ are used to simplify the result. These are

$$-\frac{\hbar^2}{2m} \nabla^2\chi(r) - \frac{e^2}{4\pi\epsilon_0 r} \chi(r) = E_{1s}\chi(r), \tag{5-14}$$

and

$$-\frac{\hbar^2}{2m} \nabla^2\chi(|\mathbf{r} - \mathbf{R}|) - \frac{e^2}{4\pi\epsilon_0|\mathbf{r} - \mathbf{R}|} \chi(|\mathbf{r} - \mathbf{R}|) = E_{1s}\chi(|\mathbf{r} - \mathbf{R}|), \tag{5-15}$$

respectively. $E_{1s}$ ($= -13.6$ eV) is the energy of an electron in a hydrogen $1s$ state. After simplification, Eq. 5-13 is

$$\langle E \rangle = E_{1s} - \frac{e^2 N^2}{4\pi\epsilon_0} \int [\chi(r) + \chi(|\mathbf{r} - \mathbf{R}|)] \left[ \frac{1}{r} \chi(|\mathbf{r} - \mathbf{R}|) + \frac{1}{|\mathbf{r} - \mathbf{R}|} \chi(r) \right] d\tau. \tag{5-16}$$

Let

$$A = \frac{e^2}{4\pi\epsilon_0} \int \chi^2(r) \frac{1}{|\mathbf{r} - \mathbf{R}|} d\tau \tag{5-17}$$

and

$$B = \frac{e^2}{4\pi\epsilon_0} \int \chi(r)\chi(|\mathbf{r} - \mathbf{R}|) \frac{1}{r} d\tau. \tag{5-18}$$

Simple changes of variable can be used to show that $(e^2/4\pi\epsilon_0)\int[\chi^2(|\mathbf{r} - \mathbf{R}|)/r]d\tau = A$ and $(e^2/4\pi\epsilon_0)\int[\chi(r)\chi(|\mathbf{r} - \mathbf{R}|)/|\mathbf{r} - \mathbf{R}|] d\tau = B$. When Eqs. 5-12, 5-17, and 5-18 are used, Eq. 5-16 becomes

$$\langle E \rangle = E_{1s} - \frac{A + B}{1 + \Delta}. \tag{5-19}$$

Since $A$ and $B$ are both positive, Eq. 5-19 clearly predicts a decrease in energy from the $1s$ atomic level. The total energy of the $H_2^+$ molecule is obtained by adding the potential energy of the proton-proton interaction to $\langle E \rangle$:

$$E_{\text{total}} = E_{1s} - \frac{A + B}{1 + \Delta} + \frac{e^2}{4\pi\epsilon_0} \frac{1}{R}. \tag{5-20}$$

$A$ and $B$ determine the extent to which the energy is lowered as the protons approach each other. $A$ is essentially the potential energy for the interaction between one of the protons and the electron when it is in an atomic state around the other proton. $B$ is a measure of the extent to which tails of neighboring atomic orbitals overlap in the interstitial region. The greater the overlap, the lower the energy and the stronger the bond.

For the atomic orbital of Eq. 5-10, Eqs. 5-12, 5-17, and 5-18 can be evaluated in closed form:

$$\Delta = \left[ 1 + \frac{R}{a_0} + \frac{R^2}{3a_0^2} \right] e^{-R/a_0}, \tag{5-21}$$

$$A = \frac{e^2}{4\pi\epsilon_0} \frac{1}{a_0} \left[ \frac{a_0}{R} - \left( \frac{a_0}{R} + 1 \right) e^{-2R/a_0} \right], \tag{5-22}$$

and

$$B = \frac{e^2}{4\pi\epsilon_0} \frac{1}{a_0} \left[ \frac{R}{a_0} + 1 \right] e^{-R/a_0}. \tag{5-23}$$

These expressions were used to plot both $\langle E \rangle$ and $E_{total}$ as functions of $R$ in Fig. 5-3.

For large proton separation, $A \to e^2/4\pi\epsilon_0 R$, $B \to 0$, and $\Delta \to 0$, so $\langle E \rangle \to E_{1s} - e^2/4\pi\epsilon_0 R$ and $E_{total} \to E_{1s}$. $\langle E \rangle$ is then the energy of an electron in a 1s atomic state, interacting with a proton far away. The system is a well-separated proton and neutral hydrogen atom, so the total energy is just that of an electron in a hydrogen 1s state. The interference term in the probability density is small since $\chi(r)$ is extremely small where $\chi(|\mathbf{r} - \mathbf{R}|)$ is not and vice versa.

As $R$ becomes small $A \to e^2/4\pi\epsilon_0 a_0$, $B \to e^2/4\pi\epsilon_0 a_0$, and $\Delta \to 1$, so $\langle E \rangle \to E_{1s} - e^2/4\pi\epsilon_0 a_0$. When the proton-proton potential energy is added to $\langle E \rangle$ to form $E_{total}$, the result becomes large and positive. $E_{total}$ passes through a minimum at about $R = 2.50a_0$. This is the predicted equilibrium separation of the protons.

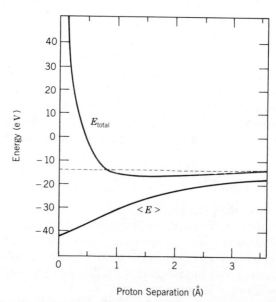

**FIGURE 5-3** Electron energy $\langle E \rangle$ and total energy as functions of proton separation for the bonding state of a hydrogen molecular ion. The 1s atomic level is indicated by a dotted line.

**EXAMPLE 5-1** For an $H_2^+$ ion with a proton separation of $2.50a_0$, use the wave function of Eq. 5-9 to calculate (a) the average electron energy, (b) the average electron potential energy, (c) the average electron kinetic energy, and (d) the total energy including the proton-proton potential energy.

**SOLUTION** For $R = 2.50a_0$, Eqs. 5-21, 5-22, and 5-23 yield $\Delta = 0.458$, $A = 1.70 \times 10^{-18}$ J $(= 10.6$ eV$)$, and $B = 1.25 \times 10^{-18}$ J $(= 7.82$ eV$)$, respectively. (a) The average electron energy is $\langle E \rangle = E_{1s} - (A + B)/(1 + \Delta) = -4.21 \times 10^{-18}$ J $(= -26.3$ eV$)$. (b) The average electron potential energy is given by

$$\langle U \rangle = \int \psi^*(\mathbf{r}) \left[ -\frac{e^2}{4\pi\epsilon_0} \frac{1}{r} - \frac{e^2}{4\pi\epsilon_0} \frac{1}{|\mathbf{r} - \mathbf{R}|} \right] \psi(\mathbf{r}) \, d\tau$$

$$= -\frac{N^2 e^2}{4\pi\epsilon_0} \left[ \int \chi^2(r) \frac{1}{r} \, d\tau + \int \chi^2(|\mathbf{r} - \mathbf{R}|) \frac{1}{|\mathbf{r} - \mathbf{R}|} \, d\tau \right] - 2N^2(A + 2B) .$$

The integrals are easily evaluated: each of the first two terms is $-N^2 e^2/4\pi\epsilon_0 a_0$ or $-1.49 \times 10^{-18}$ J $(-9.33$ eV$)$. Finally, $\langle U \rangle = -5.88 \times 10^{-18}$ J $(-36.7$ eV$)$. (c) $\langle K \rangle = \langle E \rangle - \langle U \rangle = -4.21 \times 10^{-18} + 5.88 \times 10^{-18} = 1.67 \times 10^{-18}$ J $(10.4$ eV$)$. (d) The proton-proton interaction energy is $e^2/4\pi\epsilon_0 R = 1.74 \times 10^{-18}$ J $(10.9$ eV$)$, so the total energy is $-4.21 \times 10^{-18} + 1.74 \times 10^{-18} = -2.47 \times 10^{-18}$ J $(-15.4$ eV$)$.

For comparison, the total energy at infinite separation is $-2.18 \times 10^{-18}$ J. The average electron potential energy is $-4.36 \times 10^{-18}$ J and the average electron kinetic energy is $2.18 \times 10^{-18}$ J. Note that both kinetic and potential energies decrease when the molecule is formed. ∎

Experimentally, the equilibrium separation is found to be $R = 2.0a_0$ and the total energy at this separation is found to be $E_{total} = -2.6 \times 10^{-18}$ J. Although the calculated energy behaves qualitatively in the correct manner and bonding is predicted, it is quantitatively in error. Errors occur because the approximate wave function is not sufficiently distorted from the atomic orbitals, a situation that can easily be corrected by adding other functions to the expression of Eq. 5-9. In principle, the complete set of hydrogen functions can be used and $\chi$ can be written

$$\chi(\mathbf{r}) = \sum_n A_n \Phi_n(\mathbf{r}), \tag{5-24}$$

where $\Phi_n(\mathbf{r})$ is an atomic orbital, $A_n$ is a constant, and the sum is over all atomic orbitals. In practice the sum is truncated after a few terms. A similar expression, with different constants and with atomic functions centered at $\mathbf{r} - \mathbf{R}$, is written for the second potential energy well, then these two expressions are used in Eq. 5-9. The expectation value of the energy is calculated in terms of the coefficients $A_n$, then values are assigned these coefficients so $\langle E \rangle$ has the smallest possible value. Finally, the equilibrium separation is found by searching for the minimum of $E_{total}$ as a function of $R$.

*Antibonding States.* Not all molecular states result in a lowering of energy. As an example of one which does not, consider the electron wave function

$$\psi(r) = N[\chi(r) - \chi(|\mathbf{r} - \mathbf{R}|)], \tag{5-25}$$

where $\chi(r)$ is again the hydrogen $1s$ function. Like the function given in Eq. 5-9, this is a qualitatively correct approximate solution to the time-independent Schrödinger equation.

The wave function is plotted in Fig. 5-4. Compared to the bonding function of Fig. 5-2, it is small in the region between protons and correspondingly large in the vicinity of each proton. As Fig. 5-5 shows, at any separation the electron energy is greater than the energy for the bonding state. In particular, as the protons are brought together the average potential energy decreases less for this state than for the bonding state. The electron avoids the interstitial region, where the decrease in potential energy is greatest. In addition, the kinetic energy is greater than for the bonding state, as can be seen by comparing wave function slopes in the interstitial region.

A detailed calculation, similar to the one for the bonding state, gives

$$\langle E \rangle = E_{1s} - \frac{A - B}{1 - \Delta}, \tag{5-26}$$

where $A, B$, and $\Delta$ are the same integrals as before. $E_{\text{total}}$ is again $\langle E \rangle + e^2/4\pi\epsilon_0 R$. As can be seen in Fig. 5-5, it is greater than the energy of a well-separated hydrogen atom and proton for all values of $R$. If the electron is in this state, the protons are not bound together. The state is an antibonding state.

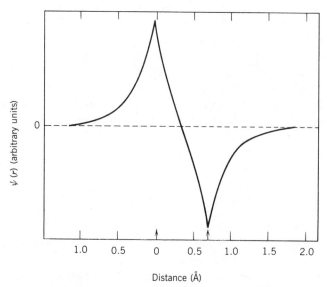

**FIGURE 5-4** Approximate wave function for the lowest energy antibonding state of an electron in a hydrogen molecular ion, along the line joining the protons.

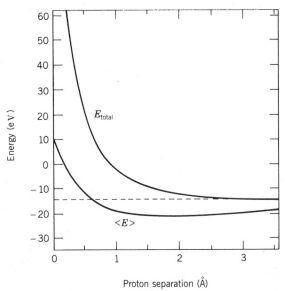

Proton separation (Å)

**FIGURE 5-5** Electron energy $\langle E \rangle$ and total energy as functions of proton separation for the lowest antibonding state of a hydrogen molecular ion. The 1s atomic level is indicated by a dotted line.

*Forces of Repulsion.* When atoms are close together they repel, rather than attract, each other. For the $H_2^+$ ion in a bonding state, repulsion is produced solely by the mutual interaction of the protons. For atoms with a larger number of electrons, core electrons contribute significantly to repulsion.

When atoms are brought close together, an electron state is created for each core state of each atom. Some have lower energy than the atomic state and are bonding states, while others have higher energy and are antibonding states. Since all core states, both bonding and antibonding, are occupied, the energy is considerably greater than if only bonding states were occupied. Note the role played by the Pauli exclusion principle. If any number of electrons could occupy the same state, all core electrons would be in bonding states and solids would collapse to extremely small volumes.

We may think of the atoms as being attracted to each other by means of the outer electrons, the majority of which occupy bonding states, and being repelled from each other by means of core electrons. Since core wave functions do not extend as far from a nucleus as wave functions for outer electrons, attraction occurs for large separations and repulsion occurs for small separations.

No analytical form can be derived from first principles for the core repulsion energy as a function of atomic separation. Approximate expressions, however, are often used. Frequently the energy of interaction for two cores is taken to have either the form

$$E_c = \frac{\beta}{R^n} \tag{5-27}$$

or the form

$$E_c = \beta\, e^{-R/\rho},\qquad(5\text{-}28)$$

where $R$ is the atomic separation. $\beta$, $n$, and $\rho$ are parameters which are chosen for a particular material by comparing measured and calculated values of properties that depend on core-core interactions. For most materials, $n$ has a value between 8 and 14, while $\rho$ has a value between $0.3a_0$ and $a_0$.

For a collection of atoms the energies must be summed. If Eq. 5-27 is used, for example, the total core-core energy for $N$ identical atoms is given by

$$E_{c\ total} = \frac{1}{2}\sum_i \sum_j{}' \frac{\beta}{R_{ij}^n},\qquad(5\text{-}29)$$

where $R_{ij}$ is the distance between atom $i$ and atom $j$. Both sums extend over atoms in the material, but terms for which $i = j$ are omitted. The prime on the second summation symbol is intended to indicate this. Each pair of atoms is included twice in the sums, once when $i$ represents the first atom of the pair and $j$ represents the second, then again when $i$ represents the second and $j$ represents the first. The factor $\frac{1}{2}$ is included to correct for double counting.

In practice, the energy of repulsion decreases rapidly as $R_{ij}$ increases and little accuracy is lost if only those pairs of atoms with the smallest separations are included. For each value of $i$ in Eq. 5-29 only a few terms in the sum over $j$ are required, representing the nearest neighbors of atom $i$. For crystals with a single atom in the primitive basis, the distributions of atoms around any given atom is the same as around any other. Equation 5-29 then reduces to

$$E_{c\ total} = \frac{Nz\beta}{2R^n},\qquad(5\text{-}30)$$

where $N$ is the number of atoms in the crystal, $z$ is the number of nearest neighbors to any atom, and $R$ is the nearest neighbor distance.

**EXAMPLE 5-2**   The total energy of two argon atoms, relative to their energy at infinite separation, is given by

$$E = -C(a_0/R)^6 + B(a_0/R)^{12},$$

where $C = 2.35 \times 10^3$ eV, $B = 1.69 \times 10^8$ eV, and $a_0$ is the Bohr radius. The first term represents the energy due to the attractive force of the outer electrons while the second term represents the energy of core-core repulsion. Calculate (a) the equilibrium separation $R_{eq}$, (b) the energy of attraction for $R = R_{eq}$, (c) the energy of repulsion for $R = R_{eq}$, and (d) the total energy for $R = R_{eq}$.

**SOLUTION**   For $R = R_{eq}$, $dE/dR = 0$ so $(6Ca_0^6/R_{eq}^7) - (12Ba_0^{12}/R_{eq}^{13}) = 0$ and $R_{eq} = (2B/C)^{1/6}a_0 = (2 \times 1.69 \times 10^8/2.35 \times 10^3)^{1/6}a_0 = 7.24a_0 = 3.83$ Å. (b) The energy of attraction is $-C(a_0/R_{eq})^6 = -1.63 \times 10^{-2}$ eV. (c) The energy of repulsion is $B(a_0/R_{eq})^{12} = 8.14 \times 10^{-3}$ eV. (d) The total energy is their sum or $-8.16 \times 10^{-3}$ eV. Although the force of repulsion exactly balances the force of attraction at the equilibrium separation, the energy associated with the attractive force dominates the energy associated with the repulsive force. ∎

## 5.3 COVALENT BONDING

Covalent bonding comes about in much the same manner as the bonding of protons in $H_2^+$. To develop a bonding wave function, we first form a linear combination of atomic orbitals for each atom. The function associated with atom $a$, for example, is

$$\chi_a(\mathbf{r}) = \sum_n A_{an} \Phi_{an}(\mathbf{r}). \tag{5-31}$$

Here $n$ runs over the atomic states, $\Phi_{an}$ is an atomic orbital, and $A_{an}$ is a constant coefficient. Similar functions are constructed for each atom, with different coefficients for different atoms. If the atoms are not identical the atomic orbitals $\Phi_{an}(\mathbf{r})$ are also different.

We consider a system of two atoms and take the wave function to be

$$\psi(r) = C_a \chi_a(\mathbf{r} - \mathbf{R}_a) + C_b \chi_b(\mathbf{r} - \mathbf{R}_b), \tag{5-32}$$

where $C_a$ and $C_b$ are constants and $\mathbf{R}_a$ and $\mathbf{R}_b$ give the atomic positions. $A_{an}$, $A_{bn}$, $C_a$, and $C_b$ are chosen so $\psi$ closely approximates a solution to the Schrödinger equation. There are a large number of solutions, all represented by the same form but differing in values of the coefficients and having different values of the energy associated with them. For the lowest energy states the coefficients have values that make overlap large.

For a covalent bond the sum in Eq. 5-31 is dominated by a linear combination of the three $p$ functions associated with the highest occupied atomic shell. For many bonds the $s$ function of the same shell is also important. Atomic orbitals are arranged so the wave function is large along lines extending radially outward from the atom and the function overlaps a similar function extending from a neighboring atom.

For an atom at the origin the $p$ functions can be written

$$\Phi_{px}(\mathbf{r}) = \frac{x}{r} f_p(r), \tag{5-33}$$

$$\Phi_{py}(\mathbf{r}) = \frac{y}{r} f_p(r), \tag{5-34}$$

and

$$\Phi_{pz}(\mathbf{r}) = \frac{z}{r} f_p(r), \tag{5-35}$$

where $f_p(r)$ is the radial function associated with the $p$ orbitals and is the same in all three expressions. It is adjusted so $\Phi_{px}$, $\Phi_{py}$, and $\Phi_{pz}$ are each normalized.

It depends only on distance from the nucleus, not on angle, and we assume it is positive in the outer reaches of the atom. The $p$ functions each have two lobes, regions of high probability density, which extend in opposite directions from the origin. Figure 5-6 is a polar diagram of the probability density associated with $\Phi_{pz}$ for the $n = 2$ shell of hydrogen and shows the lobes associated with that function, one extending in the positive $z$ direction and the other extending in the negative $z$ direction. In the outer regions $\Phi_{pz}$ is positive in one lobe and negative in the other. $\Phi_{px}$ and $\Phi_{py}$ have similar lobes along the $x$ and $y$ axes, respectively.

Lobes need not be along coordinate axes. If $\alpha$, $\beta$, and $\gamma$ are constants which obey $\alpha^2 + \beta^2 + \gamma^2 = 1$, then

$$\chi(\mathbf{r}) = \alpha\Phi_{px}(\mathbf{r}) + \beta\Phi_{py}(\mathbf{r}) + \gamma\Phi_{pz}(\mathbf{r}) \tag{5-36}$$

is a normalized function which has a positive lobe in the direction defined by the unit vector $\hat{\mathbf{n}} = \alpha\hat{\mathbf{x}} + \beta\hat{\mathbf{y}} + \gamma\hat{\mathbf{z}}$ and a negative lobe in the opposite direction. To see this, substitute Eqs. 5-33, 5-34, and 5-35 into Eq. 5-36 and note that $\alpha x + \beta y + \gamma z = \mathbf{r} \cdot \hat{\mathbf{n}}$, so $\chi(\mathbf{r}) = \mathbf{r} \cdot \hat{\mathbf{n}} f_p(r)/r$. The scalar product has largest positive value for $\mathbf{r}$ in the direction of $\hat{\mathbf{n}}$ and its largest negative value for $\mathbf{r}$ in the opposite direction.

A strong bond is generated between two atoms if the lobes from one are parallel to the lobes from the other, with lobes of the same sign overlapping. For example, two atoms might be located on the $z$ axis with both positive lobes extending in the positive $x$ direction. If the atoms are sufficiently close, both positive and negative lobes overlap. Such bonds are known as $\pi$ bonds and they are the principal means by which covalent molecules, such as $O_2$ and $N_2$, are held together.

Large overlap also occurs if the functions are arranged so that lobes of the same sign point toward each other from neighboring atoms. Such a bond, called a $\sigma$ bond, is important for many solids. The functions, however, are not simply

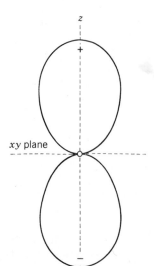

**FIGURE 5-6** Polar graph of the electron probability density associated with an atomic $p$ orbital. Along any line through the origin, the probability density is proportional to the distance from the origin to the intersection with the graph. The wave function is positive in the outer reaches of one lobe and negative in the outer reaches of the other.

linear combinations of atomic $p$ orbitals: a $\sigma$ bond is strengthened by the addition of the $s$ orbital associated with the same shell. An $s$ function is spherically symmetric so, when it is added to a $p$ function, it tends to cancel one lobe and augment the other. For example, let $f_s(r)$ be the $s$ function and suppose it is positive. Then $f_s(r) + \Phi_{pz}(\mathbf{r})$ has a major lobe in the positive $z$ direction and only a minor lobe in the negative $z$ direction. Figure 5-7 shows a polar diagram of the probability density associated with such a function.

Single atom functions that enter a $\sigma$ bond have the form

$$\chi(\mathbf{r}) = A_s f_s(r) + A_p \left[\alpha x + \beta y + \gamma z\right] f_p(r)/r, \qquad (5\text{-}37)$$

where $\alpha^2 + \beta^2 + \gamma^2 = 1$ and $A_s$ and $A_p$ are constants which, for $\chi$ to be normalized, obey $|A_s|^2 + |A_p|^2 = 1$. The ratio $|A_p|^2/|A_s|^2$ gives the amount of $p$-like character in the probability density, relative to the amount of $s$-like character. This ratio is often denoted by $n$ and, for a given value of $n$, the bond is said to be an $sp^n$ bond.

Since four linearly independent functions enter the admixture given by Eq. 5-37, up to four independent bonding wave functions can be associated with each atom, differing from each other in the values of $\alpha$, $\beta$, and $\gamma$. The most important set for solids consists of those with lobes spaced uniformly in angle. If an atom is at the center of a cube, as shown in Fig. 5-8, lobes reach outward toward four of the corners. These four corners are the positions of other atoms, and lobes associated with them point toward the central atom. The four lobes shown are in the directions $(1/\sqrt{3})(\hat{\mathbf{x}} + \hat{\mathbf{y}} + \hat{\mathbf{z}})$, $(1/\sqrt{3})(-\hat{\mathbf{x}} - \hat{\mathbf{y}} + \hat{\mathbf{z}})$, $(1/\sqrt{3})(\hat{\mathbf{x}} - \hat{\mathbf{y}} - \hat{\mathbf{z}})$, and $(1/\sqrt{3})(-\hat{\mathbf{x}} + \hat{\mathbf{y}} - \hat{\mathbf{z}})$, so the functions are

$$\chi_1(\mathbf{r}) = A_s f_s(r) + \frac{A_p}{\sqrt{3}}\,(x + y + z)\,\frac{f_p(r)}{r}, \qquad (5\text{-}38)$$

$$\chi_2(\mathbf{r}) = A_s f_s(r) + \frac{A_p}{\sqrt{3}}\,(-x - y + z)\,\frac{f_p(r)}{r}, \qquad (5\text{-}39)$$

$$\chi_3(\mathbf{r}) = A_s f_s(r) + \frac{A_p}{\sqrt{3}}\,(x - y - z)\,\frac{f_p(r)}{r}, \qquad (5\text{-}40)$$

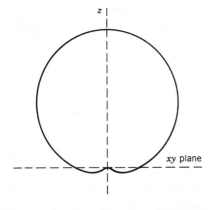

**FIGURE 5-7** Polar graph of the probability density for an $sp$ admixture, $|\phi_s + \phi_{pz}|^2$. The $s$ function, being spherically symmetric, tends to cancel one lobe of the $p$ function and enhance the other. A minor lobe extends in the negative $z$ direction but it is too small to be seen on the scale of the graph.

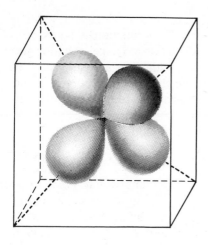

**FIGURE 5-8** Tetrahedral $sp^3$ bonds. Four lobes emanate from an atom at the center of a cube. Other atoms are at the ends of the dotted lines and lobes point from them toward the cube center.

and

$$\chi_4(\mathbf{r}) = A_s f_s(r) + \frac{A_p}{\sqrt{3}} (-x + y - z) \frac{f_p(r)}{r}. \qquad (5\text{-}41)$$

Each atom is at the center of a regular tetrahedron formed by four other atoms. This arrangement is favored by many atoms from the right side of the periodic table and leads to diamond, zinc blende, and wurzite crystal structures. If the four atoms to which the central atom is bound are identical, the four lobes must be the same except for their directions and all constants labeled $A_s$ in Eqs. 5-38, 5-39, 5-40, and 5-41 have the same value. Similarly, all constants labeled $A_p$ have the same value. Furthermore, since only a single atomic $s$ orbital is available, $4|A_s|^2 = 1$ and we may take $A_s = \frac{1}{2}$. Three atomic $p$ orbitals are available, so we may take $A_p = \sqrt{3}/2$. The ratio $|A_p|^2/|A_s|^2$ is 3 and these functions describe $sp^3$ bonds.

**EXAMPLE 5-3** Find the wave function for an $sp^3$ bond between two identical atoms separated by $R$ along the line in the direction $(1/\sqrt{3})(\hat{x} + \hat{y} + \hat{z})$.

**SOLUTION** Place the origin at one atom, then the other atom is at $\mathbf{R} = (R/\sqrt{3})(\hat{x} + \hat{y} + \hat{z})$. The wave function centered on the atom at the origin has its lobe in the direction $(1/\sqrt{3})(\hat{x} + \hat{y} + \hat{z})$ and so is the function given by Eq. 5-38. Call it $\chi_a$. The displacement vector from the second atom to the point $\mathbf{r}$ is $\mathbf{r} - \mathbf{R}$ and the lobe of the function associated with this atom is in the direction $-(1/\sqrt{3})(\hat{x} + \hat{y} + \hat{z})$. The appropriate function is

$\chi(\mathbf{r})$

$$= \frac{1}{2} f_s(|\mathbf{r} - \mathbf{R}|) - \frac{1}{2} \left[ \left( x - \frac{R}{\sqrt{3}} \right) + \left( y - \frac{R}{\sqrt{3}} \right) + \left( z - \frac{R}{\sqrt{3}} \right) \right] \frac{f_p(|\mathbf{r} - \mathbf{R}|)}{|\mathbf{r} - \mathbf{R}|}$$

$$= \frac{1}{2} f_s(|\mathbf{r} - \mathbf{R}|) - \frac{1}{2} (x + y + z - \sqrt{3}R) \frac{f_p(|\mathbf{r} - \mathbf{R}|)}{|\mathbf{r} - \mathbf{R}|},$$

where we have taken $A_s$ to be 1/2 and $A_p$ to be $\sqrt{3}/2$. The wave function is the sum, or

$$\psi(\mathbf{r}) = \frac{N}{2}\left[f_s(r) + (x + y + z)\frac{f_p(r)}{r}\right]$$
$$+ \frac{N}{2}[f_s(|\mathbf{r} - \mathbf{R}|) - (x + y + z - \sqrt{3}R)\frac{f_p(|\mathbf{r} - \mathbf{R}|)}{|\mathbf{r} - \mathbf{R}|}.$$

$N$ is the normalization constant and is given by $N^2 = 1/2(1 + S)$, where $S = \int \chi_a \chi_b \, d\tau$. ∎

Up to two electrons can have the same spatial wave function, one with spin up and the other with spin down, so the four $sp^3$ bonding states can hold up to eight electrons. Except for lead, elements of column IV of the periodic table are bound into solids by means of $sp^3$ covalent bonds. These atoms each have four electrons in outer $s$ and $p$ states so the bonding states are exactly filled.

Atoms of many binary compounds bond covalently to each other and thereby form a tetrahedral arrangement in the solid. A total of eight $s$ and $p$ electrons are contributed by each pair of atoms, and these fill the eight bonding states. Some examples are GaAs, with gallium from column III and arsenic from column V, and CdS, with cadmium from column II and sulfur from column VI.

Notice that the atomic orbitals that enter the bonding function are not necessarily those that are occupied in the ground state of the isolated atom. All column IV atoms have two $s$ electrons and two $p$ electrons in their outlet shells, but $sp^3$ bonding functions are formed with a single $s$ orbital and three $p$ orbitals for each atom. We may think of a two-step process in which an electron is first promoted from an $s$ to a $p$ state before entering the bond. Although the promotion of the electron increases the energy, bond formation lowers it and the net result is a decrease in energy.

Covalent bonds have some common characteristics. They are quite strong. Covalent materials have cohesive energies from about 3 eV to about 10 eV. Covalent bonds are directed. They occur in certain well-defined directions and, as a consequence, covalent materials are brittle and do not bend easily. Roughly speaking, the electrons are localized to regions of the bond lobes and are not free to travel though the solid. As a result, covalent materials are usually electrical insulators or semiconductors.

Some elements, such as carbon (as graphite), selenium, and tellurium, form solids with structures that can be described as stacked parallel planes. Atoms in each of the planes are strongly bonded to each other via covalent bonds, while atoms in adjacent planes are much more loosely bound to each other. The single atom functions that enter the strong bonds between atoms in the same plane have the form given by Eq. 5-37 but, for example, if the plane is the $xy$ plane then $\gamma = 0$. A situation that occurs frequently is one in which the lobes are symmetrically placed around the central atom, so the angle between adjacent lobes is 120°, as shown in Fig. 5-9. If the bonding functions are identical except for their directions, the bonds are $sp^2$ bonds.

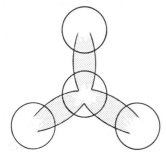

FIGURE 5-9 Schematic representation of $sp^2$ bonds in a plane, spaced equally in angle. Three bonds emanate from each atom and lines joining any atom to its nearest neighbors make angles of 120° with each other.

Atoms in different planes of these materials are attracted toward each other by means of van der Waals forces. Since bonds between atoms in the same plane are strong while interplanar bonds are weak, the separation between adjacent planes is large compared to interatomic distances in a plane. For graphite, tightly bound atoms in a plane are about 1.4 Å apart, while the separation of planes is about 3.35 Å. Interplanar bonds can be broken easily and the various layers of atom tend to slide across each other, while the structure in any layer remains intact. Graphite is often used as a dry lubricant.

## 5.4 IONIC BONDING

If two atoms joined by a covalent bond are not identical, the functions $\chi_a$ and $\chi_b$ in Eq. 5-32 are derived from different atomic orbitals. In particular, the functions $f_s$ and $f_p$ are each different for atoms $a$ and $b$. Since the system is not invariant with respect to the interchange of atoms, the constants $C_a$ and $C_b$ are also different. Although the wave function spreads into the region between atoms, the probability is higher for the electron to be nearer one atom than the other. The atoms behave, to some extent, as oppositely charged ions and attract each other electrostatically. Bonding for such pairs of atoms is described as being partially covalent and partially ionic.

In an extreme case, an electron leaves one atom and becomes bound to the other in what is nearly an atomic state. If, as a result, the electron distribution about both atoms is spherically symmetric, the bond is wholly ionic and the force of attraction can be described simply as the coulombic interaction between two oppositely charged ions.

Various measures might be devised to define the extent to which a bond is ionic. We might, for example, compare the average electron charge $e\int|\psi|^2 \, d\tau$ on the two sides of the plane that perpendicularly bisects the line joining the atoms. Alternatively, we might take the quantity $C_a^2 - C_b^2$ to measure the extent to which the electron favors atom $a$ over atom $b$. Another scheme, based on calorimetric measurements, is often used. The energy required to break a bond between two $a$ atoms in a material composed solely of $a$ atoms is measured. A similar measurement is made of the energy required to break a $b$–$b$ bond. These two bonds are wholly covalent. Finally the energy needed to break an $a$–$b$ bond is measured.

The last energy is generally greater than the average of the first two and the difference is ascribed to the partially ionic character of an $a$–$b$ bond.

Alkali halides are important examples of materials for which the bonding is almost wholly ionic. For these materials, ionic bonds account for 90% or more of the cohesive energies. Alkali metal atoms, from column IA, each have a single $s$ electron outside a core, while halide atoms, from column VII, each have a single empty $p$ state in their outer shells. In the solid, the $s$ electron is transferred and completes the outer shell of the halide atom. Both atoms become ions with completely filled shells and, for each, the average charge density is very nearly spherically symmetric. The metal ion, having lost an electron, is positive, and the halide atom, having gained an electron, is negative.

Figure 5-10 shows the total electron probability density for a typical ionic bond. For practical purposes, there are no longer any directed bonds. Unlike covalent bonding, ionic bonding does not limit the number of nearest neighbors to four. The number is limited, however, because neighbors to any atom all have charge of the same sign and repel each other. As we have seen, solid alkali halides are usually structured so there are either six or eight nearest neighbors to any given atom.

When bonding is purely ionic, the equation for the total potential energy of the system is quite simple. Since each ion can be treated as a point charge, we sum the electrostatic potential energy for all ion pairs, then add the energy of core-core repulsion. The result is

$$E = -\frac{1}{2} \sum_i \sum_j{}' \frac{q_i q_j}{4\pi\epsilon_0 |\mathbf{r}_i - \mathbf{r}_j|} + \text{(core contribution)}, \qquad (5\text{-}42)$$

where $\mathbf{r}_i$ is the position of ion $i$ and $q_i$ is its charge. The prime on the second summation indicates that the term $i = j$ is omitted and the factor $\frac{1}{2}$ is included because each bond appears twice in the sums. This expression is similar to that

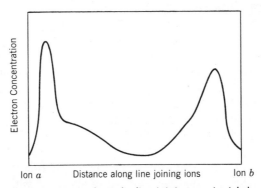

**FIGURE 5-10** Electron concentration along the line joining two ionicly bonded atoms, located at the edges of the graph. The probability density is high in the core regions and low in the interstitial region. Ion $a$ might be a halide, with filled outer $s$ and $p$ subshells, and ion $b$ might be an alkali ion, with its outer $s$ electron missing.

of Eq. 5-29 for the core-core interaction energy. Now, however, the energy of an ion pair is inversely proportional to the atomic separation, rather than to the separation raised to a high power. The interaction is long range and ions much farther away than nearest neighbors contribute significantly to the energy.

Consider a binary ionic crystal with one positive and one negative ion in its primitive basis. Suppose positive ions, each with charge $+e$, are located at $n_1\mathbf{a} + n_2\mathbf{b} + n_3\mathbf{c}$ and negative ions, each with charge $-e$, are located at $n_1\mathbf{a} + n_2\mathbf{b} + n_3\mathbf{c} + \mathbf{p}$, where $\mathbf{p}$ is a basis vector. The electrostatic energy $E_+$ of a positive ion, assumed to be at the origin, is given by

$$E_+ = -\frac{e^2}{4\pi\epsilon_0}\left[\sum_{\mathbf{R}}\frac{1}{|\mathbf{R} + \mathbf{p}|} - {\sum_{\mathbf{R}}}'\frac{1}{|\mathbf{R}|}\right], \tag{5-43}$$

where $\mathbf{R}$ is a lattice vector and each sum is over all unit cells of the crystal. The first sum accounts for interactions with ions of opposite charge, while the second accounts for interactions with ions of like charge. The prime indicates that the $\mathbf{R} = 0$ term is omitted from the second sum.

All ions, whether positive or negative, have the same potential energy, so the total electrostatic energy $E_M$ is given by

$$E_M = -\frac{Ne^2}{4\pi\epsilon_0}\left[\sum_{\mathbf{R}}\frac{1}{|\mathbf{R} + \mathbf{p}|} - {\sum_{\mathbf{R}}}'\frac{1}{|\mathbf{R}|}\right], \tag{5-44}$$

where $N$ is the number of primitive unit cells. There are $2N$ ions in the crystal but Eq. 5-43 is multiplied by $N$ rather than by $2N$ to avoid double counting the interactions. $E_M$ is called the Madelung energy, hence the subscript $M$.

Equation 5-44 is often written

$$E_M = -\frac{Ne^2}{4\pi\epsilon_0}\frac{\alpha}{R_0}, \tag{5-45}$$

where $R_0$ is the nearest-neighbor distance and $\alpha$ is a dimensionless quantity called the Madelung constant, given by

$$\alpha = R_0\left[\sum_{\mathbf{R}}\frac{1}{|\mathbf{R} + \mathbf{p}|} - {\sum_{\mathbf{R}}}'\frac{1}{|\mathbf{R}|}\right]. \tag{5-46}$$

The Madelung constant depends on the structure of the crystal but not on unit cell dimensions. For example, it is different for NaCl and CsCl but it is the same for all crystals with CsCl structures, regardless of their lattice constants. Sometimes the electrostatic energy is written in terms of a lattice constant $a$ rather than the nearest-neighbor distance $R_0$. Then $E_M = -Ne^2\alpha'/4\pi\epsilon_0 a$, where $\alpha' = \alpha a/R_0$.

Madelung constants cannot be computed by evaluating the individual sums in Eq. 5-46. We might sum the contributions of all atoms within some sphere of radius $R$, then attempt to evaluate the sum in the limit as $R$ becomes large. The number of terms, however, increases at a faster rate than $R$ and, since the denominators of the summands increase in proportion to $R$, the sums do not converge. Even if terms in the two sums are combined before summing, convergence is extremely slow unless special precautions are taken. Techniques have been developed to carry out the evaluation by grouping ions that are nearly the same distance from the origin in such a way that each group is neutral. To do this, fractions of ions are included in some groups, with the remainders in the next group. The contributions of the groups are then summed, in order of increasing distance from the origin. This technique is called the Evjen method of summation. A procedure called the Ewald method is more generally applicable, although more complicated mathematically. It is explained in Appendix C. Some values of Madelung constants, referred to nearest-neighbor distances, are*:

| | |
|---|---|
| NaCl | 1.74756 |
| CsCl | 1.76267 |
| ZnS (zinc blende) | 1.63805 |

To find the total energy of an ionic crystal, the energy of core-core repulsion must be added to the Madelung energy. If we include only nearest-neighbor interactions and take the core-core energy of two ions separated by $R$ to be proportional to $1/R^n$, then the energy of a binary crystal is given by

$$E = -\frac{e^2 N\alpha}{4\pi\epsilon_0 R} + \frac{NA}{R^n}. \tag{5-47}$$

Strictly speaking, other contributions to the energy should be included in Eq. 5-47. We will deal with one omission, the kinetic energy of the ions, in Chapters 6 and 8. For now we assume the temperature is near absolute zero and this contribution is negligible. For more precise results covalent and van der Waals contributions must be taken into account as well.

Equation 5-47 can be used to determine the equilibrium separation. The first law of thermodynamics is

$$dE = -P\, d\tau_s + T\, dS, \tag{5-48}$$

where $P$ is the pressure, $S$ is the entropy of the sample, and $\tau_s$ is the sample volume. We assume $P$ is small, an assumption that usually leads to insignificant error. At $T = 0$ the equilibrium sample volume is then determined by $dE/d\tau_s = 0$, a condition that is equivalent to $dE/dR = 0$. The derivative of Eq. 5-47 with respect to $R$ is equated to zero; then the resulting equation is solved for $R$ to obtain

$$R_{eq} = \left[\frac{4\pi\epsilon_0 nA}{\alpha e^2}\right]^{1/(n-1)} \tag{5-49}$$

for the equilibrium nearest-neighbor distance.

---

*Values are taken from the article by M. P. Tosi, referenced at the end of the chapter. This article also gives Madelung constants for other materials.

When Eq. 5-49 is substituted into Eq. 5-47, the expression for the energy becomes

$$E = -\frac{Ne^2\alpha}{4\pi\epsilon_0 R_{eq}}\left(1 - \frac{1}{n}\right). \tag{5-50}$$

For most ionic crystals $n$ is on the order of 10 and, as can be seen from this equation, the Madelung term dominates the energy. Mutual repulsion of the ions is important, of course. At equilibrium, the force of repulsion cancels the electrostatic force of attraction. Equilibrium, however, is established at a separation for which the energy of repulsion is small compared to the electrostatic energy of attraction.

In some instances, the parameters $A$ and $n$ can be determined from measurements taken on a gas; then Eq. 5-49 is used to predict $R_{eq}$. More often, $R_{eq}$ is determined from x-ray scattering data and is used, along with compressibility data, to determine the parameters $n$ and $A$ in Eq. 5-49. Other quantities, such as the cohesive energy, might be used but they are not sensitive to the values of $A$ and $n$. Table 5-1 gives values of the parameters for some alkali halides.

The isothermal compressibility $\kappa$ is defined by $\kappa = -(1/\tau_s)(\partial\tau_s/\partial P)_T$, where the subscript indicates that the temperature is held constant when the derivative is evaluated. When $dE/d\tau_s = -P$ is differentiated with respect to $\tau_s$, holding the temperature constant, the result is $d^2E/d\tau_s^2 = -(\partial P/\partial\tau)_T$, so

$$\frac{1}{\kappa} = \tau_s\left[\frac{d^2E}{d\tau_s^2}\right]_{\tau_s=\tau_{s0}}, \tag{5-51}$$

where $\tau_{s0}$ is the equilibrium sample volume. The reciprocal of the compressibility is called the bulk modulus.

To evaluate the derivative in Eq. 5-51 we need to know the relationship between the sample volume and the nearest-neighbor distance. For any sample, the volume of a unit cell is proportional to the cube of the nearest-neighbor distance,

**TABLE 5-1** Energy Parameters for Selected Alkali Halide Crystals

| Crystal | $R_0$ (Å) | $n$ | $A$ (J · m$^n$) |
|---------|-----------|-----|-----------------|
| LiF | 2.014 | 6.20 | $2.61 \times 10^{-79}$ |
| LiCl | 2.570 | 7.30 | $2.34 \times 10^{-89}$ |
| NaF | 3.317 | 6.41 | $4.98 \times 10^{-88}$ |
| NaCl | 2.820 | 8.38 | $1.77 \times 10^{-99}$ |
| KF | 2.674 | 7.39 | $4.21 \times 10^{-90}$ |
| KCl | 3.174 | 8.55 | $1.01 \times 10^{-100}$ |
| RbF | 2.815 | 8.14 | $3.85 \times 10^{-99}$ |
| CsF | 3.004 | 10.22 | $8.03 \times 10^{-117}$ |
| CsCl | 3.571 | 10.65 | $3.44 \times 10^{-120}$ |

Calculated from data given in the article by M. P. Tosi, referenced at the end of the chapter.

so we take $\tau_s = CNR^3$, where $C$ is a constant determined by the structure. For example, $R$ for the CsCl structure is $(\sqrt{3}/2)a$ so $C = a^3/R^3 = 8/3^{3/2} = 1.54$.

The first derivative of the energy can be evaluated using $dE/d\tau_s = (dE/dR)(dR/d\tau_s) = (dE/dR)/3CNR^2$ and the second derivative can be evaluated using $d^2E/d\tau_s^2 = (d^2E/dR^2)/3CNR^2$, where the condition $dE/dR = 0$ for $R = R_{eq}$ was used. $A = (\alpha e^2/4\pi\epsilon_0 n)R_{eq}^{n-1}$, from Eq. 5-49, is substituted into Eq. 5-47, then $E$ is differentiated twice with respect to $R$ and the result is divided by $3CNR^2$. Equation 5-51 becomes

$$\frac{1}{\kappa} = \frac{\alpha e^2(n-1)}{36\pi\epsilon_0 CR_{eq}^4}, \tag{5-52}$$

and when this equation is solved for $n$,

$$n = 1 + \frac{36\pi\epsilon_0 CR_{eq}^4}{\alpha e^2\kappa} \tag{5-53}$$

is obtained.

**EXAMPLE 5-4**  For sodium chloride at low temperatures the equilibrium nearest-neighbor distance is 2.79 Å and the isothermal compressibility is $3.39 \times 10^{-11}$ m$^3$/J. Use this data to calculate the energy parameters $n$ and $A$, then estimate the Madelung and core overlap contributions to the energy per unit cell.

**SOLUTION**  For NaCl $R = \frac{1}{2}a$, where $a$ is the cube edge. The volume of a cube is $a^3$ and each cube contains four ion pairs, so $a^3 = 4C(a/2)^3$ and $C = 2$. According to Eq. 5-53,

$$n = 1 + \frac{36\pi\epsilon_0 CR_{eq}^4}{\alpha e^2\kappa}$$

$$= 1 + \frac{36\pi \times 8.85 \times 10^{-12} \times 2 \times (2.79 \times 10^{-10})^4}{1.75 \times (1.60 \times 10^{-19})^2 \times 3.39 \times 10^{-11}} = 8.99.$$

According to Eq. 5-49

$$A = \frac{\alpha e^2 R_{eq}^{n-1}}{4\pi\epsilon_0 n} = \frac{1.75 \times (1.60 \times 10^{-19})^2 \times (2.79 \times 10^{-10})^{7.99}}{4\pi \times 8.85 \times 10^{-12} \times 8.99}$$

$$= 2.23 \times 10^{-105},$$

in SI units. The Madelung contribution to the energy per unit call is

$$\frac{E_M}{N} = -\frac{e^2\alpha}{4\pi\epsilon_0 R_{eq}} = -\frac{(1.60 \times 10^{-19})^2 \times 1.75}{4\pi \times 8.85 \times 10^{-12} \times 2.79 \times 10^{-10}}$$

$$= -1.44 \times 10^{-18} \text{ J}$$

and the core overlap contribution is the magnitude of $E_M/N$ divided by $n$, or $1.44 \times 10^{-18}/8.99 = 1.60 \times 10^{-19}$ J.  ∎

The crystal energy given by Eq. 5-47 is measured relative to the energy of a collection of ions at rest and far removed from each other. To calculate the cohesive energy, we must know what energy changes occur when an electron is transferred from a metal atom to a halide atom. The energy required to remove an electron from an atom is called the ionization energy, while the enegy obtained when an electron is added to an atom is called the electron affinity. To find the cohesive energy we must adjust the energy given by Eq. 5-47 by adding the ionization energy for the metal atom and subtracting the electron affinity for the halide atom.

The first ionization energy for sodium is $8.22 \times 10^{-19}$ J and the electron affinity of chlorine is $5.78 \times 10^{-19}$ J so, when we use the results of Example 5-4, the energy of NaCl is $-1.44 \times 10^{-18} + 1.60 \times 10^{-19} + 8.22 \times 10^{-19} - 5.78 \times 10^{-19} = -1.28 \times 10^{-18}$ J per primitive unit cell, relative to widely separated neutral atoms. The cohesive energy is $+1.28 \times 10^{-18}$ J/cell (7.97 eV/cell). The result obtained for NaCl is fairly typical. Cohesive energies of ionic solids are about 5 eV/atom, somewhat less than typical covalent bonds but nevertheless quite strong.

## 5.5 METALLIC BONDING

Outer electrons of atoms that form metals are loosely bound. When a solid is formed, the potential energy barrier between atoms is reduced and these electrons become free to travel throughout the solid. To a large extent, the accompanying reduction in kinetic energy is responsible for binding.

Atomic orbitals in a metal do not form directed bonds, but rather they link together to form wave functions which extend throughout the material, with nearly the same amplitude everywhere. If an electron is loosely bound in the atom then, when the solid is formed and the potential energy is reduced between atoms, the electron energy may be well above the potential energy maximum. The wave functions are then nearly plane waves in regions between atoms and the electrons are influenced only slightly by forces of the ion cores.

Metallic bonding can occur only for large aggregates of closely packed atoms, not for pairs or small groups. As we have seen, metals have face-centered cubic, hexagonal close-packed, or body-centered cubic structures, with each atom having a large number of nearest neighbors. Simple metals, such as those formed by atoms from columns IA, IIA, IB, and IIB of the periodic table, have cohesive energies in the range from 1 to 5 eV/atom, somewhat less than most covalent solids.

In addition to the simple metals, some elements on the right side of the table are also metallic. Among these are aluminum, indium, lead, and one form of tin. Solids formed from two or more of these materials are also metallic as are some, but not all, solids that are compounds of metallic and nonmetallic elements. Notable exceptions are metal oxides, which are insulators.

For some materials bonding may be intermediate between covalent and metallic. An electron wave function spreads throughout the material but there is a somewhat higher probability density along lines joining atoms. Many compounds consisting of metallic and nonmetallic elements exhibit bonding of this type. Covalent-like bonds formed by $d$ orbitals of transition metal atoms cause these metals to be tightly bound. Tungsten, with a cohesive energy of 9 eV/atom, is the most tightly bound of all metallic elements.

## 5.6 VAN DER WAALS BONDING

If the average position of the electrons in an atom does not coincide with the nucleus, the atom has an electric dipole moment and produces an electric field, even if its net charge is zero. Since an electric field exerts forces on electrons and protons in opposite directions, dipole moments are induced on other atoms. Furthermore, since the field varies with distance from the source atom, the force it exerts on the electrons of a second atom has a different magnitude than the force it exerts on the protons, so the first atom exerts a net force on the second. As we will see, the net force is one of attraction.

Consider two neutral atoms, displaced from each other by $\mathbf{R}$, as diagrammed in Fig. 5-11. Assume that, at some instant, atom 1 has dipole moment $\mathbf{p}_1$. Then the electric field produced by atom 1 at the site of atom 2 is given by

$$\mathcal{E} = \frac{1}{4\pi\epsilon_0} \frac{1}{R^3} [3\mathbf{p}_1 \cdot \hat{\mathbf{R}}\hat{\mathbf{R}} - \mathbf{p}_1], \tag{5-54}$$

where $\hat{\mathbf{R}}$ is a unit vector in the direction of $\mathbf{R}$. The field induces a dipole moment in the second atom, in the direction of the field and proportional to it. We write

$$\mathbf{p}_2 = \alpha\mathcal{E} = \frac{\alpha}{4\pi\epsilon_0 R^3} [3\mathbf{p}_1 \cdot \hat{\mathbf{R}}\hat{\mathbf{R}} - \mathbf{p}_1] \tag{5-55}$$

for the dipole moment of the second atom. Here $\alpha$ is a constant of proportionality, called the polarizability of the atom. The potential energy of a dipole $\mathbf{p}$ in an electric field $\mathcal{E}$ is given by $U = -\mathbf{p} \cdot \mathcal{E}$, so the potential energy for the interaction between the two atoms is

$$U(\mathbf{R}) = -\frac{\alpha}{(4\pi\epsilon_0)^2 R^6} [3(\mathbf{p}_1 \cdot \hat{\mathbf{R}})^2 + \mathbf{p}_1^2]. \tag{5-56}$$

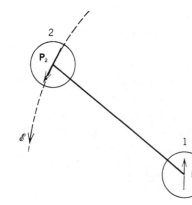

**FIGURE 5-11** Atom 1 has dipole moment $\mathbf{p}_1$ and produces an electric field at the site of atom 2. A dipole moment $\mathbf{p}_2$ is induced in atom 2 and the atoms are bound by the force of the field on the dipole. The dotted line represents a portion of an electric field line due to atom 1.

Several features of this expression are important. First, $U(\mathbf{R})$ is negative, indicating that the atoms attract each other. Second, it varies as $1/R^6$. Compared with energies associated with other bonding mechanisms, this energy falls off more rapidly with distance than any other and, for usual atomic separations, it is the weakest. Third, $U(\mathbf{R})$ is proportional to the square of the magnitude of $\mathbf{p}_1$. The dipole moment of an atom in a solid fluctuates in magnitude and direction, so its average over time vanishes. The sign of $U(\mathbf{R})$, however, does not change as $\mathbf{p}_1$ fluctuates and its time average does not vanish. The force is always one of attraction, no matter what the direction of $\mathbf{p}_1$.

Since the dipole moment $\mathbf{p}_1$ fluctuates rapidly in a random manner, we may take the van der Waals energy to be the time average of Eq. 5-56. Write $\mathbf{p}_1 \cdot \hat{\mathbf{R}} = p_1 \cos\theta$, where $\theta$ is the angle between $\mathbf{p}_1$ and $\hat{\mathbf{R}}$. If all values of $\theta$ occur with equal probability the average of $\cos^2\theta$ is $\frac{1}{2}$ and the average of $(\mathbf{p}_1 \cdot \hat{\mathbf{R}})^2$ is $\frac{1}{2}\langle p_1^2 \rangle$, where $\langle p_1^2 \rangle$ is the average of the dipole moment squared. The time average of Eq. 5-56 is

$$\langle U(\mathbf{R}) \rangle = - \frac{5\alpha \langle p_1^2 \rangle}{2(4\pi\epsilon_0)^2 R^6}. \tag{5-57}$$

All atoms attract others by means of van der Waals forces but this type force is extremely weak and if other bonding forces are present it is overshadowed by them. It is, however, the primary bonding mechanism for atoms of inert gas elements. These atoms do not have any partially filled subshells and the lowest unoccupied atomic states are well separated in energy from the highest occupied states. As a consequence, wave function distortions required for covalent or metallic bonding are energetically unfavorable. Only van der Waals bonding remains.

To find the energy of a pair of inert gas atoms we must add the energy of core repulsion to the energy given by Eq. 5-57. Usually the core-core energy is taken to be proportional to $1/R^{12}$ and the total potential energy is written

$$U_{\text{total}} = -4\epsilon \left[ \left( \frac{\sigma}{R} \right)^6 - \left( \frac{\sigma}{R} \right)^{12} \right], \tag{5-58}$$

where $\epsilon$ and $\sigma$ are parameters that depend on the polarizability and average dipole moment of an atom as well as on the extent of core overlap. Equation 5-58 gives what is known as the Lennard-Jones 6-12 potential energy for a pair of inert gas atoms.

The total potential energy of a van der Waals solid is found by summing Eq. 5-58 over all pairs of atoms. The sums converge rapidly and can be evaluated easily. Crystals of inert gas atoms have FCC structures and for them $\Sigma R^{-6} = 14.45392 R_0^{-6}$ and $\Sigma R^{-12} = 12.13188 R_0^{-12}$, where $R_0$ is the nearest-neighbor distance. If we neglect the kinetic energy of the atoms, the total energy is given by

$$E = -2N\epsilon \left[ 14.45 \left( \frac{\sigma}{R_0} \right)^6 - 12.13 \left( \frac{\sigma}{R_0} \right)^{12} \right], \tag{5-59}$$

TABLE 5-2 Lennard-Jones Parameters for Inert Gas Crystals

| Element | $R_0$ (Å) | $\epsilon$ ($10^{-23}$ J) | $\sigma$ (Å) |
|---------|-----------|---------------------------|--------------|
| Ne | 3.13 | 50 | 2.74 |
| Ar | 3.76 | 167 | 3.40 |
| Kr | 4.01 | 225 | 3.65 |
| Xe | 4.35 | 320 | 3.98 |

Data adopted from M. Born and K. Huang, *Dynamical Theory of Crystal Lattices* (London: Oxford Univ. Press, 1954).

where $N$ is the number of atoms in the crystal. Values of $\epsilon$ and $\sigma$ for inert gas crystals are given in Table 5-2.

For zero pressure and temperature, the equilibrium nearest-neighbor distance $R_{eq}$ is determined by the condition $dE/dR_0 = 0$ and for inert gas crystals this leads to

$$R_{eq} = 1.090\sigma . \tag{5-60}$$

For an FCC structure the sample volume $\tau_s$ is related to the nearest-neighbor distance by $\tau_s = NR_0^3/\sqrt{2}$ so the isothermal compressibility $\kappa$ is given by

$$\frac{1}{\kappa} = \tau_s \left. \frac{d^2E}{d\tau_s^2} \right|_{R=R_{eq}} = 97.41 \frac{\epsilon}{R_{eq}^3} . \tag{5-61}$$

Equations 5-60 and 5-61 can be used to find $\epsilon$ and $\sigma$ from x-ray and compressibility data.

**EXAMPLE 5-5** Crystalline argon has an FCC structure with a cube edge of 5.31 Å. Its isothermal compressibility is $93.8 \times 10^{-11}$ m$^2$/N. Assume the atoms interact only via van der Waals forces and compute values for the energy parameters $\epsilon$ and $\sigma$.

**SOLUTION** The nearest-neighbor distance is half a cube edge so $R_{eq} = a/\sqrt{2} = 5.31/\sqrt{2} = 3.75$ Å. From Eq. 5-60 $\sigma = R_{eq}/1.090 = 3.44$ Å. From Eq. 5-61 $\epsilon = R_{eq}^3/(97.41\kappa) = (3.75 \times 10^{-10})^3/(97.41 \times 93.8 \times 10^{-11}) = 5.77 \times 10^{-22}$ J. ∎

## 5.7 HYDROGEN BONDING

Hydrogen bonds do not fit conveniently into the categories discussed so far, although they have characteristics in common with two of them. In a hydrogen bond, a hydrogen atom forms the link that holds two other atoms together. A partially covalent, partially ionic, bond is formed between the hydrogen atom and one of the other atoms. The electron probability density is high near the other atom and the pair has an electric dipole moment which points from the other atom toward the hydrogen. Two such pairs attract each other electrostatically, with the negative ion of the second pair pulled toward the hydrogen atom of the first.

Ice is an important example. Two water molecules attract each other, the oxygen of one molecule being pulled toward one of the hydrogen atoms of the other. That is, hydrogen links two oxygen atoms in different molecules. This type bond holds the two strands of a DNA molecule together and is also responsible for the adhesion of many glues.

Hydrogen bonds are weaker than metallic bonds but are much stronger than van der Waals bonds. Typical bond energies range from 0.3 to 1 eV per atom.

## 5.8 REFERENCES

E. U. Condon and H. Odabasi, *Atomic Structure* (London: Cambridge Univ. Press, 1980).

U. Fano and L. Fano, *Physics of Atoms and Molecules* (Chicago: University of Chicago Press, 1972).

M. A. Morrison, T. L. Estle, and N. F. Lane, *Quantum States of Atoms, Molecules, and Solids* (Englewood Cliffs, NJ: Prentice-Hall, 1976).

M. O'Keeffe and A. Navrotsky (Eds), *Structure and Bonding in Crystals* (New York: Academic, 1981).

D. Park, *Introduction to the Quantum Theory* (New York: McGraw-Hill, 1964).

J. C. Phillips, *Bonds and Bands in Semiconductors* (New York: Academic, 1973).

J. C. Slater, *Quantum Theory of Matter* (New York: McGraw-Hill, 1968).

M. P. Tosi, "Cohesion of Ionic Solids in the Born Model" in *Solid State Physics* (F. Seitz and D. Turnbull, Eds.), Vol. 16, p. 1 (New York: Academic, 1964).

H. E. Zimmerman, *Quantum Mechanics for Organic Chemists* (New York: Academic, 1975).

## PROBLEMS

1. The potential energy function for the two electrons of a helium atom is

$$U(\mathbf{r}_1, \mathbf{r}_2) = \frac{e^2}{4\pi\epsilon_0} \left[ -\frac{2}{r_1} - \frac{2}{r_2} + \frac{1}{|\mathbf{r}_1 - \mathbf{r}_2|} \right],$$

where the nucleus is at the origin and the electrons are at $\mathbf{r}_1$ and $\mathbf{r}_2$. (a) Suppose the two-electron wave function is $\psi(\mathbf{r}_1, \mathbf{r}_2) = \psi_a(\mathbf{r}_1)\psi_b(\mathbf{r}_2)$, where $\psi_a$ and $\psi_b$ are normalized single particle functions. Write an expression for the average potential energy $\langle U \rangle = \int\int |\psi|^2 U \, d\tau_1 \, d\tau_2$ in terms of integrals containing the single particle functions. Do not attempt to evaluate the integrals, but show that $\langle U \rangle$ has the form $\langle U_a \rangle + \langle U_b \rangle + \langle U_{ee} \rangle$, where $\langle U_a \rangle$ is the average potential energy of an electron with wave function $\psi_a$ in the field of two protons, $\langle U_b \rangle$ is a similar quantity for an electron with wave function $\psi_b$, and $\langle U_{ee} \rangle$ is the average energy of the electron-electron interaction. Write explicit expressions for $\langle U_a \rangle$, $\langle U_b \rangle$, and $\langle U_{ee} \rangle$ in integral form.

(b) Suppose $\psi_a(\mathbf{r}_1)$ obeys the time-independent Schrödinger equation

$$-\frac{\hbar^2}{2m}\nabla_1^2\psi_a + \frac{e^2}{4\pi\epsilon_0}\left[-\frac{2}{r_1} + \int\frac{|\psi_b(\mathbf{r}_2)|^2}{|\mathbf{r}_1 - \mathbf{r}_2|}\,d\tau_2\right]\psi_a = E_a\psi_a\,.$$

Multiply by $\psi_a^*$ and integrate over the coordinates associated with $\mathbf{r}_1$. Show that $E_a = \langle K_a\rangle + \langle U_a\rangle + \langle U_{ee}\rangle$, where $\langle K_a\rangle$ is the average kinetic energy of an electron with wave function $\psi_a$. (c) Show that the total energy of the system is $E_a + E_b - \langle U_{ee}\rangle$, where $E_b$ is the energy associated with $\psi_b$ in the same way that $E_a$ is associated with $\psi_a$.

2. Consider the $sp$ admixture $\chi(\mathbf{r}) = A_s f_s(r) + A_p z f_p(r)/r$, where $A_s$ and $A_p$ are constants. Assume $f_s$, $z f_p/r$, and $\chi$ are normalized. (a) Show that $\chi$ is not an eigenfunction of the operator associated with the magnitude squared of the orbital angular momentum, but that it is an eigenfunction of the operator associated with the $z$ component of orbital angular momentum. (b) For an electron with this wave function what are the possible outcomes of a measurement of the magnitude of the orbital angular momentum? (c) With what probability does each occur?

3. Consider the $sp^3$ single atom functions $\chi_1$ and $\chi_2$, defined by Eqs. 5-38 and 5-39. (a) Show they are orthogonal to each other. That is, show that $\int\chi_1\chi_2\,d\tau = 0$ if $|A_p|^2 = 3|A_s|^2$. Recall that $s$ and $p$ functions, centered on the same atom, are automatically orthogonal. (b) What is the angle between the bond directions associated with $\chi_1$ and $\chi_2$?

4. In a crystal of identical tetrahedrally bonded atoms, $sp^3$ bonds emanating from one atom are in the direction shown in Fig. 5-8. (a) Sketch four lobes of bonding functions, each emanating from an atom that bonds to the central atom. (b) Write expressions analogous to Eqs. 5-38, 5-39, 5-40, and 5-41 for the single atom functions associated with the lobes of part a.

5. In crystalline CsCl, each cesium atom sits at the center of a cube with chlorine atoms at the corners. At equilibrium, the distance between a cesium atom and a nearest chlorine atom is 3.57 Å. Take the bonding to be purely ionic. (a) Calculate the electrostatic energy, relative to the energy at infinite separation, for the interaction between a single cesium atom and its eight nearest chlorine neighbors when all atoms are at their equilibrium sites. (b) This is a poor estimate of the energy of a cesium atom in a crystal since the ionic interaction has extremely long range. For comparison, calculate the Madelung energy per ion when all atoms are at their equilibrium sites.

6. The energy per ion for cesium chloride is nearly $-(\alpha e^2/4\pi\epsilon_0 R) + 8Ae^{-R/\rho}$, where $\alpha$ is the Madelung constant, $A = 5.64 \times 10^3$ eV, and $\rho = 0.34$ Å. Calculate the equilibrium nearest-neighbor distance. You might use a computer and a root-finding program or simply trial and error. The result is not exactly the nearest-neighbor distance given in Problem 5 since other contributions to the energy, such as that of van der Wals forces, have been omitted.

7. For sodium chloride at equilibrium, use the parameters given in Table 5-1 to estimate (a) the Madelung energy per ion, (b) the energy per ion associated with core-core repulsion, and (c) the compressibility at $T = 0$ K.

8. (a) Use values of $n$ and $A$ given in Table 5-1 to find the equilibrium nearest-neighbor separation for CsCl at $T = 0$ K. (b) Use the result to estimate the compressibility of CsCl at $T = 0$ K.

9. To investigate the influence of structure on crystal properties, suppose sodium chloride crystallizes with a zinc blende structure. Compare (a) the nearest-neighbor distance, (b) the energy per ion, and (c) the compressibility with the values these quantities have when the material has its usual NaCl structure. Take the energy of the core-core repulsion to be proportional to $1/R^n$ and use the same value of $n$ for both structures but adjust $A$ to account for different numbers of nearest neighbors.

10. The energy per atom of an argon crystal as a function of nearest-neighbor distance $R$ can be written $U(R) = -(A/R^6) + (B/R^{12})$, with $A = 1.03 \times 10^{-77}$ J$\cdot$m$^6$ and $B = 1.62 \times 10^{-134}$ J$\cdot$m$^{12}$. (a) Find the equilibrium nearest-neighbor distance, then (b) compare the van der Waals and core-core repulsion contributions to the energy at equilibrium separation. (c) Assume the van der Waals energy arises from nearest-neighbor interactions only and estimate the magnitude of the van der Waals force between two argon atoms at equilibrium separation. Crystalline argon has an FCC structure.

11. The polarizability $\alpha$ of an argon atom is given by $\alpha/4\pi\epsilon_0 = 1.62 \times 10^{-30}$ m$^3$. Consider two argon atoms interacting via the van der Waals force. Take the van der Waals contribution to the energy to be $-6.20 \times 10^{-3}$ eV/atom and the atomic separation to be 3.08 Å. Calculate the time averages (a) $\sqrt{\langle p_1^2 \rangle}$ and (b) $\sqrt{\langle \mathcal{E}^2 \rangle}$, where $\mathcal{E}$ is the magnitude of the electric field due to one of the atoms, at the site of the other. For comparison, calculate (c) the separation of an electron and proton such that their dipole moment has magnitude $\sqrt{\langle p_1^2 \rangle}$, and (d) the distance from an isolated proton to a point where the electric field has magnitude $\sqrt{\langle \mathcal{E}^2 \rangle}$.

12. Consider the van der Waals interaction of a pair of atoms with polarizability $\alpha$. Atom 1 is at the origin and has dipole moment $\mathbf{p}_1$. Atom 2 is on the $x$ axis, a distance $R$ away, and has dipole moment $\mathbf{p}_2$, induced by the electric field of atom 1. Assume $\mathbf{p}_1$ is in the $xy$ plane and makes the angle $\theta$ with the $x$ axis. (a) Find an expression, in terms of $\mathbf{p}_1, R, \alpha$, and $\theta$, for the force exerted by atom 1 on atom 2. (b) Show that the force is a central force only when $\mathbf{p}_1$ is either parallel or perpendicular to the line joining the atoms. (c) If $\mathbf{p}_1$ fluctuates randomly with time, is the time-averaged force a central force?

# Chapter 6

## ATOMIC VIBRATIONS

The triple axis neutron spectrometer at the Brookhaven high flux reactor. The reactor is in the left background, the spectrometer is just left of center.

The primary purpose of this chapter is to introduce the fundamental concepts needed to understand the vibratory motion of atoms in solids and the contributions these motions make to material properties. The most important idea is that of a normal mode of vibration, in which all atoms oscillate with the same frequency. Only atomic vibrations with certain frequencies, determined by interatomic forces, occur in any given solid. These constitute the normal mode spectrum, an important characteristic of a solid.

The concepts of normal mode displacements and frequencies are valid for both crystalline and amorphous solids. Because the atoms of a crystal are periodically placed, however, normal mode displacements for these materials have an especially simple form and are relatively easy to discuss. We will carry the discussion as far as we can in general terms, then specialize to crystals.

## 6.1 NORMAL MODES

The idea is simple. If the displacements of atoms from their equilibrium sites are small, the forces they exert on each other are proportional to their displacements, as if the atoms were connected by ideal springs. We may take the force on any atom to be a sum of terms, each proportional to the displacement of one of the atoms from its equilibrium site. This approximation leads to atomic motions that are simple harmonic.

The constants of proportionality that relate forces and displacements can, in principle, be calculated from details of the interatomic forces. We will not carry out the calculations, but instead will assume that the force constants are known. We substitute expressions for the forces into Newton's second law and generate a set of differential equations, one for each atom. We then seek solutions for which all atomic displacements have the same frequency.

Strictly speaking, quantum mechanics should be used to describe atomic motions. Both classical and quantum mechanical descriptions, however, lead to the same normal mode frequency spectrum, so we use classical mechanics for this introduction to the subject.

*Normal Modes of a Simple System.* To illustrate, consider the one-dimensional system of three atoms shown in Fig. 6-1. The atoms have equilibrium positions $x_1$, $x_2$, and $x_3$, and the displacement of atom $i$ from its equilibrium position is denoted by $u_i(t)$. We suppose only nearest-neighbor interactions are significant. The force of atom 2 on atom 1 is proportional to the difference in the displacements of those atoms from their equilibrium positions, so we write

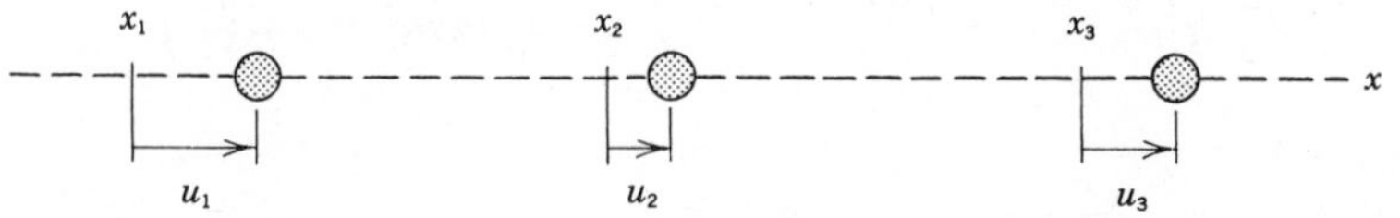

**FIGURE 6-1** A one-dimensional system of three atoms. Atom $i$ is displaced by $u_i$ from its equilibrium position $x_i$.

$F_1 = -K_1(u_1 - u_2)$, where $K_1$ is a constant. According to Newton's third law, the force of atom 1 on atom 2 is $-K_1(u_2 - u_1)$. In addition, atom 3 exerts a force $-K_2(u_3 - u_2)$ on atom 2, so the net force on this atom is $F_2 = -K_1(u_2 - u_1) - K_2(u_2 - u_3)$. Only atom 2 exerts a force on atom 3, so $F_3 = -K_2(u_3 - u_2)$. Here $K_2$ is the force constant for the interaction of atoms 2 and 3. Note that the forces tend to restore equilibrium separations. If $u_1 > u_2$, for example, the force on atom 1 is in the negative $x$ direction.

Newton's second law for each of the three atoms yields

$$m \frac{d^2 u_1}{dt^2} = -K_1(u_1 - u_2), \tag{6-1}$$

$$m \frac{d^2 u_2}{dt^2} = -K_1(u_2 - u_1) - K_2(u_2 - u_3), \tag{6-2}$$

and

$$m \frac{d^2 u_3}{dt^2} = -K_2(u_3 - u_2), \tag{6-3}$$

respectively. To find normal mode solutions we assume each displacement has the same sinusoidal time dependence. That is, we take $u_i = u_{i0}e^{-i\omega t}$ for atom $i$. Here $u_{i0}$ is independent of the time and $\omega$ is the angular frequency of the vibration. Although complex notation is used for later convenience, displacements are real. We use the real part of $u_i$ to describe a physical displacement.

The second derivative of $u_i(t)$ is $-\omega^2 u_i(t)$, so Eqs. 6-1, 6-2, and 6-3 become

$$(K_1 - m\omega^2)u_1 - K_1 u_2 = 0, \tag{6-4}$$

$$-K_1 u_1 + (K_1 + K_2 - m\omega^2)u_2 - K_2 u_3 = 0, \tag{6-5}$$

and

$$-K_2 u_2 + (K_2 - m\omega^2)u_3 = 0, \tag{6-6}$$

respectively, after some rearrangement. Since these equations are homogeneous in the displacements, a possible solution is $u_i = 0$ for all $i$. This solution is not of interest since it describes a situation in which all atoms are at rest on their equilibrium sites. Other solutions exist only if the determinant formed from the coefficients of the displacements vanishes. In detail, the displacements must all be zero unless

$$\begin{vmatrix} (K_1 - m\omega^2) & -K_1 & 0 \\ -K_1 & (K_1 + K_2 - m\omega^2) & -K_2 \\ 0 & -K_2 & (K_2 - m\omega^2) \end{vmatrix} = 0. \tag{6-7}$$

The determinant is called the secular determinant for the normal mode problem.

Equation 6-7 determines the normal mode frequencies. When the determinant is expanded, the equation becomes

$$m\omega^2[m^2\omega^4 - 2(K_1 + K_2)m\omega^2 + 3K_1 K_2]u_1 = 0. \tag{6-8}$$

The left side is a third-order polynomial in $\omega^2$, so Eq. 6-8 is satisfied by three values of $\omega^2$. Since we take the angular frequency to be positive, each of these leads to a single normal mode frequency. The normal mode frequencies are

$$\omega_1 = 0, \tag{6-9}$$

$$\omega_2 = \frac{1}{m}\left[(K_1 + K_2) - (K_1^2 + K_2^2 - K_1 K_2)^{1/2}\right]^{1/2}, \tag{6-10}$$

and

$$\omega_3 = \frac{1}{m}\left[(K_1 + K_2) + (K_1^2 + K_2^2 - K_1 K_2)^{1/2}\right]^{1/2}. \tag{6-11}$$

If $\omega$ is any one of the three normal mode frequencies, only two of the equations of motion can be considered independent and the displacements are not completely determined by Eqs. 6-9, 6-10, and 6-11. However, two ratios, $u_2/u_1$ and $u_3/u_1$, say, can be evaluated. Equation 6-4 immediately gives $u_2/u_1$; then Eq. 6-5 or 6-6 can be used to find $u_3/u_1$. Each of the ratios has different values for different normal mode frequencies.

*The Normal Mode Problem for a Solid.* The formulation given above can be broadened to include three-dimensional solids and to take into account interactions between all atoms, not just nearest neighbors. Because a large number of atoms interact with each other and because we must specify three components of the force acting on each one, the notation is necessarily complicated. We use subscripts $i$ and $j$ to distinguish one atom from another and $\alpha$ and $\beta$ to denote Cartesian components. $\alpha$, for example, may represent $x$, $y$, or $z$. In this notation, the $\alpha$ component of the net force on atom $i$ is denoted by $F_{i\alpha}$ and the $\beta$ component of the displacement of atom $j$ from its equilibrium position is denoted by $u_{j\beta}$.

Concentrate on atom $i$. A force is exerted on it if any atom, including $i$ itself, is displaced from equilibrium. Furthermore, the displacement of any atom in one direction might result in a force on atom $i$ in another direction. An expression for the force on atom $i$ in the harmonic approximation must contain a term for each component of the displacement of each atom. We write

$$F_{i\alpha} = -\sum_j \sum_\beta \Phi_{\alpha\beta}(i, j)u_{j\beta}, \tag{6-12}$$

where the first sum is over all atoms of the solid and the second is over the three Cartesian coordinates. $\Phi_{\alpha\beta}(i, j)$ is a force constant. It depends on the specific nature of the atomic interactions in the solid being considered and is different for different pairs of atoms as well as for different components of the atomic displacements. Some force constants are negative, as a comparison of Eqs. 6-1, 6-2, and 6-3 with Eq. 6-12 reveals.

The $\alpha$ component of Newton's second law for atom $i$ is

$$m_i \frac{d^2 u_{i\alpha}}{dt^2} = -\sum_j \sum_\beta \Phi_{\alpha\beta}(i, j)u_{j\beta}. \tag{6-13}$$

Equation 6-13 is representative of a set of simultaneous equations for the displacement components. Three similar equations can be written for each atom of the solid.

To find the normal mode solutions, assume $u_{i\alpha} = u_{i\alpha 0}e^{-i\omega t}$, where $u_{i\alpha 0}$ is independent of the time and $\omega$ is an angular frequency. On substitution, Eq. 6-13 becomes

$$m_i\omega^2 u_{i\alpha}(t) = \sum_j \sum_\beta \Phi_{\alpha\beta}(i,j)u_{j\beta}(t) \qquad (6\text{-}14)$$

or

$$\sum_j \sum_\beta [\Phi_{\alpha\beta}(i,j) - m_i\omega^2\delta_{ij}\delta_{\alpha\beta}]u_{j\beta}(t) = 0, \qquad (6\text{-}15)$$

where $\delta_{ij}$ and $\delta_{\alpha\beta}$ are Kronecker deltas. For example, $\delta_{ij}$ is 0 if $i \neq j$ and is 1 if $i = j$. Normal mode frequencies are solutions to the determinantal equation

$$\text{Det}[\Phi_{\alpha\beta}(i,j) - m_i\omega^2\delta_{\alpha\beta}\delta_{ij}] = 0, \qquad (6\text{-}16)$$

where the quantity on the left is the determinant formed by coefficients of the displacement components in the equations of motion. The notation may be somewhat confusing because each displacement component is labeled by two subscripts, one denoting an atom and one denoting a Cartesian component. Subscripts $i$ and $\alpha$ label determinant rows, in the order $1x$, $1y$, $1z$, $2x$, $2y$, $2z$, $3x, \ldots,$ for example. Subscripts $j$ and $\beta$, in the same order, label columns.

When the determinant is expanded, the result is a polynomial of order $3N$ in $\omega^2$, where $N$ is the number of atoms in the solid. The highest order term, proportional to $(\omega^2)^{3N}$, results from the product of the diagonal elements. The polynomial has $3N$ roots and, since only positive values of $\omega$ are accepted, the system has exactly $3N$ normal modes. Once normal mode frequencies have been found, any of them can be substituted into the set of equations represented by Eq. 6-15 and these equations can be solved for the ratios of the displacement components.

In the harmonic approximation, the set of simultaneous equations represented by Eq. 6-15 is valid for crystals and amorphous materials alike. However, it cannot be solved for amorphous solids of macroscopic size simply because there are so many atoms. Numerical techniques and high-speed computers can be used to find the normal mode frequencies for a system containing at most about 100 atoms. On the other hand, if the solid is crystalline, the complexity of the problem is reduced significantly by the periodic nature of the atomic structure, as we will soon see. First we will look in more detail at how force constants can be found from interatomic forces.

*Force Constants from the Potential Energy.* Force constants can be determined directly from interatomic forces, given as functions of atomic separations. Simply expand the forces as power series in displacement components and take the force constants to be the coefficients in the linear terms. This technique will be used later in Example 6-3.

Since the force on an atom can be determined from the potential energy of the solid, given as a function of atomic positions, force constants are directly related to the potential energy function. We now discuss the relationship.

The potential energy $U(\mathbf{r}_1, \mathbf{r}_2, \ldots, \mathbf{r}_N)$ of a solid containing $N$ atoms is a function of $3N$ independent variables, which we may take to be the Cartesian coordinates of the nuclei. In a sense, the function is misnamed. It includes the total electronic kinetic energy as well as the potential energy of electron-electron, electron-nucleus, and nucleus-nucleus interactions. In fact, it is the total energy of the solid exclusive of the atomic vibrational energy.

In terms of the potential energy function, the force on atom $i$ is given by

$$\mathbf{F}_i = -\frac{\partial U}{\partial x_i}\hat{\mathbf{x}} - \frac{\partial U}{\partial y_i}\hat{\mathbf{y}} - \frac{\partial U}{\partial z_i}\hat{\mathbf{z}}$$

$$= -\frac{\partial U}{\partial u_{ix}}\hat{\mathbf{x}} - \frac{\partial U}{\partial u_{iy}}\hat{\mathbf{y}} - \frac{\partial U}{\partial u_{iz}}\hat{\mathbf{z}}. \tag{6-17}$$

According to Eq. 6-12, the force constant $\Phi_{\alpha\beta}(i, j)$ is given by $\Phi_{\alpha\beta}(i, j) = -\partial F_{i\alpha}/\partial u_{j\beta}$, or when Eq. 6-17 is used, by

$$\Phi_{\alpha\beta}(i, j) = \frac{\partial^2 U}{\partial u_{i\alpha}\partial u_{j\beta}}. \tag{6-18}$$

Strictly speaking, we should apply the small displacement approximation to $U$. This simply means that the derivative in Eq. 6-18 is evaluated for atoms at their equilibrium sites.

In principle, Eq. 6-18 provides a means of computing force constants. First, the total energy of the solid is found for the nuclei at rest in a variety of configurations and a minimum is found. This determines an equilibrium configuration corresponding to one of the crystalline or amorphous forms of the material. Then, second derivatives of the energy with respect to displacement components are evaluated for the equilibrium configuration. Of course, the amount of work involved is far too great for the procedure to be carried out in practice, but approximations have been developed.

*Anharmonic Forces.* In writing Eq. 6-12 for the force on an atom in the harmonic approximation, we neglected terms of higher order than linear in atomic displacements. Although this approximation produces accurate results as far as normal mode frequencies are concerned, neglected terms are important for some physical phenomena.

Anharmonic terms are responsible for a continual exchange of vibrational energy between normal modes. When the atoms are in thermal equilibrium, each mode loses as much energy as it gains and there is no net change in the energy of any mode. However, when the atoms are not in thermal equilibrium, as when a temperature gradient is present, energy flows away from modes with an excess

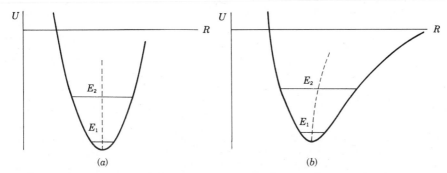

**FIGURE 6-2** (a) Harmonic and (b) anharmonic potential energy functions for the interaction of two atoms. $R$ is the atomic separation, while $E_1$ and $E_2$ represent two possible vibrational energies. For (a) an increase in energy does not result in a change in the average atomic separation, given by the midpoints of constant energy lines. For (b) an increase in energy results in an increase in average separation.

to modes with a deficit. Thus anharmonic forces are important if a solid is to reach thermal equilibrium.

Anharmonic forces are also important for the thermal expansion of a solid. The idea can be understood in terms of the potential energy functions drawn in Fig. 6-2 for two interacting atoms. In ($a$) the potential energy is quadratic and the force exerted by either atom on the other is linear in the relative displacement. On the other hand, the function shown in ($b$) is not quadratic and leads to anharmonic forces. Each of the horizontal lines shown represents a possible vibrational energy for the pair. For each energy, intersections of the horizontal line with the potential energy curve give the amplitude of vibration and the midpoint of the line gives the average atomic separation.

As the temperature increases so does the vibrational energy. Clearly, this increase leads to a change in the average separation only when the potential energy is not symmetric about its minimum. For most solids core-core interactions produce a curve that is steeper to the left of the minimum than to the right and expansion occurs as the temperature increases.

## 6.2  A MONATOMIC LINEAR CHAIN

If atomic equilibrium sites are arranged periodically, solutions to the set of equations represented by Eq. 6-15 take a particularly simple form and are relatively easy to solve. To illustrate, we consider a one-dimensional crystal with a single atom in the primitive basis.

*Equations of Motion.*  The system consists of $N$ atoms with mass $m$ and equilibrium separation $a$, as shown in Fig. 6-3. Equilibrium positions are given by $na$, where $n = 0, 1, \ldots, N - 1$. The atoms move along the $x$ axis and the position of atom $n$, for example, is given by $x_n(t) = na + u_n(t)$, where $n_n(t)$ is the

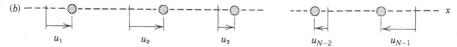

FIGURE 6-3   A one-dimensional linear chain of $N$ atoms. ($a$) Equilibrium positions, with atom $n$ at $x = na$. ($b$) The atoms displaced from their equilibrium positions at some instant of time. The displacement of atom $n$ is given by $u_n$. For the situation shown, $u_{N-2}$ and $u_{N-1}$ are negative while the other displacements shown are positive.

displacement of the atom from its equilibrium site. The equation of motion for atom $n$ is

$$m \frac{d^2u_n}{dt^2} = -\sum_i \Phi(n, i)u_i, \qquad (6\text{-}19)$$

where $\Phi(n, i)$ is the force constant for the interaction of atoms $n$ and $i$ and the sum is over all atoms of the chain. Equation 6-19 is identical to Eq. 6-13 except that subscripts specifying Cartesian components are omitted. For a normal mode vibration, $d^2u_n(t)/dt^2 = -\omega^2 u_n(t)$ and Eq. 6-19 becomes

$$m\omega^2 u_n(t) = \sum_i \Phi(n, i)u_i(t). \qquad (6\text{-}20)$$

Solutions have a particularly simple form if all atoms obey exactly the same equation of motion. As the system was described, atoms near the ends of the chain have different distributions of neighbors than atoms in the interior and so obey different equations of motion. We remedy this by assuming the $N$ atoms are part of an infinite chain, so every atom has the same distribution of neighbors. If each atom interacts with $2p$ neighbors, then the sum in Eq. 6-20 can be arranged so that

$$m\omega^2 u_n(t) = \sum_{s=-p}^{+p} \Phi(n, n + s)u_{n+s}(t). \qquad (6\text{-}21)$$

Here the atoms have been relabeled so the sum starts with the atom a distance $pa$ away from atom $n$ on one side and ends with the atom a distance $pa$ away on the other side. Some of the terms in the sum may be associated with atoms outside the actual chain, but they are included so every atom participates in the same number of interactions.

$\Phi(n, n + s)$ does not depend on $n$, only on $s$. For example, atoms with equilibrium positions $5a$ and $8a$ interact with the same force constant as atoms with equilibrium positions $10a$ and $13a$. If we write $\Phi(s)$ for $\Phi(n, n + s)$, Eq. 6-

21 becomes

$$mω^2u_n(t) = \sum_{s=-p}^{+p} Φ(s)u_{n+s}(t).$$

(6-22)

To create a model in which all atoms are equivalent, we have actually increased the number of equations to be solved. Atoms beyond the original $N$ atom chain contribute to the force on atoms in the chain and we must consider their motions. To reduce the complexity of the problem, we impose the condition

$$u_{n+N}(t) = u_n(t).$$

(6-23)

In other words, the model consists of an infinite chain composed of segments, each containing $N$ atoms. The displacement of any atom in one segment is taken to be exactly the same as the displacement of analogous atoms in other segments. Displacements that satisfy Eq. 6-23 are said to obey periodic boundary conditions.

Except for surface modes, in which only atoms near a sample surface participate, periodic boundary conditions introduce essentially no error into calculations of normal mode frequencies. If $N$ is large, the fraction of atoms near a boundary is small and periodic boundary conditions lead to a valid description of the motion of nearly every atom in a crystal. Surfaces modes will be discussed later.

*Normal Mode Displacements.* First consider a normal mode with angular frequency $ω$ and suppose no other normal modes have the same frequency. Such a mode is said to be nondegenerate. All displacements are proportional to any one of them, so we may take

$$u_1(t) = Γu_0(t),$$

(6-24)

where $Γ$ is independent of the time but may depend on $ω$. It is determined by the equations of motion.

Equation 6-24 is valid even if the structure is not crystalline. We now make use of the periodicity of the chain. Because the atoms are all equivalent, the constant of proportionality that relates the displacement of atom 2 to that of atom 1 is also $Γ$. This is, $u_2(t) = Γu_1(t)$. To understand why this is so, consider a different segment of the infinite chain, from $x = a$ to $x = Na$. The set of equations for these atoms is exactly the same as the set for the original segment. In particular, the proportionality constant relating the displacements of atom 1 and atom 0 is the same as the proportionality constant relating the displacements of atom 1 and atom 0 of the original chain. Since atom 0 of the new chain is atom 1 of the old and atom 1 of the new chain is atom 2 of the old, $u_2(t) = Γu_1(t)$. In terms of atom 0, $u_2(t) = Γ^2u_0(t)$.

We can continue with other segments and show that $u_{n+1}(t) = Γu_n(t)$. As a consequence,

$$u_n(t) = Γ^nu_0(t).$$

(6-25)

Equation 6-25 gives $u_N(t) = \Gamma^N u_0(t)$ but, according to Eq. 6-23, $u_N(t) = u_0(t)$. So $\Gamma^N = 1$ and $\Gamma$ is an $N$th root of 1. Since $e^{i2\pi k} = 1$ if $k$ is an integer, $\Gamma$ has one of the values $e^{i2\pi k/N}$. This result is usually written

$$\Gamma = e^{iqa}, \tag{6-26}$$

where $q = 2\pi k/Na$. Later we will see that a different value of $k$ is associated with each normal mode frequency.

Write $u_0(t) = Ae^{-i\omega t}$, where $A$ is a constant. Then, according to Eqs. 6-25 and 6-26,

$$u_n(t) = Ae^{i(qna-\omega t)}. \tag{6-27}$$

This is the mathematical form of a normal mode displacement for an atom in a periodic linear chain. Because we know the form in advance, we can use it to simplify greatly the solving of secular equations for normal mode frequencies.

If the same frequency is associated with more than one normal mode, the proportionality constant relating the displacements of any two selected atoms need not be the same for the translated and original chains. Nevertheless, normal mode displacements can still be written in the form given by Eq. 6-27. The proof is somewhat more complicated than for the nondegenerate case and the reader is referred to references at the end of the chapter.

As indicated by Eq. 6-27, normal mode oscillations in crystals are similar to waves traveling on a continuous string. Take $A$ to be real and compare the real part of Eq. 6-27, $u_n(t) = A\cos(qna - \omega t)$, to the expression for the displacement of a string, $y(x, t) = A\cos(qx - \omega t)$. Here $q$ is related to the wavelength $\lambda$ by $q = 2\pi/\lambda$. The functions have the same form. To obtain atomic displacements from the expression for the string displacement, the latter is evaluated with $x = na$ and the propagation constant $q$ is limited to the values $2\pi k/Na$. We conclude that a normal mode oscillation is a traveling sinusoidal wave with wavelength given by $\lambda = 2\pi/q = Na/k$.

*Normal Mode Frequencies.* Equation 6-27 is substituted into Eq. 6-22 to obtain

$$m\omega^2 = \sum_s \Phi(s)\, e^{iqsa}, \tag{6-28}$$

after division by $Ae^{i(qna-\omega t)}$. Equation 6-28 gives the dispersion relationship for normal modes of a chain: the normal mode frequency as a function of propagation constant. To use the equation, simply substitute one of the allowed values of $q$ and solve for $\omega$. Note the enormous reduction in the complexity of the problem. We need solve only one equation, not $N$. However, it must be solved a large number of times, once for each allowed value of $q$.

As an example, consider a chain in which each atom interacts only with its nearest neighbors. Take the force of atom $n + 1$ on atom $n$ to be $\gamma(u_{n+1} - u_n)$ and the force of atom $n - 1$ atom on atom $n$ to be $\gamma(u_{n-1} - u_n)$. Equation 6-28 then becomes

$$\begin{aligned} m\omega^2 &= -\gamma[e^{-iqa} + e^{iqa} - 2] \\ &= 2\gamma[1 - \cos(qa)] = 4\gamma\sin^2(\tfrac{1}{2}qa), \end{aligned} \tag{6-29}$$

where the identities $e^{iqa} + e^{-iqa} = 2\cos(qa)$ and $\cos(qa) = 1 - 2\sin^2(\tfrac{1}{2}qa)$ were used. The normal mode frequencies are given by

$$\omega(q) = \left[\frac{4\gamma}{m}\right]^{1/2} |\sin(\tfrac{1}{2}qa)|. \tag{6-30}$$

and are plotted in Fig. 6-4.

We compare the dispersion relationship given by Eq. 6-30 with similar curve for waves traveling with speed $v$ on a continuous string. For the latter, $\omega = vq$ so $\omega(q)$ is a straight line.

The dispersion curve for the chain is linear for $q$ near zero, just like the curve for a continuous string. Near $q = 0$ the wavelength is long compared to the interatomic spacing and neighboring atoms have nearly the same displacement from equilibrium at the same time. The granularity of the medium does not influence the dispersion curve at long wavelengths.

Sound waves have frequencies in the linear portion of the dispersion curve. The speed of sound in our one-dimensional crystal is the slope of $\omega(q)$ in the limit $q \to 0$. In this limit $\sin(\tfrac{1}{2}qa)$ can be replaced by $\tfrac{1}{2}qa$ and Eq. 6-30 becomes $\omega = (\gamma a^2/m)^{1/2}q$ for $q$ positive. Thus the speed of sound is $(\gamma a^2/m)^{1/2}$. Since $m/a$ is the linear mass density and $\gamma a$ can be interpreted as the tension in the chain, this is consistent with the result given in elementary texts for the speed of a wave on a string.

As $q$ becomes large, $\omega(q)$ deviates significantly from a straight line. When $qa = \pm\pi$, it reaches a maximum given by

$$\omega_{max} = \left[\frac{4\gamma}{m}\right]^{1/2}. \tag{6-31}$$

No normal modes have angular frequency greater than $\omega_{max}$. On the other hand, $\omega$ is not bounded for a continuous string. Later we will see what happens if a vibration is excited with angular frequency greater than $\omega_{max}$.

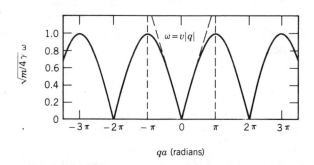

FIGURE 6-4  The dispersion relationship $\omega = \sqrt{(4\gamma/m)}\,|\sin(\tfrac{1}{2}qa)|$ for a monatomic linear chain with nearest-neighbor interactions. Here $\omega$ is the angular frequency, $q$ is the propagation constant, $\gamma$ is the force constant, and $a$ is the equilibrium atomic separation. Propagation constants are usually limited to the range $-\pi/a$ to $+\pi/a$. Dotted lines show the dispersion relationship $\omega = v|q|$ for a continuous string.

As the following example shows, characteristics of the dispersion curve, such as the speed of sound and the maximum frequency, can be used to gain information about the force constants.

**EXAMPLE 6-1** The speed of sound in a certain linear monatomic chain is $1.08 \times 10^4$ m/s. If the mass of each atom is $6.81 \times 10^{-26}$ kg and the atomic separation at equilibrium is 4.85 Å, find (a) the force constant and (b) the maximum normal mode angular frequency. Assume only nearest-neighbor atoms interact.

**SOLUTION** (a) Since $v = \sqrt{(\gamma a^2/m)}$, $\gamma = v^2 m/a^2 = (1.08 \times 10^4)^2 \times 6.81 \times 10^{-26}/(4.85 \times 10^{-10})^2 = 33.8$ J/m². (b) $\omega_{max} = \sqrt{(4\gamma/m)} = (4 \times 33.8/6.81 \times 10^{-26})^{1/2} = 4.46 \times 10^{13}$ rad/s. Note that $\omega_{max}$ is well above the angular frequency of audible sound. ∎

*Limitations on the Propagation Constant.* The dispersion curve for our one-dimensional crystal is periodic in the propagation constant: it repeats in intervals of $2\pi/a$. Atomic displacements are also periodic in $q$. In particular, displacements are the same for normal modes with propagation constants $q$ and $q + 2\pi\hbar/a$, where $\hbar$ is any integer. For the mode with propagation constant $q + 2\pi\hbar/a$,

$$u_n(t) = Ae^{i(qna + 2\pi n\hbar - \omega t)} = Ae^{i(qna - \omega t)}, \tag{6-32}$$

which is, of course, the same as the displacement for the mode with propagation constant $q$. The second equality follows from the identity $e^{i2\pi n\hbar} = 1$.

Figure 6-5 shows a graph of the displacement for two continuous strings carrying waves with propagation constants that differ by $2\pi/a$. Clearly the waves are different. However, string displacements at points given by $x = na$, where $n$ is an integer, are the same for the two waves. String displacements differ between these points but this difference has no meaning for a crystal.

Two modes with propagation constants which differ by $2\pi/a$ are physically the same and only one can be included in a list of normal modes. Duplication of modes is avoided if we restrict the propagation constant to any range of width $2\pi/a$. Usually the smallest propagation constant possible for any given mode is used, so $q$ is restricted to the range

$$-\frac{\pi}{a} \leq q < +\frac{\pi}{a}. \tag{6-33}$$

All values in this range differ by less than $2\pi/a$. Only one of the end points may be included, but which one is immaterial. Dotted lines mark the limits of the range in Fig. 6-4.

Precisely $N$ allowed values of the propagation constant are in the range from $-\pi/a$ to $+\pi/a$. Allowed values are given by $2\pi\hbar/Na$, so, according to Eq. 6-33, $\hbar$ ranges from $-\frac{1}{2}N$ to $+\frac{1}{2}N$, an interval that contains exactly $N$ integers. There is one allowed propagation vector for each atom in the chain.

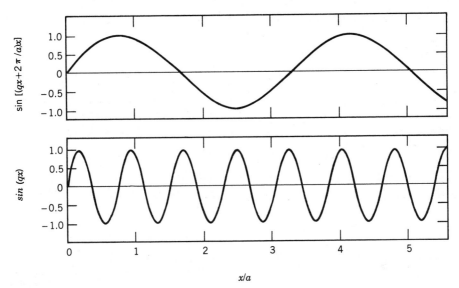

**FIGURE 6-5** The two functions $\sin(qx)$ and $\sin[(q + 2\pi/a)x]$ have the same values for $x = na$, where $n$ is any integer. The waves represent the same normal mode. To plot the graphs, $a$ was chosen to be $\lambda/0.3$, where $\lambda$ is the wavelength in the upper diagram.

In general, the number of normal modes equals the number of degrees of freedom of the system. Our linear chain of $N$ atoms has $N$ degrees of freedom and $N$ normal modes, the same as the number of possible propagation constants. For this system, a different propagation constant is associated with each mode and $q$ may be used to label the modes.

**EXAMPLE 6-2** Consider vibrational waves traveling in a one-dimensional monatomic crystal in which equilibrium sites are $a = 5.00$ Å apart. (a) If the chain contains $6.00 \times 10^8$ atoms, find the allowed values of the propagation constant. (b) Suppose the length of the chain is doubled by adding atoms with the same equilibrium spacing. What then are the allowed values of the propagation constant? In both cases assume the propagation constant for a given wave has the smallest magnitude possible for that wave.

**SOLUTION** (a) Allowed values of the propagation constant are given by $q = 2\pi k/Na = 2\pi k/(6.00 \times 10^8 \times 5.00 \times 10^{-10}) = 20.9k$ m$^{-1}$. Values of $k$ range from $-\frac{1}{2}N = -3.00 \times 10^8$ to $+\frac{1}{2}N = +3.00 \times 10^8$ so there are $6.00 \times 10^8$ allowed propagation constants, equally spaced between $-6.27 \times 10^9$ m$^{-1}$ and $+6.27 \times 10^9$ m$^{-1}$. (b) If $N = 1.20 \times 10^9$ atoms, then $q = 10.5k$ m$^{-1}$. Values of $k$ again range from $-6.00 \times 10^8$ to $+6.00 \times 10^8$, but now there are $1.20 \times 10^9$ allowed propagation constants. Changing the length of the chain by adding atoms with the same separation does not change the limits on the propagation constant. It simply squeezes more allowed values between the same limits. ∎

*High-Frequency Excitations.* If atomic vibrations are excited by forcing surface atoms to oscillate with an angular frequency greater than $\omega_{max}$, atoms in the interior vibrate but the vibrations are severely attenuated with distance into the sample.

We consider a chain in which only nearest neighbors interact. Equation 6-29 is valid, but $q$ must be complex to satisfy it. Substitute $q = \alpha + i\beta$, where $\alpha$ and $\beta$ are real, into Eq. 6-29 and make use of the identity $\sin(A + iB) = \sin(A)\cos(iB) + \cos(A)\sin(iB) = \sin(A)\cosh(B) + i\cos(A)\sinh(B)$ to find

$$m\omega^2 = 4\gamma[\sin(\tfrac{1}{2}\alpha a)\cosh(\tfrac{1}{2}\beta a) + i\cos(\tfrac{1}{2}\alpha a)\sinh(\tfrac{1}{2}\beta a)]^2. \qquad (6\text{-}34)$$

Since $\omega$ is real, $\cos(\tfrac{1}{2}\alpha a)\sinh(\tfrac{1}{2}\beta a) = 0$. If $\sinh(\tfrac{1}{2}\beta a) = 0$, then $\beta = 0$, $q$ is real, and $\omega$ must be less than $\omega_{max}$. We reject this solution and select $\alpha = \pi/a$ so $\cos(\tfrac{1}{2}\alpha a) = 0$. Then $\sin(\tfrac{1}{2}\alpha a) = 1$ and Eq. 6-34 becomes $m\omega^2 = 4\gamma\cosh^2(\tfrac{1}{2}\beta a)$, or

$$\beta = \frac{2}{a}\cosh^{-1}\left[\frac{m\omega^2}{4\gamma}\right]^{1/2} = \frac{2}{a}\cosh^{-1}\left[\frac{\omega}{\omega_{max}}\right]. \qquad (6\text{-}35)$$

For a given $\omega$, two values of $\beta$ satisfy Eq. 6-35, one positive and one negative. If the chain lies along the positive $x$ axis, from $x = 0$ to $x = (N - 1)a$ and the atom at $x = 0$ is set into oscillation, we use the positive value. Then the displacement of atom $n$ is given by

$$u_n(t) = Ae^{i(qna - \omega t)} = Ae^{i(n\pi + i\beta na - \omega t)}$$
$$= (-1)^n Ae^{-\beta na}e^{-i\omega t}. \qquad (6\text{-}36)$$

Each atom vibrates with angular frequency $\omega$ but the amplitude decreases exponentially from atom to atom along the chain. For $\omega$ near $\omega_{max}$, $\beta$ is nearly zero and oscillations extend far into the chain. However, $\beta$ increases rapidly as $\omega$ increases and, even for $\omega$ only slightly above $\omega_{max}$, relatively few atoms vibrate with significant amplitude. In any event, the excitation is not a traveling wave.

## 6.3 A DIATOMIC LINEAR CHAIN

Important features of some vibrational spectra are illustrated by a linear chain with two atoms per primitive unit cell. A cell of the chain shown in Fig. 6-6 has length $a$ and contains one atom of mass $M$ and one of mass $m$. Equilibrium positions of mass $M$ atoms are given by $na$, while those of mass $m$ atoms are given by $(n + \tfrac{1}{2})a$, where $n$ is an integer. Without loss of generality we take

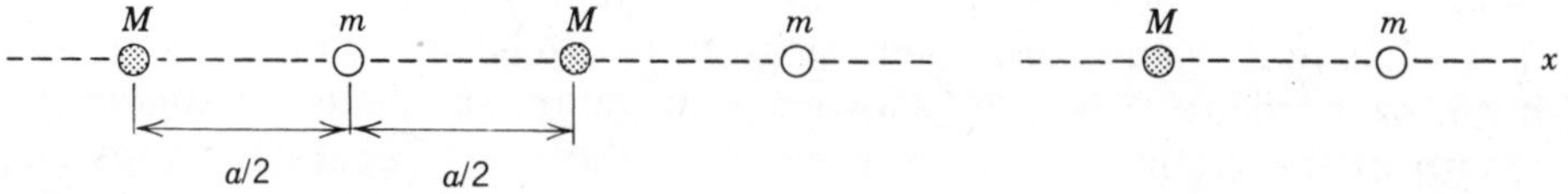

**FIGURE 6-6** A diatomic linear chain. A primitive unit cell has length a and contains two atoms, one of mass M and the other of mass m.

$M > m$. We assume only nearest neighbors interact and take the force constant to be $\gamma$.

*Normal Mode Frequencies.*   Represent the displacements from equilibrium of the two atoms in unit cell $n$ by $u_{Mn}(t)$ and $u_{mn}(t)$, respectively. In the harmonic approximation, Newton's second law yields

$$M\frac{d^2u_{Mn}(t)}{dt^2} = \gamma[u_{mn} + u_{mn-1} - 2u_{Mn}] \tag{6-37}$$

for the atom of mass $M$ and

$$m\frac{d^2u_{mn}(t)}{dt^2} = \gamma[u_{Mn} + u_{Mn+1} - 2u_{mn}] \tag{6-38}$$

for the atom of mass $m$ in the same cell. All atoms of mass $M$ are equivalent by virtue of translational symmetry and their displacements from equilibrium sites are related to each other in the same way as displacements for a monatomic chain. Consequently,

$$u_{Mn}(t) = A_M e^{i(qna-\omega t)}. \tag{6-39}$$

Displacements from equilibrium sites of mass $m$ atoms are similarly related to each other, so

$$u_{mn}(t) = A_m e^{i[qna+(1/2)qa-\omega t]}. \tag{6-40}$$

The factor $e^{iqa/2}$ need not be included but the mathematics is somewhat less complicated if it is. Mass $M$ and mass $m$ atoms are not equivalent, so we may not assume $A_M = A_m$. Note that $n$ labels a unit cell and not an atom.

When Eqs. 6-39 and 6-40 are substituted into Eqs. 6-37 and 6-38, the results are

$$-M\omega^2 A_M = 2\gamma[A_m\cos(\tfrac{1}{2}qa) - A_M] \tag{6-41}$$

and

$$-m\omega^2 A_m = 2\gamma[A_M\cos(\tfrac{1}{2}qa) - A_m], \tag{6-42}$$

respectively. We divided by $e^{i(qna-\omega t)}$ to obtain Eq. 6-41 and by $e^{i(qna+qa/2-\omega t)}$ to obtain Eq. 6-42. We also used the identity $e^{iqa/2} + e^{-iqa/2} = 2\cos(\tfrac{1}{2}qa)$. Equations 6-41 and 6-42 are two linear, homogeneous algebraic equations in the unknowns $A_M$ and $A_m$. One solution is $A_M = A_m = 0$ but we reject it because it describes a situation for which no vibrations exist. For other solutions the determinant formed from the coefficients of $A_M$ and $A_m$ vanishes. That is,

$$\begin{vmatrix} 2\gamma - M\omega^2 & -2\gamma\cos(\tfrac{1}{2}qa) \\ -2\gamma\cos(\tfrac{1}{2}qa) & 2\gamma - m\omega^2 \end{vmatrix} = 0 \tag{6-43}$$

or, when the determinant is expanded,

$$(2\gamma - M\omega^2)(2\gamma - m\omega^2) - 4\gamma^2\cos^2(\tfrac{1}{2}qa) = 0. \tag{6-44}$$

This equation is quadratic in $\omega^2$ and has the two solutions

$$\omega^2 = \gamma \frac{M + m}{Mm} \pm \gamma \left[ \left( \frac{M + m}{Mm} \right)^2 - \frac{4}{Mm} \sin^2(\tfrac{1}{2}qa) \right]^{1/2}. \qquad (6\text{-}45)$$

Normal mode frequencies are given by the positive square root of the expression on the right.

As for a monatomic chain, waves with propagation constants $q$ and $q + 2\pi \hbar / a$ are identical and $q$ is limited to a range of values with width $2\pi/a$ to avoid duplication. Usually $q$ is restricted to the range from $-\pi/a$ to $+\pi/a$ so the propagation constant for any wave has the smallest magnitude appropriate for that wave.

Allowed values of $q$ are the same as for the monatomic chain. Periodic boundary conditions take the form $u_{M\ n+N}(t) = u_{Mn}(t)$ and $u_{m\ n+N}(t) = u_{mn}(t)$ and lead to the condition $q = 2\pi\hbar/Na$, where $\hbar$ is an integer. In the reduced zone scheme, $\hbar$ is limited to values in the range from $-\tfrac{1}{2}N$ to $+\tfrac{1}{2}N$ and the number of allowed values exactly equals the number of unit cells. Two normal modes correspond to each value, so there are $2N$ normal modes in all, one for each degree of freedom.

Figure 6-7 shows the dispersion relationships given by Eq. 6-45. The upper branch corresponds to the positive sign, while the lower branch corresponds to the negative sign. Because frequencies in the upper branch are usually in or near the optical region of the electromagnetic spectrum, it is called the optical branch. The lower branch is called the acoustical branch. If a vibration is excited with a frequency in the gap between the branches or above the optical branch, it is attenuated so that only atoms near the source oscillate with significant amplitude.

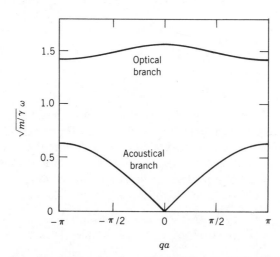

**FIGURE 6-7** The dispersion relationship for a diatomic linear chain with mass ratio $M/m = 5$. Two normal mode frequencies are associated with each value of $q$, one in the optical branch and one in the acoustical branch. In general for a linear chain, the number of branches equals the number of atoms in a primitive unit cell.

Sometimes optical and acoustical branches are plotted in separate regions of the graph, as shown in Fig. 6-8. Here the acoustical branch is plotted from $-\pi/a$ to $+\pi/a$, while the optical branch is plotted in two sections, from $-2\pi/a$ to $-\pi/a$ and from $+\pi/a$ to $+2\pi/a$. This form is known as the extended zone scheme, while the form shown in Fig. 6-7 is known as the reduced zone scheme. One scheme can be transformed into the other by translating some sections of the graph by $2\pi/a$, either to the left or to the right.

Maximum and minimum angular frequencies of the optical branch can be found by setting $q = 0$ and $q = \pi/a$, respectively, in Eq. 6-45. They are

$$\omega_{op\ max} = \left[ 2\gamma \frac{M + m}{Mm} \right]^{1/2} \tag{6-46}$$

and

$$\omega_{op\ min} = \left[ \frac{2\gamma}{m} \right]^{1/2}. \tag{6-47}$$

Their ratio is $(\omega_{op\ max})/(\omega_{op\ min}) = [1 + (m/M)]^{1/2}$. The branch is narrow for small $m/M$ and widens as $m/M$ tends toward 1.

The maximum frequency of the acoustical branch occurs for $q = \pi/a$ and is given by

$$\omega_{ac\ max} = \left[ \frac{2\gamma}{M} \right]^{1/2}. \tag{6-48}$$

The difference $\omega_{op\ min} - \omega_{ac\ max}$ gives the width of the gap between the branches at $q = \pi/a$. The gap narrows as the masses become more alike and disappears

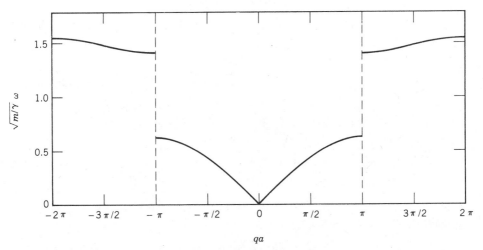

**FIGURE 6-8**  The dispersion relationship for the diatomic linear chain of Fig. 6-7, plotted in the extended zone scheme. The optical branch is in two sections: $-2\pi/a < q < -\pi/a$ and $\pi/a < q < +2\pi/a$. Figure 6-7 is obtained when each section is translated by a reciprocal lattice vector.

entirely for $m = M$. The chain is then monatomic with a unit cell length of $\frac{1}{2}a$. When the dispersion curve is plotted in the extended zone scheme the result is identical to the central region of Fig. 6-4, except that $q$ ranges from $-2\pi/a$ to $+2\pi/a$, as it should for the new cell size.

The dispersion curve for the acoustical branch is nearly linear in the region of small $q$. In that region $\sin(\frac{1}{2}qa)$ may be replaced by $\frac{1}{2}qa$ and the square root that appears in Eq. 6-45 may be expanded as a power series in $qa$. To lowest order the result is

$$\omega^2 \approx \frac{\gamma a^2}{2(M + m)} q^2 , \tag{6-49}$$

so the speed of sound is

$$v = \left[ \frac{\gamma a^2}{2(M + m)} \right]^{1/2} . \tag{6-50}$$

Since the linear mass density is $(M + m)/a$, this expression agrees with the corresponding expression for a continuous string with tension $\frac{1}{2}\gamma a$.

**EXAMPLE 6-3** A linear chain is made up of sodium and chlorine atoms, alternately placed with neighboring equilibrium positions separated by 2.81 Å. A sodium atom has a mass of $3.82 \times 10^{-26}$ kg and a chlorine atom has a mass of $5.89 \times 10^{-26}$ kg. Suppose the atoms interact electrostatically, with each sodium atom having a net charge of $+e$ and each chlorine atom having a net charge of $-e$, where e is the charge on a proton. Assume only nearest neighbors interact and estimate the maximum and minimum angular frequencies in each branch of the vibration spectrum.

**SOLUTION** The magnitude of the force between two neighboring atoms is given by

$$F = \frac{e^2}{4\pi\epsilon_0 r^2} ,$$

where $r$ is their separation. Write $r = \frac{1}{2}a + x$, where $x$ is the difference in the displacements of the atoms from their equilibrium sites. The binomial theorem gives $r^{-2} = (\frac{1}{2}a + x)^{-2} \approx (\frac{1}{2}a)^{-2} - 2(\frac{1}{2}a)^{-3}x$, to first order in $x$. So

$$\gamma = \frac{2e^2}{4\pi\epsilon_0(\frac{1}{2}a)^3} = \frac{(1.60 \times 10^{-19})^2}{4\pi \times 8.85 \times 10^{-12}} \frac{2}{(2.81 \times 10^{-10})^3} = 20.7 \text{ J/m}^2 .$$

For the acoustical branch, the minimum angular frequency is 0 and the maximum angular frequency is

$$\omega_{ac\ max} = \sqrt{2\gamma/M} = (2 \times 20.7/5.89 \times 10^{-26})^{1/2} = 2.65 \times 10^{13} \text{ rad/s} .$$

For the optical branch

$$\omega_{op\ min} = \sqrt{2\gamma/m} = (2 \times 20.7/3.82 \times 10^{-26})^{1/2} = 3.29 \times 10^{13} \text{ rad/s}$$

and

$$\omega_{op\ max} = \sqrt{2\gamma(M + m)/Mm}$$

$$= [2 \times 20.7 \times (5.89 \times 10^{-26} + 3.82 \times 10^{-26})$$

$$\div 5.89 \times 10^{-26} \times 3.82 \times 10^{-26}]^{1/2}$$

$$= 4.23 \times 10^{13}\ rad/s\ .$$

These are actually poor estimates of the branch limits. The electrostatic interaction is long range and to obtain more accurate results we must include interactions between atoms with greater separations than nearest neighbors. ∎

*Displacement Amplitudes.* Equations 6-41 and 6-42 determine the amplitude ratio $A_m/A_M$ for each normal mode frequency. According to the first of these

$$\frac{A_m}{A_M} = \frac{2\gamma - M\omega^2}{2\gamma \cos(\frac{1}{2}qa)}$$

$$= \frac{1 - \dfrac{M}{m} \pm \left[ \left(\dfrac{M + m}{m}\right)^2 - 4\dfrac{M}{m} \sin^2(\frac{1}{2}qa) \right]^{1/2}}{2 \cos(\frac{1}{2}qa)}, \quad (6\text{-}51)$$

where the second equality follows when Eq. 6-45 is used to substitute for $\omega^2$. The upper sign is used for the optical branch and the lower sign is used for the acoustical branch.

We investigate the acoustical branch first. For $q = 0$, Eq. 6-51 gives $A_m/A_M = 1$. As is typical of an acoustical branch in the long wavelength limit, all atoms move together with the same amplitude. For $q = \pi/a$, Eq. 6-51 is indeterminant but the L'Hôpital rule gives zero for the limit. $A_m = 0$ while $A_M$ is arbitrary, indicating that only the heavy atoms oscillate. If $A_M$ remains the same as $q$ increases from zero, $A_m$ decreases until it vanishes at $q = \pi/a$.

For the optical branch at $q = 0$, Eq. 6-51 yields $A_m/A_M = -M/m$. The negative sign indicates that the atoms of a unit cell move in opposite directions. Since the amplitude ratio is inversely proportional to the corresponding mass ratio, the center of mass of each cell remains fixed during the motion. In detail, the $x$ coordinate $x_c$ of the center of mass for cell $n$ is given by

$$(M + m)x_c = Mna + m(n + \tfrac{1}{2})a + MA_M e^{i(qna - \omega t)}$$

$$+ mA_m e^{i[qna + (1/2)qa - \omega t]} \quad (6\text{-}52)$$

and, for $q = 0$ and $mA_m = -MA_M$, this expression reduces to $(M + m)x_c = Mna + m(n + \frac{1}{2})a$, the same as the expression for stationary atoms.

If the two atoms of the basis are oppositely charged, long wavelength optical modes are easily excited by an oscillating electric field. The forces of the field on the two atoms have the same magnitude and are oppositely directed, so they do not change the motion of the center of mass. This is an important mechanism

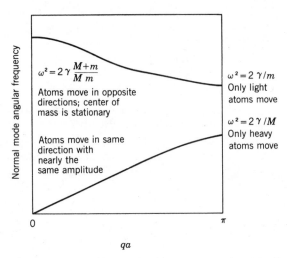

**FIGURE 6-9** Summary of normal mode angular frequencies and amplitude ratios at special points on the dispersion curves for a diatomic linear chain.

for the excitation of optical branch vibrations in ionic solids. We will discuss it in more detail later.

For short wavelength optical modes ($q \to \pi/a$), Eq. 6-51 gives $A_m/A_M \to \infty$. That is, $A_M = 0$, $A_m$ is arbitrary, and only the light atoms oscillate. If $A_m$ remains the same as $q$ increases from zero, $A_M$ decreases in magnitude until it vanishes at $q = \pi/a$. A summary of dispersion curve features for a diatomic chain is shown in Fig. 6-9.

## 6.4  ATOMIC VIBRATIONS IN THREE DIMENSIONS

The concepts introduced in the last two sections are easily generalized to three dimensions. In the harmonic approximation the normal modes are again plane waves, but normal mode amplitudes and propagation constants are vectors. The propagation vector $\mathbf{q}$ for a wave is in the direction of wave propagation and has magnitude $2\pi/\lambda$, where $\lambda$ is the wavelength. The amplitude $\mathbf{A}$ is along the direction of particle motion.

*Monatomic Crystals.*   We first consider a crystal with one atom per primitive unit cell. When the atoms vibrate in a normal mode with propagation vector $\mathbf{q}$ and frequency $\omega$, the displacement of atom j from its equilibrium position $\mathbf{R}_j$ is described by

$$\mathbf{u}_j(t) = \mathbf{A}e^{i(\mathbf{q}\cdot\mathbf{R}_j - \omega t)},\tag{6-53}$$

where $\mathbf{A}$ is the vector amplitude. When Eq. 6-53 is substituted into Eq. 6-15, the result is

$$\sum_j \sum_\beta [\Phi_{\alpha\beta}(i,j) - m\omega^2 \delta_{ij}\delta_{\alpha\beta}] A_\beta \, e^{i\mathbf{q}\cdot\mathbf{R}_j} = 0,\tag{6-54}$$

after division by $e^{-i\omega t}$. The first sum is over all atoms but, in practice, it is limited to a relatively small number of neighbors of atom $i$. The second sum is over the coordinate labels $x$, $y$, and $z$.

Two atoms with equilibrium positions $\mathbf{R}_1$ and $\mathbf{R}_1 + \mathbf{R}_s$ interact with the same force constants as two atoms with equilibrium positions $\mathbf{R}_2$ and $\mathbf{R}_2 + \mathbf{R}_s$. The force constants are identical because the atoms in the pairs have the same equilibrium separation. In general, $\Phi_{\alpha\beta}(i, j)$ depends on $\mathbf{R}_i$ and $\mathbf{R}_j$ only in the combination $\mathbf{R}_j - \mathbf{R}_i$. Let $\mathbf{R}_s = \mathbf{R}_j - \mathbf{R}_i$ and write $\Phi_{\alpha\beta}(s)$ for $\Phi_{\alpha\beta}(i, j)$. Then multiply Eq. 6-54 by $e^{-i\mathbf{q} \cdot \mathbf{R}_i}$ to obtain

$$\sum_s \sum_\beta [\Phi_{\alpha\beta}(s) - m\omega^2 \delta_{s0}\delta_{\alpha\beta}]A_\beta \, e^{i\mathbf{q} \cdot \mathbf{R}_s} = 0. \tag{6-55}$$

The quantity $D_{\alpha\beta}(\mathbf{q})$, defined by

$$D_{\alpha\beta}(\mathbf{q}) = \frac{1}{m} \sum_s \Phi_{\alpha\beta}(s) \, e^{i\mathbf{q} \cdot \mathbf{R}_s} \tag{6-56}$$

is called the dynamical matrix for the system, and in terms of it, Eq. 6-55 is

$$\sum_\beta [D_{\alpha\beta}(\mathbf{q}) - \omega^2 \delta_{\alpha\beta}]A_\beta = 0. \tag{6-57}$$

The three equations represented by Eq. 6-57, corresponding to the three possible values of $\alpha$, form a set of simultaneous equations in the three unknowns $A_x, A_y,$ and $A_z$. Nontrivial solutions exist only if the determinant formed from the coefficients of the amplitude components vanishes. That is, the amplitude of the motion vanishes unless

$$\begin{vmatrix} D_{xx} - \omega^2 & D_{xy} & D_{xz} \\ D_{yx} & D_{yy} - \omega^2 & D_{yz} \\ D_{zx} & D_{zy} & D_{zz} - \omega^2 \end{vmatrix} = 0. \tag{6-58}$$

Equation 6-58 is solved for the normal mode angular frequencies $\omega$ as functions of $\mathbf{q}$. Since the left side is a third-degree polynomial in $\omega^2$, Eq. 6-58 yields three positive values of $\omega$ for each $\mathbf{q}$.

Dispersion curves are displayed as in Fig. 6-10. The angular frequency is plotted as a function of the magnitude of $\mathbf{q}$ for a selected direction. Similar graphs are drawn for other directions of $\mathbf{q}$. The three branches correspond to the three solutions of Eq. 6-58 for each value of $\mathbf{q}$.

Once a normal mode frequency is found, the three equations represented by Eq. 6-57 can be solved for the ratios of the amplitude components: $A_y/A_x$ and $A_z/A_x$, for example. These ratios determine the direction of motion of the atoms. For $\mathbf{q}$ along one of the symmetry axes of a crystal, two of the modes correspond to transverse waves, with $\mathbf{A}$ perpendicular to $\mathbf{q}$. These modes have the same frequency and are said to be degenerate. The third normal mode corresponds to a longitudinal wave, with $\mathbf{A}$ parallel to $\mathbf{q}$. For most directions of $\mathbf{q}$, however, the three frequencies are all different and $\mathbf{A}$ is neither parallel nor perpendicular to $\mathbf{q}$, so the wave cannot be classified as longitudinal or transverse.

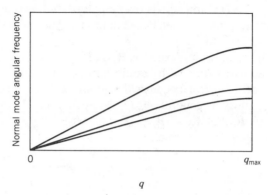

**FIGURE 6-10** Dispersion curves for a three-dimensional crystal with one atom in the primitive basis. Three modes are associated with each value of **q**. The curves are for **q** in a low symmetry direction. For **q** in a high-symmetry direction, the two lowest curves are degenerate and correspond to transverse modes while the highest curve corresponds to a longitudinal mode.

For each branch, $\omega(\mathbf{q})$ approaches zero linearly with **q** and the slopes of the dispersion curves give the speeds of sound. Near $\mathbf{q} = 0$ the waves are either longitudinal or transverse, regardless of propagation direction. Since force constants for shear are usually smaller than force constants for compression, transverse waves are slower than longitudinal waves.

*Limitations on the Propagation Vector.* We consider a monatomic sample with $N^3$ primitive unit cells, arranged so there are $N$ cells along **a**, $N$ cells along **b**, and $N$ cells along **c**. This sample is part of an infinite crystal.

The argument to show that Eq. 6-53 gives the correct form for an atomic displacement proceeds like the analogous argument for a monatomic chain. Because the crystal is invariant when it is translated by $\mathbf{R}_n = n\mathbf{a}$,

$$\mathbf{u}_n(t) = \Gamma_a^n \, \mathbf{u}_0(t), \tag{6-59}$$

where $\Gamma_a$ is a constant. Periodic boundary conditions, $\mathbf{u}_{n+N}(t) = \mathbf{u}_n(t)$, lead to the condition $\Gamma_a^N = 1$, so $\Gamma_a = e^{i2\pi\hbar/N}$, where $\hbar$ is an integer. For atoms separated by $\mathbf{R}_n = n\mathbf{b}$ and for atoms separated by $\mathbf{R}_n = n\mathbf{c}$, the displacements are related by expressions like that of Eq. 6-59, with multiplying factors given by $\Gamma_b = e^{i2\pi\ell/N}$ and $\Gamma_c = e^{i2\pi\ell/N}$, respectively. Here $\hbar$ and $\ell$ are integers. Thus the displacement of the atom at $\mathbf{R} = n_1\mathbf{a} + n_2\mathbf{b} + n_3\mathbf{c}$ is

$$\mathbf{u} = \Gamma_a^{n_1} \Gamma_b^{n_2} \Gamma_c^{n_3} \mathbf{u}_0 = e^{i2\pi(n_1\hbar + n_2\hbar + n_3\ell)/N} \mathbf{u}_0. \tag{6-60}$$

The exponential factor in Eq. 6-60 can be written in terms of **R** by using the result $\mathbf{R} \cdot (\hbar\mathbf{A} + \hbar\mathbf{B} + \ell\mathbf{C}) = 2\pi(n_1\hbar + n_2\hbar + n_3\ell)$, where **A**, **B**, and **C** are fundamental reciprocal lattice vectors. The form given by Eq. 6-53 is obtained, with

$$\mathbf{q} = \frac{\hbar}{N}\mathbf{A} + \frac{\hbar}{N}\mathbf{B} + \frac{\ell}{N}\mathbf{C}. \tag{6-61}$$

These are the allowed propagation vectors for normal mode waves in a three dimensional crystal.

Two modes with propagation vectors differing by a reciprocal lattice vector are physically identical. They have the same frequency and the same atomic displacement ratios. Consider the mode with propagation vector $\mathbf{q} + \mathbf{G}$, where $\mathbf{G}$ is any reciprocal lattice vector. Since $\mathbf{G} \cdot \mathbf{R}_n$ is a multiple of $2\pi$, the displacement of the atom in cell $n$ is

$$\mathbf{u}_n(t) = \mathbf{A}\, e^{i(\mathbf{q} \cdot \mathbf{R}_n + \mathbf{G} \cdot \mathbf{R}_n - \omega t)} = \mathbf{A}\, e^{i(\mathbf{q} \cdot \mathbf{R}_n - \omega t)}, \tag{6-62}$$

the same as the displacement for the mode with propagation vector $\mathbf{q}$. Clearly, Eq. 6-15 yields the same set of normal mode frequencies for these two propagation vectors.

In a listing of distinct normal modes, no two propagation vectors may differ by a reciprocal lattice vector. One method of assuring this is to limit $\hbar$, $\mathfrak{k}$, and $\ell$ in Eq. 6-61 to the range from $-\frac{1}{2}N$ to $+\frac{1}{2}N$. There are exactly $N$ integers in this range so there are $N^3$ distinct propagation vectors, exactly the same as the number of unit cells in the sample. Since there are three normal modes for each propagation vector and one atom for each unit cell, the number of normal modes equals the number of degrees of freedom: three times the number of atoms. A mode can be designated by giving its propagation vector and branch.

*Brillouin Zones.* A propagation vector $\mathbf{q}$ defines a point in reciprocal space, the space spanned by the reciprocal lattice. When $\hbar$, $\mathfrak{k}$, and $\ell$ in Eq. 6-61 are limited to the range from $\frac{1}{2}N$ to $+\frac{1}{2}N$, $\mathbf{q}$ is in what is known as the Brillouin zone.* Figure 6-11 shows the construction of a Brillouin zone for a two-dimensional crystal. One reciprocal lattice point is chosen as the origin and reciprocal lattice vectors are drawn from the origin to other points. Then the perpendicular bisectors of these vectors are constructed. The smallest geometrical figure in area, centered at the origin and bounded by the perpendicular bisectors, is the Brillouin zone.

A similar process is carried out for a three-dimensional crystal. Now the perpendicular bisectors are planes and the Brillouin zone is a three-dimensional region. It has the smallest volume of all figures centered at the origin and bounded by planes that perpendicularly bisect reciprocal lattice vectors. It is exactly analogous to the Wigner-Seitz unit cell of a direct lattice. Clearly, no two points within the zone are separated by a reciprocal lattice vector.

Points on a Brillouin zone surface can be located easily. In Fig. 6-12, $\mathbf{q}$ is a vector from the zone center to a point on the plane that perpendicularly bisects the reciprocal lattice vector $\mathbf{G}$. The projection of $\mathbf{q}$ on $\mathbf{G}$, given by $\mathbf{q} \cdot \mathbf{G}/|\mathbf{G}|$, must be half the magnitude of $\mathbf{G}$, so $\mathbf{q} \cdot \mathbf{G}/|\mathbf{G}| = \frac{1}{2}|\mathbf{G}|$ or

$$\mathbf{q} \cdot \mathbf{G} - \tfrac{1}{2}|\mathbf{G}|^2 = 0. \tag{6-63}$$

---

*This region is sometimes designated the *first* Brillouin zone to distinguish it from other regions of reciprocal space, called the second and higher Brillouin zones. We will not explicitly use zones other than the first.

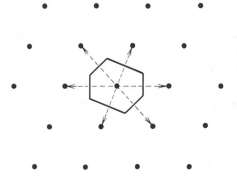

**FIGURE 6-11** Construction of a Brillouin zone for a two-dimensional lattice. The dots represent reciprocal lattice points and the dotted lines are reciprocal lattice vectors from the central point to neighboring points. The zone is the smallest figure enclosed by the perpendicular bisectors of the vectors. For a three-dimensional lattice, the perpendicular bisectors are planes and the zone is a three-dimensional figure with the smallest volume bounded by them.

To find the zone boundary, first pick a direction for **q**. Then, for each reciprocal lattice vector, solve Eq. 6-63 for the magnitude of **q**. The smallest magnitude obtained gives a point on the zone surface and the vector **G** used to obtain the point gives the normal to the surface. In practice, Eq. 6-63 need be solved for only a few reciprocal lattice vectors, the shortest ones nearby **q** in direction.

Brillouin zones for cubic lattices are shown in Fig. 6-13. The zone for a simple cubic direct lattice with cube edge $a$ is a cube with edge $2\pi/a$. For a body-centered cubic direct lattice, the reciprocal lattice is face-centered cubic. The zone has 12 faces, each of which bisects one of the lines from a cube corner to a neighboring face center. For a face-centered cubic direct lattice, the reciprocal lattice is body-centered cubic. Each of the hexagonal faces bisects a line from the cube center to a cube corner and each of the square faces bisects a line that joins the body center to body centers of adjacent cubes.

The Brillouin zone for a given crystal depends only on the lattice, not on the basis. Cesium chloride has a simple cubic direct lattice and so has a simple cubic Brillouin zone. Both sodium chloride and copper have Brillouin zones appropriate for a face-centered cubic direct lattice.

**EXAMPLE 6-4** (a) Draw the Brillouin zone for a two-dimensional hexagonal lattice with hexagon edge $a$. (b) What is the length of the largest propagation vector to the zone?

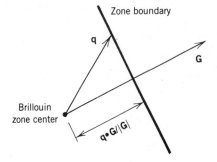

**FIGURE 6-12** A Brillouin zone boundary bisects the reciprocal lattice vector **G**. The projection on **G** of a vector **q**, from the center of the zone to the boundary, is $\mathbf{q} \cdot \mathbf{G}/|\mathbf{G}|$. This must be half the magnitude of **G**, so $\mathbf{q} \cdot \mathbf{G} = \frac{1}{2}|\mathbf{G}|^2$.

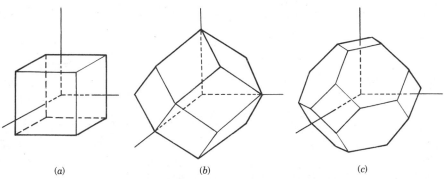

**FIGURE 6-13** Brillouin zones for cubic direct lattices: (a) simple cubic, (b) body-centered cubic, and (c) face-centered cubic. Each plane in (b) perpendicularly bisects a reciprocal lattice vector from a cube corner to one of the 12 nearest face centers. In (c) the square faces are perpendicular bisectors of vectors that join a cube center to the six nearest cube centers. The hexagonal faces are perpendicular bisectors of vectors from a cube center to the eight nearest cube corners.

**SOLUTION** (a) Orient a Cartesian coordinate system as shown in Fig. 6-14a. Then $\mathbf{a} = \frac{1}{2}a(\hat{x} + \sqrt{3}\hat{y})$ and $\mathbf{b} = a\hat{x}$. $\mathbf{A} = (4\pi/\sqrt{3}a)\hat{y}$ and $\mathbf{B} = (2\pi/\sqrt{3}a)(\sqrt{3}\hat{x} - \hat{y})$ are fundamental reciprocal lattice vectors. Reciprocal lattice points are drawn in (b) and $O$ is taken as the origin. The perpendicular bisectors of six reciprocal lattice vectors, form $O$ to other reciprocal lattice points, form the Brillouin zone. No other perpendicular bisectors lie inside this region. The zone is hexagonal with edge $4\pi/3a$. (b) Six vectors from the origin to the boundary of the zone are larger than any others. One of

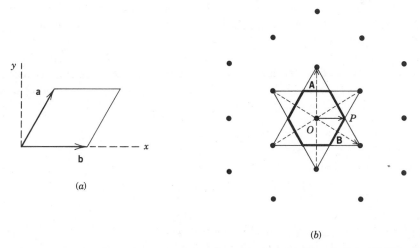

**FIGURE 6-14** (a) The unit cell of a two-dimensional hexagonal direct lattice. (b) The corresponding reciprocal lattice. $O$ is the origin and dotted lines connect it to nearby reciprocal lattice points. Perpendicular bisectors form the Brillouin zone, shown by heavy lines. $OP$ is the longest propagation vector in the zone.

them, for example, is from $O$ to $P$ on the diagram. They all have length $4\pi/3a$, the same as a hexagon edge. ∎

**EXAMPLE 6-5**  For a face-centered cubic direct lattice find the length of the largest propagation vector in the Brillouin zone that is (a) parallel to a cube edge; (b) parallel to a cube body diagonal; (c) parallel to a cube face diagonal.

**SOLUTION**  Place a coordinate system with axes along three mutually perpendicular cube edges and take the fundamental vectors of the direct lattice to be $a\hat{x}$, $a\hat{y}$, and $a\hat{z}$. Then $\mathbf{A} = (2\pi/a)(\hat{x} + \hat{y} + \hat{z})$, $\mathbf{B} = (2\pi/a)(\hat{x} + \hat{y} - \hat{z})$, and $\mathbf{C} = (2\pi/a)(\hat{x} - \hat{y} + \hat{z})$ are fundamental vectors of the reciprocal lattice. (a) Take $\mathbf{q} = q\hat{x}$. As shown in Fig. 6-13, the $x$ axis intersects the square face of the Brillouin zone that perpendicularly bisects $\mathbf{G} = \mathbf{B} + \mathbf{C} = (4\pi/a)\hat{x}$. Thus $\mathbf{q} \cdot \mathbf{G} = 4\pi q/a$, $|\mathbf{G}|^2 = 16\pi^2/a^2$, and Eq. 6-63 becomes $(4\pi q/a) = 8\pi^2/a^2$, so $q = 2\pi/a$. (b) Take $\mathbf{q} = (q/\sqrt{3})(\hat{x} + \hat{y} + \hat{z})$. A line in this direction intersects the hexagonal zone face that bisects $\mathbf{G} = \mathbf{A}$. Thus $\mathbf{q} \cdot \mathbf{G} = 2\pi\sqrt{3}q/a$ and $|\mathbf{G}|^2 = 12\pi^2/a^2$, so $q = \sqrt{3}\pi/a$. (c) Take $\mathbf{q} = (q/\sqrt{2})(\hat{x} + \hat{y})$. A line in this direction meets the zone boundary at the midpoint of the intersection of two hexagonal faces. We may take $\mathbf{G}$ to be $\mathbf{A}$ or $\mathbf{B}$. In either case, $\mathbf{q} \cdot \mathbf{G} = (2\pi\sqrt{2}q/a)$ and $|\mathbf{G}|^2 = 12\pi^2/a^2$, so $q = 3\pi/\sqrt{2}a$. ∎

*Symmetries of Dispersion Relationships.*   $\omega(\mathbf{q})$ is invariant under all the symmetry operations of the point group of the crystal. To be more specific, suppose a symmetry operation of the point group is applied to the propagation vector $\mathbf{q}$ and the result is the vector $\mathbf{q}'$. Then $\omega(\mathbf{q}') = \omega(\mathbf{q})$. This is easy to understand. Consider two crystals which are exactly the same except that the second is rotated relative to the first. If $\mathbf{q}$ rotates to $\mathbf{q}'$, then $\omega(\mathbf{q}')$ for the second crystal is exactly the same as $\omega(\mathbf{q})$ for the first. Now suppose the rotation is one of the symmetry elements of the crystal. Then the second crystal is precisely the same as the first and $\omega(\mathbf{q}')$ for the second must be the same as $\omega(\mathbf{q}')$ for the first. Thus $\omega(\mathbf{q}) = \omega(\mathbf{q}')$ for either crystal. Note that $\omega(\mathbf{q})$ has the symmetry of the crystal, not the lattice alone.

Even if the crystal does not have inversion symmetry, $\omega(-\mathbf{q}) = \omega(\mathbf{q})$. If $t$ is replaced by $-t$ in Eq. 6-13, the equation remains the same. Thus $\mathbf{u}_j = Ae^{i(\mathbf{q} \cdot \mathbf{R}_j + \omega t)}$ is a normal mode displacement if $\mathbf{u}_j = Ae^{i(\mathbf{q} \cdot \mathbf{R}_j - \omega t)}$ is. The real part of the former expression is identical to the real part of $Ae^{i(-\mathbf{q} \cdot \mathbf{R} - \omega t)}$, the displacement for a normal mode with propagation vector $-\mathbf{q}$.

Advantage is taken of these symmetry properties when normal mode frequencies are calculated and plotted. Equation 6-57 need not be solved for both $\omega(\mathbf{q})$ and $\omega(\mathbf{q}')$ if $\mathbf{q}$ and $\mathbf{q}'$ are related by a symmetry operation of the crystal point group or if $\mathbf{q}' = -\mathbf{q}$. Equation 6-57 is solved only for propagation vectors in a sector of the zone, chosen so that each vector outside the sector is related to a vector inside by a symmetry operation of the point group. Considerable effort is saved if the point group is large.

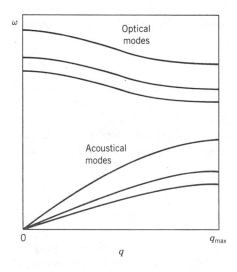

**FIGURE 6-15** Schematic representation of vibration dispersion curves for a three-dimensional crystal with two atoms in its primitive basis. Three acoustical modes and three optical modes are associated with each propagation vector. If $\mathbf{q}$ is along a high-symmetry direction the two lowest curves of each set are identical.

*Crystals with Multiatom Bases.* If $\mathbf{R}_n + \mathbf{p}_i$ is the equilibrium position of atom $i$ in cell $n$ and $\mathbf{u}_{ni}$ is the displacement of that atom from its equilibrium position, then

$$\mathbf{u}_{ni}(t) = \mathbf{A}_i\, e^{i[\mathbf{q}\cdot(\mathbf{R}_n+\mathbf{p}_i)-\omega t]} \tag{6-64}$$

for the normal mode with propagation vector $\mathbf{q}$ and angular frequency $\omega$. The subscript $i$ on the amplitude denotes an atom of the basis, not a cell. For any normal mode, the vibration amplitudes of different atoms in the basis are usually different, just as they are for the one dimensional diatomic chain of Section 6.3. Periodic boundary conditions again lead to propagation vectors given by Eq. 6-61. Atomic displacements are identical for modes with propagation vectors $\mathbf{q}$ and $\mathbf{q} + \mathbf{G}$, where $\mathbf{G}$ is a reciprocal lattice vector, so $\mathbf{q}$ may be restricted to the Brillouin zone.

Normal mode frequencies are found by substituting Eq. 6-64 into the equations of motion for the atoms and setting the secular determinant equal to zero. There are three equations for each atom in the basis, one for each Cartesian component of the amplitude. Although the idea is simple, the analysis is necessarily quite complicated, so we will simply state the results rather than give the details.

If there are $p$ atoms in the basis, the secular determinant is a polynomial of order $3p$ in $\omega^2$, so $3p$ normal modes are associated with each propagation vector $\mathbf{q}$ and the dispersion curve has $3p$ branches. Three of these are acoustical branches, for which the frequency tends to zero with the propagation vector. The other $3p - 3$ are optical branches, for which it does not. The various branches are shown schematically in Fig. 6-15.

*Phonons.* The energy of a harmonic oscillator is quantized[*]: it has one of the values $(n + \tfrac{1}{2})\hbar\omega$, where $n$ is zero or a positive integer and $\omega$ is the angular

[*]See, for example, R. Eisberg and R. Resnick, *Quantum Physics* (New York: Wiley, 1985).

frequency of vibration. No other value is possible. In a similar manner, normal mode energies of solids are also quantized and

$$E = [n + \tfrac{1}{2}]\hbar\omega \tag{6-65}$$

gives the energy of a mode with angular frequency $\omega$. Here the energy is measured relative to the energy of a rigid translation, for which the angular frequency vanishes. Each atom always has some vibrational energy: a normal mode with angular frequency $\omega$ has an energy of least $\tfrac{1}{2}\hbar\omega$, called the zero-point energy.

Because possible energy levels of a mode with angular frequency $\omega$ are equally spaced with an interval of $\hbar\omega$, a particle interpretation can be given normal mode energies. The particles are called phonons and each phonon associated with a mode has energy $\hbar\omega$. In Eq. 6-65, $n$ gives the number of phonons and $n\hbar\omega$ gives the total phonon energy of a mode with angular frequency $\omega$. Note that all phonons associated with the same mode have the same energy but phonons associated with modes of different frequency have different energies.

The momentum $\mathbf{p}$ of a free electron is related to the propagation vector of its wave by $\mathbf{p} = \hbar\mathbf{k}$. By analogy, $\hbar\mathbf{q}$ is called the crystal momentum of a phonon with propagation vector $\mathbf{q}$. It is not the momentum of the vibrating atoms, which is zero. It does, however, enter into conservation laws in a manner similar to a true momentum. When we study inelastic neutron scattering, for example, we will see that the sum of the neutron momentum and the phonon crystal momentum is conserved during interactions of neutrons with atoms of the solid.

## 6.5 SURFACE VIBRATIONS

Normal modes of vibration are also associated with sample surfaces. Force constants linking neighbors near the surface are different from those linking neighbors in the interior. In addition, impurity atoms may be adsorbed on the surface and these have masses and force constants that differ from those associated with host atoms. As a consequence, we expect vibration frequencies of surface modes to be different from those of bulk modes.

Qualitative understanding of surface related modes can be obtained from a simple model, called the Rosenzweig model.* A (001) surface of a simple cubic structure is considered. Equilibrium positions on the surface form a square lattice with edge $a$, where $a$ is the edge of a primitive unit cell of the bulk material. Nearest-neighbor interactions are described by the force constant $\gamma$, second-neighbor interactions by the force constant $\tfrac{1}{2}\gamma$, and other interactions are neglected. Nearest-neighbor forces have noncentral parts and second neighbors are assumed to interact only if the components of their displacements along the line of wave propagation differ.

To find dispersion relations for bulk modes, periodic boundary conditions are imposed and the normal mode problem is solved. For waves traveling parallel

---

*Details of this model can be found in G. Armand and P. Masri, *J. Vacuum Sci. Technol.* **9**:705, 1971.

to a (010) plane, normal mode frequencies are

$$\omega_1^2 = \omega_2^2 = 4(\gamma/M)[2 - \cos(q_x a) - \cos(q_z a)] \qquad (6\text{-}66)$$

and

$$\omega_3^2 = 4(\gamma/M)[4 - \cos(q_x a) - \cos(q_z a) - 2\cos(q_x a)\cos(q_z a)], \quad (6\text{-}67)$$

where the first equation holds for the two transverse branches and the second holds for the longitudinal branch. $M$ is the mass of an atom. Angular frequencies are plotted in Fig. 6-16 as functions of $q_x$. Shaded regions mark all angular frequencies given by Eqs. 6-66 and 6-67 as $q_x$ and $q_z$ take on all allowed values. Angular frequencies for transverse modes lie between the curves labeled $\omega_1(0)$ and $\omega_1(\pi)$, while angular frequencies for longitudinal modes lie between curves labeled $\omega_3(0)$ and $\omega_3(\pi)$. The arguments are limiting values of $q_z a$. The region labeled *ABC* represents a gap in the bulk normal mode spectrum.

When the surface is taken into account, bulk mode frequencies change slightly, but more importantly for us, new modes appear. They are waves traveling parallel to the surface and are characterized by the vibration of only those atoms within a few atomic layers of the surface. The amplitudes of these waves decrease in a nearly exponential fashion with distance from the surface.

Atomic positions are periodic in the $xy$ plane so $q_x$ and $q_y$ still represent components of a propagation vector for surface waves. The ratio of displacements for two atoms separated by na along the $x$ axis, for example, is $e^{iq_x na}$. For surface waves, dispersion relations are found as functions of these two com-

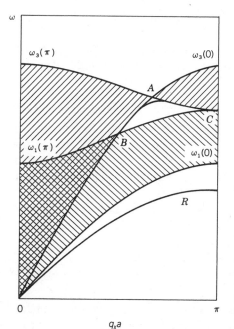

**FIGURE 6-16** Normal mode angular frequencies for a cubic crystal with a (001) surface. Bulk modes are represented by the shaded regions. The curves labeled $\omega_1(0)$ and $\omega_1(\pi)$ mark the limits of transverse bulk modes while those labeled $\omega_3(0)$ and $\omega_3(\pi)$ mark the limits of longitudinal bulk modes. No bulk modes correspond to the region *ABC*. *R* is the dispersion curve for Rayleigh surface waves, just below the curve for transverse bulk modes. Some surface waves have frequencies in the gap, just below *A*. (From G. Armand, L. Dobrzynski, and P. Masri, *J. Vacuum Sci. Technol.* **9**: 705, 1971. Used with permission.)

ponents and $\omega$ is plotted as a function of the magnitude of $\mathbf{q}$ for some direction parallel to the surface. In Fig. 6-16 $\omega$ has been plotted for $\mathbf{q}$ in the $x$ direction.

Some surface waves have frequencies that are nearly identical to those of some bulk modes. Others have frequencies in gaps of the bulk spectrum. In Fig. 6-16, for example, a series of modes appears in the forbidden region denoted by *ABC*. In addition, a new branch to the spectrum, consisting of what are known as Rayleigh waves, appears below the transverse branches. Near $\mathbf{q} = 0$, this dispersion curve has nearly the same slope as a transverse branch, but near the Brillouin zone boundary it bends more sharply.

The extent of the forbidden gap, the frequencies of the surface modes, and the speed of Rayleigh waves are all sensitive to the force constants and masses used in the calculation and are different for different materials. Nevertheless, the results are valid in a qualitative fashion. That is, we expect all samples to show evidence of surface modes with frequencies in forbidden gaps and below the acoustical branch.

## 6.6 INELASTIC NEUTRON SCATTERING

In Chapter 4 we discussed the scattering of waves from stationary atoms. When the atoms are in motion, the scattered intensity pattern is different and the difference allows us to experimentally determine vibrational dispersion curves. Since neutrons make better probes of the bulk vibration spectrum than x rays or electrons, we will discuss scattering in terms of those particles. The general principles, however, hold for other incident particles. Inelastic low-energy electron scattering is used to study surface modes, for example.

*Conservation Laws.* Aside from elastic scattering events, the most likely interaction between a neutron and the atoms of a solid is one in which a single quantum of energy is exchanged with one of the normal vibrational modes. If the normal mode has angular frequency $\omega$, then the law of conservation of energy is

$$E' - E = \pm\hbar\omega, \tag{6-68}$$

where $E$ and $E'$ are the energies of the neutron before and after scattering, respectively. The plus sign is used if the neutron absorbs a phonon and the minus sign is used if the neutron emits a phonon. Since neutrons with energies on the order of phonon energies are used, the fractional change in energy on scattering is large and easily detected.

To derive the relationship between wave vectors, we consider a crystal with a primitive basis of one atom. According to Eq. 4-12, the scattered neutron wave is then proportional to

$$S = \sum_j e^{-i\Delta\mathbf{s}\cdot\mathbf{r}_j}, \tag{6-69}$$

where the sum is over all atoms, assumed to have the same atomic form factor. Atom $j$ is at $\mathbf{r}_j$ and $\Delta\mathbf{s} = \mathbf{s}' - \mathbf{s}$ is the change on scattering of the neutron propagation vector. We used an expression like this to discuss scattering from stationary atoms. Now, however, we take $\mathbf{r}_j = \mathbf{R}_j + \mathbf{u}_j(t)$, where $\mathbf{R}_j$ is a lattice vector and $\mathbf{u}_j$ is the displacement of atom $j$ from its lattice site. Since $\mathbf{u}_j$ is small, we may approximate $e^{-i\Delta s \cdot \mathbf{u}_j}$ by $1 - i\Delta s \cdot \mathbf{u}_j$ and write

$$S = \sum_j e^{-i\Delta s \cdot \mathbf{R}_j} [1 - i\Delta s \cdot \mathbf{u}_j]. \tag{6-70}$$

Suppose the atoms are vibrating in a normal mode with angular frequency $\omega$ and wave vector $\mathbf{q}$. Since the displacement is real we take $\mathbf{u}_j$ to be sum of Eq. 6-53 and its complex conjugate. We also assume the amplitude $\mathbf{A}$ is real and write

$$\mathbf{u}_j = \mathbf{A}[e^{i(\mathbf{q} \cdot \mathbf{R}_j - \omega t)} + e^{-i(\mathbf{q} \cdot \mathbf{R}_j - \omega t)}]. \tag{6-71}$$

When Eq. 6-71 is substituted into Eq. 6-70, the result is

$$S = \sum_j e^{-i\Delta s \cdot \mathbf{R}_j} - i\,\Delta s \cdot \mathbf{A}\, e^{-i\omega t} \sum_j e^{i(\mathbf{q} - \Delta s) \cdot \mathbf{R}_j}$$
$$- i\,\Delta s \cdot \mathbf{A}\, e^{i\omega t} \sum_j e^{-i(\mathbf{q} + \Delta s) \cdot \mathbf{R}_j}. \tag{6-72}$$

A sum of the form $\sum e^{i\alpha \cdot \mathbf{R}_j}$ has a large magnitude only if $\alpha$ is a reciprocal lattice vector. The first sum in Eq. 6-72 is large if $\Delta s = \mathbf{G}$, where $\mathbf{G}$ is a reciprocal lattice vector. This is the condition for elastic scattering. The second sum is large if $\Delta s = \mathbf{q} + \mathbf{G}$, corresponding to the absorption of a phonon by the neutron. The third is large if $\Delta s = -\mathbf{q} + \mathbf{G}$, corresponding to phonon emission. Thus

$$\Delta s = \pm\mathbf{q} + \mathbf{G}, \tag{6-73}$$

for inelastic scattering involving a normal mode with propagation vector $\mathbf{q}$.

Equation 6-73 replaces the usual conservation of momentum law for interactions of particles. The quantities $\hbar s$ and $\hbar s'$ give the momentum of the neutron before and after scattering, respectively, so $\hbar \Delta s$ is the change in neutron momentum. The quantity $\hbar \mathbf{q}$ is the phonon crystal momentum. According to Eq. 6-73, the sum of neutron momentum and phonon crystal momentum may not be conserved: it may change by $\hbar \mathbf{G}$. This term insures $\mathbf{q}$ is in the Brillouin zone.

*Analysis of Data.* Triple-axis neutron spectrometers, like that diagrammed in Fig. 6-17, are widely used for inelastic scattering experiments. The incident beam, usually from a nuclear reactor, contains particles with a wide variety of energies. Elastic scattering from crystal A is used to create a monochromatic beam and, by varing the orientation of this crystal, particles with a given energy are selected. The collimated monochromatic beam is then scattered from the sample and the emerging beam strikes an analyzing crystal, where it is again elastically scattered. Finally, the beam enters a detector, where the number of neutrons is counted. Neutrons with different wavelengths and hence different energies are detected as the angle $\theta_B$ is changed. The analyzing crystal and detector can be moved as

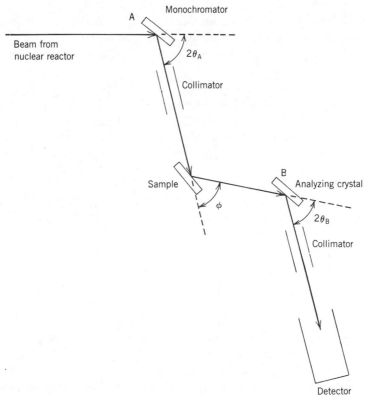

**FIGURE 6-17** The geometry of a triple axis neutron spectrometer. Incident particles with a particular energy are selected by fixing $\theta_A$, while scattered particles with a particular energy are selected for detection by fixing $\theta_B$. For a given scattering angle $\phi$, $\theta_B$ is varied until an inelastic peak occurs in the counting rate. $\phi$ is varied to sample interactions with many normal modes.

a unit in an arc around the sample so neutrons scattered at various angles can be detected.

The orientation of crystal A determines the propagation vector of the incident wave. The angle $2\theta_A$ is measured and, for first order diffraction, $\lambda = 2d \sin \theta_A$, where $d$ is a lattice plane spacing. The magnitude of the propagation vector is found using $s = 2\pi/\lambda$. Its direction, of course, is along the beam. Once $\lambda$ is known the energy of an incident neutron can be calculated.

For a given scattering angle $\phi$, the number of neutrons received by the detector per unit time is measured as a function of the angle $\theta_B$. Neutrons are detected for every angle but, for most angles, these have been scattered in multiphonon events and their number is small. A peak in the counting rate is obtained when conditions for single phonon scattering are met for a normal mode. The value of $\theta_B$ at a peak is used to compute the energy and propagation vector of the scattered neutrons. The sign of $E' - E$ is used to decide if a phonon was absorbed or emitted, then Eq. 6-68 is used to find $\omega$ and Eq. 6-73 is used to find $\mathbf{q} + \mathbf{G}$ or $-\mathbf{q} + \mathbf{G}$, as appropriate.

To proceed, reciprocal lattice vectors for the sample must be known. They are determined prior to the inelastic scattering experiment, perhaps by analysis of the pattern for elastic x-ray scattering. Given $\mathbf{q} + \mathbf{G}$ or $-\mathbf{q} + \mathbf{G}$, a reciprocal lattice vector $\mathbf{G}$ is chosen so that $\mathbf{q}$ lies in the Brillouin zone. Both the frequency and propagation vector for a normal mode are then known. To investigate other modes, the experiment is repeated at other scattering angles.

Inelastic neutron scattering can also be used to study normal modes of amorphous solids. Although wave vectors are not associated with normal mode vibrations, peaks occur in the counting rate when the change in neutron energy matches a phonon energy. The vibration spectrum is found by searching out inelastic scattering peaks.

## 6.7  ELASTIC CONSTANTS

The harmonic nature of interatomic forces for small atomic displacements has implications for the macroscopic behavior of solids: the compression or shear of a solid is proportional to the force applied to its surface, provided the elastic limit is not exceeded. The force per unit area is described by a stress tensor and the extent to which a solid distorts is described by a strain matrix. These two matrices are proportional to each other, the constants of proportionality being the elastic constants. Important mechanical properties of a solid, such as its compressibility, depend on these constants.

*Stress and Strain.*  Figure 6-18a shows a portion of a solid, sufficiently large that it contains many primitive unit cells but sufficiently small that we may treat

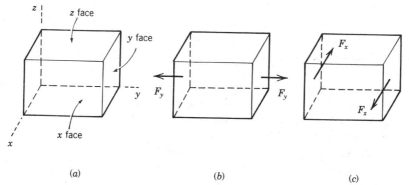

(a)  (b)  (c)

**FIGURE 6-18**  (a) The coordinate system used to describe stress. Each face of the sample is labeled by the axis perpendicular to it. (b) The stress element $T_{yy}$: forces along the y axis are applied to y faces and $T_{yy} = F_y/A_y$, where $A_y$ is the face area. If $T_{yy}$ is positive an elongation is produced. (c) The stress element $T_{xy}$: forces parallel to the x axis are applied to the y faces and $T_{xy} = F_x/A_y$. A shear is produced. Rotation is prevented by application of the stress $T_{yx} = T_{xy}$.

it as infinitesimal as far as macroscopic phenomena are concerned. A coordinate system is shown and each face is labeled according to the coordinate axis perpendicular to it. For example, areas of front and back faces are denoted by $A_x$, areas of left and right faces by $A_y$, and areas of top and bottom faces by $A_z$. An area is taken to be positive if the outward normal unit vector is in the positive direction along the appropriate axis and negative if it is in the negative direction.

To stress a solid, forces are applied in such a way that both the net force and net torque vanish. Two possibilities are shown in Fig. 6-18b and c. In the first case, forces that are equal in magnitude but opposite in direction are applied perpendicularly to opposite faces, tending to elongate the sample. In the second case, equal and opposite forces are applied parallel to opposite faces, tending to shear it. If one of the faces is part of the sample surface, stresses on that face are applied by means of an external agent. If the face is an imaginary boundary between two parts of the sample, stresses are applied by neighboring portions of the sample.

Stress is described by means of a second rank tensor $T_{\alpha\beta}$. The first index gives the direction of the force and the second labels the surface to which it is applied. The magnitude of a stress element is the ratio of the force to the area of the surface. For example, in Fig. 6-18b the stress is $T_{yy} = F_y/A_y$ and in (c) the stress is $T_{xy} = F_x/A_y$. Other elements of the stress tensor are defined in a similar way. Diagonal elements of the stress tensor are positive if they tend to elongate the sample and negative if they tend to compress it. A negative diagonal element represents pressure. Note that the stress $T_{xy}$, acting alone, represents a net torque on the sample. This torque is exactly canceled if stress $T_{yx}$ also acts and $T_{yx} = T_{xy}$. We always assume $T_{\alpha\beta} = T_{\beta\alpha}$.

Deformations of a solid are described by a strain matrix, with elements $e_{\alpha\beta}$. Suppose $\mathbf{r}$ gives the position of some matter in the unstrained sample of Fig. 6-18 and suppose further that the matter moves to $\mathbf{r}' = \mathbf{r} + \mathbf{u}$ when the sample is strained. Different parts of a strained solid are displaced differently, so the displacement $\mathbf{u}$ is a function of the original position of the matter. Its derivatives with respect to the original coordinates form the elements of the strain matrix. In particular, $e_{\alpha\alpha} = \partial u_\alpha/\partial r_\alpha$, with $\alpha = x, y,$ or $z$. Off-diagonal elements are defined by $e_{\alpha\beta} = (\partial u_\alpha/\partial r_\beta) + (\partial u_\beta/\partial r_\alpha)$. Diagonal elements $e_{xx}$, $e_{yy}$, and $e_{zz}$ describe a compression or elongation of the solid, while the other elements describe shears. The definition assures that $e_{\alpha\beta} = e_{\beta\alpha}$.

For uniform compression along the $x$ axis, $e_{xx}$ is a constant and the point originally at $x, y, z$ moves to $x(1 + e_{xx}), y, z$. If, on the other hand, $e_{xy}$ is constant and all other strain elements vanish, the material has a uniform shear and the point originally at $x, y, z$ moves to $x + \frac{1}{2}e_{xy}y, y + \frac{1}{2}e_{xy}x$. These distortions are illustrated in Fig. 6-19.

When a solid is strained its volume may change. Suppose the dimensions of the sample were originally $\ell_x$, $\ell_y$, and $\ell_z$, along the $x, y,$ and $z$ axes, respectively. If the sample is uniformly strained, its dimensions become $\ell_x(1 + e_{xx})$, $\ell_y(1 + e_{yy})$, and $\ell_z(1 + e_{zz})$, respectively. The volume of the strained sample is $\ell_x\ell_y\ell_z(1 + e_{xx})(1 + e_{yy})(1 + e_{zz})$, which may be approximated by $\ell_x\ell_y\ell_z(1 + e_{xx} + e_{yy} + e_{zz})$ to first order in the strain. The fractional change of volume on

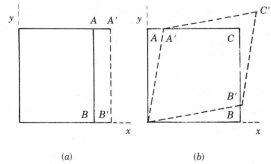

**FIGURE 6-19** The $xy$ plane in a uniformly strained solid. In (a) $e_{xx}$ is the only nonvanishing strain element, while in (b) $e_{xy}$ is the only nonvanishing element. When the solid is strained, point $A$ goes to $A'$, $B$ goes to $B'$, and $C$ goes to $C'$.

straining is $\delta\tau_s/\tau_s = e_{xx} + e_{yy} + e_{zz}$. Pure shear deformations do not change the volume in first order.

Although stress and strain matrices each have nine elements, the conditions $T_{\alpha\beta} = T_{\beta\alpha}$ and $e_{\alpha\beta} = e_{\beta\alpha}$ reduce the number of independent elements in each to six. To simplify the notation they are often numbered according to the scheme $xx \to 1$, $yy \to 2$, $zz \to 3$, $yz \to 4$, $xz \to 5$, and $xy \to 6$. The fractional volume change, for example, is $e_1 + e_2 + e_3$.

Hooke's law, which is valid as long as the elastic limit is not exceeded, gives the relationship between stress and strain. In the notation described in the last paragraph,

$$T_i = \sum_{j=1}^{6} c_{ij} e_j, \tag{6-74}$$

where $c_{ij}$ are the elastic constants of the solid. There are 36 elastic constants, but energy considerations can be used to show that $c_{ij} = c_{ji}$, so only 21 are independent.

If a solid is crystalline with symmetry axes or planes, the number of independent elastic constants is less than 21. For example, if the crystal has cubic symmetry, $c_{11} = c_{22} = c_{33}$, $c_{12} = c_{13} = c_{23}$, $c_{44} = c_{55} = c_{66}$, and all other elements vanish. We may take the three independent elastic constants for such a crystal to be $c_{11}$, $c_{12}$, and $c_{44}$.

The equality of $c_{11}$, $c_{22}$, and $c_{33}$ follows immediately from the equivalence of the [100], [010], and [001] directions for a cube. To show that $c_{15} = 0$, we suppose the crystal is deformed so that only $e_5$ does not vanish. Then $T_1 = c_{15}e_5$. Now rotate the crystal through 180° about the $z$ axis. Because $r_x$ and $u_x$ change sign while $r_z$ and $u_z$ do not, $e_5$ changes sign. The sign of $T_1$, on the other hand, does not change. Since the crystal is cubic, $c_{15}$ does not change either, so it must vanish. Similar arguments can be given to demonstrate the other equalities given above.

All directions are equivalent in amorphous and polycrystalline materials. Symmetry arguments can be used to show that all the equalities which hold for cubic crystals also hold for these materials and, in addition, $c_{11} = c_{12} + 2c_{44}$. For these materials $c_{12}$ and $c_{44}$ are usually denoted by $\lambda$ and $\mu$, respectively, and are called Lamé's constants. In terms of them, $c_{11} = \lambda + 2\mu$.

**EXAMPLE 6-6** Young's modulus is often used to characterize the elasticity of an isotropic sample, such as an amorphous or polycrystalline solid. The sample is subjected to uniaxial stress, with forces applied normally to one pair of opposite surfaces, and the fractional elongation or compression of the sample dimension along the line of the force is measured. The modulus is the ratio of the applied stress to the fractional change in dimension. That is $Y = T/e_1$, where $T$ is the stress for compressional forces in the $x$ direction. Find an expression for Young's modulus in terms of Lamé's constants.

**SOLUTION** Let $T_1 = T$ and take all other stress elements to be 0. As $i$ successively takes on the values 1 through 6, Eq. 6-74 yields $T = (\lambda + 2\mu)e_1 + \lambda e_2 + \lambda e_3$, $0 = \lambda e_1 + (\lambda + 2\mu)e_2 + \lambda e_3$, $0 = \lambda e_1 + \lambda e_2 + (\lambda + 2\mu)e_3$, $0 = \mu e_4$, $0 = \mu e_5$, and $0 = \mu e_6$. These are solved simultaneously for the strain elements, with the result $e_1 = (\lambda + \mu)T/[\mu(3\lambda + 2\mu)]$, $e_2 = e_3 = -\lambda T/[2\mu(3\lambda + 2\mu)(\lambda + \mu)]$, $e_4 = e_5 = e_6 = 0$. From the first of these, $Y = \mu(3\lambda + 2\mu)/(\lambda + \mu)$. ∎

When a uniaxial stress is applied, sample dimensions perpendicular to the direction of the force also change. For the situation described in the example, the quantity $\sigma = -e_2/e_1$ is a measure of this change. It is called Poisson's ratio and, in terms of Lamé's constants, is given by $\lambda/[2(\lambda + \mu)]$.

*Compressibility.* When uniform pressure is applied to a sample its volume decreases. The compressibility, defined by $\kappa = -(1/\tau_s)(\partial\tau_s/\partial P)_T$, measures the fractional change in sample volume. Here $P$ is the pressure and $T$ is the temperature. In Chapter 5 we evaluated this quantity for ionic and van der Waals solids in terms of forces between atoms. We now take a more macroscopic view and evaluate it in terms of the elastic constants.

If the pressure is not too great we may approximate the compressibility by $\kappa = -(1/P)(\delta\tau_s/\tau_s)$. Uniform pressure is described by $T_1 = T_2 = T_3 = -P$ and $T_4 = T_5 = T_6 = 0$. For a cubic crystal Eq. 6-74 yields

$$-P = c_{11}e_1 + c_{12}e_2 + c_{12}e_3, \tag{6-75}$$

$$-P = c_{12}e_1 + c_{11}e_2 + c_{12}e_3, \tag{6-76}$$

$$-P = c_{12}e_1 + c_{12}e_2 + c_{11}e_3, \tag{6-77}$$

$$0 = c_{44}e_4, \tag{6-78}$$

$$0 = c_{44}e_5, \tag{6-79}$$

and

$$0 = c_{44}e_6 \qquad (6\text{-}80)$$

as $i$ successively takes on the values 1 through 6. The sum of the first three relationships gives $-3P = (c_{11} + 2c_{12})(e_1 + e_2 + e_3)$, so the fractional change in volume is $\delta\tau_s/\tau_s = e_1 + e_2 + e_3 = -3P/(c_{11} + 2c_{12})$ and

$$\kappa = \tfrac{1}{3}(c_{11} + 2c_{12}). \qquad (6\text{-}81)$$

For an isotropic solid $c_{11} = \lambda + 2\mu$ and $c_{12} = \lambda$, so $\kappa = 3/(3\lambda + 2\mu)$. Many handbooks list the bulk modulus $1/\kappa$.

## 6.8   REFERENCES

For excellent introductions to the subject of atomic vibrations, see:

L. Brillouin, *Wave Propagation in Periodic Structures* (New York: Dover, 1953).
A. K. Ghatak and L. S. Kothari, *Lattice Dynamics* (Reading, MA: Addison-Wesley, 1972).

For more advanced treatments, see:

M. Born and K. Huang, *Dynamical Theory of Crystal Lattices* (London: Oxford Univ. Press, 1954).
J. T. Devreese, V. E. Van Doren, and P. E. Van Camp (Eds.), *Ab Initio Calculation of Phonon Spectra* (New York: Plenum, 1983).
A. A. Maradudindin, E. W. Montroll, G. H. Weiss, and I. P. Ipatova, *Theory of Lattice Dynamics in the Harmonic Approximation* (New York: Academic, 1963).
S. S. Mitra, "Vibrational Spectra of Solids" in *Solid State Physics* (F. Seitz and D. Turnbull, Eds.), Vol. 13, p. 1 (New York: Academic, 1962).

Vibrations in amorphous materials are discussed in:

M. F. Thorpe, "Phonons in Amorphous Soilds" in *Physics of Structurally Disordered Solids*, (S. S. Mitra, Ed.) (New York: Plenum, 1974).
J. F. Vetelino and S. S. Mitra, "Dynamics of Structurally Disordered Solids" in *Physics of Structurally Disordered Solids* (S. S. Mitra, Ed.) (New York: Plenum, 1974).

Elasticity is treated from a macroscopic point of view in many intermediate mechanics texts. Among them are:

J. Norwood, Jr., *Intermediate Classical Mechanics* (Englewood Cliffs, NJ: Prentice-Hall, 1979).
A. Sommerfeld, *Mechanics of Deformable Bodies* (New York: Academic, 1950).
K. R. Symon, *Mechanics* (Reading, MA: Addison-Wesley, 1971).

An advanced treatment of elastic constants can be found in:

H. B. Huntington, "The Elastic Constants of Crystals" in *Solid State Physics* (F. Seitz and D. Turnbull, Eds.), Vol. 7, p. 213 (New York: Academic, 1958).

## PROBLEMS

1. Consider a linear monatomic chain of $N$ atoms with equilibrium separation $a$, vibrating in a normal mode with propagation constant $q = 2\pi \ell / Na$. Show that, at any instant of time, (a) the average displacement of the atoms from their equilibrium sites and (b) the total momentum of the atoms are both zero. (c) Show that these quantities do not necessarily vanish if $q = 0$. The atoms then move together, retaining their relative positions.

2. A linear monatomic chain of $N$ atoms with mass $m$ and equilibrium separation $a$ vibrates in a normal mode with propagation constant $q$, angular frequency $\omega$, and amplitude $A$. Only nearest neighbors interact and the force constant is $\gamma$. Assume periodic boundary conditions and represent the displacements and velocities by real, not complex, functions. (a) Show that the classical kinetic energy of the atom with equilibrium position $na$ is $K_n = \frac{1}{2} m \omega^2 A^2 \sin^2(qna - \omega t)$. (b) Show that, except for an additive constant, the total potential energy is $U = \frac{1}{2} \gamma \Sigma (u_{n+1} - u_n)^2$, where the sum is over the atoms. Show further that $U = \frac{1}{2} \gamma \Sigma (2u_n^2 - u_n u_{n+1} - u_n u_{n-1})$. (c) Show that the total energy is $E = 2\gamma N A^2 \sin^2(\frac{1}{2} qa)$.

3. Classical statistical mechanics predicts that each normal mode has energy $k_B T$ when a solid is in thermal equilibrium at temperature $T$. Here $k_B$ is the Boltzmann constant. Consider a linear monatomic chain of $5.0 \times 10^7$ atoms, with equilibrium separation 6.5 Å. The mass of each atom is $1.8 \times 10^{-26}$ kg and only nearest neighbors interact. Take the force constant to be 9.7 N/m. Use classical mechanics to estimate: (a) the amplitude of vibration at $T = 300$ K for the mode with the maximum frequency; (b) the amplitude of vibration at $T = 300$ K for a mode with a frequency that is 1% of the maximum frequency; and (c) the temperature for which the amplitude of a mode with the maximum frequency is 1% of the equilibrium atomic spacing.

4. A linear chain consists of $2N$ atoms of mass $M$, with equilibrium spacing $\frac{1}{2} a$. Each atom interacts only with its nearest neighbors. The force constant for the interaction between the atom at $na$ and the atom at $na + \frac{1}{2} a$ is $\gamma_1$ while the force constant for the interaction between the atom at $na$ and the atom at $na - \frac{1}{2} a$ is $\gamma_2$. Take $\gamma_1 > \gamma_2$. Each cell of length $a$ contains two nonequivalent atoms. (a) Derive an expression for the normal mode angular frequency as a function of propagation constant $q$. (b) Find the normal mode frequencies for $q = 0$ and for $q = \pi / a$. Indicate which modes belong to the acoustical branch and which belong to the optical branch. (c) Sketch the dispersion curves. (d) How does the answer to part b change if $\gamma_2 > \gamma_1$?

5. The magnitude of force between two argon atoms is approximated by

$$F(r) = 24\epsilon \left| 2 \frac{\sigma^{12}}{R^{13}} - \frac{\sigma^6}{R^7} \right|,$$

where $R$ is their separation, $\epsilon = 0.0104$ eV, and $\sigma = 3.40$ Å. Consider a linear chain of argon atoms with only nearest-neighbor interactions. (a)

Find the equilibrium separation, assuming that the force between two neighboring atoms vanishes for that separation. (b) Find the force constant in the harmonic approximation for two neighboring atoms. (c) The mass of an argon atom is $6.64 \times 10^{-26}$ kg. Calculate the speed of sound in the linear chain. (d) Calculate the maximum normal mode angular frequency for the chain.

6. In many cases experimental dispersion curves are closely reproduced if ion cores and outer electrons are assumed to have different displacements. Consider a monatomic linear chain with equilibrium separation $a$. Suppose each ion core interacts only with its own outer electrons and that the force is proportional to the displacement of the electron center of mass from the nucleus. The force constant is $\gamma_2$. Neighboring electron distributions interact with force constant $\gamma_1$. Let $u_{in} = Ae^{i(qna-\omega t)}$ be the displacement of ion core $n$ and $u_{en} = Be^{i(qna-\omega t)}$ be the displacement of the center of mass of the outer electrons associated with ion $n$. (a) Show that

$$-\omega^2 M u_{in} = \gamma_2(u_{en} - u_{in})$$

and

$$-\omega^2 m u_{en} = \gamma_2(u_{in} - u_{en}) + \gamma_1(u_{en-1} + u_{en+1} - 2u_{en}).$$

Here $M$ is the ion mass and $m$ is the total mass of the outer electrons associated with an ion. (b) Solve for the normal mode angular frequency associated with propagation constant $q$, then show that

$$\omega^2 = \frac{4\gamma_1}{M} \frac{\sin^2(\tfrac{1}{2}qa)}{1 + \dfrac{4\gamma_1}{\gamma_2}\sin^2(\tfrac{1}{2}qa)}$$

in the limit as $m \to 0$ (the electrons are much lighter than the ions). (c) Find expressions for the speed of sound and the maximum angular frequency. Compare their values with those for rigid atom motion.

7. Consider a diatomic linear chain with only nearest-neighbor interactions. In the $q = 0$ optical mode atoms of a cell move in such as way that their center of mass remains at rest. In the $q = \pi/a$ optical mode only the light atoms move, and in the $q = \pi/a$ acoustical mode only the heavy atoms move. Consider the same chain but suppose nearest neighbors interact with force constant $\gamma_1$ and next nearest neighbors interact with force constant $\gamma_2$. (a) Which of the above statements are still true? (b) Does the width of the acoustical branch depend on $\gamma_2$? (c) Does the width of the optical branch depend on $\gamma_2$? (d) Does the gap between the two branches at $q = \pi/a$ depend on $\gamma_2$?

8. A two-dimensional crystal has a square lattice and one atom in its primitive basis. Take the mass of an atom to be $M$ and the edge of a primitive unit cell to be $a$. Assume only nearest and next nearest neighbors interact and, in each case, suppose the force is along the line joining the interacting atoms. In the harmonic approximation the force constants are $\gamma_1$ for near-

est-neighbor interactions and $\gamma_2$ for next-nearest-neighbor interactions. (a) Assume the atoms move in the plane of the lattice and develop the secular equation for the normal mode frequencies as functions of the propagation vector $\mathbf{q}$. (b) Find an expression for the frequencies of normal modes that propagate parallel to an edge of the square cell. For each mode determine if the wave is transverse, longitudinal, or neither. (c) Repeat for propagation along a cell diagonal. (d) For each of the two propagation directions considered, compare the speeds of sound for the two branches of the spectrum.

9. Consider the two-dimensional crystal of Problem 8. Describe the boundaries of the Brillouin zone. Use the results of Problem 8 to find normal mode angular frequencies for a propagation vector that extends from the zone center to (a) a zone edge center and (b) a zone corner. (c) For each of the waves find the wavelength in terms of the square edge $a$.

10. The secular equation derived in Problem 8 provides a nice illustration of symmetry displayed by normal mode frequencies as functions of the propagation vector $\mathbf{q}$. Orient a Cartesian coordinate system so its axes are parallel to edges of the square unit cell. Show that the secular equation does not change if the following replacements are made: (a) $\mathbf{q} \rightarrow -\mathbf{q}$ (180° rotation); (b) $q_x \rightarrow -q_x$ (mirror reflection); (c) $q_y \rightarrow -q_y$ (mirror reflection); and (d) $q_x \rightarrow q_y, q_y \rightarrow -q_x$ (90° rotation). These are all operations in the point group of a square.

11. Consider the Brillouin zone for a face-centered cubic direct lattice with cube edge $a$. Find the components of a vector (a) from the zone center to the center of a square face; (b) from the zone center to the center of a hexagonal face; and (c) from the zone center to a point where two hexagonal faces and one square face meet.

12. Find an expression in terms of the cube edge $a$ for the longest propagation vector in the Brillouin zone of a body-centered cubic crystal. Suppose the shortest wavelength associated with any propagation vector in the Brillouin zone is $\lambda = 3.75$ Å. Calculate the edge length of the direct lattice cubic unit cell and the edge length of the reciprocal lattice cubic unit cell.

13. Atoms of mass $m$ form a simple cubic structure with cube edge $a$. Only nearest neighbors interact and, for any pair of neighbors, the force exerted by atom 2 on atom 1 is

$$\mathbf{F} = (\gamma_c - \gamma_a)(\mathbf{u}_2 - \mathbf{u}_1) \cdot \hat{\mathbf{R}}\,\hat{\mathbf{R}} + \gamma_a(\mathbf{u}_2 - \mathbf{u}_1),$$

where $\gamma_c$ and $\gamma_a$ are constants and $\hat{\mathbf{R}}$ is a unit vector in the direction of the equilibrium displacement of atom 2 from atom 1. The term proportional to $\gamma_c$ is the harmonic approximation to a central force while the two terms proportional to $\gamma_a$, taken together, describe a bond bending force that tends to keep the line joining the atoms parallel to its equilibrium direction. (a) Find expressions for the dynamical matrix elements and, in particular, show that this matrix is diagonal. (b) Find an expression for the normal mode angular frequency as a function of the propagation vector. (c) Take $m =$

$1.5 \times 10^{-25}$ kg, $a = 3.6$ Å, $\gamma_c = 16$ N/m, and $\gamma_a = 5.8$ N/m. Find the normal mode angular frequencies for propagation vector $\mathbf{q} = (\pi/a)(3\hat{\mathbf{x}} + 2\hat{\mathbf{y}} + 4\hat{\mathbf{z}})$, where the coordinate axes are along cube edges. (d) For each of the normal mode angular frequencies found in part c, find the direction of particle motion. (e) Are the normal mode frequencies for the propagation vector of part c greater or less than they would be if $\gamma_a$ were 0?

14. Atoms in a certain monatomic linear chain have an equilibrium separation of 4.85 Å and their maximum normal mode frequency is $4.46 \times 10^{13}$ rad/s. If a $5.75 \times 10^{13}$ rad/s vibration is excited in the chain, how far from the source is the amplitude 10% of the amplitude at the source?

15. Atoms in a linear monatomic chain have mass $m = 6.44 \times 10^{-25}$ kg and equilibrium separation $a = 4.85$ Å. Only nearest neighbors interact and the force constant is 15.0 N/m. (a) Find the energy and crystal momentum of a phonon associated with the maximum frequency normal mode. (b) Suppose a 2.5-Å neutron absorbs one of these phonons. Find the fractional changes in its energy and momentum magnitude. Assume the change in momentum is within the Brillouin zone.

16. A beam of 3.50-Å neutrons is incident normally on a cube face of a monatomic simple cubic crystal, with cube edge 4.25 Å. Some neutrons that are forward scattered in single phonon events exit the crystal along a cube body diagonal and have a wavelength of 2.33 Å. What is the frequency and wave vector of the mode with which they interacted? Were phonons absorbed or emitted by these neutrons?

17. A cube with edge $a$ is distorted so that the point originally at $x, y, z$ moves to $[(1 - \gamma)x + \gamma y]$, $[\gamma x + (1 - \gamma)y]$, $z$, where $\gamma \ll 1$. The origin is at a cube corner and the coordinate axes are aligned with the original cube edges. (a) Find the coordinates of each of the cube corners after the distortion. Make a sketch of the distorted cube. (b) What are the elements of the strain matrix? (c) To first order in $\gamma$, find the length after distortion of each of the cube edges originally in the $xy$ plane. (d) To first order in $\gamma$, what is the fractional change in the volume of the cube? (e) Assume the material is isotropic and find the stress, in terms of $\gamma$ and Lamé's constants, required to produce the distortion.

18. A cubic crystal is uniformly stressed in such a way that all elements of the stress tensor are zero except $T_1$ and $T_4$. (a) Find the elements of the strain matrix in terms of $T_1$, $T_4$, and the elastic constants. (b) Assume $\partial u_i / \partial r_j = \partial u_j / \partial r_i$ and find a general expression for the position of the material originally at $x, y, z$, relative to an origin fixed in the sample.

19. Young's modulus and Poisson's ratio for a certain brass sample are found to be $Y = 2.48 \times 10^{10}$ N/m² and $\sigma = 0.462$, respectively. Brass has cubic symmetry. Which elastic constants can be determined from this data and what are their values? What experiment(s) should be performed to find the values of the other elastic constants?

# Chapter 7

## ELECTRON STATES

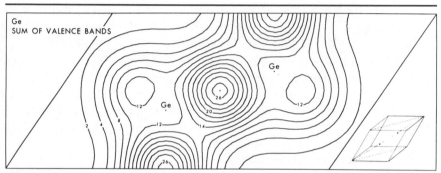

Lines of constant electron density near two atoms in a germanium crystal. Note the electron buildup in the region between the atoms.

Nearly all properties of materials depend, either directly or indirectly, on the electron system. This chapter contains the fundamentals: descriptions of wave functions and energy levels for electrons in solids. Electron wave functions and energy levels are found by solving the time-independent Schrödinger equation, Eq. 5-4, usually with the Hartree or, in more ambitious programs, the Hartree-Fock potential energy function. Brief descriptions of these potential energy functions and the self-consistent technique to which they give rise were given in Chapter 5. There the system was quickly reduced to one with only two atoms to illustrate some of the fundamentals of atomic bonding. We now consider large collections of atoms.

## 7.1  QUALITATIVE RESULTS

Figure 7-1 is a highly schematic diagram of an energy spectrum for an ideal crystal, with all atoms on periodically placed sites and with surface effects neglected. The dominant feature is the grouping of levels, with groups separated by gaps in the spectrum. Each of the shaded regions actually contains a large number of discrete levels, on the order of the number of atoms in the solid, but the levels are so closely spaced they cannot be shown individually. Shaded regions are wider at the top of the diagram than at the bottom, indicating that groups of levels span greater ranges at higher energies than at lower.

The main features of the energy spectrum can be explained qualitatively in terms of atomic orbital overlap, discussed in Chapter 5. For a collection of well-separated atoms, the allowed energies are the atomic levels and the wave functions are the corresponding atomic orbitals. As the atoms are brought close together to form a solid, orbitals centered on different atoms overlap and each atomic level splits, just as the $1s$ level for hydrogen splits into bonding and antibonding levels when an $H_2^+$ molecule is formed. For a solid, however, the number of levels is much larger than two.

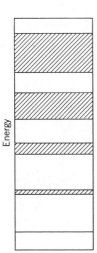

**FIGURE 7-1**  Energy levels for an electron bound in a crystalline solid. Shaded regions indicate groups of allowed energy values. Each group actually consists of a large number of discrete levels, too close together to distinguish on the diagram. The groups are narrow for low energy and wide for high energy. No electron may have energy in a gap.

The extent of orbital overlap depends on the energies of the states involved and on the potential energy function. Figure 7-2 shows schematically the potential energy of an electron along a line of atoms in a crystal. At a surface (not shown) it rises steeply to zero, the value selected for the potential energy of an electron outside the solid. Inside, the function displays the same periodicity as the crystal. Except near a surface, it has the same value at any two points separated by a lattice vector.

Near each nucleus the potential energy forms a deep well, essentially the same as that around the nucleus of an isolated atom. In interstitial regions it is the sum of contributions of many atoms and so is lower than the potential energy outside the solid.

Core electrons have energies that are deep in the potential energy wells near nuclei. Their wave functions are nearly atomic orbitals and have exponential-like tails outside the classical turning points (the ends of the horizontal lines representing energy levels in Fig. 7-2). Between atoms the energy is far below the potential energy, so the tails drop steeply toward zero and the probability density is extremely small in interstitial regions. As a result, the splitting of levels is small. The lowest level in Fig. 7-1, for example, is shown as a single line at essentially the same energy as the lowest atomic level. Two states, one spin up and the other spin down, are associated with this level for each atom in the solid.

Wave functions for electrons with higher energy drop off less sharply than those for deeper lying electrons, so overlap of neighboring atomic orbitals is greater at higher energies than at lower. As a consequence, the spread in energy levels increases with energy. Nevertheless, splitting is small for nearly all core

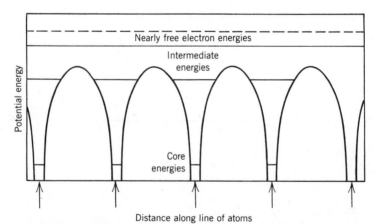

FIGURE 7-2 Electron potential energy along a line of atoms in a crystal. The dotted line indicates the energy for an electron outside the solid. Positions of nuclei are marked by arrows and representative allowed energy levels are shown as solid horizontal lines. Core states have energies deep in the potential wells near each nucleus. Nearly free electrons have energies well above the potential energy maxima.

states and gaps occur in this energy region because the atomic spectrum is discrete. Gap widths are determined primarily by separations of atomic levels, mediated by the splitting that accompanies overlap.

We now turn to the region near the potential energy maxima. It is important because it contains the highest occupied levels of insulators and semiconductors at low temperatures. Electrons with energy in this region can no longer be considered core electrons. Neighboring atomic orbitals link together to form wave functions that extend with significant amplitude throughout the solid. Except quite close to a nucleus these functions do not bear a close resemblance to atomic orbitals. In this energy region, the width of a group of levels is usually greater than the separation of atomic levels. Not only do orbitals associated with any atom spread far into the interstitial region, but near a nucleus each wave function is a mixture of several atomic orbitals. For example, both $s$ and $p$ functions of the outer shell usually enter the admixture with large amplitudes. Gaps appear but their positions are not simply related to the $s$ and $p$ energy levels of an electron in an isolated atom.

Wave functions for electrons with energies well above the potential energy maxima are nearly plane waves and the probability density for any one of them is nearly uniform throughout most of the solid. Ion cores exert only weak forces on these electrons and have less influence on wave functions than at lower energies. The forces do, however, give rise to gaps. Later we will investigate in detail how they come about.

At low temperatures metals are the only defect-free solids to have significant numbers of nearly free electrons, as electrons with energies in this region are called. The formation of an aggregate lowers the potential energy in interstitial regions so it is below the energy of an electron in the outermost occupied shell of an isolated metal atom. As a consequence, outermost electrons in atoms become nearly free in solids. On the other hand, states well above the potential energy maxima in semiconductors are occupied only at high temperatures or if the solid contains certain impurities.

Additional energy levels, some perhaps in gaps, are associated with structural defects, surfaces as well as impurities, vacancies, and interstitials. Electron energy spectra of amorphous materials are also different from that shown in Fig. 7-1. Levels appear at energies corresponding to gaps in the spectrum of the crystalline form of the same material and boundaries of groups of levels are considerably blurred.

## 7.2 ELECTRON STATES IN CRYSTALS

Like calculations of normal mode frequencies and displacements, calculations of electron wave functions and energy levels simplify considerably when the ion cores form an ideal crystal and periodic boundary conditions are applied. The chief reason is that the mathematical form of the wave functions is determined by the translational symmetry of the crystal and is known in advance. The Schrödinger equation need be integrated for only a single unit cell.

*Bloch Functions.* In an ideal crystal, without boundaries and with atoms at sites prescribed by lattice and basis vectors, electron wave functions have the form

$$\psi(\mathbf{r}) = e^{i\mathbf{k}\cdot\mathbf{r}}u(\mathbf{k}, \mathbf{r}), \tag{7-1}$$

where $\mathbf{k}$ is a real constant vector and $u(\mathbf{r})$ is a function that is periodic with the same periodicity as the lattice. That is, $u(\mathbf{k}, \mathbf{r} + \mathbf{R}) = u(\mathbf{k}, \mathbf{r})$ for any lattice vector $\mathbf{R}$. The vector $\mathbf{k}$ and the function $u(\mathbf{k}, \mathbf{r})$ may be different for different electron states in the same crystal, but in each case Eq. 7-1 is satisfied. A function having this form is called a Bloch function.

Equation 7-1 follows directly from the periodicity of the crystal. Since all unit cells are equivalent, the probability density is the same at analogous points in all cells. Since $|\psi(\mathbf{r} + \mathbf{R})|^2 = |\psi(\mathbf{r})|^2$, $\psi(\mathbf{r} + \mathbf{R})$ and $\psi(\mathbf{r})$ differ at most by a multiplicative factor with magnitude 1. That is,

$$\psi(\mathbf{r} + \mathbf{R}) = e^{i\alpha(\mathbf{R})}\psi(\mathbf{r}), \tag{7-2}$$

where $\alpha(\mathbf{R})$ is real and may depend on $\mathbf{R}$ but not on $\mathbf{r}$.

$\alpha(\mathbf{R})$ must be linear in $\mathbf{R}$. To understand this, consider two lattice translations, $\mathbf{R}_1$ and $\mathbf{R}_2$, and evaluate $\psi(\mathbf{r} + \mathbf{R}_1 + \mathbf{R}_2)$ in two ways. First, treat $\mathbf{R}_1 + \mathbf{R}_2$ as a single translation and write $\psi(\mathbf{r} + \mathbf{R}_1 + \mathbf{R}_2) = e^{i\alpha(\mathbf{R}_1 + \mathbf{R}_2)}\psi(\mathbf{r})$. Second, treat $\mathbf{R}_1$ and $\mathbf{R}_2$ as two separate translations and write $\psi(\mathbf{r} + \mathbf{R}_1 + \mathbf{R}_2) = e^{i\alpha(\mathbf{R}_2)}\psi(\mathbf{r} + \mathbf{R}_1) = e^{i\alpha(\mathbf{R}_2)}e^{i\alpha(\mathbf{R}_1)}\psi(\mathbf{r})$. Since the two results must be the same, $\alpha(\mathbf{R}_1 + \mathbf{R}_2) = \alpha(\mathbf{R}_1) + \alpha(\mathbf{R}_2)$. Because it is linear in $\mathbf{R}$, $\alpha(\mathbf{R})$ can be written in the form $AR_x + BR_y + CR_z + D$, where $A, B, C$, and $D$ are independent of $\mathbf{R}$. The constant term $D$ leads to a phase factor that is the same for all cells and can be ignored. The other terms can be written as the scalar product of $\mathbf{R}$ and a constant vector $\mathbf{k}$. Thus $\alpha(\mathbf{R}) = \mathbf{k}\cdot\mathbf{R}$ and

$$\psi(\mathbf{r} + \mathbf{R}) = e^{i\mathbf{k}\cdot\mathbf{R}}\psi(\mathbf{r}), \tag{7-3}$$

for any lattice vector $\mathbf{R}$. This equation is known as Bloch's theorem.

Equation 7-3 implies that $\psi(\mathbf{r})$ has the Bloch form, given by Eq. 7-1. For any function $\psi(\mathbf{r})$ we can define $u(\mathbf{k}, \mathbf{r})$ so that $\psi(\mathbf{r}) = e^{i\mathbf{k}\cdot\mathbf{r}}u(\mathbf{k}, \mathbf{r})$. We wish to show that Eq. 7-3 implies $u(\mathbf{k}, \mathbf{r})$ is periodic. When $e^{i\mathbf{k}\cdot\mathbf{r}}u(\mathbf{k}, \mathbf{r})$ is substituted for $\psi(\mathbf{r})$ in Eq. 7-3, the result is

$$e^{i\mathbf{k}\cdot(\mathbf{r}+\mathbf{R})}u(\mathbf{k}, \mathbf{r} + \mathbf{R}) = e^{i\mathbf{k}\cdot\mathbf{R}}e^{i\mathbf{k}\cdot\mathbf{r}}u(\mathbf{k}, \mathbf{r}) \tag{7-4}$$

or

$$u(\mathbf{k}, \mathbf{r} + \mathbf{R}) = u(\mathbf{k}, \mathbf{r}). \tag{7-5}$$

So $u(\mathbf{k}, \mathbf{r})$ has the periodicity of the lattice.

Strictly speaking, Bloch functions are valid electron wave functions only if the crystal is infinite. Near the surface of a finite crystal the structure is different from that in the interior, so $|\psi(\mathbf{r} + \mathbf{R})|^2$ is different from $|\psi(\mathbf{r})|^2$ if $\mathbf{r}$ is near a surface and $\mathbf{r} + \mathbf{R}$ is not, or vice versa. In the interior of a macroscopic crystal, however, Bloch functions are excellent approximations to true wave functions and we will use them to discuss bulk properties of materials. Wave functions for points near surfaces are considered separately.

When multiplied by the time-dependent factor, a Bloch function becomes

$$\Psi(\mathbf{r}, t) = e^{i(\mathbf{k} \cdot \mathbf{r} - \omega t)} u(\mathbf{k}, \mathbf{r}). \tag{7-6}$$

This represents a traveling wave propagating in the direction of $\mathbf{k}$. Its amplitude is modulated from point to point within a unit cell but is the same for equivalent points in all cells.

If $u(\mathbf{k}, \mathbf{r})$ is independent of $\mathbf{r}$, $\Psi(\mathbf{r}, t)$ is a plane wave and the electron has momentum $\mathbf{p} = \hbar \mathbf{k}$. However, a plane wave is a valid wave function and the electron has a definite momentum only if the potential energy function is independent of $\mathbf{r}$. Then the net force on the electron vanishes and momentum is conserved. In a crystal, on the other hand, the potential energy function depends on $\mathbf{r}$, so electron momentum is not conserved.

The quantity $\hbar \mathbf{k}$ is conserved for an electron in an ideal crystal, even though ion cores exert forces on it. Once in a state with a particular value of $\hbar \mathbf{k}$, an electron remains in that state and retains that value unless it is acted on by forces that do not have the periodicity of the crystal. The quantity $\hbar \mathbf{k}$ is called the electron crystal momentum and plays the same role in electron interactions as $\hbar \mathbf{q}$ does in phonon interactions. It is used to label electron states and enters into nearly all discussions dealing with the influence of external forces on the electron system.

*Energy Bands.* A differential equation for $u(\mathbf{k}, \mathbf{r})$ is obtained by substituting Eq. 7-1 into the Schrödinger equation. Consider $\nabla^2 \psi(\mathbf{k}, \mathbf{r})$ first. The product rule for differentiation is used to obtain

$$\frac{\partial^2 \psi}{\partial x^2} = \frac{\partial^2}{\partial x^2} \left[ e^{i(k_x x + k_y y + k_z z)} u \right]$$

$$= \left[ \frac{\partial^2 u}{\partial x^2} + 2ik_x \frac{\partial u}{\partial x} - k_x^2 u \right] e^{i\mathbf{k} \cdot \mathbf{r}}. \tag{7-7}$$

Similarly,

$$\frac{\partial^2 \psi}{\partial y^2} = \left[ \frac{\partial^2 u}{\partial y^2} + 2ik_y \frac{\partial u}{\partial y} - k_y^2 u \right] e^{i\mathbf{k} \cdot \mathbf{r}} \tag{7-8}$$

and

$$\frac{\partial^2 \psi}{\partial z^2} = \left[ \frac{\partial^2 u}{\partial z^2} + 2ik_z \frac{\partial u}{\partial z} - k_z^2 u \right] e^{i\mathbf{k} \cdot \mathbf{r}}. \tag{7-9}$$

When summed, Eqs. 7-7, 7-8, and 7-9 produce

$$\nabla^2 \psi = [\nabla^2 u + 2i\mathbf{k} \cdot \nabla u - k^2 u] e^{i\mathbf{k} \cdot \mathbf{r}}, \tag{7-10}$$

so the time-independent Schrödinger equation becomes

$$-\frac{\hbar^2}{2m} \nabla^2 u - i \frac{\hbar^2}{m} \mathbf{k} \cdot \nabla u + \frac{\hbar^2 k^2}{2m} u + Uu = Eu, \tag{7-11}$$

once each term is divided by $e^{i\mathbf{k} \cdot \mathbf{r}}$.

Equation 7-11 is to be solved for $u(\mathbf{k}, \mathbf{r})$. Since $u(\mathbf{k}, \mathbf{r})$ is periodic, a solution need be obtained for only one unit cell and, to do this, boundary conditions are placed on $u$ rather than on $\psi$. Since $\psi$ and its slope are continuous, $u(\mathbf{k}, \mathbf{r})$ and its slope at any point on the surface of a unit cell are related, respectively, to $u(\mathbf{k}, \mathbf{r})$ and its slope at other points on the surface displaced from the first by lattice vectors. Specifically, if $\mathbf{r}$ and $\mathbf{r} + \mathbf{R}$ are two points on the surface of a cell, then

$$u(\mathbf{k}, \mathbf{r}) = u(\mathbf{k}, \mathbf{r} + \mathbf{R}) \tag{7-12}$$

and

$$\hat{\mathbf{n}}_1 \cdot \nabla u(\mathbf{k}, \mathbf{r}) = -\hat{\mathbf{n}}_2 \cdot \nabla u(\mathbf{k}, \mathbf{r} + \mathbf{R}). \tag{7-13}$$

Here $\hat{\mathbf{n}}_1$ is the unit outward normal to the surface at $\mathbf{r}$ and $\hat{\mathbf{n}}_2$ is the unit outward normal to the surface at $\mathbf{r} + \mathbf{R}$. The first condition assures the continuity of $\psi$ and the second assures the continuity of its normal derivative. The minus sign appears in Eq. 7-13 because $\hat{\mathbf{n}}_1$ and $\hat{\mathbf{n}}_2$ are in opposite directions.

Functions $u(\mathbf{k}, \mathbf{r})$ that satisfy Eqs. 7-11, 7-12, and 7-13 exist only if the energy $E$ has certain discrete values. These are the allowed energy levels for states with propagation vector $\mathbf{k}$. The symbol $E_n(\mathbf{k})$ is used to denote an energy level. Here

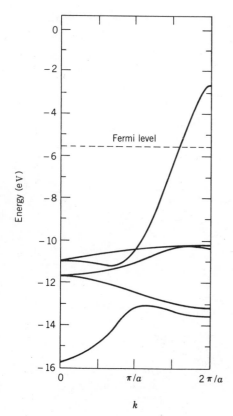

**FIGURE 7-3** Electron energy bands for crystalline copper. The propagation vector is in the [100] direction. Bands associated with core states are not shown. The potential energy is zero at points outside the crystal. The Fermi level is explained in Chapter 8. (From E. C. Snow and J. T. Waber, *Phys. Rev.*; **157**:570, 1967. Used with permission.)

$n$ is an integer, called the band index, that labels levels in order of increasing energy: $E_1(\mathbf{k})$ is the lowest energy level associated with propagation vector $\mathbf{k}$, $E_2(\mathbf{k})$ is the next highest, and so on. The band index and propagation vector are also used to specify the wave function: $\psi_n(\mathbf{k}, \mathbf{r})$ denotes a wave function with band index $n$ and $u_n(\mathbf{k}, \mathbf{r})$ denotes its periodic part.

The collection of all levels with the same band index is called an energy band. Bands are usually displayed as in Fig. 7-3. A direction is selected for $\mathbf{k}$ and the energy is plotted as a function of the magnitude of $\mathbf{k}$. To give a more complete picture, of course, similar plots must be shown for other directions of $\mathbf{k}$.

Energy levels for any given propagation vector are separated by as much as a few electron volts. In Fig. 7-3, for example, energy levels for $k = \pi/2a$ are at about $-10.6$, $-11.0$, $-11.8$, $-12.0$, and $-14.9$ eV. These separations give rise to the gaps in the energy spectrum, but a gap does not necessarily occur between every pair of adjacent bands. Specifically, no gap occurs if the minimum energy of one band is lower than the maximum energy of the band with the next lower index. Figure 7-4 illustrates the point.

*Properties of Energy Bands.* An energy band displays the periodicity of the reciprocal lattice. That is,

$$E_n(\mathbf{k} + \mathbf{G}) = E_n(\mathbf{k}), \tag{7-14}$$

where $\mathbf{G}$ is any reciprocal lattice vector. In addition, $\psi_n(\mathbf{k} + \mathbf{G}, \mathbf{r})$ represents the same state as $\psi_n(\mathbf{k}, \mathbf{r})$. To show this, we first note that the function $e^{i\mathbf{G}\cdot\mathbf{r}}$ has the periodicity of the direct lattice. Since $\mathbf{G}\cdot\mathbf{R}$ is a multiple of $2\pi$, $e^{i\mathbf{G}\cdot(\mathbf{r}+\mathbf{R})} = e^{i\mathbf{G}\cdot\mathbf{r}}$ for any direct lattice vector $\mathbf{R}$. For the wave function in band $n$ associated with propagation vector $\mathbf{k} + \mathbf{G}$, we write

$$\psi_n(\mathbf{k} + \mathbf{G}, \mathbf{r}) = e^{i(\mathbf{k}+\mathbf{G})\cdot\mathbf{r}}u_n(\mathbf{k} + \mathbf{G}, \mathbf{r}) = e^{i\mathbf{k}\cdot\mathbf{r}}u'_n(\mathbf{k} + \mathbf{G}, \mathbf{r}), \tag{7-15}$$

where $u'_n(\mathbf{k} + \mathbf{G}, \mathbf{r}) = e^{i\mathbf{G}\cdot\mathbf{r}}u_n(\mathbf{k} + \mathbf{G}, \mathbf{r})$. Clearly, $u'_n(\mathbf{k} + \mathbf{G}, \mathbf{r})$ has the periodicity

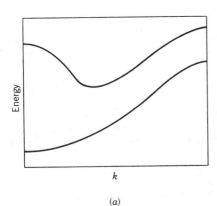
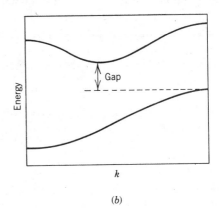

(a)  (b)

**FIGURE 7-4** (a) The minimum of the second band is below the maximum of the first. For each propagation vector the second band energy is above the first band energy, but the energy spectrum has no gap. Compare with the situation shown in (b), for which the two bands are separated by a gap.

of the lattice. As substitution of Eq. 7-15 into the time-independent Schrödinger equation shows, $u_n'(\mathbf{k} + \mathbf{G}, \mathbf{r})$ obeys precisely the same differential equation as $u_n(\mathbf{k}, \mathbf{r})$. Furthermore, $u_n'(\mathbf{k} + \mathbf{G}, \mathbf{r})$ and $u_n(\mathbf{k}, \mathbf{r})$ obey the same boundary conditions. They are therefore the same function, except perhaps for an unimportant multiplicative factor with magnitude 1. Since $u_n'(\mathbf{k} + \mathbf{G}, \mathbf{r}) = u_n(\mathbf{k}, \mathbf{r})$, $\psi_n(\mathbf{k} + \mathbf{G}, \mathbf{r}) = \psi_n(\mathbf{k}, \mathbf{r})$. Wave functions for the two propagation vectors are identical. In addition, exactly the same set of energy levels is obtained, so $E_n(\mathbf{k} + \mathbf{G}) = E_n(\mathbf{k})$.

To avoid redundancy in the list of states, propagation vectors are usually confined to the Brillouin zone. Then no two of them can differ by a reciprocal lattice vector. Sometimes, however, bands are plotted in an extended zone scheme, just as the vibration spectrum of a diatomic chain was plotted in Fig. 6-8.

$E_n(\mathbf{k})$ has the symmetry of the crystal point group. That is, $E_n(\mathbf{k}') = E_n(\mathbf{k})$ if $\mathbf{k}'$ is related to $\mathbf{k}$ by one the symmetry operations that leaves the crystal invariant. The proof is exactly the same as that used in Section 6.4 to show normal mode frequencies display the symmetry of the crystal point group.

In addition, $E_n(-\mathbf{k}) = E_n(\mathbf{k})$, whether the crystal has a center of inversion or not. Strictly speaking, the potential energy function should include spin-orbit interactions, small corrections that are spin dependent. When these are included, $E_n(-\mathbf{k}) \neq E_n(\mathbf{k})$ for electrons with the same spin. But $E_n(-\mathbf{k})$ for a spin-down electron is the same as $E_n(\mathbf{k})$ for a spin-up electron and, similarly, $E_n(-\mathbf{k})$ for a spin-up electron is the same as $E_n(\mathbf{k})$ for a spin-down electron.

*Allowed Propagation Vectors.* Boundary conditions applied to a wave function lead to quantization of electron crystal momentum. To find the allowed values of crystal momentum we use periodic boundary conditions, similar to those used in the study of normal modes. More realistically, an electron wave function tends toward zero outside a solid, but, except for extremely simple models, this boundary condition is difficult to apply. When it is applied, it also leads to quantization of crystal momentum and allowed values are only insignificantly different from those produced by periodic boundary conditions.

Assume a crystal is composed of $N^3$ unit cells, arranged so $N$ of them are along each crystal axis. Then periodic boundary conditions are expressed by

$$\psi(\mathbf{r} + N\mathbf{a}) = \psi(\mathbf{r} + N\mathbf{b}) = \psi(\mathbf{r} + N\mathbf{c}) = \psi(\mathbf{r}), \qquad (7\text{-}16)$$

where $\mathbf{a}$, $\mathbf{b}$, and $\mathbf{c}$ are primitive vectors of the lattice. Substitute the Bloch form into these expressions and use $u_n(\mathbf{k}, \mathbf{r} + N\mathbf{a}) = u_n(\mathbf{k}, \mathbf{r} + N\mathbf{b}) = u_n(\mathbf{k}, \mathbf{r} + N\mathbf{c}) = u_n(\mathbf{k}, \mathbf{r})$ to obtain $e^{iN\mathbf{k} \cdot \mathbf{a}} = 1$, $e^{iN\mathbf{k} \cdot \mathbf{b}} = 1$, and $e^{iN\mathbf{k} \cdot \mathbf{c}} = 1$. These relations are satisfied by

$$\mathbf{k} = \frac{h}{N}\mathbf{A} + \frac{k}{N}\mathbf{B} + \frac{\ell}{N}\mathbf{C}, \qquad (7\text{-}17)$$

where $\mathbf{A}$, $\mathbf{B}$, and $\mathbf{C}$ are fundamental vectors of the reciprocal lattice and $h$, $k$, and $\ell$ are integers. If $\mathbf{k}$ is restricted to the Brillouin zone, then $h$, $k$, and $\ell$ are each restricted to the range from $-\frac{1}{2}N$ to $+\frac{1}{2}N$. Within their respective ranges there are exactly $N$ values of each integer, so there are $N^3$ allowed propagation vectors in all. Allowed propagation vectors are the same for electrons and normal vibration modes.

Because **k** is quantized we must reinterpret band diagrams slightly. The diagram remains as drawn, but each of the curves is really a series of discrete points, one point on each curve for each allowed propagation vector. Since the number of points on each curve is extremely large, perhaps $10^8$ or more, the points are quite close together and the curves may often be treated as continuous.

We are now in a position to enumerate the states in a band. The number of distinct propagation vectors equals the number of unit cells in the crystal and, for any specified propagation vector and band index, there are two states, one with spin up and one with spin down. The total number of states in a band therefore equals twice the number of unit cells in the crystal. If the crystal is enlarged by adding more cells, the number of states in a band increases but the maximum and minimum energies in the band do not. More allowed energy values crowd between the same limits.

**EXAMPLE 7-1** Show that for each band there are $2N^3\tau/(2\pi)^3$ states per unit volume of reciprocal space. Here $\tau$ is the volume of a direct lattice unit cell.

**SOLUTION** A Brillouin zone is a primitive unit cell of the reciprocal lattice, so its volume is $|\mathbf{A} \cdot (\mathbf{B} \times \mathbf{C})| = (2\pi)^3/|\mathbf{a} \cdot (\mathbf{b} \times \mathbf{c})| = (2\pi)^3/\tau$. To obtain this result, use Eqs. 4-21, 4-22, and 4-23 to substitute for $\mathbf{A}$, $\mathbf{B}$, and $\mathbf{C}$, respectively; then use the vector identity $(\mathbf{V}_1 \times \mathbf{V}_2) \cdot (\mathbf{V}_3 \times \mathbf{V}_4) = (\mathbf{V}_1 \cdot \mathbf{V}_3)(\mathbf{V}_2 \cdot \mathbf{V}_4) - (\mathbf{V}_1 \cdot \mathbf{V}_4)(\mathbf{V}_2 \cdot \mathbf{V}_3)$. There are $N^3$ propagation vectors in the Brillouin zone and two states in each band for each propagation vector, so for each band there are $2N^3\tau/(2\pi)^3$ states per unit volume of reciprocal space. ■

## 7.3 TIGHTLY BOUND ELECTRONS

There are two idealized situations for which electron wave functions can be expressed in a simple manner and an energy band calculation can be carried out with relative ease. One of these occurs when the energies of interest are deep within the potential energy wells at nuclei and the other occurs when the energies of interest are far above the maxima of the potential energy function. The first case is discussed in this section and the second is discussed in the next.

*Wave Functions for Tightly Bound Electrons.* The calculation is quite similar to that for covalent bonding functions, discussed in Section 5.3. We suppose the crystal basis contains only one atom and all atoms are at lattice points. We also suppose the wave function near any nucleus is very nearly an atomic orbital centered on that nucleus.

The atomic orbital $\chi(\mathbf{r} - \mathbf{R})$ centered on the nucleus at $\mathbf{R}$ satisfies the time-independent Schrödinger equation

$$-\frac{\hbar^2}{2m}\nabla^2\chi(\mathbf{r} - \mathbf{R}) + U_a(\mathbf{r} - \mathbf{R})\chi(\mathbf{r} - \mathbf{R}) = E_a(\mathbf{r} - \mathbf{R}),\qquad(7\text{-}18)$$

where $U_a(\mathbf{r} - \mathbf{R})$ is the potential energy for an electron in an isolated atom and $E_a$ is the atomic energy level associated with $\chi$. $\chi(\mathbf{r})$ decreases in an exponentiallike manner in regions for which $U_a(\mathbf{r}) > E_a$, so it is small far from the nucleus.

An electron wave function for tightly bound electrons in a crystal is approximated by

$$\psi(\mathbf{k}, \mathbf{r}) = \Gamma \sum_{\mathbf{R}} e^{i\mathbf{k}\cdot\mathbf{R}}\chi(\mathbf{r} - \mathbf{R}),\qquad(7\text{-}19)$$

where the sum is over all atoms of the crystal and $\Gamma$ is a constant determined by the normalization condition. Note that if $\mathbf{r}$ is near any nucleus, $\psi$ is nearly the atomic orbital centered on that nucleus, multiplied by a constant factor. If $\mathbf{r}$ is near $\mathbf{R}'$, for example, $\chi(\mathbf{r} - \mathbf{R}) \approx 0$ for all $\mathbf{R}$ except $\mathbf{R} = \mathbf{R}'$ and $\psi(\mathbf{k}, \mathbf{r}) \approx \Gamma e^{i(\mathbf{k}\cdot\mathbf{R}')}\chi(\mathbf{r} - \mathbf{R}')$.

The factors $e^{i\mathbf{k}\cdot\mathbf{R}}$ in Eq. 7-19 were chosen so $\psi(\mathbf{k}, \mathbf{r})$ has the Bloch form. If $\mathbf{R}'$ is any lattice vector, then

$$\psi(\mathbf{k}, \mathbf{r} + \mathbf{R}') = \Gamma \sum_{\mathbf{R}} e^{i\mathbf{k}\cdot\mathbf{R}}\chi(\mathbf{r} + \mathbf{R}' - \mathbf{R}).\qquad(7\text{-}20)$$

$\mathbf{R}'' = \mathbf{R} - \mathbf{R}'$ is another lattice vector, so we may replace $\mathbf{R}$ by $\mathbf{R}'' + \mathbf{R}'$ and use $\mathbf{R}''$ as the summation variable. When this is done

$$\psi(\mathbf{k}, \mathbf{r} + \mathbf{R}') = \Gamma \sum_{\mathbf{R}''} e^{i\mathbf{k}\cdot(\mathbf{R}''+\mathbf{R}')}\chi(\mathbf{r} - \mathbf{R}'')$$

$$= \Gamma e^{i\mathbf{k}\cdot\mathbf{R}'} \sum_{\mathbf{R}''} e^{i\mathbf{k}\cdot\mathbf{R}''}\chi(\mathbf{r} - \mathbf{R}'') = e^{i\mathbf{k}\cdot\mathbf{R}'}\psi(\mathbf{k}, \mathbf{r}).\qquad(7\text{-}21)$$

For this proof to be valid, the sample must be part of a much larger crystal so each atom has the same distribution of neighbors around it. Then each term in the sum over $\mathbf{R}''$ that appears in Eq. 7-21 is identical to a term in the sum over $\mathbf{R}$ that appears in Eq. 7-19.

The constant $\Gamma$ is determined by the condition $\int \psi^*\psi\, d\tau = 1$, where the integral is over the volume of the sample. To evaluate it, write

$$\psi^*(\mathbf{k}, \mathbf{r}) = \Gamma^* \sum_{\mathbf{R}'} e^{-i\mathbf{k}\cdot\mathbf{R}'}\chi^*(\mathbf{r} - \mathbf{R}').\qquad(7\text{-}22)$$

Then

$$\int \psi^*(\mathbf{k}, \mathbf{r})\psi(\mathbf{k}, \mathbf{r})\, d\tau$$

$$= |\Gamma|^2 \sum_{\mathbf{R}'} \sum_{\mathbf{R}} e^{i\mathbf{k}\cdot(\mathbf{R}-\mathbf{R}')} \int \chi^*(\mathbf{r} - \mathbf{R}')\chi(\mathbf{r} - \mathbf{R})\, d\tau.\qquad(7\text{-}23)$$

The sum over $\mathbf{R}$ may be considered to be a sum over the neighbors of the atom at $\mathbf{R}'$ and, since every atom has the same distribution of neighbors around it, the value of the sum is the same for every value of $\mathbf{R}'$. We evaluate the sum for $\mathbf{R}' = 0$ and multiply the result by the number of atoms $N$ in the sample. If $\int \chi^*(\mathbf{r})\chi(\mathbf{r} - \mathbf{R})\,d\tau = B(\mathbf{R})$, then

$$1 = |\Gamma|^2 N \sum_{\mathbf{R}} e^{i\mathbf{k} \cdot \mathbf{R}} B(\mathbf{R}), \tag{7-24}$$

and we may take

$$\Gamma = \left[ N \sum_{\mathbf{R}} e^{i\mathbf{k} \cdot \mathbf{R}} B(\mathbf{R}) \right]^{-1/2}. \tag{7-25}$$

*Tight Binding Bands.* Substitute the wave function, given by Eq. 7-19, into the Schrödinger equation for an electron in a crystal and use Eq. 7-18 to replace $-(\hbar^2/2m)\nabla^2\chi(\mathbf{r} - \mathbf{R})$ with $E_a\chi(\mathbf{r} - \mathbf{R}) - U_a(\mathbf{r} - \mathbf{R})\chi(\mathbf{r} - \mathbf{R})$. After slight rearrangement, the result can be written

$$\sum_{\mathbf{R}} e^{i\mathbf{k} \cdot \mathbf{R}} [U(\mathbf{r}) - U_a(\mathbf{r} - \mathbf{R})]\chi(\mathbf{r} - \mathbf{R}) = [E(\mathbf{k}) - E_a] \sum_{\mathbf{R}} e^{i\mathbf{k} \cdot \mathbf{R}} \chi(\mathbf{r} - \mathbf{R}). \tag{7-26}$$

Finally, multiply by $\psi^*(\mathbf{k}, \mathbf{r})$, given by Eq. 7-22, and integrate over the volume of the sample, to obtain

$$E(\mathbf{k}) = E_a + |\Gamma|^2 \sum_{\mathbf{R}'} \sum_{\mathbf{R}} e^{i\mathbf{k} \cdot (\mathbf{R} - \mathbf{R}')} \int \chi^*(\mathbf{r} - \mathbf{R}')[U(\mathbf{r}) - U_a(\mathbf{r} - \mathbf{R})]\chi(\mathbf{r} - \mathbf{R})\,d\tau. \tag{7-27}$$

Again the sum over $\mathbf{R}$ is the same for each value of $\mathbf{R}'$, so we multiply the $\mathbf{R}' = 0$ term by the number of atoms $N$. Let

$$A(\mathbf{R}) = \int \chi^*(\mathbf{r})[U_a(\mathbf{r} - \mathbf{R}) - U(\mathbf{r})]\chi(\mathbf{r} - \mathbf{R})\,d\tau \tag{7-28}$$

and use Eq. 7-25 to substitute for $|\Gamma|^2$. Then Eq. 7-27 becomes

$$E(\mathbf{k}) = E_a - \frac{\sum A(\mathbf{R})e^{i\mathbf{k} \cdot \mathbf{R}}}{\sum B(\mathbf{R})e^{i\mathbf{k} \cdot \mathbf{R}}}, \tag{7-29}$$

where both sums are over atoms of the crystal.

The $\mathbf{R} = 0$ term dominates each sum of Eq. 7-29. Integrals in all other terms contain the product $\chi^*(\mathbf{r})\chi(\mathbf{r} - \mathbf{R})$ of orbitals centered on different atoms and are small. In the denominator we retain only the $\mathbf{R} = 0$ term, which has the value 1. We denote the $\mathbf{R} = 0$ term in the numerator by $\alpha$. Specifically, $\alpha = \int \chi(\mathbf{r})[u_a(\mathbf{r}) - U(\mathbf{r})]\chi(\mathbf{r})\,d\tau$. Then, to first order in small quantities,

$$E(\mathbf{k}) = E_a - \alpha - \sum_{\mathbf{R}}{}' A(\mathbf{R})e^{i\mathbf{k} \cdot \mathbf{R}}, \tag{7-30}$$

where the prime indicates that the $\mathbf{R} = 0$ term is omitted from the sum.

Because $U_a(\mathbf{r}) > U(\mathbf{r})$ for all values of $\mathbf{r}$, $\alpha$ is positive and the second term of Eq. 7-30 represents a lowering of energy from the atomic level. The mechanism is like that discussed in Chapter 5 in connection with bonding: the average potential energy associated with an atomic orbital is less in a crystal than in an isolated atom. If the atomic separation is decreased, $U_a(\mathbf{r}) - U(\mathbf{r})$ increases and the band as a whole moves downward in energy.

The third term depends on the overlap of orbitals centered on different atoms. $A(\mathbf{R})$ is large if both $\chi(\mathbf{r})$ and $\chi(\mathbf{r} - \mathbf{R})$ are large in interstitial regions, where $U(\mathbf{r})$ differs greatly from $U_a(\mathbf{r} - \mathbf{R})$. The sum describes the spread of levels to form a band: it has different values for different propagation vectors. If there are $N$ atoms, then $N$ levels are formed, corresponding to the $N$ values of $\mathbf{k}$. Some are greater than $E_a - \alpha$ while others are less.

If $\mathbf{k} = 0$, for example, then $\chi(\mathbf{r})$ and $\chi(\mathbf{r} - \mathbf{R})$ enter the wave function in the combination $\chi(\mathbf{r}) + \chi(\mathbf{r} - \mathbf{R})$, like a bonding function. On the other hand, if $\mathbf{k} \cdot \mathbf{R} = \pi$ they enter in the combination $\chi(\mathbf{r}) - \chi(\mathbf{r} - \mathbf{R})$, like an antibonding function. Other possibilities exist, corresponding to other values of $\mathbf{k} \cdot \mathbf{R}$. For different values, atomic orbitals link differently and give rise to different electron probability distributions. Both the average potential energy and average kinetic energy are different for electrons with different propagation vectors. If the atomic separation decreases, $A(\mathbf{R})$ increases and the width of the band, measured from the lowest to the highest energy, also increases. Thus large overlap produces a wide band.

**EXAMPLE 7-2** (a) Find an expression for the energies of a tight binding band for a crystal with a simple cubic lattice and a basis of one atom. Assume the atomic orbital $\chi(\mathbf{r})$ is real and spherically symmetric and take $A(\mathbf{R})$ to be zero except for nearest neighbors. (b) Find expressions for the minimum and maximum energies in the band.

**SOLUTION** (a) Take the unit cell to be a cube with edge $a$ and place a Cartesian coordinate system with axes parallel to cube edges. Each atom has nearest neighbors at $\pm a\hat{\mathbf{x}}$, $\pm a\hat{\mathbf{y}}$, and $\pm a\hat{\mathbf{z}}$. Since $\chi$ is spherically symmetric, the integral for $A(\mathbf{R})$ has the same value for all nearest neighbor pairs. If $A = \int \chi^*(\mathbf{r})[U_a(\mathbf{r} - \mathbf{R}) - U(\mathbf{r})]\chi(\mathbf{r} - \mathbf{R}) \, d\tau$ for nearest neighbors, then

$$E(\mathbf{k}) = E_a - \alpha$$
$$- A[e^{ik_x a} + e^{-ik_x a} + e^{ik_y a} + e^{-ik_y a} + e^{ik_z a} + e^{-ik_z a}]$$
$$= E_a - \alpha - 2A[\cos(k_x a) + \cos(k_y a) + \cos(k_z a)] \, .$$

(b) Since the Brillouin zone is a cube with edge $2\pi/a$, $k_x$, $k_y$, and $k_z$ each range from $-\pi/a$ to $+\pi/a$. If $A$ is positive the minimum energy occurs for $\mathbf{k} = 0$ and is $E_a - \alpha - 6A$. The maximum energy occurs for $k_x = k_y = k_z = \pi/a$ and is $E_a - \alpha + 6A$. The band width is $12A$. ∎

## 7.4  NEARLY FREE ELECTRONS

*Free Electrons.*  For electrons with energies well above the potential energy maxima, the potential energy function does not influence wave functions except at points quite close to ion cores. As a first approximation we replace the actual potential energy function by its volume average, $U_0$. $U_0$ does not depend on $\mathbf{r}$, so we may use it to replace the actual potential energy function only if forces on electrons can be ignored. Later we will assume weak forces act and modify the model accordingly.

The time-independent Schrödinger equation,

$$-\frac{\hbar^2}{2m}\,\nabla^2\psi(r) + U_0\psi(r) = E\psi(r) \tag{7-31}$$

has solutions of the form

$$\psi = Ae^{i\mathbf{k}'\cdot\mathbf{r}}, \tag{7-32}$$

where $\mathbf{k}'$ is a propagation vector. A prime is used to reserve the symbol $\mathbf{k}$, without a prime, for propagation vectors in the Brillouin zone. The constant $A$ is determined by the normalization condition $\int\psi^*\psi\,d\tau = 1$, where the integral is over the volume of the sample. Since $\psi^*\psi = |A|^2$, we may take $A$ to be $1/\sqrt{\tau_s}$ where $\tau_s$ is the sample volume, and write

$$\psi = \frac{1}{\sqrt{\tau_s}}\,e^{i\mathbf{k}'\cdot\mathbf{r}}. \tag{7-33}$$

A wave function of the form given by Eq. 7-33 is a plane traveling wave with wavelength $\lambda = 2\pi/k'$. We have implicitly applied boundary conditions appropriate to an infinite solid. The wave exists throughout all space and is not, for example, reflected or attenuated at boundaries. When we must consider a finite sample we take it to be a portion of an infinite solid and continue to use plane waves.

Since $\nabla^2\psi = -(k')^2\psi$, Eq. 7-31 gives

$$E = U_0 + \frac{\hbar^2}{2m}\,(k')^2 \tag{7-34}$$

for the relationship between the electron energy and propagation vector. An electron with a wave function given by Eq. 7-33 has momentum $\mathbf{p} = \hbar\mathbf{k}'$, so Eq. 7-34 can be written $E = U_0 + p^2/2m$. The first term is the potential energy while the second is the kinetic energy.

Although free electron wave functions do not depend in any way on the structure of the solid, they can be written in the form of Bloch functions. Consider any crystal and suppose the set of reciprocal lattice vectors are known. For every propagation vector $\mathbf{k}'$ there is a reciprocal lattice vector $\mathbf{G}$ such that $\mathbf{k} = \mathbf{k}' + \mathbf{G}$ is a vector in the Brillouin zone. The selection of $\mathbf{G}$ is illustrated in Fig. 7-5: for any $\mathbf{k}'$ the appropriate reciprocal lattice vector is the one from the reciprocal lattice point nearest $\mathbf{k}'$ to the origin. In terms of a propagation

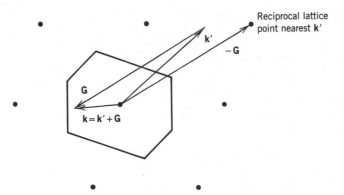

**FIGURE 7-5** Dots represent reciprocal lattice points and the Brillouin zone boundary is drawn with heavy lines. $\mathbf{k}'$ is an arbitrary vector in reciprocal space, with the point at $-\mathbf{G}$ nearest its tip. $\mathbf{k} = \mathbf{k}' + \mathbf{G}$ is a vector in the zone.

vector $\mathbf{k}$ in the zone, the wave function and energy are given by

$$\psi(\mathbf{k}, \mathbf{r}) = \frac{1}{\sqrt{\tau_s}} e^{i\mathbf{G}\cdot\mathbf{r}} e^{i\mathbf{k}\cdot\mathbf{r}} \qquad (7\text{-}35)$$

and

$$E(\mathbf{k}) = U_0 + \frac{\hbar^2}{2m} |\mathbf{k} + \mathbf{G}|^2, \qquad (7\text{-}36)$$

respectively. Note that $e^{i\mathbf{G}\cdot\mathbf{r}}$ has the periodicity of the lattice, so $\psi(\mathbf{k}, \mathbf{r})$ is a Bloch function.

Figure 7-6 illustrates free electron bands for a one-dimensional crystal with cell length $a$. In $(a)$ the energy is plotted as a function of the propagation constant $k'$. Dotted lines mark boundaries of the Brillouin zone. To draw $(b)$, each region outside the zone is translated by a multiple of $2\pi/a$ and thereby brought into the zone. The result is very much like a standard band diagram in that a large number of discrete energy levels exist for each propagation vector in the zone. It differs from a band diagram for a real crystal in that there are no gaps in the energy spectrum. In the next section we will see how energy gaps come about.

**EXAMPLE 7-3** A certain simple cubic crystal has cube edge $a = 5.7$ Å. Calculate the four lowest free electron energies if the wave vector $\mathbf{k}$ in the reduced zone scheme has magnitude $\pi/2a$ and is normal to a cube face.

**SOLUTION** Orient a Cartesian coordinate system with its axes parallel to cube edges and take $\mathbf{k} = (\pi/2a)\hat{\mathbf{x}}$. Reciprocal lattice vectors have the form $\mathbf{G} = h(2\pi/a)\hat{\mathbf{x}} + k(2\pi/a)\hat{\mathbf{y}} + l(2\pi/a)\hat{\mathbf{z}}$ so, for $U_0 = 0$, the energy levels are

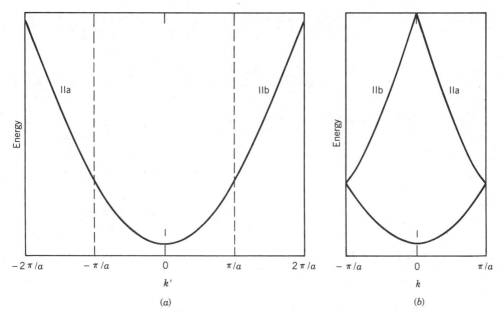

**FIGURE 7-6** (a) Energy as a function of propagation constant for free electrons in a one-dimensional solid. The dotted lines mark Brillouin zone boundaries for a crystal with a direct lattice cell of length $a$. (b) The same energy spectrum as shown in (a) but plotted in the reduced zone scheme. The portion of the curve to the right of $\pi/a$ has been translated left by $2\pi/a$ and the portion to the left of $-\pi/a$ has been translated right by $2\pi/a$.

given by

$$E(\mathbf{k}) = \frac{\hbar^2}{2m}\left[\left(\frac{\pi}{2a} + \frac{2\pi h}{a}\right)^2 + \left(\frac{2\pi k}{a}\right)^2 + \left(\frac{2\pi \ell}{a}\right)^2\right]$$

$$= \frac{2\hbar^2\pi^2}{ma^2}\left[\left(\frac{1}{4} + h\right)^2 + k^2 + \ell^2\right].$$

Select integer values for $h$, $k$, and $\ell$ to obtain the lowest four values. For (000), $E = 4.59 \times 10^{-20}$ J (0.287 eV); for $(\overline{1}00)$, $E = 4.14 \times 10^{-19}$ J (2.59 eV); for (010), $E = 7.81 \times 10^{-19}$ J (4.88 eV); and for (100), $E = 1.15 \times 10^{-18}$ J (7.18 eV).   ■

*Nearly Free Electrons.*   We now suppose the electrons interact weakly with the ion cores of the solid. Neither electron wave functions or energy levels are much different from those for strictly free electrons. Since a wave function for a nearly free electron is nearly a plane wave, we expand the periodic part $u(\mathbf{k}, \mathbf{r})$ of a Bloch function in a Fourier series.

Any continuous function $f(\mathbf{r})$ with the periodicity of the lattice has a Fourier series of the form

$$f(\mathbf{r}) = \sum_{\mathbf{G}} A(\mathbf{G})\, e^{i\mathbf{G}\cdot\mathbf{r}}, \tag{7-37}$$

where $A(\mathbf{G})$ is independent of $\mathbf{r}$ and the sum is over all reciprocal lattice vectors. If the periodic function $f(\mathbf{r})$ is known, the Fourier coefficients $A(\mathbf{G})$ can be found using

$$A(\mathbf{G}) = \frac{1}{\tau} \int f(\mathbf{r}) \, e^{-i\mathbf{G} \cdot \mathbf{r}} \, d\tau , \qquad (7\text{-}38)$$

where the integral is over the volume $\tau$ of a unit cell. To show Eq. 7-38 is valid, multiply both sides of Eq. 7-34 by $e^{-i\mathbf{G}' \cdot \mathbf{r}}$, integrate over the volume of a unit cell, and use the identity

$$\int e^{i(\mathbf{G} - \mathbf{G}') \cdot \mathbf{r}} \, d\tau = 0 \qquad (7\text{-}39)$$

if $\mathbf{G}' \neq \mathbf{G}$. If $\mathbf{G}' = \mathbf{G}$, the integral yields $\tau$. The only term in the sum that survives is the $\mathbf{G}' = \mathbf{G}$ term and Eq. 7-38 follows immediately. See Appendix B for details.

If we write

$$u_n(\mathbf{k}, \mathbf{r}) = \sum_{\mathbf{G}} u(\mathbf{G}) \, e^{i\mathbf{G} \cdot \mathbf{r}} \qquad (7\text{-}40)$$

for the periodic part of the wave function, the wave function itself is given by

$$\psi(\mathbf{k}, \mathbf{r}) = \sum_{\mathbf{G}} u(\mathbf{G}) \, e^{i(\mathbf{k} + \mathbf{G}) \cdot \mathbf{r}} , \qquad (7\text{-}41)$$

where $\mathbf{k}$ is a propagation vector in the Brillouin zone. Equation 7-41 can be used to represent the wave function of any electron in a crystal. If all but a few of the coefficients are small, it is appropriate for a nearly free electron.

The potential energy function also has the periodicity of the lattice, so its Fourier series has the form

$$U(\mathbf{r}) = \sum_{\mathbf{G}'} U(\mathbf{G}') \, e^{i\mathbf{G}' \cdot \mathbf{r}} . \qquad (7\text{-}42)$$

$U(0)$, corresponding to $\mathbf{G}' = 0$, is the volume average of the potential energy, which we denote by $U_0$. In addition, $U(\mathbf{r})$ is real, so $U^*(\mathbf{G}) = U(-\mathbf{G})$.

Substitute Eqs. 7-41 and 7-42 into the time-independent Schrödinger equation and make use of $\nabla^2 e^{i(\mathbf{k} + \mathbf{G}) \cdot \mathbf{r}} = -|\mathbf{k} + \mathbf{G}|^2 e^{i(\mathbf{k} + \mathbf{G}) \cdot \mathbf{r}}$. Cancel the common factor $e^{i\mathbf{k} \cdot \mathbf{r}}$ and rearrange the terms to obtain

$$\sum_{\mathbf{G}} \left[ E - U_0 - \frac{\hbar^2}{2m} |\mathbf{k} + \mathbf{G}|^2 \right] u(\mathbf{G}) e^{i\mathbf{G} \cdot \mathbf{r}}$$

$$= \sum_{\mathbf{G}} \sum_{\mathbf{G}'} u(\mathbf{G}) U(\mathbf{G}') e^{i(\mathbf{G} + \mathbf{G}') \cdot \mathbf{r}} , \qquad (7\text{-}43)$$

where the prime on the summation symbol indicates that the $\mathbf{G}' = 0$ term is omitted. It has been transposed to the left side. Now multiply by $e^{-i\mathbf{G}'' \cdot \mathbf{r}}$ and integrate over a unit cell. According to Eq. 7-39, the result is zero for every term on the left except the one for which $\mathbf{G} = \mathbf{G}''$ and for every term on the right except those for which $\mathbf{G} + \mathbf{G}' = \mathbf{G}''$. We choose to retain the sum over $\mathbf{G}'$. For a particular $\mathbf{G}'$ the only terms to survive are those for which $\mathbf{G} = \mathbf{G}'' - \mathbf{G}'$, so Eq. 7-43 becomes

$$\left[ E - U_0 - \frac{\hbar^2}{2m} |\mathbf{k} + \mathbf{G}''|^2 \right] u(\mathbf{G}'') = \sum_{\mathbf{G}}{}' u(\mathbf{G}'' - \mathbf{G}')U(\mathbf{G}'). \quad (7\text{-}44)$$

A set of simultaneous equations for the unknowns $u(\mathbf{G})$ is generated by taking $\mathbf{G}''$ to be each of the reciprocal lattice vectors in turn. Allowed energies for a given propagation vector $\mathbf{k}$ are determined by equating to zero the determinant formed by the coefficients of the unknowns. Practical considerations limit the number of terms in the sum and the number of equations in the set to several hundred. Instead of carrying out the procedure described above, we will use an approximation technique.

Suppose the dependence of $U(\mathbf{r})$ on $\mathbf{r}$ is so weak that $U(\mathbf{G})$ is small for every reciprocal lattice vector except $\mathbf{G} = 0$. Since the wave function then differs only slightly from a free electron wave function, we suppose only one of the coefficients $u(\mathbf{G})$ is large while the others are small. For illustrative purposes we examine a wave function that is nearly $e^{i\mathbf{k}\cdot\mathbf{r}}$ and, to do this, we assume all coefficients $u(\mathbf{G})$ are small except $u(0)$. Later we will need to change this assumption.

In the sum on the right side of Eq. 7-44, neglect all terms that are products of small quantities. The only term that is not negligible is the $\mathbf{G}' = \mathbf{G}''$ term, so the equation becomes

$$\left[ E - U_0 - \frac{\hbar^2}{2m} |\mathbf{k} + \mathbf{G}''|^2 \right] u(\mathbf{G}'') = u(0)U(\mathbf{G}'') \quad (7\text{-}45)$$

for $\mathbf{G}'' \neq 0$ and

$$\left[ E - U_0 - \frac{\hbar^2}{2m} k^2 \right] u(0) = 0 \quad (7\text{-}46)$$

for $\mathbf{G}'' = 0$. The right side of the last equation vanishes since the sum in Eq. 7-44 does not contain a $\mathbf{G}'' = 0$ term. Equation 7-46 is satisfied if $E = U_0 + (\hbar^2 k^2 / 2m)$. In this approximation, the energy levels are exactly the same as the free electron levels.

The expression for the energy is substituted into Eq. 7-45 and the result is solved for $u(\mathbf{G}'')$:

$$u(\mathbf{G}'') = \frac{2mU(\mathbf{G}'')}{\hbar^2[k^2 - |\mathbf{k} + \mathbf{G}''|^2]} u(0) \quad (7\text{-}47)$$

for all $\mathbf{G}''$ except $\mathbf{G}'' = 0$. To find the wave function, this expression is substituted into Eq. 7-41 and $u(0)$ is chosen so the wave function is normalized. Although energy levels are the same as free electron levels, wave functions are not. The periodic part $u(\mathbf{k}, \mathbf{r})$ is no longer constant but varies within a unit cell to reflect the influence of interactions with ion cores.

The calculation just carried out fails if $\mathbf{k}$ is near a Brillouin zone boundary.

Suppose, for example, it is near the plane that bisects $G_1$. Then $\mathbf{k}$ is nearly $\frac{1}{2}G_1$ and $k^2$ is nearly $|\mathbf{k} - G_1|^2$. According to Eq. 7-47, $u(-G_1)$ is large, in contradiction to the original assumption that it is not.

The method can be corrected easily. In Eq. 7-44, we assume both $u(0)$ and $u(-G_1)$ are large while other Fourier coefficients $u(G)$ are small. The most significant terms in the sum on the right side of the equation are the $\mathbf{G'} = \mathbf{G''}$ term, $u(0)U(\mathbf{G''})$, and the $\mathbf{G'} = \mathbf{G''} + G_1$ term, $u(-G_1)U(\mathbf{G''} + G_1)$. Other terms are products of small quantities.

Equation 7-44 becomes

$$\left[ E - U_0 - \frac{\hbar^2}{2m} k^2 \right] u(0) = u(-G_1)U(-G_1) \tag{7-48}$$

for $\mathbf{G''} = 0$ and

$$\left[ E - U_0 - \frac{\hbar^2}{2m} |\mathbf{k} - G_1|^2 \right] u(-G_1) = u(0)U(G_1) \tag{7-49}$$

for $\mathbf{G''} = -G_1$. These are two simultaneous homogeneous equations for $u(0)$ and $u(-G_1)$. The energy is determined by the condition

$$\begin{vmatrix} E - U_0 - \dfrac{\hbar^2}{2m} k^2 & - U(G_1) \\[2mm] -U(-G_1) & E - U_0 - \dfrac{\hbar^2}{2m} |\mathbf{k} + G_1|^2 \end{vmatrix} = 0. \tag{7-50}$$

When the determinant is expanded, Eq. 7-50 becomes

$$\left[ E - U_0 - \frac{\hbar^2}{2m} k^2 \right]\left[ E - U_0 - \frac{\hbar^2}{2m} |\mathbf{k} - G_1|^2 \right] - |U(G_1)|^2 = 0, \tag{7-51}$$

where we have used $U(-G_1) = U^*(G_1)$. This is a quadradic equation for $E$ and has the two solutions

$$E(\mathbf{k}) = U_0 + \frac{\hbar^2}{4m} [k^2 + |\mathbf{k} - G_1|^2]$$
$$+ \tfrac{1}{2}[(\hbar^2/2m)^2(|\mathbf{k} - G_1|^2 - k^2)^2 + 4|U(G_1)|^2]^{1/2}. \tag{7-52}$$

Figure 7-7 shows the two bands described by Eq. 7-52, plotted for $\mathbf{k}$ parallel to $G_1$. They should be compared with the free electron bands shown in Fig. 7-6b. Far from the zone boundary the two diagrams are alike, but near the boundary the lower curve is below the free electron curve while the upper curve is above it. A gap appears at the zone boundary. The split in energy at the zone boundary can be found by setting $\mathbf{k} = \frac{1}{2}G_1$ in Eq. 7-52. The higher and lower energies are $E = U_0 + (\hbar^2 G_1^2/8m) + |U(G_1)|$ and $E = U_0 + (\hbar^2 G_1^2/8m) - |U(G_1)|$, respectively, so the gap at the zone boundary is $2|U(G_1)|$, twice the Fourier coefficient of the potential energy function. In the limit as $U(G_1)$ vanishes, both energies have the value $U_0 + (\hbar^2 G_1^2/8m)$, the appropriate free electron energy.

Expressions for the wave functions can also be found. Equation 7-48 is solved for $u(-G_1)$ in terms of $u(0)$ and one of the expressions given by Eq. 7-52 is

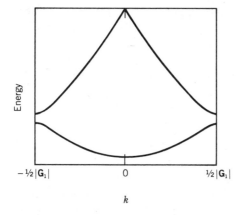

**FIGURE 7-7** Energy as a function of propagation vector magnitude for nearly free electrons. The propagation vector is taken to be parallel to reciprocal lattice vector $\mathbf{G}_1$, which is bisected by a Brillouin zone boundary. The curves are similar to those of Fig. 7-6$b$ but gaps occur at $\mathbf{k} = \pm\frac{1}{2}\mathbf{G}_1$. The gap width depends on the Fourier component $U(\mathbf{G}_1)$ of the potential energy function.

substituted for the energy. At the zone boundary the wave function for the lower energy state is

$$\psi(\mathbf{k}, \mathbf{r}) = e^{i\mathbf{k}\cdot\mathbf{r}}\, u(0)[1 - e^{-i\mathbf{G}_1\cdot\mathbf{r}}] \tag{7-53}$$

while the wave function for the upper energy state is

$$\psi(\mathbf{k}\cdot\mathbf{r}) = e^{i\mathbf{k}\cdot\mathbf{r}}\, u(0)[1 + e^{-i\mathbf{G}_1\cdot\mathbf{r}}] . \tag{7-54}$$

In each case $u(0)$ is found by requiring that the wave function be normalized.

Equations 7-53 and 7-54 help us qualitatively understand the origin of the energy gap. The Fourier coefficients $U(\mathbf{G}_1)$ and $U(-\mathbf{G}_1)$ of the potential energy function lead to a spatial dependence of the form

$$U(\mathbf{r}) = U(\mathbf{G}_1)e^{i\mathbf{G}_1\cdot\mathbf{r}} + U(-\mathbf{G}_1)e^{-i\mathbf{G}_1\cdot\mathbf{r}} = 2U(\mathbf{G}_1)\cos(\mathbf{G}_1\cdot\mathbf{r}), \tag{7-55}$$

where we have assumed $U(\mathbf{G}_1)$ is real. The true potential energy function has other Fourier components but the two used in Eq. 7-55 are the ones responsible for the gap. The probability density for the lower energy state,

$$|\psi|^2 = 2|u(0)|^2[1 - \cos(\mathbf{G}_1\cdot\mathbf{r})] , \tag{7-56}$$

is large at places where the potential energy is large and negative. On the other hand, the probability density for the higher energy state,

$$|\psi|^2 = 2|u(0)|^2[1 + \cos(\mathbf{G}_1\cdot\mathbf{r})] , \tag{7-57}$$

is large where the potential energy is large and positive. Electrons in the two states also have different average kinetic energies. These differences account for the gap.

Equations 7-53 and 7-54 are often interpreted as resulting from the Bragg diffraction of the electron wave. A free electron wave in a crystal undergoes scattering and the propagation vector for the scattered wave differs from that for the incident wave by a reciprocal lattice vector. Both the incident and scattered waves are included in $\psi$.

The analysis can be modified easily to deal with nearly free electron states of higher energy than those considered above. Suppose $\mathbf{G}_2$ is any reciprocal lattice vector and $G_1$ is again a reciprocal lattice vector that is bisected by the Brillouin zone boundary. The two free electron bands with energy given by $(\hbar^2/2m)|k + G_2|^2$ and $(\hbar^2/2m)|k - G_1 - G_2|^2$, respectively, are degenerate for $k = \frac{1}{2}G_1$. To find the gap when weak forces act, assume $u(G_2)$ and $u(G_1 + G_2)$ are large while other Fourier coefficients are small, then carry out the analysis as before.

**EXAMPLE 7-4** A certain simple cubic structure has a cube edge of 4.85 Å. Take the zero of energy to be at the bottom of the lowest free electron band. (a) Assume the electrons are completely free and calculate the energy of the lowest energy state with propagation vector at the center of a Brillouin zone face. (b) Suppose $U(G_1) = 0.24$ eV, where $G_1$ is the reciprocal lattice vector perpendicular to the Brillouin zone face of part a. Calculate the energy of the two lowest nearly free electron states with the propagation vector used in part a.

**SOLUTION** (a) At the zone face center $k = \pi/a$ and $E = \hbar^2 k^2/2m =$ $(1.05 \times 10^{-34})^2(\pi/4.85 \times 10^{-10})^2/(2 \times 9.11 \times 10^{-31}) = 2.54 \times 10^{-19}$ J (1.59 eV). (b) There are now two distinct levels, one $|U(G_1)|$ below the free electron level, at 1.35 eV, and one $|U(G_1)|$ above the free electron level, at 1.83 eV. ∎

## 7.5  ELECTRON ENERGY CALCULATIONS

*Advanced Techniques.* Expansions of wave functions as linear combinations of atomic orbitals and as Fourier series can both be used, in principle, to solve for electron wave functions and energy levels in solids. Most electrons of interest, however, have energies that are neither high above potential energy maxima nor deep in potential energy wells. If atomic orbitals are used, a large number must be included in the sum to give a reasonable approximation to the wave function in interstitial regions. If Fourier series are used, an enormous number of terms are required to reproduce wave functions near nuclei, where the potential energy function changes rapidly with position.

Apparently, wave functions of interest can be represented best by linear combinations of atomic orbitals near a nucleus and by linear combinations of plane waves in the interstitial regions. Most band structure calculations make use of functions that are compromises between plane waves and atomic orbitals. Each of these functions is nearly a plane wave in interstitial regions but other terms are added to obtain the correct functional behavior near nuclei. A trial wave function is constructed as a linear combination of these special functions and the Schrödinger equation is used to find the coefficients. Calculations using special functions such as these have been highly successful in obtaining energy levels to three significant figures or better. The reader is referred to references listed at the end of this chapter for details.

Linear combinations of both atomic orbitals and linear combinations of plane waves are used to obtain wave functions and energy levels for electrons in amorphous solids. Wave functions for electrons in these materials, however, do not have the Bloch form, so the calculations are much more formidable. Some are described in references listed at the end of this chapter.

*Examples of Band Structures.* Energy bands of interest for many metals are above potential energy maxima and resemble nearly free electron bands. The potential energy function is uniform over a large fraction of the unit cell volume and a significant force acts on an electron only when it is quite near a nucleus. Although the force of an ion core on an electron is strong at short ranges, its effect on the wave function is small. Except near Brillouin zone boundaries bands remain nearly parabolic, with the form $E = Ak^2$, although the value of $A$ may differ somewhat from $\hbar^2/2m$. Bands bend near boundaries to form gaps.

Figure 7-8 shows the band structure of potassium for three directions of **k**. These curves are typical of all alkali metals, which come the closest of all materials to the predictions of the nearly free electron model.

The presence of atomic $d$ states with energies near the free electron bands complicates the band structure. Look again at Fig. 7-3, which shows the band structure of copper. The lowest band shown starts out near **k** $= 0$ like a free electron band. At about $-13$ eV the electron energy is roughly that of an atomic $d$ state in copper and the $d$ component of the wave function becomes significant. From then on the band loses its free electron character and the energy becomes much less dependent on **k**, like a tight binding band. The reverse happens for one of the higher energy curves. Near **k** $= 0$ the associated wave functions have large $d$ components and the curve starts out like a tight binding band; then, near the center of the diagram, it changes character to become like a free electron band. If there were no $d$ states in this energy region, the two curves would link together to form a single nearly free electron band.

Wave functions for the remaining bands shown all have large $d$ components. If spin is ignored, an atomic $d$ subshell consists of five $d$ states and, as a consequence, five different wave functions can be constructed for each value of **k**. Their energies are shown in the diagram along with nearly free electron energies. Although there are six bands in all, two of the $d$ bands are degenerate for the direction of **k** chosen, so the diagram shows only five distinct curves.

The band structure for a typical covalent solid, silicon, is shown in Fig. 7-9. Wave functions associated with the lowest bands shown, those with energy below 0 on the diagram, are formed primarily of outer shell $s$ and $p$ functions and are similar to covalent bonding functions. All states of these bands are occupied when the crystal has the lowest possible total energy. There are four valence bands, as they are called, and they are distinct for the direction of **k** in ($b$) but two are degenerate for the directions in ($a$) and ($c$). Higher energy bands, in the region of the potential energy maxima and above, are completely empty when the total crystal energy is a minimum. Some states in these bands are occupied at elevated temperatures and the electrons in them contribute to electrical conduction; hence, they are known as conduction bands. An energy

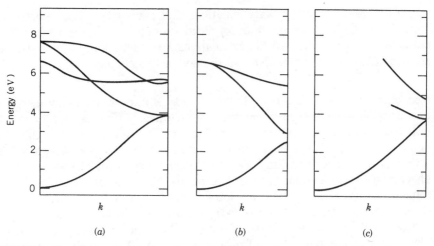

**FIGURE 7-8** Electron band structure of potassium for energies near the 4s atomic level: (a) **k** in the [100] direction; (b) **k** in the [110] direction; and (c) **k** in the [111] direction. Some curves are incomplete. The bands shown are quite similar in form to those predicted by the nearly free electron model. (From F. S. Ham, *Phys. Rev.* **128:**82, 1962. Used with permission.)

gap exists between these two sets of bands and, as we will see, it plays an important part in determining the electrical properties of silicon.

*Effective Mass.* Over much of the Brillouin zone electron energies in some bands can be approximated by an expression of the form

$$E_n(\mathbf{k}) = A|\mathbf{k} - \mathbf{k}_0|^2, \tag{7-58}$$

where $A$ is a constant and $\mathbf{k}_0$ is the propagation vector associated with the minimum energy state of the band. For a free electron band $A = \hbar^2/2m$, but for other bands it has a different value. By analogy with free electrons, $A$ is usually replaced by $\hbar^2/2m^*$, where $m^*$ is called the electron effective mass. The effective mass is not usually the same as the mass of a free electron because electrons interact with ion cores of the crystal. The stronger those interactions, the more tightly electrons are bound to atoms and the larger the effective mass. A band with $m^* > m$ is flatter than a free electron band.

The definition of effective mass can be generalized so it is valid for every band, even those that do not have the form given in Eq. 7-58. The generalization takes the form of a tensor called the reciprocal effective mass tensor, defined by

$$\left[\frac{1}{m^*}\right]_{ij} = \frac{1}{\hbar^2} \frac{\partial^2 E(\mathbf{k})}{\partial k_i \partial k_j}, \tag{7-59}$$

where $i$ and $j$ represent Cartesian coordinates. A reciprocal effective mass tensor is defined for each electron state.

Elements of the tensor give the curvature of the band at a point in reciprocal

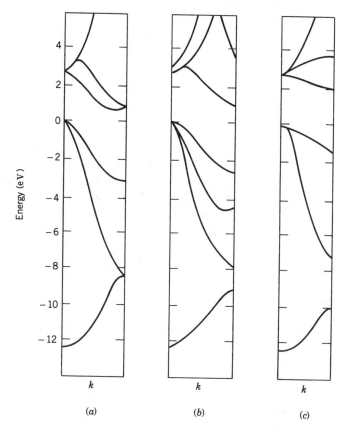

**FIGURE 7-9**  Electron band structure of crystalline silicon for energies near the 3s and 3p atomic levels: (a) **k** in the [100] direction; (b) **k** in the [110] direction; and (c) **k** in the [111] direction. Bands below $E = 0$ are valence bands and are associated with bonding. Higher bands are conduction bands and are important for the electrical properties of silicon. A gap exists between the valence and conduction bands. (From A. Zunger and M. L. Cohen, *Phys. Rev.* **B** 20:4082, 1979. Used with permission.)

space and they may be positive, negative, or zero. If electron energies are given by Eq. 7-58, then the reciprocal effective mass tensor reduces to a diagonal tensor with each diagonal element given by $(2A/\hbar^2)$. The effective mass is then said to be a scalar.

Since they give the curvature, reciprocal effective mass tensors are useful for describing bands. As we will see in Chapter 9, they are also needed to describe the reaction of an electron to an electromagnetic field and so are useful for the study of electron motion in applied fields.

**EXAMPLE 7-5**  Derive expressions for elements of the reciprocal effective mass tensor for the tight binding band of Example 7-2. Find limiting values for states near **k** = 0.

**SOLUTION**   Differentiate $E = E_a - \alpha - 2A[\cos(k_x a) + \cos(k_y a) + \cos(k_z a)]$ twice with respect to $k_x$ and divide by $\hbar^2$ to find

$$(1/m^*)_{xx} = (2Aa^2/\hbar^2)\cos(k_x a) .$$

Similarly,

$$(1/m^*)_{yy} = (2Aa^2/\hbar^2)\cos(k_y a)$$

and

$$(1/m^*)_{zz} = (2Aa^2/\hbar^2)\cos(k_z a) .$$

All other elements vanish. For $\mathbf{k}$ nearly 0, each of the cosine functions may be replaced by 1 so each diagonal element is $2Aa^2/\hbar^2$. In this limit the effective mass is a scalar and its value is given by $m^* = \hbar^2/2Aa^2$. For a tightly bound electron, the overlap integral $A$ is small and the effective mass is large. If overlap increases and the wave function spreads over a larger volume, the effective mass decreases. ∎

## 7.6   SURFACE STATES

Crystal surfaces have only small influence on the bulk wave functions and energy levels we have been studying. They do, however, give rise to additional electron states. Wave functions associated with surface states extend over the entire surface, but they are attenuated quite sharply with distance into the sample. Energies of some surface states coincide with electron energies for the bulk material while energies of others lie in gaps between bands.

If atoms on the surface form a two-dimensional lattice, then each electron wave function satisfies a two-dimensional Bloch theorem: $\psi(\mathbf{r} + \mathbf{R}) = e^{i\mathbf{k}\cdot\mathbf{R}}\psi(\mathbf{r})$, where $\mathbf{R}$ is a surface lattice vector and the propagation vector $\mathbf{k}$ is parallel to the surface. Atomic positions do not form a lattice along lines extending into the interior and no analogous theorem relates the wave function at a surface point to the wave function at any interior point. Each wave function tends to zero at large distances on either side of the surface.

As a consequence, calculations of electron wave functions and energies are formidable. Only one unit cell of the surface lattice need be considered, but the Schrödinger equation must be integrated over a region that extends reasonably far on either side of the surface and so contains a large number of atoms. The width of the region is limited by practical considerations of computation time and computer memory size.

In one model calculation the crystal is replaced by a slab, on the order of 10 atomic layers thick.[*] The slab is bounded by two parallel infinite planes, one coinciding with the sample surface and the other in the interior of the sample. A structure that is periodic in three dimensions is constructed mathematically

---

[*]Details of this model can be found in Kai-Ming Ho, B. N. Harmon, and S. H. Liu, *Phys. Rev. Lett.* **44:**1531, 1980.

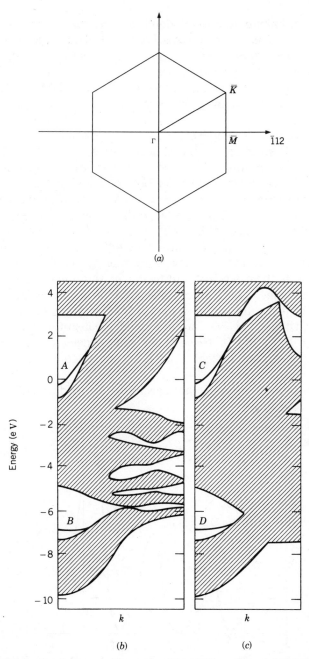

**FIGURE 7-10** Energy levels for a gold sample with a (111) surface. (a) The two-dimensional Brillouin zone for the surface lattice, showing directions of the propagation constant for which energies are plotted in (b) and (c). In (b) and (c), energies for the bulk material are indicated by shaded regions and energies for surface states are indicated by lines labeled A, B, C, and D. (From S. H. Liu, C. Hinnen, C. Nguyen van Huong, N. R. De Tacconi, and K. M. Ho, *J. Electroanal. Chem.* **176**:325, 1984. Used with permission.)

by arranging an infinite number of identical slabs, one on top of another with narrow spaces between adjacent slabs. For this structure, one face of a unit cell coincides with the two-dimensional unit cell of the surface lattice, but the unit cell dimension in the direction perpendicular to the surface is the combined thickness of one slab and one gap. If the primitive basis of the surface contains one atom, the unit cell for the model contains one atom for each atomic layer in a slab.

Figure 7-10 shows an example of results for a gold sample with a (111) surface. Gold has an FCC structure so the surface lattice is hexagonal and the two-dimensional Brillouin zone associated with it is the hexagon shown in ($a$). Energies are plotted in ($b$) and ($c$) for both bulk and surface states. For bulk states, the propagation vector can be written $\mathbf{k}_\parallel + \mathbf{k}_\perp$, where these vectors are respectively parallel and perpendicular to the surface. Energies are plotted as functions of the magnitude of $\mathbf{k}_\parallel$ for $\mathbf{k}_\parallel$ in each of the two directions indicated in ($a$). For each value of $\mathbf{k}_\parallel$ there are many different levels, corresponding to different values of $\mathbf{k}_\perp$, and these levels are represented by shaded regions on the diagram. Note the existence of forbidden gaps in the energy spectrum.

Allowed energy levels for surface states are represented by lines in Fig. 7-10$b$ and $c$. Some of these are within shaded regions and are not distinguished on the diagrams. Others, however, are in forbidden gaps of the bulk material and are labeled $A$, $B$, $C$, and $D$. These states can be detected experimentally by observing the absorption of light which is accompanied by electron transitions into surface states.

Figure 7-11 shows the probability density for the surface state with $\mathbf{k}_\parallel = 0$,

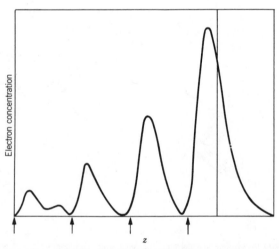

**FIGURE 7-11** Calculated probability density for the electron surface state with propagation vector at the center of the two-dimensional Brillouin zone, plotted along a line perpendicular to the surface. The sample is gold with a (111) surface. The vertical line indicates the boundary and the arrows indicate atomic layers in the material. (From S. H. Liu, C. Hinnen, C. Nguyen van Huong, N. R. De Tacconi, and K. M. Ho; *J. Electroanal. Chem.* **176**:325, 1984. Used with permission.)

plotted along a line perpendicular to the surface. The surface is indicated by a vertical line, with the sample to the left and vacuum to the right, while positions of atomic layers are indicated by arrows. In the sample, the probability density oscillates but the amplitude of oscillation decreases with distance from the surface and is quite small after only a few atomic layers. Note the exponential-like decay outside the sample.

## 7.7 REFERENCES

**Elementary:**

R. H. Bube, *Electrons in Solids* (New York: Academic, 1981).
C. M. Hurd, *Electrons in Metals* (New York: Wiley, 1975).
L. Pincherle, *Electronic Energy Bands in Solids* (London: MacDonald, 1971).

**Advanced:**

J. Callaway, *Energy Band Theory* (New York: Academic, 1964).
J. R. Reitz, "Methods of the One-Electron Theory of Solids" in *Solid State Physics* (F. Seitz and D. Turnbull, Eds.), Vol. 1, p. 1 (New York: Academic, 1955).
J. M. Ziman, "The Calculation of Bloch Functions" in *Solid State Physics* (H. Ehrenreich, F. Seitz, and D. Turnbull, Eds.), Vol. 26, p. 1 (New York: Academic, 1971).

**Books and articles on specialized computational techniques:**

D. W. Bullett, "The Renaissance and Quantitative Development of the Tight-Binding Method" in *Solid State Physics* (H. Ehrenreich, F. Seitz, and D. Turnbull, Eds.), Vol. 35, p. 129 (New York: Academic, 1980).
J. O. Dimmock, "The Calculation of Electronic Energy Bands by the Augmented Plane Wave Method" in *Solid State Physics* (H. Ehrenreich, F. Seitz, and D. Turnbull, Eds.), Vol. 26, p. 103 (New York: Academic, 1971).
F. S. Ham, "The Quantum Defect Method" in *Solid State Physics* (F. Seitz and D. Turnbull, Eds.), Vol. 1, p. 127 (New York: Academic, 1955).
W. A. Harrison, *Pseudopotentials in the Theory of Metals* (New York: Benjamin, 1966).
V. Heine, "The Pseudopotential Concept" in *Solid State Physics* (H. Ehrenreich, F. Seitz, and D. Turnbull, Eds.), Vol. 24, p. 1; (New York: Academic, 1970).
V. Heine, "Electronic Structure from the Point of View of the Local Atomic Environment" in *Solid State Physics* (H. Ehrenreich, F. Seitz, and D. Turnbull, Eds.), Vol. 35, p. 1 (New York: Academic, 1980).
T. Loucks, *Augmented Plane Wave Method* (New York: Benjamin, 1967).
T. O. Woodruff, "The Orthogonalized Plane-Wave Method" in *Solid State Physics* (F. Seitz and D. Turnbull, Eds.), Vol. 4, p. 367 (New York: Academic, 1957).

## PROBLEMS

1. (a) Give the details of an argument that shows $E_n(\mathbf{k})$ exhibits all the symmetry elements of the crystalline point group. Consider solutions to Eq. 7-11 for a crystal before and after it is rotated, for example. (b) For a crystal with a cubic structure and spherically symmetric atoms, find propagation

vectors for all states that, by virtue of symmetry, have the same energy as the state with $\mathbf{k} = (\pi/8a)[110]$.

2.  In the tight binding approximation, energies for an $s$ band in a simple cubic crystal are given by Eq. 7-30, with $\alpha$ and $A$ both positive. Rank the following points in the Brillouin zone in order of increasing energy: (a) the center of a face, (b) the center of an edge, (c) a point halfway between the center and a corner, along a body diagonal, (d) a point halfway between a face center and a corner, along a face diagonal, and (e) a corner.

3.  For the band given in Example 7-2, prove that: (a) $E(\mathbf{k} + \mathbf{G}) = E(\mathbf{k})$, where $\mathbf{G}$ is any reciprocal lattice vector; (b) $E(\mathbf{k}_1) = E(\mathbf{k}_2)$, where $\mathbf{k}_1$ and $\mathbf{k}_2$ are related by a rotation of $\pi/2$ about the [001] axis; and (c) $E(-\mathbf{k}) = E(\mathbf{k})$.

4.  (a) Derive an expression for the electron energy as a function of propagation vector $\mathbf{k}$ in the tight binding approximation for an FCC structure with cube edge $a$. Assume wave functions can be constructed from $s$ atomic orbitals and that only nearest-neighbor overlap is important. Express your answer in terms of the overlap integrals $\alpha$ and $A$, defined in the text. Assume they are positive and $A$ is the same for all nearest neighbors. (b) For what value(s) of $\mathbf{k}$ is the energy a maximum? (c) For what value(s) of $\mathbf{k}$ is the energy a minimum? (d) In terms of $\alpha$ and $A$, what range of energy does the band span? (e) Sketch the energy as a function of the magnitude of $\mathbf{k}$ for $\mathbf{k}$ along a line from the center of the Brillouin zone to the center of a hexagonal face. (f) Sketch the energy as a function of the magnitude of $\mathbf{k}$ for $\mathbf{k}$ along a line from the center of the zone to the center of a square face.

5.  (a) For the energy band found in Problem 4, use the power series expansion of a cosine function to find an expression for $E(\mathbf{k})$ valid for $\mathbf{k}$ near 0. Retain terms that are quadratic in the components of $\mathbf{k}$, but no higher order terms. (b) Use the result of part a to show that the effective mass is a scalar for $\mathbf{k}$ near 0 and find an expression its value. What happens to the effective mass if the atoms are brought closer together? (c) For $\mathbf{k}$ close to the center of a hexagonal zone face, let $\mathbf{k} = \mathbf{k}_0 - \delta\mathbf{k}$, where $\mathbf{k}_0$ is the center of the face, and show that, for $\delta\mathbf{k}$ small,

$$E \approx E_a - \alpha - A(a^2\delta k_x\delta k_y + a^2\delta k_x\delta k_z + a^2\delta k_y\delta k_z),$$

where $\alpha$ and $A$ are the usual overlap integrals. (d) Find expressions for the elements of the reciprocal effective mass tensor for a state with propagation vector at the center of a hexagonal zone face.

6.  As an example of a calculation of overlap integrals used in the tight binding approximation, consider two protons a distance $R$ apart on the $z$ axis. The total potential energy of an electron is

$$U(\mathbf{r}) = -\frac{e^2}{4\pi\epsilon_0}\left[\frac{1}{r} + \frac{1}{|\mathbf{r} - \mathbf{R}|}\right],$$

where $\mathbf{R} = R\hat{z}$. Take the atomic orbital to be the $1s$ wave function for hydrogen, $\chi(r) = (1/\sqrt{\pi})(1/a_0)^{3/2}e^{-r/a_0}$, and the atomic potential energy to be $U_a(r) = -(e^2/4\pi\epsilon_0)(1/r)$. Here $a_0$ is the Bohr radius, 0.529 Å. (a) Evaluate the overlap integrals $\alpha$ and $A(\mathbf{R})$, defined in the text. (b) Estimate the width of the tight binding band for a collection of hydrogen atoms that form a simple cubic structure with cube edge $a_0$. (c) What is the width if the cube edge is $2a_0$?

7. A simple cubic crystal structure has cube edge $a$. Tell whether each of the following propagation vectors $\mathbf{k}$ is in the Brillouin zone. If it is not, find a reciprocal lattice vector $\mathbf{G}$ such that $\mathbf{k} + \mathbf{G}$ is in the zone. In each case, find the electron energy in the free electron approximation. (a) $(3\pi/a)[100]$. (b) $(3\pi/a)[111]$. (c) $(5\pi/a)[110]$ (d) $(7\pi/2a)[121]$.

8. Rubidium has a body-centered cubic structure with a cube edge of 5.585 Å. Suppose the gap between the lowest and next lowest nearly free electron band is 0.857 eV at the center of the zone face in the [110] direction. Assume the gap is produced by a single Fourier component of the potential energy function and take the effective mass to be the free electron mass. Calculate the energies, relative to the bottom of the lower band, of states in the two bands with propagation vectors at the center of the zone face in the [110] direction.

9. Both sodium and cesium have BCC structures. The cube edge for sodium is 4.225 Å while the cube edge for cesium is 6.045 Å. Consider the lowest nearly free electron band for each of these materials and take $\mathbf{k}$ to be in the [110] direction. $E(\mathbf{k})$ for sodium is nearly parabolic to the zone boundary and the gap there may be neglected. The width of the band is 4.19 eV. $E(\mathbf{k})$ for cesium bends in a manner typical of a nearly free electron band. The width of the band is 1.67 eV and the gap at the zone boundary is 1.16 eV. Estimate the effective masses of electrons near $\mathbf{k} = 0$ in the lowest nearly free electron bands of these two crystals.

10. Calcium has an FCC structure with a cube edge of 5.58 Å. Assume the bands of interest are free electron bands. (a) Find the maximum energy for the lowest free electron band, relative to the energy at the bottom of the band. What are the propagation vectors of the states associated with this energy? (b) Find the minimum energy of the second free electron band, relative to the bottom of the first band. What are the propagation vectors of the states associated with this energy? (c) Now use the nearly free electron model to estimate the magnitude of the Fourier component of the potential energy function required for the second band to be entirely above the first. Which Fourier components should be increased to do this?

11. Consider a nearly free electron band for which the wave functions are nearly $Ae^{i(\mathbf{k}+\mathbf{G}_2)\cdot\mathbf{r}}$ for $\mathbf{k}$ near the center of the Brillouin zone. Here $\mathbf{G}_2$ is a reciprocal lattice vector. As $\mathbf{k}$ approaches $\frac{1}{2}\mathbf{G}_1$, what Fourier component of the potential energy is most significant for the determination of the energy

gap? Assume all other components can be neglected and find an expression for the gap at the center of that face.

12. Many energy band calculations are based on what is known as a muffin-tin potential energy function. Imagine a sphere centered on each atom. Within any one of the spheres the potential energy is assumed to be spherically symmetric, while outside all spheres it is assumed to be constant. As an example, consider a crystal with a simple cubic structure (cube edge $a$) and take the sphere radius to be $R = \frac{1}{2}a$. Suppose the potential energy inside a sphere is given by $U(r) = -(e^2/4\pi\epsilon_0)(1/r)$, where $r$ is the radial distance from the sphere center, and suppose it is the constant $U_0$ at points outside all spheres. (a) Find an algebraic expression for the Fourier component $U(\mathbf{G})$ corresponding to the reciprocal lattice vector $\mathbf{G}$. The integral of $e^{-i\mathbf{G}\cdot\mathbf{r}}$ over the region outside the sphere can be evaluated as the difference of the integrals over the unit cell and over the sphere. The integral over the sphere can easily be evaluated if the $z$ axis is placed along $\mathbf{G}$ and spherical coordinates are used. (b) Use the nearly free electron approximation to estimate the energy gap between the first and second bands at the center of a Brillouin zone face. Take $U_0 = -(e^2/4\pi\epsilon_0)(1/R)$ and $a = 6.0$ Å.

13. The conduction band of silicon has a minimum for $\mathbf{k}$ about 0.85 of the distance from the zone center to the center of the square zone face in the [100] direction. Near the minimum, the energy is given by

$$E = \frac{\hbar^2}{2m}\left[\frac{(k_x - k_0)^2}{0.916} + \frac{k_y^2 + k_z^2}{0.191}\right].$$

(a) Show that the minimum energy occurs for $\mathbf{k} = k_0\hat{\mathbf{x}}$. (b) Evaluate the elements of the reciprocal effective mass tensor for an electron at this band minimum. (The conduction band of silicon has five other minima, related to this one by cubic symmetry.)

# Chapter 8

# THERMODYNAMICS OF PHONONS AND ELECTRONS

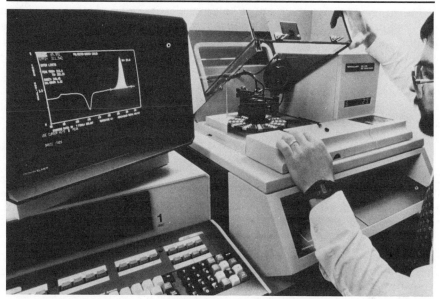

A modern calorimeter. The graph on the screen shows the rate of energy absorption as a function of temperature.

Properties of materials are temperature dependent because electrons occupy different states and vibrational modes have different energies at different temperatures. Imagine a solid in contact with a thermal reservoir at absolute temperature $T$. According to statistical mechanics, the probability that the solid is in a state with energy $E$ is proportional to $e^{-E/k_B T}$, where $k_B$ is Boltzmann's constant ($k_B = 1.381 \times 10^{-23}$ J/K $= 8.62 \times 10^{-5}$ eV/K). At high temperatures the solid has greater probability of occupying high-energy states than it does at low temperatures.

The energy $E$ includes interactions between all particles and is rather difficult to work with. Fortunately, thermodynamic calculations can be simplified because the total energy may be approximated by a sum of independent contributions, one for each normal mode and each electron state. As a consequence, we need not consider the solid as a whole but rather may treat each vibrational mode and each electron state as an independent system in contact with the reservoir. In the next section we use this approach to examine the energy of a normal mode. Electrons are discussed in later sections.

## 8.1  ATOMIC VIBRATIONS

*Phonon Statistics.*  A normal mode of angular frequency $\omega$ with $n$ phonons has energy $(n + \frac{1}{2})\hbar\omega$. Each phonon has energy $\hbar\omega$ and $\frac{1}{2}\hbar\omega$ is the zero-point energy of the vibration. The probability that the mode has $n$ phonons at temperature $T$ is given by

$$P_n = Ae^{-\beta(n+\frac{1}{2})\hbar\omega}, \tag{8-1}$$

where $\beta = 1/k_B T$ and $A$ is a constant. When Eq. 8-1 is summed over zero and all positive integers, the result must be 1. Terms in the sum form a simple geometric progression, so it can be written in closed form:

$$\sum_{n=0}^{\infty} P_n = Ae^{-\beta\hbar\omega/2} \sum_{n=0}^{\infty} e^{-\beta n\hbar\omega} = \frac{Ae^{-\beta\hbar\omega/2}}{1 - e^{-\beta\hbar\omega}}. \tag{8-2}$$

Thus

$$A = \frac{1 - e^{-\beta\hbar\omega}}{e^{-\beta\hbar\omega/2}}. \tag{8-3}$$

The average energy $\langle E \rangle$ of a mode in thermal equilibrium is found by summing all possible energy values, each weighted by the probability of its occurrence:

$$\langle E \rangle = A \sum_{n} (n + \tfrac{1}{2})\hbar\omega e^{-\beta[n+(1/2)]\hbar\omega}. \tag{8-4}$$

This can be evaluated by noting that

$$\sum_{n} (n + \tfrac{1}{2})\hbar\omega e^{-\beta[n+(1/2)]\hbar\omega} = -\frac{d}{d\beta}\left[\sum_{n} e^{-\beta[n+(1/2)]\hbar\omega}\right]$$

$$= -\frac{d}{d\beta}\frac{e^{-\beta\hbar\omega/2}}{1 - e^{-\beta\hbar\omega}}$$

$$= \hbar\omega\frac{e^{-3\beta\hbar\omega/2}}{(1 - e^{-\beta\hbar\omega})^2} + \frac{1}{2}\hbar\omega\frac{e^{-\beta\hbar\omega/2}}{1 - e^{-\beta\hbar\omega}}. \tag{8-5}$$

Equation 8-4 becomes

$$\langle E \rangle = \frac{\hbar\omega}{e^{\beta\hbar\omega} - 1} + \tfrac{1}{2}\hbar\omega , \tag{8-6}$$

after Eq. 8-3 is used to substitute for $A$. Equation 8-6 is equivalent to $\langle E \rangle = (\langle n \rangle + \tfrac{1}{2})\hbar\omega$, where

$$\langle n \rangle = \frac{1}{e^{\beta\hbar\omega} - 1} \tag{8-7}$$

is the average number of phonons in the mode at temperature $T$.

Figure 8-1 shows $\langle n \rangle$ as a function of $\beta\hbar\omega$. A mode for which $\hbar\omega \gg k_B T$ has few phonons and the energy in the mode is small, in spite of the high energy of each phonon. In the limit of large $\beta\hbar\omega$, the average values are

$$\langle n \rangle \longrightarrow e^{-\beta\hbar\omega} \tag{8-8}$$

and

$$\langle E \rangle \longrightarrow \hbar\omega e^{-\beta\hbar\omega} + \tfrac{1}{2}\hbar\omega , \tag{8-9}$$

respectively. $\langle n \rangle$ tends toward zero and $\langle E \rangle$ tends toward its zero-point value. On the other hand, a mode for which $\hbar\omega \ll k_B T$ has a large number of phonons and contributes significantly to the vibrational energy. To see this, substitute the power series expansion $e^{\beta\hbar\omega} \approx 1 + \beta\hbar\omega + \cdots$ into Eqs. 8-6 and 8-7 and obtain

$$\langle n \rangle \longrightarrow \frac{1}{\beta\hbar\omega} = \frac{k_B T}{\hbar\omega} \tag{8-10}$$

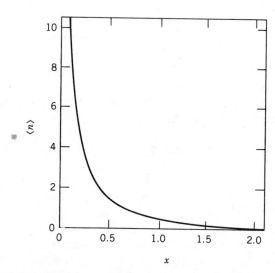

**FIGURE 8-1** The thermodynamic average number of phonons associated with a normal mode as a function of $x = \beta\hbar\omega$. Only modes for which $\hbar\omega < k_B T$ have an appreciable number of phonons.

and

$$\langle E \rangle \longrightarrow \frac{1}{\beta} = k_B T . \tag{8-11}$$

The zero-point contribution to $\langle E \rangle$ is small compared to $k_B T$ and is neglected.

At sufficiently high temperatures, $k_B T \gg \hbar\omega$ for every mode and each contributes $k_B T$ to the total energy. This is also the classical result. Classically, however, the energy of a mode is $k_B T$ at *all* temperatures. At low temperature, only low-frequency modes have energy $k_B T$. Energies of high-frequency modes are less and, according to Eqs. 8-8 and 8-9, the highest frequency modes are effectively frozen.

*Density of Modes.*   To find the total vibrational energy $E_T$ for a solid at temperature $T$, Eq. 8-6 is summed over all normal modes. If $\omega_s$ is the angular frequency of mode $s$,

$$E_T = \sum_s \frac{\hbar\omega_s}{e^{\beta\hbar\omega_s} - 1} + \frac{1}{2} \sum_s \hbar\omega_s . \tag{8-12}$$

The sum contains three terms for each atom in the solid.

Since all solids have a large number of normal modes, closely spaced in frequency, the frequency spectrum can be treated as continuous and the sum in Eq. 8-12 can be replaced by an integral. A density of modes function $g(\omega)$ is defined so $g(\omega)\,d\omega$ gives the number of modes with angular frequency between $\omega$ and $\omega + d\omega$ and the total energy is given by

$$E_T = \int \frac{\hbar\omega}{e^{\beta\hbar\omega} - 1} g(\omega)\,d\omega + \frac{1}{2} \int \hbar\omega g(\omega)\,d\omega , \tag{8-13}$$

where each integral is over the entire frequency spectrum. This equation is valid for both crystalline and amorphous solids. For a crystal $g(\omega) = 0$ for $\omega$ in a gap.

In practice, the density of modes for a crystal is evaluated by using dispersion relationships to calculate a large number of normal mode frequencies, with propagation vectors selected randomly from within the Brillouin zone. The range of possible frequencies is divided into intervals of width $\Delta\omega$ and a histogram is plotted showing the number of modes obtained in each interval. If $\Delta\omega$ is small and the number of frequencies obtained is large, the histogram approximates $g(\omega)\,\Delta\omega$.

Another method can be used. For any given branch of the spectrum, tips of propagation vectors associated with modes having the same angular frequency lie on a surface in reciprocal space, as illustrated in Fig. 8-2. Imagine two such surfaces, corresponding to angular frequencies $\omega$ and $\omega + \Delta\omega$, respectively. If $\Delta\omega$ is sufficiently small, the contribution of the branch to $g(\omega)\,\Delta\omega$ is just the number of modes with propagation vectors between these surfaces. Since allowed propagation vectors are distributed uniformly in reciprocal space, the

**FIGURE 8-2** A schematic representation of constant frequency surfaces in reciprocal space. A Brillouin zone is shown and each contour is defined by propagation vectors such that $\omega(\mathbf{q})$ = constant. Different surfaces are associated with different values of $\omega$.

number of modes between the surfaces is easy to calculate, in principal. It is the product of the number of propagation vectors per unit volume of reciprocal space and the volume between the two surfaces. As shown in Example 7-1 the former quantity is $N\tau/(3\pi)^3$, where $N$ is the number of primitive unit cells in the sample and $\tau$ is the volume of a cell.

**EXAMPLE 8-1** Derive an expression for the density of modes in the low-frequency limit for a single acoustic branch of a crystal with $N$ primitive unit cells. Assume the speed of sound is independent of propagation direction.

**SOLUTION** Since $\omega = qv$, a surface of constant frequency is a sphere with radius $q = \omega/v$. It encloses the volume $(4\pi/3)q^3 = (4\pi/3)(\omega/v)^3$ in reciprocal space. A sphere with radius $q + dq$ encloses the volume $(4\pi/3)(q + dq)^3 \simeq (4\pi/3)(q^3 + 3q^2\,dq)$ and the volume between the spheres is $4\pi q^2\,dq = 4\pi(\omega^2/v^3)\,d\omega$, where $\omega = qv$ was used. This volume is multiplied by $N\tau/(2\pi)^3$ to obtain $g(\omega) = (N\tau/2\pi^2v^3)\omega^2$. ∎

A general expression for the density of modes can be developed for a crystal. The difference in frequency of two modes in the same branch with propagation vectors $\mathbf{q}$ and $\mathbf{q} + d\mathbf{q}$, respectively, is

$$d\omega = \frac{\partial\omega}{\partial q_x}\,dq_x + \frac{\partial\omega}{\partial q_y}\,dq_y + \frac{\partial\omega}{\partial q_z}\,dq_z = \nabla_q\omega\cdot d\mathbf{q}, \qquad (8\text{-}14)$$

where $\nabla_q\omega = (\partial\omega/\partial q_x)\hat{\mathbf{x}} + (\partial\omega/\partial q_y)\hat{\mathbf{y}} + (\partial\omega/\partial q_z)\hat{\mathbf{z}}$, in Cartesian coordinates. First suppose the two propagation vectors are on the same constant frequency surface. Then $d\mathbf{q}$ is tangent to the surface and $d\omega = 0$. The scalar product $\nabla_q\omega\cdot d\mathbf{q}$ vanishes but neither $\nabla_q\omega$ nor $d\mathbf{q}$ is zero, so they must be perpendicular to each other. That is, $\nabla_q\omega$ is perpendicular to the surface of constant angular frequency $\omega$. Now suppose $\mathbf{q}$ and $d\mathbf{q}$ are on different surfaces separated by $dq_\perp$, as shown in Fig. 8-3. Then $d\omega = |\nabla_q\omega|dq_\perp$ and the volume between the surfaces

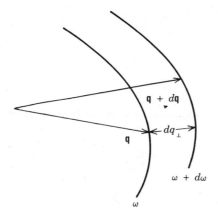

**FIGURE 8-3** Two constant frequency surfaces in reciprocal space. The difference in angular frequency can be written $d\omega = \nabla_q\omega \cdot d\mathbf{q}$ and, since $\nabla_q\omega$ is a vector normal to the surface, $d\omega = |\nabla_q\omega|dq_\perp$.

is $\oint dq_\perp dS = d\omega \oint |\nabla_q\omega|^{-1}dS$, where the integral is over either surface. The branch contributes

$$g(\omega) = \frac{N\tau}{(2\pi)^3} \oint \frac{dS}{|\nabla_q\omega|} \qquad (8\text{-}15)$$

to the density of modes. Equation 8-15 is summed over branches of the spectrum to calculate the total density of modes.

For low-frequency acoustical modes, Eq. 8-15 yields the result found in Example 8-1. For these modes $|\nabla_q\omega| = v$, a constant, and $\oint dS/|\nabla_q\omega| = 4\pi q^2/v = 4\pi\omega^2/v^3$. This is multiplied by $N\tau/(2\pi)^3$ to obtain the result given in the example. If all three acoustical branches are taken into account, the low-frequency density of modes is

$$g(\omega) = \frac{N\tau}{2\pi^2} \left[ \sum (1/v_s^3) \right]\omega^2, \qquad (8\text{-}16)$$

where the sum is over acoustical branches.

A typical density of modes curve is shown in Fig. 8-4. Near $\omega = 0$ it is proportional to $\omega^2$, but from midfrequency on the general trend is downward. The reason can be seen in plots of constant energy surfaces, such as Fig. 8-2. Above midfrequency, constant $\omega$ surfaces cut the Brillouin zone boundary and the area between nearby surfaces decreases strongly as the frequency increases. In spite of the trend, the density of modes has one or more fairly sharp spikes at high frequencies. These arise because, near a zone boundary, the dispersion curve is flat along a line normal to the boundary. Thus $|\nabla_q\omega|$ is extremely small or vanishes close to a boundary, so $g(\omega)$ is large. $|\nabla_q\omega|$ vanishes at other points in the zone and some peaks on the diagram correspond to those points.

*The Debye Approximation.* Sometimes Eq. 8-16 is used to approximate the density of modes over the entire acoustical spectrum. In the Debye approximation, as it is called, structure in the density of modes is neglected and the speeds of sound are assumed to be isotropic. Nevertheless, the approximation

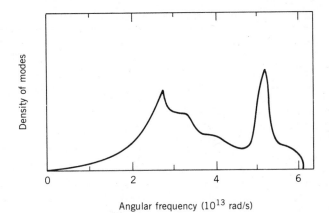

Angular frequency ($10^{13}$ rad/s)

**FIGURE 8-4** A typical density of modes function. For low frequencies the function is well represented by the Debye approximation, but high-frequency portions show complicated structure.

often produces reasonable results, particularly for temperatures that are so low that only modes in the linear portions of the dispersion curves have significant energy.

When the Debye approximation is used, the upper limit of integrals involving the density of modes is not taken to be the actual maximum frequency for acoustical modes but rather a frequency $\omega_m$ chosen so Eq. 8-16 predicts the correct number of modes. If there are $N$ unit cells in the sample, then $\omega_m$ satisfies

$$3N = \int_0^{\omega_m} g(\omega)\, d\omega. \tag{8-17}$$

When the integration is carried out and the result is solved for $\omega_m$,

$$\omega_m = \left[\frac{18\pi^2}{\tau} \frac{1}{\Sigma(1/v_s^3)}\right]^{1/3}, \tag{8-18}$$

is obtained. The sum is over the three acoustical branches.

The angular frequency $\omega_m$ can be used to characterize the density of modes for a crystal. However, the Debye temperature $T_D$, defined by

$$T_D = \frac{\hbar\omega_m}{k_B}, \tag{8-19}$$

is usually used instead. In terms of $T_D$,

$$g(\omega) = \frac{9N\hbar^3}{k_B^3 T_D^3}\, \omega^2. \tag{8-20}$$

The Debye temperature is used to distinguish between high- and low-temperature regions for a given solid. If $T > T_D$ we expect all modes to have energy $k_B T$. If $T < T_D$ we expect high-frequency modes to be frozen.

Debye temperatures of some solids are listed in Table 8-1. Values are some-

**TABLE 8-1**  Debye Temperatures for Selected Elements

| Element | $T$ (K) | Element | $T$ (K) |
|---|---|---|---|
| Aluminum | 428 | Lanthanum | 142 |
| Cesium | 38 | Lead | 105 |
| Carbon | | Lithium | 344 |
|   Diamond | 2230 | Magnesium | 400 |
|   Graphite | 420 | Manganese | 410 |
| Chlorine | 115 | Nickel | 450 |
| Cobalt | 445 | Osmium | 500 |
| Copper | 343 | Potassium | 91 |
| Germanium | 370 | Scandium | 360 |
| Gold | 165 | Sodium | 158 |
| Iodine | 106 | Silver | 225 |
| Iron | 467 | Vanadium | 380 |
| Krypton | 72 | Zinc | 327 |

Source: *American Institute of Physics Handbook*, 3rd ed. (New York: McGraw-Hill, 1972).

times determined experimentally by measuring the speeds of sound and using Eqs. 8-18 and 8-19. More often, however, heat capacity data are used. The relevant expression will be derived later.

*Total Vibrational Energy.*    We are now in a position to evaluate Eq. 8-13 in both high- and low-temperature limits. We consider a crystal with $N$ primitive unit cells and one atom in its primitive basis. We select the zero of energy so the total contribution of zero-point motions vanishes.

At high temperature $\beta\hbar\omega \ll 1$ for all frequencies in the spectrum and we may replace $e^{\beta\hbar\omega}$ by the first two terms in its power series expansion, $1 + \beta\hbar\omega$. The integrand of Eq. 8-13 then becomes $g(\omega)/\beta$ and, since the integral of $g(\omega)$ gives the total number of normal modes,

$$E_T = \frac{3N}{\beta} = 3Nk_B T . \tag{8-21}$$

This is the classical result: each mode contributes $k_B T$ to the total energy. The same result is obtained as long as $g(\omega)$ predicts the correct number of modes, regardless of its dependence on $\omega$.

To obtain the low-temperature form, we use the Debye approximation for the density of modes. When $g(\omega)$ from Eq. 8-20 is substituted into Eq. 8-13, the result is

$$E_T = \frac{9N\hbar^4}{k_B^3 T_D^3} \int_0^{\omega_m} \frac{\omega^3}{e^{\beta\hbar\omega} - 1} \, d\omega = \frac{9Nk_B T^4}{T_D^3} \int_0^{\beta\hbar\omega_m} \frac{x^3}{e^x - 1} \, dx , \tag{8-22}$$

where $x = \beta\hbar\omega$. For very low temperatures $\beta\hbar\omega_m \gg 1$ and we may replace the upper limit by $\infty$. The value of the integral is then $\pi^4/15$, so

$$E_T = \frac{3\pi^4 N}{5} \frac{k_B T^4}{T_D^3}.$$ (8-23)

At low temperatures the total vibrational energy is proportional to $T^4$ and inversely proportional to $T_D^3$. A simple qualitative argument gives the same result. We suppose each mode with propagation vector inside the reciprocal space sphere corresponding to $\omega = k_B T/\hbar$ contributes $k_B T$ to the total energy and modes with propagation vectors outside the sphere do not contribute. The number of contributing modes is proportional to $q^3 = \omega^3/v^3$ and this in turn is proportional to $T^3$. The total number of modes equals the number with propagation vectors in a sphere corresponding to $\omega_m = k_B T_D/\hbar$. This is proportional to $T_D^3$, so the fraction of modes that contribute to the total energy is $(T/T_D)^3$. Multiplication by $k_B T$ gives the energy per mode of the system.

## 8.2 ELECTRONS AT THE ABSOLUTE ZERO OF TEMPERATURE

At the absolute zero of temperature, electrons of a solid are distributed among the states in such a way that their total energy is as low as possible: if the solid contains $N$ electrons, the $N$ states lowest in energy are occupied and all higher states are unoccupied. As the temperature increases, electrons are excited to states that are unoccupied at $T = 0$ K, leaving empty states behind. In this and following sections we will determine the thermodynamic probability that each state is occupied and identify the highest occupied states at $T = 0$ K. Electrons in these states and states that are nearby in energy are the most important for material properties.

*The Fermi Energy.* The electron system is characterized by its Fermi energy, a quantity that lies between the energies of the highest occupied state and the lowest unoccupied state at $T = 0$ K. In principle, the Fermi energy for any solid can be found by counting states, starting with the lowest in energy and continuing until the number of states counted equals the number of electrons in the solid. The Fermi energy is between the energy of the last state counted and the energy of the next highest. It is usually indicated on energy spectra and band diagrams by a line called the Fermi level. See, for example, Fig. 7-3.

Crystalline metals can be distinguished from insulators by the position of the Fermi level relative to gaps in the electron energy spectrum. The Fermi level of an insulator falls in a gap while the Fermi level of a metal falls in a band, as illustrated in Fig. 8-5. As we will see, a completely occupied band cannot contribute to the electrical current. If the Fermi level is in a gap, all bands are either completely filled or completely empty, so no current is produced when an electric field is turned on. By way of contrast, if the Fermi level is in a band, that band is only partially occupied and electrons in it contribute to the current in an electric field. Later we will add a third category. Defect-free semiconductors are insulators at low temperatures but become conductors at high temperatures

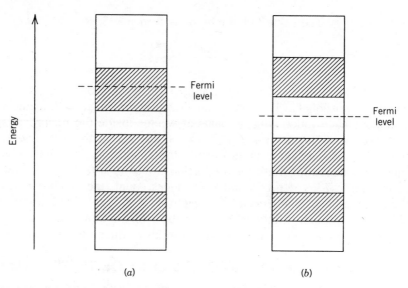

**FIGURE 8-5** The relationship between the Fermi level and the energy spectrum for (a) a crystalline metal and (b) a crystalline insulator at $T = 0$ K. Allowed energies are indicated by shading. For a metal the Fermi level lies in a band, while for an insulator it is in a gap.

or when they contain appropriate impurities. Both these situations result in partially filled bands.

Since the number of states in a band equals twice the number of primitive unit cells in the sample, a crystal with an odd number of electrons per primitive unit cell is metallic. The Fermi energy must fall within a band. On the other hand, a crystal with an even number of electrons per primitive unit cell is not necessarily an insulator at $T = 0\,K$, even though the number of electrons exactly equals the number of states in an integer number of bands. If two bands overlap in the neighborhood of the Fermi level, a gap does not occur there and the crystal is metallic. The alkaline earths, for example, have an even number of electrons per unit cell but they are not insulators because they have overlapping bands near the Fermi energy.

*Density of States.* To count states, we treat the electron energy as a continuous function of the propagation vector and make use of a density of states function, similar to the density of modes function for a vibration spectrum. The density of states $\rho(E)$ is defined so $\rho(E)\,dE$ gives the number of electron states with energy in the interval from $E$ to $E + dE$. Figure 8-6 gives some examples.

Propagation vectors are uniformly distributed in reciprocal space, so the density of states is the product of the number of states per unit volume of reciprocal space and the volume between the constant energy surfaces for energies $E$ and $E + dE$. In general,

$$\rho(E) = \frac{2\tau_s}{(2\pi)^3} \oint \frac{dS}{|\nabla_k E|}, \tag{8-24}$$

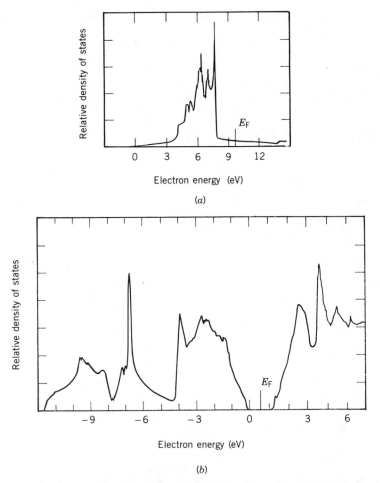

FIGURE 8-6   (a) Electron density of states for copper. The dotted line indicates the Fermi level. Many *d* states crowded into a small energy interval give rise to the peaks just below the Fermi energy. (b) Electron density of states for silicon. The gap between the valence and conduction bands is evident. ((a) from O. Jepson, D. Glotzel, and A. R. Mackintosh, *Phys. Rev.* **B 23**:2684, 1981. Used with permission. (b) from D. J. Stukel, T. C. Collins, and R. N. Euwema, in *Electronic Density of States* (L. H. Bennett, Ed.), National Bureau of Standards Special Publication 323; 1971. Reprinted courtesy of National Bureau of Standards.)

where the integral is over the constant energy surface for energy $E$ and

$$\nabla_k E(k) = \frac{\partial E(\mathbf{k})}{\partial k_x} \, \hat{\mathbf{x}} + \frac{\partial E(\mathbf{k})}{\partial k_y} \, \hat{\mathbf{y}} + \frac{\partial E(\mathbf{k})}{\partial k_z} \, \hat{\mathbf{z}} . \qquad (8\text{-}25)$$

Equation 8-24 is quite similar to Eq. 8-15: $\omega$ is replaced by $E$, $\mathbf{q}$ is replaced by $\mathbf{k}$, and $N\tau$ is replaced by the sample volume $\tau_s$. A factor of 2 appears in Eq. 8-24 to account for the two possible values of the $z$ component of spin.

For free electrons the constant energy surface for energy $E$ is a sphere with radius $k = (2mE/\hbar^2)^{1/2}$ and volume $(4\pi/3)k^3 = (4\pi/3)(2mE/\hbar^2)^{3/2}$. The number of states per unit volume of reciprocal space is $\tau_s/4\pi^3$, so $(\tau_s/3\pi^2)(2mE/\hbar^2)^{3/2}$ states have propagation vectors inside the sphere. The density of states is the derivative of this quantity with respect to $E$, or

$$\rho(E) = \frac{\sqrt{2}\tau_s m^{3/2}}{\pi^2 \hbar^3} E^{1/2} . \tag{8-26}$$

**EXAMPLE 8-2** Verify that Eq. 8-24 gives Eq. 8-26 for the free electron density of states.

**SOLUTION** Relative to the bottom of the band, the energy is given by $E(\mathbf{k}) = \hbar^2 k^2/2m$, so

$$\nabla_k E = \frac{\hbar^2}{2m} \left[ \frac{\partial k^2}{\partial k_x} \hat{\mathbf{x}} + \frac{\partial k^2}{\partial k_y} \hat{\mathbf{y}} + \frac{\partial k^2}{\partial k_z} \hat{\mathbf{z}} \right] = \frac{\hbar^2}{m} [k_x \hat{\mathbf{x}} + k_y \hat{\mathbf{y}} + k_z \hat{\mathbf{z}}] = \frac{\hbar^2}{m} \mathbf{k} .$$

A constant energy surface is a sphere with radius $k = (2mE/\hbar^2)^{1/2}$ and $|\nabla_k E|$ is constant on it. So

$$\oint \frac{dS}{|\nabla_k E|} = \frac{1}{|\nabla_k E|} \oint dS = \frac{4\pi m}{\hbar^2 k} k^2 = \frac{4\pi mk}{\hbar^2}$$

and

$$\rho(E) = \frac{\tau_s}{4\pi^3} \frac{4\pi mk}{\hbar^2} = \frac{\sqrt{2}\tau_s m^{3/2}}{\pi^2 \hbar^3} E^{1/2} . \qquad ■$$

*Calculation of the Fermi Energy.* The Fermi energy $E_F$ is obtained by solving

$$n = \frac{1}{\tau_s} \int_{E_0}^{E_F} \rho(E) \, dE , \tag{8-27}$$

where $n$ is the electron concentration and $E_0$ is the energy of the lowest electron state. The integral gives the number of states with energy between that of the lowest state and the Fermi energy. Since all these states and no others are occupied it gives the number of electrons in the solid. In practice, the integration need not be carried out over the entire energy spectrum. Usually the concentration of electrons in core states is subtracted from the left side of Eq. 8-27, then $E_0$ is interpreted as the energy of the lowest state outside the core. Equation 8-27 is valid for all solids, crystalline or amorphous.

**EXAMPLE 8-3** (a) Find an expression for the Fermi energy of a metal for which all bands except one are either completely filled or completely empty and the partially filled band is free electron-like, with a scalar effective mass $m^*$. (b) The assumptions made in part a are valid for sodium. Take the cube edge to be 4.225 Å, assume each atom contributes one electron to the partially filled band, and calculate the Fermi energy for sodium relative to the bottom of the band. Take the effective mass to be the mass of a free electron.

**SOLUTION** (a) For the partially filled band

$$\rho(E) = \frac{\sqrt{2}\tau_s(m^*)^{3/2}}{\pi^2\hbar^3} (E - E_0)^{1/2},$$

where $E_0$ is the minimum energy in the band. If $n$ is the concentration of electrons in that band, then

$$n = \frac{\sqrt{2}(m^*)^{3/2}}{\pi^2\hbar^3} \int_{E_0}^{E_F} (E - E_0)^{1/2} \, dE = \frac{2\sqrt{2}(m^*)^{3/2}}{3\pi^2\hbar^3} (E_F - E_0)^{3/2}.$$

This equation is solved for the Fermi energy:

$$E_F = E_0 + \frac{\hbar^2}{2m^*} (3\pi^2 n)^{2/3}.$$

The Fermi energy depends on the concentration of electrons, so it does not change if the sample size is increased by adding more material with the same electron concentration. (b) Sodium has two atoms per cubic unit cell and each contributes one electron to the free electron band, so $n = 2/a^3 = 2/(4.225 \times 10^{-19})^3 = 2.65 \times 10^{28}$ electrons/$m^3$ and

$$E_F = E_0 + \frac{(1.05 \times 10^{-34})^2}{2 \times 9.11 \times 10^{-31}} (3\pi^2 \times 2.65 \times 10^{28})^{2/3}$$

$$= E_0 + 5.15 \times 10^{-19} \text{ J} = E_0 + 3.22 \text{ eV}.$$

The Fermi energy is 3.22 eV above the minimum of the free electron band. ∎

*Fermi Surfaces of Metals.* The Fermi surface of a crystalline metal is the constant energy surface in reciprocal space corresponding to the Fermi energy. In the free electron approximation a Fermi surface can also be defined for an amorphous metal. On the other hand, the Fermi level of an insulator lies within a gap, so a Fermi surface does not exist.

Once the energy bands and Fermi energy of a metal have been determined, the Fermi surface can be generated as the locus of all propagation vectors **k** that satisfy $E(\mathbf{k}) = E_F$. Consider, for example, free electrons and take $E_0 = 0$. Points that satisfy $\hbar^2 k^2/2m^* = E_F$ are on the surface of a sphere with radius $k_F = (2m^* E_F/\hbar^2)^{1/2}$. Since $E_F = (\hbar^2/2m^*)(3\pi^2 n)^{2/3}$,

$$k_F = (3\pi^2 n)^{1/3}. \tag{8-28}$$

If the Fermi sphere is not entirely within the Brillouin zone, we translate parts outside by reciprocal lattice vectors so they lie inside.

Figure 8-7 illustrates the evolution of the Fermi surface in the free electron approximation as the electron concentration increases. We take the sample to

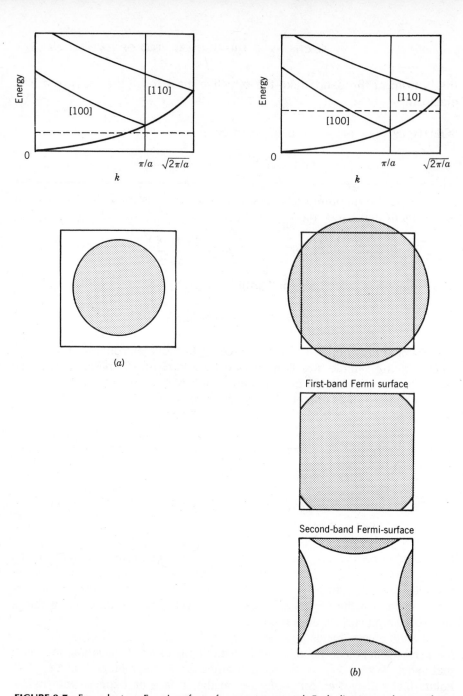

**FIGURE 8-7** Free electron Fermi surfaces for a square crystal. Each diagram at the top shows free electron bands for **k** in [100] and [110] directions, drawn on the same graph. The Fermi level is shown as a dotted line. Lower diagrams show the Brillouin zone and Fermi surface. (*a*) One electron per primitive unit cell. The Fermi level is within the first band and the Fermi surface is within the Brillouin zone. (*b*) Two electrons per primitive unit cell. The Fermi level is in two bands. The second diagram from top shows the Fermi surface in the extended zone scheme, while the other diagrams show the first and second band surfaces, respectively, in the reduced zone scheme. The second-band surface is obtained from the extended zone surface by translating those parts that lie outside the zone by reciprocal lattice vectors. Occupied states are indicated by shading.

be a two-dimensional square crystal. Diagram *a* shows the Fermi surface if the concentration is one electron per primitive unit cell. At the top is a plot of the band structure, with curves for **k** along a square edge and along a square diagonal, drawn on the same graph. The Fermi level is shown as a dotted line. The Fermi circle, shown in the second diagram down, is entirely within the Brillouin zone.

If there are two electrons per primitive unit cell, the diagrams are like those in *b*. As the top diagram indicates, the Fermi surface has two sheets, one in the first band and one in the second. The second diagram down shows the Fermi circle and the last two show the first- and second-band Fermi surfaces, respectively, in the reduced zone scheme. Portions of the circle outside the zone have been translated by reciprocal lattice vectors so they lie inside. Occupied states are indicated by shading.

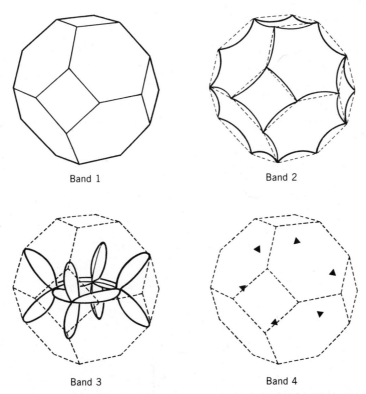

Band 1                Band 2

Band 3                Band 4

**FIGURE 8-8** The Fermi surface of aluminum. The various branches can be constructed by translating portions of a sphere from outside the zone to inside. The first band is completely filled, while the second band surface encloses empty states; filled states are between the surface and the zone boundary. Filled states in the third band are enclosed by the surface shown. The fourth band has small pockets of filled states. Zones shown for the third and fourth bands have been shifted relative to zones used for the first and second bands; the zone center for the last two diagrams is at the center of a square face of the zone for the first two. (From W. A. Harrison, *Electronic Structure and the Properties of Solids* (San Francisco, Freeman, 1980). Used with permission.)

As these diagrams imply, the Fermi surface can be quite complicated even though the electrons are nearly free. As an example, the Fermi surface for aluminum is depicted in Fig. 8-8. For this metal, each primitive unit cell contributes three electrons to nearly free electron bands. The first band is completely occupied and portions of the Fermi surface are in the second, third, and fourth bands. Surfaces for the second and third bands are nearly portions of spheres, translated to the Brillouin zone.

A weak potential energy causes the Fermi sphere to bulge toward nearby zone boundaries. Figure 8-9a shows a nearly free electron energy band and, for com-

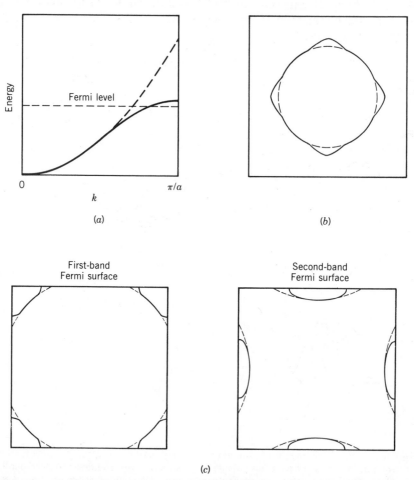

**FIGURE 8-9** (a) Free (dotted curve) and nearly free (solid curve) electron bands, drawn on the same graph. (b) Fermi surfaces for free (dotted curve) and nearly free (solid curve) electrons. When electrons interact with atoms of the crystal, the Fermi surface bulges toward nearby zone boundaries. (c) The effect of a weak potential on a two-band Fermi surface. Each surface bends toward the zone boundary.

parison, a parabolic band. If the Fermi level is in the region where the nearly free electron curve deviates from a parabola, the Fermi surface is closer to the zone boundary than the corresponding surface for completely free electrons. The effect on a one-band Fermi surface is shown in *b*, while the effect on a two-band Fermi surface is shown in *c*. Surfaces in both bands bend toward nearby zone boundaries.

The presence of *d* bands has the same effect. For example, the Fermi level of copper intersects the nearly free electron energy curve for **k** in the [100] and [110] directions but no intersections occur for **k** in the [111] direction. The Fermi surface is shown in Fig. 8-10. Its body is nearly spherical but there are necks that extend from the central region to the centers of the hexagonal faces, in ⟨111⟩ directions.

We have suggested two criteria for distinguishing a metal from an insulator. In terms of bonding, the potential energy function for an electron in a metal is weak and wave functions for outer electrons spread throughout the solid to produce a nearly uniform probability density. In terms of band structure, a metal has partially filled bands at $T = 0$ K or, equivalently, it has a Fermi surface. The relationship between these two ways of looking at a metal is illustrated by the two-band, nearly free electron Fermi surface of Fig. 8-9c. For the situation depicted there, states in both the first and second bands are partially occupied and the material is clearly metallic. First-band states in corners of the zone and second-band states in the center of the zone are empty. Now suppose the magnitude of the potential energy increases so electrons are more tightly bound to ion cores. Then the energy bands bend more strongly toward zone boundaries and the first-band Fermi surface moves outward toward zone corners while the second-band surface moves outward toward zone face centers. More states in the first band and fewer states in the second are occupied.

When the potential energy becomes sufficiently great, the first-band surface disappears into zone corners and the second-band surface disappears into zone face centers. All second-band states now have energy greater than first-band states and a gap appears in the energy spectrum. The solid has become an insulator.

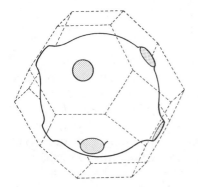

**FIGURE 8-10** The Fermi surface of copper. Necks extend from the central sphere to the centers of hexagonal Brillouin zone faces. Regions of contact with the faces are shaded. (Redrawn from C. Kittel, *Introduction to Solid State Physics*, 5th ed. (New York: Wiley, 1976). Used with permission.)

## 8.3 ELECTRONS AT TEMPERATURES ABOVE ABSOLUTE ZERO: METALS

*The Fermi-Dirac Distribution Function.* When the temperature of a solid is increased from $T = 0$ K, electrons receive a portion of the energy absorbed by the solid and are promoted from states below the Fermi energy to previously unoccupied states above. For a solid in thermal equilibrium at temperature $T$, the probability that an electron state with energy $E$ is occupied is given by the Fermi-Dirac distribution function

$$f(E) = \frac{1}{e^{\beta(E-\eta)} + 1},$$
(8-29)

where $\beta = 1/k_B T$ and $\eta$ is a quantity called the chemical potential of the system. It will be discussed in detail soon. All states with the same energy are occupied with the same probability.

The Fermi-Dirac distribution function follows from the Pauli exclusion principle and the laws of statistical mechanics. To derive it, we treat each electron state as an independent system in equilibrium with a thermal reservoir at temperature $T$. Since the number of particles in each state changes with temperature, we also suppose the state can exchange electrons with a particle reservoir. The energy of each particle in the reservoir is the chemical potential $\eta$. If $s$ electrons leave the reservoir and enter a state with energy $E$, assumed to be empty at first, the increase in energy is $s(E - \eta)$ and, according to statistical mechanics, the probability this occurs is proportional to $e^{-\beta s(E-\eta)}$. Take $A$ to be the constant of proportionality and write $P_s = Ae^{-\beta s(E-\eta)}$ for the probability the state is occupied by $s$ electrons. Of course, $s$ is either 0 or 1 since electrons obey the Pauli exclusion principle.

Since the state is either occupied by a single electron or else is unoccupied, $P_0 + P_1 = 1$. So

$$A = \frac{1}{1 + e^{-\beta(E-\eta)}}$$
(8-30)

and

$$P_1 = \frac{e^{-\beta(E-\eta)}}{1 + e^{-\beta(E-\eta)}} = \frac{1}{e^{-\beta(E-\eta)} + 1}.$$
(8-31)

This is the Fermi-Dirac distribution function. It is also the average number of electrons in a state with energy $E$ at temperature $T$. Even in thermal equilibrium, electrons continually make transitions into and out of the state and $f(E)$ may also be interpreted as the fraction of time the state is occupied. Its value is between 0 and 1.

The pool of electrons in a solid forms a particle reservoir and, to conform to the discussion above, we may think of an electron as temporarily having energy $\eta$ as it passes from one state to another. A reservoir that exchanges electrons freely with a sample is different from the actual situation in one important aspect, however. If the chemical potential associated with the reservoir is fixed, the number of electrons in the sample is temperature dependent. According to Eq. 8-29, the total electron concentration is

$$n = \frac{1}{\tau_s} \sum \frac{1}{e^{\beta(E-\eta)} + 1},$$ (8-32)

where the sum is over all electron states and $\tau_s$ is the sample volume. Clearly, $n$ is a function of temperature if $\eta$ is not. For each temperature, however, we select a value for the chemical potential so the sum on the right equals the actual electron concentration in the sample. Then $\eta$ is temperature dependent and $n$ is not. In most cases, $n$ is known and Eq. 8-32 is solved for $\eta$, then Eq. 8-29 is used to find the probability any state is occupied.

The Fermi-Dirac distribution function for $T = 0$ K is plotted in Fig. 8-11a. For states with $E < \eta$, the exponent in Eq. 8-29 is negative and, as $\beta$ becomes large, the exponential vanishes. In the low-temperature limit, $f(E) \simeq 1$ for these states. On the other hand, the exponent is positive for states with $E > \eta$ and, for these states, $f(E) \to 0$ as $T \to 0$. At the absolute zero of temperature, states with energy less than the chemical potential are guaranteed to be occupied, while states

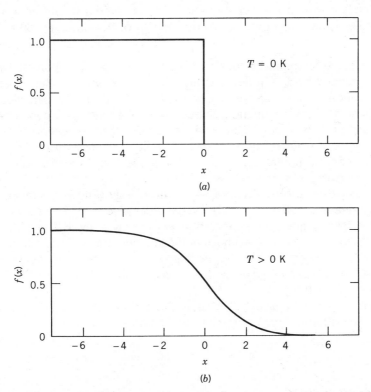

**FIGURE 8-11**  The Fermi distribution function (a) for $T = 0$ K and (b) for $T > 0$ K, plotted as a function of $x = \beta(E - \eta)$. The curves differ only in a region a few $k_B T$ wide near $E = \eta$.

with energy greater than the chemical potential are guaranteed to be unoccupied. We identify $\eta$ at $T = 0$ with the Fermi energy.*

The Fermi-Dirac distribution function for $T > 0$ K is plotted in Fig. 8-11$b$. For $E = \eta$, the function has the value 0.5 for every temperature above absolute zero. For energies much different from the chemical potential, the function tends exponentially toward 0 or 1. The following example indicates quantitatively how the function depends on temperature.

**EXAMPLE 8-4** (a) Find the occupation probability for a state 0.110 eV below the chemical potential at $T = 50, 300$, and 1200 K. (a) Repeat the calculation for a state 0.110 eV above the chemical potential.

**SOLUTION** For $T = 50$ K, $\beta = 1/(k_B T) = 1/(8.62 \times 10^{-5} \times 50) = 2.32 \times 10^2$ $eV^{-1}$. For $T = 300$ K, $\beta = 38.7$ $eV^{-1}$ and for $T = 1200$ K, $\beta = 9.67$ $eV^{-1}$. (a) At 50 K, $f(E) = [e^{\beta(E-\eta)} + 1]^{-1} = [e^{-232 \times 0.110} + 1]^{-1} = 1.00$. At 300 K, $f(E) = 0.986$ and at 1200 K, $f(E) = 0.744$. This state is essentially always occupied at 50 K. At 300 K it is empty a small fraction of the time and at 1200 K it is empty about one-fourth of the time. (b) At 50 K, $f(E) = [e^{+232 \times 0.110} + 1]^{-1} = 8.24 \times 10^{-12} = 0$. At 300 K, $f(E) = 1.41 \times 10^{-2}$ and at 1200 K, $f(E) = 0.256$. This state is essentially unoccupied at 50 K, occupied a small fraction of the time at 300 K, and occupied about one-fourth of the time at 1200 K. ∎

The example is somewhat misleading since the chemical potential is temperature dependent. $E - \eta$ depends on $T$ but, as we will see, the dependence is not strong and the example can be used as a guide to the behavior of the Fermi-Dirac distribution function. Notice that, as the temperature increases, the occupation probability for a state with energy less than the chemical potential decreases from 1 while the occupation probability for a state with energy greater than the chemical potential increases from 0. Electrons are thermally excited from states with $E < \eta$ to states with $E > \eta$.

The occupation probability of a state with energy $E$ does not differ greatly from its value at $T = 0$ unless $k_B T$ is on the order of $|E - \eta|$. Significant thermal excitation takes place over a range of energies only a few $k_B T$ wide, centered on $E = \eta$. For example, the width $\Delta E$ of the function from $f(E) = 0.99$ to $f(E) = 0.01$ is about $4.6 k_B T$. Even at the melting point of a solid, this range is small compared to the total range of electron energies: at 300 K, $k_B T = 0.0259$ eV and at 1200 K, $k_B T = 0.104$ eV.

To solve for the chemical potential, each band is treated as continuous and Eq. 8-32 is written in terms of the density of states:

$$n = \frac{1}{\tau_s} \int_{E_0}^{\infty} f(E)\rho(E) \, dE , \qquad (8-33)$$

---

*In insulator and semiconductor physics the term Fermi energy is used for the chemical potential at all temperatures.

where $E_0$ is the energy of the lowest electron state in the solid. Usually $n$ is taken to be the concentration of electrons in states outside the core and $E_0$ is taken to be the minimum energy of those electrons. This procedure is valid because $f(E)$ is essentially 1 for core electrons. We now consider the calculation of the chemical potential for a simple model of a metal.

*The Chemical Potential of a Metal.*   The integral in Eq. 8-33 can easily be evaluated as a power series in the temperature $T$. It can be written

$$n = \int_{E_0}^{\infty} h'(E)f(E) \, dE \, , \tag{8-34}$$

where

$$h(E) = \int_{E_0}^{E} \rho(E) \, dE \tag{8-35}$$

and $h'$ is the derivative of $h$ with respect to $E$. Integrals of this form are considered in Appendix D and are shown to have the expansion

$$n = h(\eta) + \frac{\pi^2}{6} (k_B T)^2 h''(\eta) + \cdots \, , \tag{8-36}$$

where $h''$ is the second derivative of $h$. We need only the first two terms. When Eq. 8-35 is used to substitute for $h$, Eq. 8-36 becomes

$$n = \frac{1}{\tau_s} \int_{E_0}^{\eta} \rho(E) \, dE + \frac{\pi^2}{6\tau_s} (k_B T)^2 \rho'(\eta) \, . \tag{8-37}$$

This equation is to be solved for $\eta$.

At $T = 0$ K, electrons fill all states with energy below $E_F$, with none left over, so

$$n = \frac{1}{\tau_s} \int_{E_0}^{E_F} \rho(E) \, dE \, . \tag{8-38}$$

Substitute this into Eq. 8-37 and use

$$\int_{E_0}^{\eta} \rho \, dE - \int_{E_0}^{E_F} \rho \, dE = \int_{E_F}^{\eta} \rho \, dE \tag{8-39}$$

to obtain

$$\frac{1}{\tau_s} \int_{E_F}^{\eta} \rho(E) \, dE + \frac{\pi^2}{6\tau_s} (k_B T)^2 \rho'(\eta) = 0 \, . \tag{8-40}$$

Since $\eta$ is not much different from the Fermi energy $E_F$, we approximate $\rho(E)$ by $\rho(E_F)$ in the integral and $\rho'(\eta)$ by $\rho'(E_F)$ in the last term. Then

$$\frac{1}{\tau_s} (\eta - E_F)\rho(E_F) + \frac{\pi^2}{6\tau_s} (k_B T)^2 \rho'(E_F) = 0 \, , \tag{8-41}$$

and

$$\eta = E_F - \frac{\pi^2}{6}(k_B T)^2 \frac{\rho'(E_F)}{\rho(E_F)}. \qquad (8\text{-}42)$$

If the density of states is greater for energies just above the Fermi level than for energies just below, then $\rho'(E_F)$ is positive and the chemical potential decreases from $E_F$ as the temperature increases. Similarly, the chemical potential increases with temperature if $\rho'(E_F)$ is negative.

Equation 8-42 can be understood qualitatively. With an increase in temperature, the central portion of the Fermi-Dirac distribution function broadens and occupation probabilities for states above $E_F$ increase while occupation probabilities for states below $E_F$ decrease. Suppose $\rho'(E_F)$ is positive. Then if the chemical potential did not change, the increase in the number of electrons above $E_F$ would be greater than the decrease in the number below. A slight downward shift in the chemical potential is required to conserve electrons. The following example demonstrates the size of the shift.

**EXAMPLE 8-5** (a) Find an expression for the chemical potential as a function of temperature if the Fermi level lies in a single free electron band. (b) Estimate the chemical potential relative to the Fermi energy for sodium. First take the temperature to be 50 K, then 390 K. The latter temperature is close to the melting point.

**SOLUTION** (a) The density of states for a free electron band is proportional to $E^{1/2}$, so $\rho'(E_F)/\rho(E_F) = 1/2E_F$ and, according to Eq. 8-42, $\eta = E_F - (\pi^2/12)(k_B T)^2/E_F$. (b) At 50 K, $k_B T = 4.31 \times 10^{-3}$ eV and at 390 K, $k_B T = 3.36 \times 10^{-2}$ eV. In Example 8-3, we obtained $E_F = 3.22$ eV for the Fermi energy of sodium, relative to the bottom of the free electron band, so at 50 K, $\eta - E_F = -(\pi^2/12)(4.31 \times 10^{-3})^2/3.22 = -4.74 \times 10^{-6}$ eV and at 390 K, $\eta - E_F = -2.88 \times 10^{-4}$ eV. These deviations of the chemical potential from the Fermi energy are quite small compared to the Fermi energy itself. ∎

*Total Electron Energy.* The total energy of an electron system in thermal equilibrium at temperature $T$ is given by

$$E_T = \sum E f(E), \qquad (8\text{-}43)$$

where the sum is over all electron states. Each term is the product of the energy of a state and the average number of electrons occupying the state. The density of states $\rho(E)$ can be used to write the total energy as an integral:

$$E_T = \int_{E_0}^{\infty} E\rho(E)f(E)\, dE, \qquad (8\text{-}44)$$

where $E_0$ is the lowest electron energy for the solid.

Equation 8-44 has the same form as the integral in Eq. 8-34, with

$$h(E) = \int_{E_0}^{E} E\rho(E)\, dE. \tag{8-45}$$

According to Eq. 8-36,

$$E_T = \int_{E_0}^{\eta} E\rho(E)\, dE + \frac{\pi^2}{6} (k_B T)^2 [\rho(\eta) + \eta\rho'(\eta)]. \tag{8-46}$$

Since the difference between $\eta$ and $E_F$ is small, we write

$$\int_{E_0}^{\eta} E\rho(E)\, dE = \int_{E_0}^{E_F} E\rho(E)\, dE + \int_{E_F}^{\eta} E\rho(E)\, dE$$

$$= \int_{E_0}^{E_F} E\rho(E)\, dE + E_F\rho(E_F)[\eta - E_F]$$

$$= \int_{E_0}^{E_F} E\rho(E)\, dE - \frac{\pi^2}{6}(k_B T)^2 E_F\rho'(E_F), \tag{8-47}$$

where Eq. 8-42 was used to substitute for $\eta - E_F$. Replace $\eta$ by $E_F$ in the brackets of Eq. 8-46. Then

$$E_T = \int_{E_0}^{E_F} E\rho(E)\, dE + \frac{\pi^2}{6}(k_B T)^2\rho(E_F). \tag{8-48}$$

The first term of Eq. 8-48 represents the total electron energy at $T = 0$ K, while the second represents the additional energy received by the electrons as the temperature increases. This term is sometimes called the thermal energy of the electrons.

We can qualitatively understand the quadratic dependence of $E_T$ on $T$. Suppose the temperature is raised from 0 K to $T$. The electrons most affected are those within a few $k_B T$ of the Fermi energy and their number is proportional to $k_B T$. On average, they each receive energy $k_B T$ so the total energy is proportional to $T^2$. On the other hand, if electrons were classical particles, then each would receive energy $k_B T$ and the total energy would be proportional to $T$. The actual thermal energy is much less than the classical value.

## 8.4 ELECTRONS AT TEMPERATURES ABOVE ABSOLUTE ZERO: SEMICONDUCTORS

*The Chemical Potential.*  For simplicity we consider a pure crystal with a single conduction band and a single valence band. At $T = 0$ K the valence band is completely filled, the conduction band is completely empty, and the chemical potential is in the gap between these bands.

The integration range in Eq. 8-33 is divided into two parts, corresponding to the valence and conduction bands, respectively. The total concentration is written $n = n_c + n_v$, where $n_c$ is the concentration of electrons in the conduction band and $n_v$ is the concentration of electrons in the valence band. For a pure insulator or semiconductor the total number of electrons is just sufficient to fill the valence band so

$$n_c + n_v = N_v,\qquad(8\text{-}49)$$

where $N_v$ is the number of states per unit volume in the valence band.

Contributions of a valence band to material properties are usually discussed in terms of holes rather than electrons. Details are given in Chapter 9. For now we simply state that the number of holes is the same as the number of empty states, so the hole concentration $p_v$ in the valence band is given by $p_v = N_v - n_v$ and Eq. 8-49 becomes

$$n_c = p_v.\qquad(8\text{-}50)$$

A hole is created in the valence band for each electron thermally excited to the conduction band.

For electrons in the conduction band

$$n_c = \frac{1}{\tau_s}\int_{E_c}^{\infty}\rho_c(E)f(E)\,dE = \frac{1}{\tau_s}\int_{E_c}^{\infty}\frac{\rho_c(E)}{e^{\beta(E-\eta)}+1}\,dE$$

$$\approx \frac{1}{\tau_s}\int_{E_c}^{\infty}\rho_c(E)e^{-\beta(E-\eta)}\,dE,\qquad(8\text{-}51)$$

where $\rho_c(E)$ is the conduction band density of states and $E_c$ is the energy at the bottom of the band. Since $f(E)$ is extremely small for energies at the top of the band and above, the upper limit of the integral is taken to be $\infty$.

To obtain the last form given in Eq. 8-51, we assumed $e^{\beta(E-\eta)} \gg 1$ for all energies in the band. The assumption can be justified by precise numerical calculations, which indicate that the value of $\eta$ is near the center of the gap between the valence and conduction bands. Since $E - \eta$ is about half an electron volt and $\beta$ is about 40 eV$^{-1}$ at room temperature, the assumption is well justified. Even for $T = 2000$ K, it leads to error in only the third significant figure.

We treat an isotropic parabolic band, with

$$\rho_c(E) = \frac{\sqrt{2}\tau_s(m_e)^{3/2}}{\pi^2\hbar^3}(E - E_c)^{1/2}\qquad(8\text{-}52)$$

where $m_e$ is the electron effective mass. The integral in Eq. 8-51 can be evaluated, with the result

$$n_c = N_c^* e^{-\beta(E_c-\eta)},\qquad(8\text{-}53)$$

where

$$N_c^* = 2\left[\frac{m_e k_B T}{2\pi\hbar^2}\right]^{3/2}.\qquad(8\text{-}54)$$

$N_c^*$ is called the effective number of states per unit volume for the conduction band. Equation 8-53 has a nice interpretation. We replace the true conduction band spectrum by a single level at $E_c$ and suppose there are $N_c^*$ states per unit volume associated with it. Then Eq. 8-53 gives the electron concentration if $E_c - \eta \gg k_B T$.

For states in the valence band we write

$$f(E) = 1 - \frac{1}{e^{-\beta(E-\eta)} + 1},  \tag{8-55}$$

which is algebraically identical to Eq. 8-29. The concentration of electrons in the valence band is given by

$$n_v = \frac{1}{\tau_s} \int_{-\infty}^{E_v} \rho(E) f(E) \, dE = N_v - \int_{-\infty}^{E_v} \frac{\rho(E)}{e^{-\beta(E-\eta)} + 1} \, dE,  \tag{8-56}$$

where $E_v$ is the energy at the top of the band and the first term arises from the integral of $\rho_v(E)$ over the band. The lower limit is taken to be $-\infty$ since the integrand is extremely small for energies at the bottom of the band and below.

A comparison of $n_v = N_v - p_v$ with Eq. 8-56 reveals that the hole concentration is

$$p_v = \int_{-\infty}^{E_v} \frac{\rho_v(E)}{e^{-\beta(E-\eta)} + 1} \, dE \approx \int_{-\infty}^{E_v} \rho_v(E) e^{\beta(E-\eta)} \, dE,  \tag{8-57}$$

where the last form is valid because $\eta - E \gg k_B T$ for all states in the band.

We take the density of states to be

$$\rho_v(E) = \frac{\sqrt{2}\tau_s(m_h)^{3/2}}{\pi^2 \hbar^3} (E_v - E)^{1/2},  \tag{8-58}$$

where $m_h$ is the effective mass for the valence band. It is called the hole effective mass, hence the subscript. Equation 8-57 becomes

$$p_v = N_v^* e^{\beta(E_v - \eta)},  \tag{8-59}$$

where

$$N_v^* = 2 \left[ \frac{m_h k_B T}{2\pi \hbar^2} \right]^{3/2}  \tag{8-60}$$

is the effective number of states per unit volume in the valence band.

When Eqs. 8-53 and 8-59 are substituted into Eq. 8-50 the result is

$$N_c^* e^{-\beta(E_c - \eta)} = N_v^* e^{\beta(E_v - \eta)}.  \tag{8-61}$$

Thus

$$\eta = \frac{E_c + E_v}{2} + \frac{k_B T}{2} \ln \left[ \frac{N_v^*}{N_c^*} \right].  \tag{8-62}$$

At $T = 0$ K, $\eta = \frac{1}{2}(E_c + E_v)$. The Fermi level is at the midpoint of the gap between the valence and conduction bands. As the temperature increases, it moves away from the gap center and, just as for a metal, its value is adjusted so Eq. 8-33 predicts the same total number of electrons for all temperatures. If

$N_v^* > N_c^*$, for example, then the density of states in the valence band is greater than the density of states in the conduction band and $\eta$ increases with $T$.

The concentration of electrons in the conduction band is given by $n_c = N_c^* e^{-\beta(E_c - \eta)}$ and, when Eq. 8-62 is used for $\eta$, the result is

$$n_c = [N_c^* N_v^*]^{1/2} e^{-\beta E_g/2}, \tag{8-63}$$

where $E_g = E_c - E_v$ is the energy gap between the valence and conduction bands. Equation 8-63 also gives the concentration of holes in the valence band. Since $N_v^*$ and $N_c^*$ both depend weakly on the temperature, the temperature dependence of $n_c$ is determined chiefly by the exponential factor in Eq. 8-63. If the natural logarithm of the concentration is plotted as a function of $\beta$, the result is nearly a straight line with slope $-\frac{1}{2}E_g$.

**EXAMPLE 8-6**  In a simplified model, assume germanium has a single valence band and a single conduction band, with a gap of 0.670 eV. The effective masses are $m_h = 0.370m_0$ and $m_e = 0.550m_0$, respectively, where $m_0$ is the free electron mass. Calculate: (a) the Fermi energy relative to the top of the valence band; (b) the chemical potential at 300 K, relative to the Fermi energy; (c) the occupation probability at 300 K for a state at the bottom of the conduction band; and (d) the probability at 300 K that a state at the top of the valence band is empty; (e) the electron concentration in the conduction band at 300 K.

**SOLUTION**  (a) Take $E_v = 0$ and $E_c = 0.670$ eV. Then $E_F = \frac{1}{2}(E_v + E_c) = \frac{1}{2}(0 + 0.670) = 0.335$ eV. (b) Once Eqs. 8-54 and 8-60 are substituted into Eq. 8-62, that equation becomes $\eta - E_F = \frac{3}{4}k_B T \ln(m_h/m_e) = \frac{3}{4} \times 8.62 \times 10^{-5} \times 300 \times \ln(0.370/0.550) = -7.69 \times 10^{-3}$ eV. (c) Since $E_c - \eta \gg k_B T$, the probability that a state with energy $E_c$ is occupied is $f(E_c) = e^{-\beta(E_c - \eta)} = \exp[-(0.670 - 0.335 + 0.008)/8.62 \times 10^{-5} \times 300] = 1.74 \times 10^{-6}$. (d) Since $\eta - E_v \gg k_B T$, the probability the state with energy $E_v$ is empty is $1 - f(E_v) = e^{\beta(E_v - \eta)} = \exp[(0 - 0.335 + 0.008)/8.62 \times 10^{-5} \times 300] = 3.22 \times 10^{-6}$. (e) $N_c^* = 2(m_e k_B T/2\pi\hbar^2)^{3/2} = 2[0.550 \times 9.11 \times 10^{-31} \times 1.38 \times 10^{-23} \times 300/2\pi(1.05 \times 10^{-34})^2]^{2/3} = 1.04 \times 10^{25}$ states/m$^3$ and $n_c = N_c^* f(E) = 1.04 \times 10^{25} \times 1.74 \times 10^{-6} = 1.80 \times 10^{19}$ electrons/m$^3$. ∎

Whether a solid is an insulator or semiconductor depends on the magnitude of the gap between valence and conduction bands. Although the distinction is not sharp, solids with gaps less than about 2.5 eV are usually considered semiconductors while solids with gaps greater than 3 eV are usually considered insulators.

*Doped Semiconductors.*  Many semiconductor devices depend for their operation on the electron distribution when small amounts of certain impurities have been added. Such semiconductors are said to be doped. Impurity atoms replace host atoms in the crystal structure and, because an impurity atom has a different number of electrons than a host atom, the electron distribution in the solid is changed.

Atoms with five electrons in their outer shells, like phosphorous, arsenic, and antimony, are often used as dopants. At $T = 0$ K, four electrons per impurity occupy valence band states while the fifth is in a state that is localized near the impurity. Its wave function is like an atomic orbital, but with considerably greater extent because the potential energy is lower in the solid than in the isolated atom. Its energy is in the gap between valence and conduction bands, as indicated schematically in Fig. 8-12a. Each level in the diagram represents many states, one for each impurity atom. If the impurity concentration is high, so neighboring orbitals overlap, the single level shown on the diagram is broadened into a band.

At the absolute zero of temperature all valence band states and one localized state for each impurity atom are filled, leaving conduction bands empty. As the temperature increases, electrons from impurity levels are thermally promoted to conduction bands. Because these impurities donate electrons to conduction bands, they are called donors, and because semiconductors with donors have more electrons in conduction bands than holes in valence bands, they are called $n$-type semiconductors.

Atoms with three electrons in their outer shells, like gallium and indium, are called acceptors. They also contribute electron states with energies in the gap, as indicated in Fig. 8-12b, but the states are empty at the absolute zero of temperature. As the temperature increases, electrons are promoted to these states from valence bands, so semiconductors doped with acceptor impurities

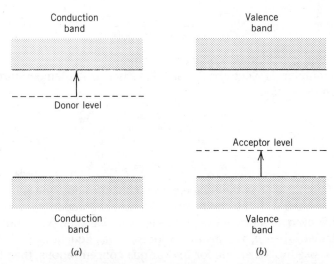

**FIGURE 8-12** (a) A donor level in the gap between valence and conduction bands of a semiconductor. At $T = 0$ K the level is filled, but at higher temperatures electrons are promoted from the donor level to the conduction band. (b) An acceptor level. At $T = 0$ K the level is empty, but at higher temperatures electrons are promoted from the valence band to the acceptor level, thereby creating holes in the valence band.

have more holes in valence bands than electrons in conduction bands and are called p type.

A great many different types of impurity atoms act as donors or acceptors. Some are given for silicon in Fig. 8-13, along with associated energy levels. Many states are associated with each level in the diagram, one for each impurity atom.

We wish to compute the concentration of electrons in the conduction band and holes in the valence band when impurities are present. If each impurity introduces two or more states with the same energy, the Fermi-Dirac distribution function must be modified. The modification arises because occupation numbers for such states are not independent of each other.

Consider first a donor atom with $g$ states corresponding to energy $E_d$, in the gap. An electron has equal probability of occupying any one of the $g$ states but, once one state is occupied, the occupation probability for the others is dramatically decreased by the mutual repulsion of electrons. We assume that either all states are empty or one is filled. The sum of the probabilities for the two possibilities is $A[1 + ge^{-\beta(E_d - \eta)}]$. This must be 1, so $A = [1 + ge^{-\beta(E_d - \eta)}]^{-1}$ and the concentration of electrons in donor states is

$$n_d = N_d \frac{ge^{-\beta(E_d - \eta)}}{1 + ge^{-\beta(E_d - \eta)}} = \frac{N_d}{(1/g)e^{\beta(E_d - \eta)} + 1}, \tag{8-64}$$

where $N_d$ is the concentration of donors in the sample.

Now consider acceptors that contribute $g$ states each, all with energy $E_a$. We assume either one of the states is empty or all are filled. The sum of the probabilities is $A[g + e^{-\beta(E_a - \eta)}]$ and the concentration of electrons in acceptor states is

$$n_a = N_a \frac{e^{-\beta(E_a - \eta)}}{g + e^{-\beta(E_a - \eta)}} = \frac{N_a}{ge^{\beta(E_a - \eta)} + 1}, \tag{8-65}$$

where $N_a$ is the concentration of acceptor atoms.

Since the number of electrons remains the same as the temperature changes, Eq. 8-33 can be written

$$n_c + n_v + n_d + n_a = N_v + N_d \tag{8-66}$$

or

$$n_c + n_d + n_a = p_v + N_d. \tag{8-67}$$

The left side of Eq. 8-66 is the total electron concentration, the sum of concentrations associated with the conduction, valence, and impurity levels. The total is just sufficient to fill all valence band states and one state per donor. Equation 8-67 is obtained when $p_v$ is substituted for $N_v - n_v$. Equations 8-51, 8-57, 8-64, and 8-65 are used to substitute for the various concentrations, then Eq. 8-67 is solved for the chemical potential. Finally, the same equations are used, along with the value found for $\eta$, to obtain the various electron concentrations. For many situations of interest the chemical potential is close to an impurity level or to a band edge and the approximations made to obtain the final expressions

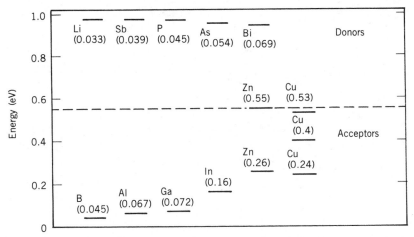

**FIGURE 8-13** Impurity levels in silicon. Energies given in parentheses are measured from the bottom of the conduction band for donors and from the top of the valence band for acceptors. Zn and Cu atoms can accept more than one electron each. (From S. M. Sze, *Physics of Semiconductor Devices* (New York: Wiley, 1981). Used with permission.)

in Eqs. 8-51 and 8-57 are not valid. Numerical techniques must be used to obtain a value for $\eta$.

Figure 8-14 shows the results for a typical $n$-type semiconductor. At low temperature, the chemical potential has a value between the donor energy and the bottom of the conduction band. This is reasonable since the donor levels are filled and the conduction band is empty. As the temperature increases, the chemical potential decreases to a value close to the midpoint of the gap between the valence and conduction bands. Essentially all donor states have then been emptied and excitation from the valence band dominates. In this temperature region, the chemical potential behaves as if there were no donors and the material is said to be intrinsic. Equations 8-62 and 8-63 are then valid.

The second graph shows the natural logarithm of the concentration of electrons in the conduction band as a function of $\beta$. At low temperature, a sharp rise occurs as electrons are promoted from donor states, then $n_c$ remains constant over the wide middle portion of the curve. Donor states have been emptied but the temperature is not yet sufficiently high for excitation from the valence band to be significant. Finally, at high temperature, promotion from the valence band takes place and the sample becomes intrinsic. This portion of the graph is a straight line with a slope of $-\frac{1}{2}E_g$.

If the number of donors is increased, the number of electrons in the conduction band increases, as we might expect. We might not expect the number of holes in the valence band to decrease but it does. If $|E_c - \eta|$ and $|E_v - \eta|$ are both large compared to $k_B T$, then electron and hole concentrations are given by the last expressions of Eqs. 8-51 and 8-57, respectively, and the product of the concentrations is

$$n_c p_v = N_c^* N_v^* e^{-\beta(E_c - E_v)} = N_c^* N_v^* e^{-\beta E_g}, \qquad (8\text{-}68)$$

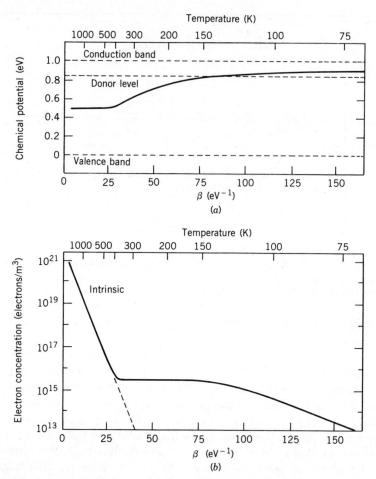

**FIGURE 8-14** (a) Chemical potential as a function of $\beta$ for an $n$-type semiconductor. Dotted lines indicate band edges and the donor level. (b) Electron concentration in the conduction band as a function of $\beta$. The vertical scale is logarithmic. The dotted line gives the concentration for the intrinsic material. To make the plots, the number of states per unit volume in the valence and conduction bands were each taken to be $5.0 \times 10^{21}$ states/m$^3$ and the number of donor impurities was taken to be $3.0 \times 10^{15}$ donors/m$^3$.

independently of the chemical potential. Impurities change the value of the chemical potential and the individual concentrations, but at any temperature the product $n_c p_v$ has the value it would have if the semiconductor were pure, provided the chemical potential is far from the band edges. With the addition of donors, for example, $n_c$ increases and $p_v$ decreases but $n_c p_v$ remains the same. Of course, if high-impurity concentrations drive the chemical potential close to a band edge or into a band, then Eq. 8-68 is not valid.

Equation 8-68 is often used in conjunction with Eq. 8-67 to solve for $n_c$ and $p_v$ when all donor levels are empty ($n_d = 0$) or all acceptor levels are filled ($n_a = N_a$).

**EXAMPLE 8-7** A semiconductor is doped with $N_d$ donors and, at the temperature of interest, all donor states are empty. Take the number of electrons in the conduction band to be $n_i$ for the intrinsic material at the same temperature and develop an expression for the number of electrons in the conduction band of the doped semiconductor.

**SOLUTION** In Eq. 8-67 set $n_d = 0$ and $n_a = 0$. The equation then becomes $n_c = p_v + N_d$. When $p_v = n_i^2/n_c$, from Eq. 8-68, is substituted for $p_v$, the result is $n_c^2 - N_d n_c - n_i^2 = 0$. So

$$n_c = \frac{N_d + [N_d^2 + 4n_i^2]^{1/2}}{2}.$$

Since $n_c$ must be positive, we use the positive root. Note that if $N_d \gg n_i$ then $n_c = N_d$. This situation prevails in the temperature region for which the curve of Fig. 8-14 is flat. If $n_i \gg N_d$, then $n_c = n_i$. This occurs at high temperatures, for which the semiconductor is intrinsic. ∎

## 8.5 HEAT CAPACITY

The heat capacity $C$ of a solid is found by measuring the heat $\Delta Q$ absorbed by the solid when its temperature increases by a small amount $\Delta T$. Specifically,

$$C = \frac{\Delta Q}{\Delta T}. \tag{8-69}$$

As the temperature changes, many types of energy transfer may occur. For example, the sample may do work on its environment. The magnetization or electric polarization of the sample may change, producing changes in the energy stored in the electromagnetic field. Consequently, every material has several different heat capacities, differing from each other in the quantities that are held constant during the temperature increase.

We consider a sample for which only vibrational and electron energies change. Such a sample is characterized by a heat capacity at constant volume $C_\tau = (dQ/dT)_\tau$ and a heat capacity at constant pressure $C_P = (dQ/dT)_P$, where the subscript identifies the quantity held constant during the temperature increase. The two are related by*

$$C_P - C_\tau = \frac{T\tau_s\alpha^2}{\kappa}, \tag{8-70}$$

where $\tau_s$ is the sample volume, $\alpha = (1/\tau_s)(\partial\tau_s/\partial T)_P$ is the coefficient of volume expansion, and $\kappa = -(1/\tau_s)(\partial\tau_s/\partial P)_T$ is the compressibility.

Since it is difficult to hold the volume of a sample constant while the temperature increases, $C_\tau$ is not usually measured. On the other hand, since normal

---

*See, for example, Chapter 11 of M. W. Zemansky, *Heat and Thermodynamics* (New York: McGraw-Hill, 1957).

mode frequencies and electron energy levels depend on interatomic spacings, $C_P$ is difficult to deal with theoretically. As a consequence, $C_P$ is measured and $C_\tau$ is calculated. Equation 8-70 is used to compare experimental and theoretical results. For most solids the difference between these two heat capacities is small except at high temperature.

According to the first law of thermodynamics, the heat $dQ$ absorbed by the sample, the increase $dE_T$ in the total energy of the sample, and the work $P\ d\tau_s$ done by the sample are related by

$$dQ = dE_T + P\ d\tau .$$
(8-71)

For a constant volume process the last term is zero so the heat capacity at constant volume is

$$C_\tau = \left[\frac{dQ}{dT}\right]_\tau = \left[\frac{\partial E_T}{\partial T}\right]_\tau .$$
(8-72)

To calculate $C_\tau$, we evaluate the derivative of the total energy with respect to temperature. Since the sample volume is constant, only the numbers of phonons in individual normal modes and the numbers of electrons in individual energy levels change, not the normal mode frequencies and energy levels themselves.

Figure 8-15 shows $C_\tau$ as a function of temperature for a typical solid. At high temperature it is independent of the temperature but at low temperature it decreases rapidly as the temperature decreases. For most materials at low temperature, $C_\tau$ is nearly proportional to $T^3$. As we will see, this behavior arises from the excitation of atomic vibrations. For metals and semiconductors with large concentrations of electrons in conduction bands, $C_\tau$ becomes proportional to $T$ at extremely low temperatures. This behavior arises from the excitation of electrons. We discuss phonon and electron contributions separately.

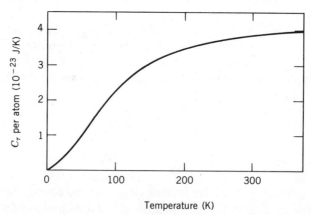

**FIGURE 8-15** Heat capacity per atom at constant volume as a function of temperature $T$ for a solid. It is proportional to $T^3$ for most materials at low temperature and it rises to a high-temperature value of $3k_B$ (= $4.14 \times 10^{-23}$ J/K). For metals at extremely low temperature it is proportional to $T$.

*Phonon Contribution to the Heat Capacity.* An expression for the total phonon energy as a function of temperature was given in Eq. 8-13. When it is differentiated with respect to the temperature, the result is

$$C_\tau = \frac{\hbar^2}{k_\mathrm{B} T^2} \int \frac{\omega^2 e^{\beta \hbar \omega}}{(e^{\beta \hbar \omega} - 1)^2} \, g(\omega) \, d\omega , \qquad (8\text{-}73)$$

where $g(\omega)$ is the density of modes and the integral is over the entire frequency spectrum. Note that zero-point motion does not contribute to $C_\tau$, but it does contribute to $C_P$.

If the temperature is so high that $\beta \hbar \omega \ll 1$ for all normal mode frequencies, the exponential may be replaced by 1 in the numerator and by $1 + \beta \hbar \omega$ in the denominator. The integrand is then independent of $\omega$ and, since $\int g(\omega) \, d\omega = 3N$, where $N$ is the number of atoms in the solid,

$$C_\tau = 3N k_\mathrm{B} . \qquad (8\text{-}74)$$

At high temperature the heat capacity is independent of the temperature and is proportional to the number of atoms in the solid. This is known as the Dulong-Petit law. It is valid for any material, as long as $\beta \hbar \omega \gg 1$ for all vibrational frequencies.

High-frequency modes are frozen at low temperature. Only acoustical modes contribute to the heat capacity and, for them, the Debye approximation to the density of states may be used. When the expression given by Eq. 8-20 is substituted for $g(\omega)$ in Eq. 8-73 and $x$ is substituted for $\beta \hbar \omega$, the result is

$$C_\tau = 9N k_\mathrm{B} \left[ \frac{T}{T_\mathrm{D}} \right]^3 \int_0^{x_\mathrm{m}} \frac{x^4 e^x}{(e^x - 1)^2} \, dx , \qquad (8\text{-}75)$$

where $x_\mathrm{m} = \beta \hbar \omega_\mathrm{m} = T_\mathrm{D}/T$, $T_\mathrm{D}$ is the Debye temperature, and $N$ is now the number of unit cells in the sample. Since $T \ll T_\mathrm{D}$, the upper limit may be replaced by $+\infty$. Then the value of the integral is $12\pi^4/45$ and

$$C_\tau = \frac{12\pi^4 N}{5} \, k_\mathrm{B} \, \frac{T^3}{T_\mathrm{D}^3} . \qquad (8\text{-}76)$$

This is the well-known Debye $T^3$ law for the phonon contribution to the low-temperature heat capacity of a solid. It can also be obtained by differentiating Eq. 8-23.

Equation 8-76 is often used to obtain the Debye temperature. The heat capacity is measured and other contributions are subtracted to obtain the phonon contribution, then the equation is solved for $T_\mathrm{D}$. If the heat capacity deviates from the $T^3$ law, the value obtained for $T_\mathrm{D}$ is temperature dependent and the extent of the dependence measures the validity of the Debye approximation.

Optical modes make contributions when $k_\mathrm{B} T$ reaches a value corresponding to optical phonon energies. These are sometimes taken into account by means of the Einstein approximation. All modes in a branch are assumed to have the same angular frequency $\omega_\mathrm{E}$, called the Einstein frequency, and the contribution of the branch to the total energy is

$$E = \frac{N\hbar\omega_E}{e^{\beta\hbar\omega_E} - 1}. \tag{8-77}$$

$N$ is again the number of unit cells in the sample. In this approximation, an optical branch contributes

$$\left[\frac{\partial E}{\partial T}\right]_\tau = \frac{N\hbar^2\omega_E^2}{k_B T^2} \frac{e^{\beta\hbar\omega_E}}{(e^{\beta\hbar\omega_E} - 1)^2} \tag{8-78}$$

to the heat capacity. Note that the contribution decreases exponentially in the low-temperature limit.

High-frequency acoustical modes become significant at temperatures somewhat below those for which optical modes are important. Since the density of modes for this region of the spectrum cannot be represented by a simple function of $\omega$, numerical techniques must be used to evaluate the integral in Eq. 8-73.

**EXAMPLE 8-8** At low temperature the phonon contribution to the heat capacity can be written $aNT^3$, where $a$ is a constant. Take the Debye temperature for germanium to be 363 K and calculate the value of $a$. Then calculate the phonon contribution per unit cell to the heat capacity at 10, 20, and 30 K, assuming the Debye approximation holds.

**SOLUTION** By comparison with Eq. 8-76, $a = 12\pi^4 k_B/5T_D^3 = 12\pi^4 \times 1.38 \times 10^{-23}/5 \times 363^3 = 6.60 \times 10^{-29}$ J/K⁴. At 10 K, $C_\tau/N = 6.60 \times 10^{-26}$ J/K; at 20 K, $C_\tau/N = 5.28 \times 10^{-25}$ J/K; and at 30 K, $C_\tau/N = 1.78 \times 10^{-24}$ J/K. ∎

*Electron Contribution to the Heat Capacity of a Metal.* An expression for the electron contribution to $C_\tau$ at low temperature is found by differentiating Eq. 8-48 with respect to temperature. The result is

$$C_\tau = \frac{\pi^2}{3} \rho(E_F) k_B^2 T, \tag{8-79}$$

where $\rho(E_F)$ is the density of states at the Fermi level.

Recall that the number of electrons affected by a change in temperature is proportional to $k_B T$, and as a result, the total electron energy is proportional to $T^2$. Thus the electron contribution to the heat capacity is proportional to $T$. If electrons were classical each would receive energy $k_B T$ as the temperature is raised from 0 K to $T$. The heat capacity would be independent of the temperature and would be very much larger than the values actually observed.

The total heat capacity of a metal is the sum of electron and phonon contributions:

$$C_\tau = \gamma T + aT^3, \tag{8-80}$$

where $\gamma$ and $a$ are constants. They are evaluated from experimental data by plotting $C_\tau/T$ as a function of $T^2$ and drawing the best straight line through the

**TABLE 8-2**  Values of the Electronic Constant $\gamma$ for Selected Elements

| Element | $\gamma$ (mJ/mol $\cdot$ K$^2$) | Element | $\gamma$ (mJ/mol $\cdot$ K) |
|---|---|---|---|
| Aluminum | 1.35 | Manganese | 14 |
| Antimony | 0.112 | Mercury | 1.79 |
| Barium | 2.7 | Nickel | 7.1 |
| Cadmium | 0.69 | Niobium | 7.79 |
| Cesium | 3.2 | Platinum | 6.8 |
| Chromium | 1.40 | Potassium | 2.1 |
| Copper | 0.688 | Silver | 0.650 |
| Indium | 1.6 | Sodium | 1.4 |
| Lanthanum | 10 | Yttrium | 10.2 |
| Lead | 3.0 | Zinc | 0.65 |
| Magnesium | 1.3 | Zirconium | 2.80 |

Source: American Institute of Physics Handbook, 3rd ed. (New York: McGraw-Hill, 1972).

points. The slope of the line gives $a$ and its intercept gives $\gamma$. The value of $\gamma$ can be used, via Eq. 8-79, to find the electron density of states $\rho(E_F)$ at the Fermi energy. As they appear in Eq. 8-80, $C_\tau$, $\gamma$, and $a$ are proportional to the number of primitive unit cells in the sample. Experimentally determined values are usually given per atom or per mole. Some are listed in Table 8-2.

**EXAMPLE 8-9**  Use Eq. 8-79 to calculate the coefficient $\gamma$ for 1 mol of sodium. Electrons in only one band, a free electron band, are affected by the increase in temperature. Take the cube edge to be 4.225 Å.

**SOLUTION**  If the zero of energy is at the bottom of the free electron band, $\rho(E_F) = (\sqrt{2}N\tau m^{3/2}/\pi^2\hbar^3)E_F^{1/2}$. If $a$ is the cube edge, then $\tau = a^3/2$ since sodium has a BCC structure. Take $N$ to be Avogadro's number (6.02 $\times$ 10$^{23}$ atoms/mol) and $E_F$ to be 3.22 eV or 5.15 $\times$ 10$^{-19}$ J, relative to the bottom of the band. When these values re substituted into

$$\gamma = \frac{k_B^2}{3} \frac{\sqrt{2}(N/2)a^3 m^{3/2}}{\hbar^3} E_F^{1/2}$$

the result is 1.10 $\times$ 10$^{-3}$ J/K$^2$.  ∎

## 8.6  REFERENCES

In addition to references listed at the end of the last two chapters, the following contain sections on the thermodynamics of phonons and electrons:

C. Kittel, Thermal Physics (New York: Wiley, 1969).
P. G. Klemens, "Thermal Conductivity and Lattice Vibrational Modes" in Solid State Physics (F. Seitz and D. Turnbull, Eds.), Vol. 8, p. 110 (New York: Academic, 1959).
L. A. Girifalco, Statistical Physics of Materials (New York: Wiley, 1973).

F. Mandl, *Statistical Physics* (New York: Wiley, 1973).

F. Reif, *Fundamentals of Statistical And Thermal Physics* (New York: McGraw-Hill, 1965).

F. W. Sears and G. L. Salinger, *Thermodynamics, Kinetic Theory, and Statistical Thermodynamics* (Reading, MA: Addison-Wesley, 1975).

K. Stowe, *Introduction to Statistical Mechanics and Thermodynamics* (New York: Wiley, 1984).

Some modern physics texts contain material pertinent to this chapter:

R. Eisberg and R. Resnick, *Quantum Physics* (New York: Wiley, 1985).

J. D. McGervey, *Introduction to Modern Physics* (New York: Academic, 1971).

R. L. Sproull and W. A. Phillips, *Modern Physics* (New York: Wiley, 1980).

For examples of Fermi surfaces, see:

A. P. Cracknell, *The Fermi Surfaces of Metals* (New York: Barnes & Noble, 1971).

W. A. Harrison and W. B. Webb (Eds.), *The Fermi Surface* (New York: Wiley, 1960).

## PROBLEMS

1. For many solids, $\omega_1 = 6.00 \times 10^8$ rad/s is a low angular frequency and $\omega_2 = 6.00 \times 10^{13}$ rad/s is a high angular frequency in an acoustical branch. For a vibrational mode associated with each of these angular frequencies, determine: (a) the temperature for which the mode has a single phonon; (b) the fractional increase in temperature $\Delta T/T$ required to double the number of phonons from 1 to 2; and (c) the number of phonons and the energy associated with the mode at 50 K and at 300 K.

2. Unlike the number of electrons, the number of phonons in a solid is not conserved as the temperature changes. Consider a three-dimensional sample with one atom in the basis and approximate the dispersion relationship by $\omega = vq$, where $v$ is the speed of sound. To simplify the calculation, assume the speed of sound is the same for all directions of $\mathbf{q}$ and assume all three branches of the spectrum have the same maximum angular frequency $\omega_m$. (a) Show that the total number of phonons is given by

$$n_T = \frac{9\tau_s}{2\pi^2 v^3} \left[\frac{k_B T}{\hbar}\right]^3 \int_0^{x_m} \frac{x^2 \, dx}{e^x - 1},$$

where $x_m = \hbar\omega_m/k_B T$. (b) Show that, for $k_B T \ll \hbar\omega_m$, the total number of phonons is proportional to $T^3$.

3. When a solid is in thermal equilibrium, the energy of each normal mode fluctuates. The square of the standard deviation $S$ is given by $S^2 = \langle E^2 \rangle - \langle E \rangle^2$, where

$$\langle E^2 \rangle = \frac{\sum n^2 \hbar^2 \omega^2 e^{-\beta n\hbar\omega}}{\sum e^{-\beta n\hbar\omega}}$$

for a mode of frequency $\omega$. Each sum runs from $n = 0$ to $n = \infty$. Show that: (a) $S^2 = \hbar^2\omega^2 e^{\beta\hbar\omega}[e^{\beta\hbar\omega} - 1]^{-2}$; (b) the contribution to the constant volume heat capacity is $S^2/k_B T^2$; (c) the fractional standard deviation $S/\langle E \rangle$ is to $e^{\beta\hbar\omega}$ as $T$ becomes small; and (d) the fractional standard deviation tends to 1 as $T$ becomes large.

4. Consider a linear monatomic chain of $N$ atoms, each of which interacts only with its nearest neighbors. (a) Show that $(Na/2\pi)\,dq$ modes have propagation constant between $q$ and $q + dq$. Here $a$ is the equilibrium separation of the atoms. (b) Show that the density of modes is given by $g(\omega) = (2N/\pi)(\omega_m^2 - \omega^2)^{-1/2}$, where $\omega_m$ is the maximum normal mode angular frequency. (c) Show that the total phonon energy is proportional to $T^2$ at low temperatures.

5. For each of the following solids, use the Debye temperature to estimate the average speed of sound: (a) potassium (BCC with $a = 5.225$ Å; $T_D = 89$ K); (b) magnesium (HCP with $a = 3.21$ Å, $c = 5.21$ Å; $T_D = 450$ K); and (c) germanium (diamond with $a = 5.658$ Å; $T_D = 363$ K). Take $(1/v_{ave})^3 = \frac{1}{3}\sum(1/v_s)^3$, where the sum is over the three acoustical branches.

6. The maximum frequency $\omega_m$ used in the Debye approximation can be obtained in another way. The Brillouin zone is replaced by a sphere of equal volume in reciprocal space and a linear, isotropic dispersion relationship $\omega = vq$ is assumed. Here $v$ is an average of the speeds of sound for the three acoustical branches. Show that: (a) the radius of the Debye sphere is $q_D = (6\pi^2/\tau)^{1/3}$, where $\tau$ is the volume of a direct lattice unit cell; (b) there are $3N$ normal modes with propagation vectors inside the Debye sphere; and (c) the maximum frequency of any mode with propagation vector in the sphere is $\omega_m = (6\pi^2 v^3/\tau)^{1/3}$. (d) To obtain agreement with Eq. 8-18, how should the average speed of sound be related to the speeds for the individual branches?

7. A tight-binding $s$ band for electrons in a simple cubic lattice with cube edge $a$ is given by

$$E(\mathbf{k}) = E_a - \alpha - 2\gamma[\cos(k_x a) + \cos(k_y a) + \cos(k_z a)],$$

where $E_a$ is the atomic level and $\alpha$ and $\gamma$ are positive overlap integrals. Draw intersections of Brillouin zone boundaries with the $k_z = 0$ plane and, for each of the following values of the energy, plot on the same diagram intersections of the constant energy surface with that plane: (a) $E_a - \alpha - 4\gamma$; (b) $E_a - 2\gamma$; and (c) $E_a - \alpha - \gamma$. If possible, use a computer.

8. Consider nearly free electron bands for a simple cubic crystal with cube edge $a = 5.80$ Å. For each state assume the only important Fourier component of the potential energy function is the one corresponding to the nearest Brillouin zone face. Take this component to have magnitude 0.133 eV in each case. (a) Draw the (001) cross section of the Brillouin zone and plot the constant energy surface for $E = 0.666$ eV, relative to the bottom of the lowest nearly free electron band. There is only one branch, associated

with the lowest band. (b) Plot the constant energy surface for $E = 1.28$ eV. There are two branches. If possible, use a computer.

9. Aluminum has a face-centered cubic structure with cube edge $a = 4.05$ Å. Each atom contributes 3 electrons to nearly free electron bands and the Fermi level is 12.0 eV above the bottom of the lowest nearly free electron band. Assume the free electron model with an effective mass is valid and calculate (a) the radius of the Fermi sphere and (b) the effective mass for electrons in aluminum. (c) Compare the volume of the Fermi sphere to the volume of the Brillouin zone.

10. Strontium has an FCC structure with cube edge $a = 6.08$ Å. Each atom contributes 2 electrons to nearly free electron bands. Assume the electrons are absolutely free. Calculate (a) the radius of the Fermi sphere and (b) the Fermi energy. (c) The Fermi sphere intersects only the hexagonal faces of the Brillouin zone surface. At $T = 0$ K, what fraction of states in the lowest free electron band are occupied?

11. Suppose electron bands for a simple cubic crystal can be approximated by free electron bands. For each of the following situations calculate the fraction of states that are occupied at $T = 0$ K in the lowest free electron band: (a) each atom contributes 1 electron to the free electron bands; (b) each atom contributes 2 electrons; and (c) each atom contributes 3 electrons.

12. Consider a nearly free electron band and investigate the constant energy surface for energy $E = 2.60$ eV above the bottom of the band. Suppose a Brillouin zone face is normal to the $x$ direction at $k_x = 0.780 \times 10^{10}$ m$^{-1}$ and the gap between the first and second bands is 0.150 eV there. Assume the gap is due to a single Fourier component $U(\mathbf{G})$ of the potential energy function. (a) First assume the electrons are absolutely free and calculate the radius of the constant energy sphere. (b) The intersection of the constant energy sphere and the Brillouin zone face is a circle. Calculate its radius. (c) When the potential energy function is taken into account, the sphere distorts but the intersection of the constant energy surface with the zone face is still a circle for each band. Find the radii. (d) For what value of $U(\mathbf{G})$ does the second band surface disappear?

13. (a) Consider an electron state that is occupied with probability 0.95 at temperature $T$. Derive an expression for $E - \eta$, its energy relative to the chemical potential. Evaluate the expression for $T = 100, 300$, and $1200$ K. (b) Repeat the calculation for a state that is occupied with probability 0.05.

14. Consider free electron states for two metals with the same crystal structure, the same size unit cells, and the same electron concentration. Each has a single occupied free electron band, but the effective mass for one of them is $m_1^*$ while the effective mass for the other is $m_2^* = 2m_1^*$. Take the zero of energy to be at the bottom of the free electron band in each case. Find the ratio of: (a) their Fermi energies; (b) their densities of states at the Fermi energy; (c) their chemical potentials for $T > 0$, each measured rel-

ative to the Fermi energy; and (d) the thermal energies of their free electrons for $T > 0$.

15. A certain semiconductor has a diamond structure with a cube edge of 5.4 Å. It has parabolic conduction and valence bands, characterized by effective masses $m_e = 0.88m_0$ and $m_h = 0.42m_0$, respectively. The gap between the valence and conduction bands is 0.82 eV. Assume the material is pure and calculate: (a) the Fermi energy; (b) the chemical potential at 300 K, relative to the Fermi energy; (c) the number of electrons per unit volume in the conduction band at 300 K; and (d) the number of electrons per unit volume in the valence band at 300 K. (e) By numerical substitution into Eqs. 8-60 and 8-61 show that $n_c = p_v$ at 300 K.

16. Two semiconductors have the same crystal structure and the same size unit cell. Each has a single parbolic valence band, characterized by the same effective mass $m_h$. Each also has a single parabolic conduction band, characterized by the same effective mass $m_e$. The gap between the valence and conduction band is $E_{g1} = 0.65$ eV for the first semiconductor and $E_{g2} = 2E_{g1}$ for the second. Assume $|E - \eta| \gg k_B T$ for all states. Compare: (a) their Fermi energies, each measured from the top of the valence band; (b) changes in their chemical potentials as the temperature is raised from 0 to 300 K; and (c) the numbers of conduction band electrons at 300 K.

17. Consider an insulator with a single conduction band and a single valence band, both parabolic. The densities of states are given by Eqs. 8-52 and 8-58. Show that the total energy per unit volume of all electrons in the conduction band is given by $n(E_c + \tfrac{3}{2}k_B T)$ and the total energy per unit volume of all electrons in the valence band is given by $E_f - p(E_v - \tfrac{3}{2}k_B T)$, where $E_f$ is the energy per unit volume of a full valence band.

18. For an $n$-type semiconductor with a single donor level in the gap near the bottom of the conduction band, sketch the Fermi-Dirac distribution function and mark possible positions of the top of the valence band, the bottom of the conduction band, and the donor level: (a) when the donor states are partially filled and the valence band is essentially filled, (b) when the donor states are essentially empty and the valence band is essentially filled, and (c) when the donor states are essentially empty and the valence band is partially filled.

19. Near a conduction band minimum, electron band energies for silicon, germanium, and some other semiconductors can be written

$$E(k) = \hbar^2 \left[ \frac{k_x^2}{2m_1} + \frac{k_y^2}{2m_2} + \frac{k_z^2}{2m_3} \right],$$

where $m_1$, $m_2$, and $m_3$ are effective mass parameters. (a) Show that the contribution of states around one such minimum to the density of states is

$$\rho(E) = \frac{\sqrt{2}\tau_s (m_d^*)^{3/2}}{\pi^2 \hbar^3} E^{1/2},$$

where $m_d^* = (m_1 m_2 m_3)^{1/3}$. This quantity is called the density of states effective mass. *Hint:* Calculate the number of states with **k** inside a constant energy ellipsoid, then differentiate with respect to $E$. The number of states is $(\tau_s/4\pi^3) \int dk_x \, dk_y \, dk_z$. If $k_x' = (m/m_1)^{1/2}k_x$, $k_y' = (m/m_2)^{1/2}k_y$, and $k_z' = (m/m_3)^{1/2}k_z$, then the integral is over the volume of a sphere with radius $k' = [(k_x')^2 + (k_y')^2 + (k_z')^2]^{1/2} = (2mE/\hbar^2)^{1/2}$. (b) Silicon has six equivalent minima and germanium has four, so the expression for $\rho(E)$ must be multiplied by 6 and 4, respectively, for these materials. This is done by defining the density of states effective mass as $m_d^* = (m_1 m_2 m_3)^{1/3}s^{2/3}$, where $s$ is the number of equivalent minima. For silicon $m_1 = 0.9163m_0$ and $m_2 = m_3 = 0.1905m_0$, where $m_0$ is the free electron mass. Calculate the effective number of states per unit volume in the silicon conduction band at 300 K.

20. When pure, a certain semiconductor has $n_i$ electrons per unit volume in the conduction band. Show that, when it is doped with $N_d$ donors per unit volume, the electron concentration in the conduction band is $n = n_i + N_d/2$ if $N_d \ll n_i$. What then is the hole concentration in the valence band? Assume all donors are singly ionized.

21. Pure germanium at 300 K has $2.5 \times 10^{19}$ electrons/m³ in its conduction band. A germanium crystal is doped with $7.6 \times 10^{18}$ acceptor atoms/m³. Assume each impurity has accepted one electron and calculate the concentration of electrons in the conduction band and the concentration of holes in the valence band at 300 K.

22. Consider a semiconducting sample at 300 K, with $2.70 \times 10^{16}$ states/m³ in the valence band and $3.50 \times 10^{16}$ states/m³ in the conduction band. The gap between the bands is 1.06 eV. The sample is doped with $4.50 \times 10^{11}$ donor atoms/m³ and the chemical potential is 0.760 eV above the top of the valence band. Assume the degeneracy of the impurity level is 2. (a) How many electrons per unit volume are in the conduction band? (b) How many valence band states per unit volume are empty? (c) How many electrons per unit volume are in impurity states? (d) Where is the donor energy level, relative to the top of the valence band?

23. Use the two level model to estimate the temperature at which an $n$ type semiconductor becomes intrinsic. Take the effective concentration of states in the conduction band to be $N_c^*$, the effective concentration of states in the valence band to be $N_v^*$, and the concentration of donors to be $N_d$. Assume each donor introduces a single state in the gap and neglect the temperature dependence of $N_c$ and $N_v^*$. (a) Show that the concentration of electrons in the conduction band differs from the intrinsic value by 10% when $n_i = 5.24n_D$ and (b) that this occurs at temperature $T$ given by

$$T = \frac{E_g}{k_B \ln \left[ \dfrac{N_c^* N_v^*}{27.4 N_d^2} \right]},$$

where $E_g$ is the gap between the valence and conduction bands. (c) Take $N_c^* = N_v^* = 1.04 \times 10^{25}$ states/m$^3$, $E_g = 0.67$ eV, and $N_d = 2.71 \times 10^{19}$ donors/m$^3$, then evaluate the expression for the temperature.

24. For each of the two normal mode angular frequencies given in Problem 1, find the contribution of the mode to the constant volume heat capacity at 50 K and at 300 K. Give your answers in J/K and as multiples of $k_B$.

25. For each of the solids listed in Problem 5, use the Debye approximation to estimate the phonon contribution to the constant volume heat capacity per primitive unit cell at 10 K and at 25 K.

26. Use the Debye approximation and the free electron model to compare phonon and electron contributions to the constant volume heat capacity for potassium at 0.1, 1, and 10 K. First calculate the electronic constant $\gamma$ and compare the value with that given in Table 8-2. Potassium has a BCC structure with a cube edge of 5.225 Å and its Debye temperature is 89 K. Each atom contributes 1 electron to the lowest free electron band. Take the effective mass to be the same as the mass of a free electron.

27. For (a) potassium and (b) lanthanum, use data given in Tables 8-1 and 8-2 to estimate the temperature at which the phonon and electron contributions to the constant volume heat capacity are equal. (c) For each of these metals, use the value of $\gamma$ from Table 8-2 to estimate the density of electron states per atom at the Fermi level.

# Chapter 9

## ELECTRICAL AND THERMAL CONDUCTION

A variable temperature research dewar used to measure electrical, thermal, magnetic, and optical properties as functions of temperature. The sample is mounted in the tail, behind the window, and wires run up the tail center to measuring instruments. Liquid helium for cooling is held in the tank which forms the upper portion of the dewar.

- 9.1 ELECTRON DYNAMICS

- 9.2 THE BOLTZMANN TRANSPORT EQUATION

- 9.3 ELECTRICAL CONDUCTION

- 9.4 THERMAL CONDUCTION

- 9.5 SCATTERING

When an electric field or temperature gradient is applied to a solid, charge and energy flow. Metals are excellent electrical and thermal conductors and, for them, electrons are the predominant carriers of both charge and energy. Thermal conduction in most semiconductors and all insulators, on the other hand, is overwhelmingly due to phonon flow. Exceptions are semiconductors that have been heavily doped so the electron or hole concentration is extremely high.

In this chapter we study the fundamental processes of conduction: the acceleration of electrons in an electric field and the diffusion of both electrons and phonons in a temperature gradient. Results are used to develop expressions for the electrical and thermal conductivities of materials. Scattering of electrons and phonons from defects and from vibrating atoms plays an important role in both electrical and thermal conduction. It is taken into account in a phenomenological manner and discussed qualitatively in the last section.

## 9.1 ELECTRON DYNAMICS

*Electron Velocity.* For purposes of discussing macroscopic electric current we take the momentum of an electron with wave function $\psi$ to be the average

$$\langle \mathbf{p} \rangle = -i\hbar \int \psi^* \nabla \psi \, d\tau , \qquad (9\text{-}1)$$

where the integral is over the volume of the sample. Equation 9-1 simplifies considerably if the sample is a crystal. We treat a one-dimensional crystal, for which

$$\langle \mathbf{p} \rangle = -i\hbar \int \psi^* \frac{\partial \psi}{\partial x} \, dx . \qquad (9\text{-}2)$$

and

$$\psi = e^{ikx} u(k, x) , \qquad (9\text{-}3)$$

where $u(k, x)$ is periodic. Since $d\psi/dx = (iku + \partial u/\partial x)e^{ikx}$ and $\int u^* u \, dx = 1$,

$$\langle \mathbf{p} \rangle = \hbar k - i\hbar \int u^* \frac{\partial u}{\partial x} \, dx . \qquad (9\text{-}4)$$

Note that $\langle \mathbf{p} \rangle = \hbar k$ if $u$ is independent of $x$.

The differential equation obeyed by $u(k, x)$ can be used to write the integral of Eq. 9-4 in compact form. In one dimension, Eq. 7-11 is

$$-\frac{\hbar^2}{2m} \frac{\partial^2 u}{\partial x^2} - i\frac{\hbar^2 k}{m} \frac{\partial u}{\partial x} + \frac{\hbar^2 k^2}{2m} u + U(x)u = E(k)u . \qquad (9\text{-}5)$$

When this is differentiated with respect to $k$ and then multiplied by $u^*$, the result is

$$-\frac{\hbar^2}{2m} u^* \frac{\partial^3 u}{\partial k \, \partial^2 x} - i\frac{\hbar^2}{m} u^* \frac{\partial u}{\partial x} - i\frac{\hbar^2 k}{m} u^* \frac{\partial^2 u}{\partial k \, \partial x} + \frac{\hbar^2 k}{m} u^* u$$

$$+ \frac{\hbar^2 k^2}{2m} u^* \frac{\partial u}{\partial k} + U u^* \frac{\partial u}{\partial k} = \frac{dE}{dk} u^* u + E u^* \frac{\partial u}{\partial k} . \qquad (9\text{-}6)$$

The complex conjugate of Eq. 9-5 is multiplied by $\partial u/\partial k$ and the result subtracted from Eq. 9-6 to obtain

$$
-\frac{\hbar^2}{2m}\left[u^*\frac{\partial^3 u}{\partial k\,\partial x^2} - \frac{\partial^2 u^*}{\partial x^2}\frac{\partial u}{\partial k}\right] - i\frac{\hbar^2 k}{m}\left[u^*\frac{\partial^2 u}{\partial k\,\partial x} + \frac{\partial u^*}{\partial x}\frac{\partial u}{\partial k}\right]
$$

$$
- i\frac{\hbar^2}{m}u^*\frac{\partial u}{\partial x} + \frac{\hbar^2 k}{m}u^* u = \frac{dE}{dk}u^* u. \quad (9\text{-}7)
$$

This equation is integrated over the sample. The first two terms are exact derivatives: the first is the derivative of $(\hbar^2/2m)[u^*(\partial^2 u/\partial k\,\partial x) - (\partial u^*/\partial x)(\partial u/\partial k)]$ and the second is the derivative of $(i\hbar^2 k/m)u^*(\partial u/\partial k)$. The indefinite integrals are periodic and each has the same value at the upper and lower limits, so the definite integrals vanish. Thus

$$
- i\frac{\hbar^2}{m}\int u^*\frac{\partial u}{\partial x}\,dx + \frac{\hbar^2 k}{m} = \frac{dE}{dk}. \quad (9\text{-}8)
$$

When this result is used to replace the integral in Eq. 9-4, that equation becomes

$$
\langle p \rangle = \frac{m}{\hbar}\frac{\partial E}{\partial k}, \quad (9\text{-}9)
$$

When a similar derivation is carried out for three dimensions, the average momentum of an electron with propagation vector $\mathbf{k}$ in band $n$ is found to be

$$
\langle \mathbf{p} \rangle = \frac{m}{\hbar}\nabla_k E_n(\mathbf{k}), \quad (9\text{-}10)
$$

where $\nabla_k E_n$ is the gradient in reciprocal space of the electron energy. The electron velocity is taken to be

$$
\mathbf{v}_n(\mathbf{k}) = \frac{\langle \mathbf{p} \rangle}{m} = \frac{1}{\hbar}\nabla_k E_n(\mathbf{k}). \quad (9\text{-}11)
$$

Since $E_n = \hbar\omega$, where $\omega$ is the angular frequency of the electron wave function, the particle velocity is identical to the group velocity of the wave, given by $\nabla_k\omega$.

Since $\nabla_k E_n$ is perpendicular to a constant energy surface associated with band $n$, so is $\mathbf{v}_n(\mathbf{k})$. If the energy increases with distance from the center of the Brillouin zone, $\mathbf{v}_n$ is directed away from the center. For a one-dimensional crystal $v_n(k) = (1/\hbar)(dE_n/dk)$ and the direction of the velocity is indicated by the sign of the slope of the band curve $E_n(k)$. For the lower band shown in Fig. 9-1, the electron velocity is positive for $k$ positive, while for the upper it is negative. Near a Brillouin zone boundary the energy is nearly independent of $k$ and the electron velocity is small.

**EXAMPLE 9-1** Find an expression for the velocity of a nearly free electron in a one-dimensional crystal with cell length $a$. Assume only the Fourier components $U(G)$ and $U(-G)$ of the potential energy function are significant. Evaluate the expression for $a = 4.5 \times 10^{-10}$ m, $G = -2\pi/a$, $U(G) = 1.7$ eV, and $k = -\frac{1}{4}G$.

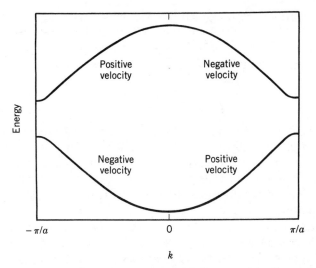

**FIGURE 9-1** Two electron energy bands for a one-dimensional crystal. The sign of the electron velocity is determined by the slope of $E(k)$. At the Brillouin zone boundary $dE/dk = 0$ and the velocity vanishes. Electron velocities for a three-dimensional crystal are proportional to the gradient of the energy in reciprocal space.

**SOLUTION**  For a one-dimensional crystal, Eq. 7-50 becomes

$$E(k) = \frac{\hbar^2}{4m}(2k^2 + 2kG + G^2) - \frac{1}{2}\left[\left(\frac{\hbar^2}{2m}\right)^2(2kG + G^2)^2 + 4|U(G)|^2\right]^{1/2},$$

where we have taken $U_0$ to be 0 and have assumed the electron is in the lower band. The electron velocity is

$$v = \frac{1}{\hbar}\frac{dE}{dk} = \frac{\hbar}{2m}(2k + G) - \frac{\hbar^3}{4m^2}\frac{G^2(2k + G)}{\left[\left(\frac{\hbar^2}{2m}\right)^2(2k + G)^2G^2 + 4|U(G)|^2\right]^{1/2}}.$$

For $k = -\frac{1}{4}G$,

$$v = \frac{\hbar G}{4m} - \frac{\hbar^3 G^3}{4m^2}\left[\left(\frac{\hbar^2 G^2}{2m}\right)^2 + 16|U(G)|^2\right]^{-1/2}.$$

For the values given, $\hbar G = -2\pi\hbar/a = -2\pi \times 1.05 \times 10^{-34}/4.5 \times 10^{-10} = -1.47 \times 10^{-24}$ kg · m/s and $|U(G)| = 1.7 \times 1.60 \times 10^{-19} = 2.72 \times 10^{-19}$ J. So,

$$v = -\frac{1.47 \times 10^{-24}}{4 \times 9.11 \times 10^{-31}}$$
$$+ \frac{(1.47 \times 10^{-24})^3}{4(9.11 \times 10^{-31})^2}\left[\frac{(1.47 \times 10^{-24})^4}{4(9.11 \times 10^{-31})^2} + 16(2.72 \times 10^{-19})^2\right]^{-1/2}$$
$$= 1.91 \times 10^5 \text{ m/s}.$$

This is a typical velocity for an electron in the middle of a nearly free electron band. ∎

*Electron Acceleration.* An applied electric field causes an electron to change states. If the solid is a crystal and the field is constant and weak, the electron remains in the same band. If, in addition, the field is uniform, each electron changes states in such a way that its crystal momentum obeys

$$\frac{d\hbar\mathbf{k}}{dt} = -e\mathcal{E}, \tag{9-12}$$

where $\mathcal{E}$ is the applied field and $e$ is the magnitude of the charge on an electron. This equation is often likened to Newton's second law. The right side is the electrical force on the electron, while the left is the rate of change of crystal momentum. We stress, however, that $\hbar\mathbf{k}$ is a crystal momentum and not a true momentum. The two are the same only for free electrons and, in many instances, are quite different.

The electron wave function reflects the behavior described above. It is now time dependent and we denote it by $\Psi(\mathbf{r}, t)$. If the electron is initially in a state with crystal momentum $\hbar\mathbf{k}_0$, then the spatial part of $\Psi(\mathbf{r}, 0)$ is $\psi_n(\mathbf{k}_0, r)$. At a later time $t$, the spatial part of $\Psi(\mathbf{r}, t)$ is nearly $\psi_n(\mathbf{k}', r)$, where $\mathbf{k}' = \mathbf{k}_0 - e\mathcal{E}t/\hbar$. This result, which is equivalent to Eq. 9-12, can be obtained from the time-dependent Schrödinger equation, with $+e\mathcal{E} \cdot \mathbf{r}$ included in the potential energy function. The derivation is somewhat complicated but the result is important for studies of material properties. Details can be found in Appendix E.

According to Eq. 9-12, the crystal momentum changes uniformly. If an electron is initially in a state with crystal momentum $\hbar\mathbf{k}_0$, then at a later time $t$ it is in a state with crystal momentum $\hbar\mathbf{k}_0 - e\mathcal{E}t$, in the same band. This change is illustrated in Fig. 9-2. A circle marks the state that is occupied by the electron

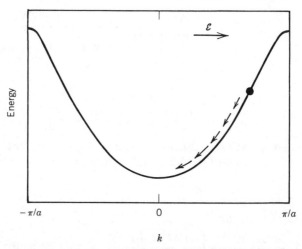

**FIGURE 9-2** An electron energy band for a one-dimensional crystal. The circle marks a state initially occupied by an electron. When an electric field is turned on in the direction shown, the electron makes transitions as indicated by arrows to other states in the band. For the band shown, its energy decreases until it passes $k = 0$.

and, as time increases, it moves along the band, covering equal intervals of $\mathbf{k}$ in equal times. The change in crystal momentum is opposite in direction to the applied field.

If no other external forces act, the electron eventually occupies a state with $\mathbf{k}$ on a Brillouin zone boundary, if only for an instant. In the next instant, it moves in accordance with Eq. 9-12 to a state with $\mathbf{k}$ outside the zone. The new state is identical to a state with $\mathbf{k}$ inside, a reciprocal lattice vector away, so in the reduced zone scheme an electron in a state with $\mathbf{k}$ on a zone boundary next makes a transition to a state with $\mathbf{k}$ on the opposite boundary.

As an electron changes states, its velocity changes, simply because different velocities are associated with different states. The change in velocity is not necessarily uniform in time even though the change in propagation vector is. To obtain the acceleration, we differentiate the velocity with respect to time and observe that it depends on time through $\mathbf{k}$. If we use $dk_j/dt = -(e/\hbar)\mathcal{E}_j$ and $v_i = (1/\hbar)(\partial E/\partial k_i)$, where subscripts denote Cartesian components, then component $i$ of the acceleration is

$$a_i = \frac{dv_i(\mathbf{k})}{dt} = \sum_j \frac{\partial v_i}{\partial k_j}\frac{dk_j}{dt} = \frac{1}{\hbar}\sum_j \frac{\partial v_i}{\partial k_j}(-e\mathcal{E}_j)$$

$$= \frac{1}{\hbar^2}\sum_j \frac{\partial^2 E}{\partial k_i \partial k_j}(-e\mathcal{E}_j) = \sum_j \left[\frac{1}{m^*}\right]_{ij}(-e\mathcal{E}_j), \qquad (9\text{-}13)$$

where

$$\left[\frac{1}{m^*}\right]_{ij} = \frac{1}{\hbar^2}\frac{\partial^2 E(k)}{\partial k_i \partial k_j} \qquad (9\text{-}14)$$

is an element of the reciprocal effective mass tensor, introduced in Section 7.5. As Eq. 9-13 shows, the electron acceleration is not necessarily in the direction of the applied force. For example, if $\mathcal{E}$ is in the $x$ direction, the acceleration has a nonvanishing $y$ component if $(1/m^*)_{xy} \neq 0$. The electron moves to a new state and acquires a new velocity, with direction determined by the shape of the band and ultimately by interactions with ion cores.

If the effective mass is a scalar, Eq. 9-13 becomes

$$m^*\mathbf{a} = -e\mathcal{E}, \qquad (9\text{-}15)$$

indicating that the acceleration is parallel to the field. Even in this case, the acceleration may not be in the same direction as the applied force. Near a maximum in a band, for example, $d^2E/dk^2$ is negative and the acceleration is opposite to the force. Such an electron has a negative effective mass.

As an electron changes states, its energy may increase for some transitions and decrease for others. The rate of energy change is given by $dE/dt = \nabla_k E \cdot (d\mathbf{k}/dt)$ and, once Eqs. 9-11 and 9-12 are used, this expression becomes $dE/dt = -(e/\hbar)\nabla_k E \cdot \mathcal{E} = -e v \cdot \mathcal{E}$, a result that is also true classically. Both energy increasing and energy decreasing transitions are evident in the diagram of Fig. 9-2.

*Holes.* Properties of a nearly full band are usually ascribed to holes rather than electrons. Although the number of holes in a band equals the number of empty states, a hole is not simply an empty state. In particular, a single hole in an otherwise full band must account for the collective properties of all electrons in the band.

To find the crystal momentum of a hole in a given band, assume the band is completely filled except for the state with crystal momentum $\hbar\mathbf{k}_{empty}$. The crystal momentum $\hbar\mathbf{k}_h$ of the hole is not the crystal momentum associated with the empty state, but rather, it is the total crystal momentum of all electrons in the band:

$$\hbar\mathbf{k}_h = \sum_{\substack{occupied \\ states}} \hbar\mathbf{k} = \sum_{\substack{all \\ states}} \hbar\mathbf{k} - \hbar\mathbf{k}_{empty} = -\hbar\mathbf{k}_{empty}. \qquad (9\text{-}16)$$

The sum over all states vanishes because the total crystal momentum of a completely filled band is zero: for every propagation vector $\mathbf{k}$ in the Brillouin zone, $-\mathbf{k}$ is also in the zone and both are included in the sum. The band can be described as occupied by electrons except for the state with crystal momentum $\hbar\mathbf{k}_{empty}$, or as having a single hole with crystal momentum $-\hbar\mathbf{k}_{empty}$.

In a uniform applied electric field, the crystal momenta of all electrons change at the same rate. As illustrated in Fig. 9-3a, the vector $\mathbf{k}_{empty}(t)$, used to designate the state that is empty at time $t$, changes in exactly the same way as the propagation vector of an electron. On the other hand, $\mathbf{k}_h$ behaves quite differently. Differentiate Eq. 9-16 and use Eq. 9-12 to obtain

$$\frac{d\hbar\mathbf{k}_h}{dt} = +e\mathcal{E}. \qquad (9\text{-}17)$$

Figure 9-3b and c illustrate the dynamics. According to Eq. 9-17, the crystal momentum of a hole changes in the same way as that of a particle with charge $+e$. That is, electrons in a band with one empty state behave collectively in an electric field exactly like a single positive charge.

The velocity of a hole is the group velocity of the wave associated with the empty state and so is the same as the velocity an electron would have if it occupied that state. That is, if the state with wave vector $\mathbf{k}$ is empty,

$$\mathbf{v}_h = \frac{1}{\hbar} \nabla_k E_n(\mathbf{k}). \qquad (9\text{-}18)$$

The effective mass for a hole is found by differentiating Eq. 9-18 with respect to time. Component $i$ of the acceleration is

$$a_i = \frac{dv_{hi}}{dt} = -\frac{1}{\hbar^2} \sum_j \frac{\partial^2 E(\mathbf{k})}{\partial k_i \partial k_j} e\mathcal{E}_j = \sum_j \left[\frac{1}{m_h^*}\right]_{ij} e\mathcal{E}_j, \qquad (9\text{-}19)$$

where $(1/m_h^*)_{ij} = -(1/\hbar^2)[\partial^2 E(\mathbf{k})/\partial k_i \partial k_j]$ is an element of the hole reciprocal effective mass tensor. Each element is the negative of the corresponding element of the electron reciprocal effective mass tensor for the empty state. The negative

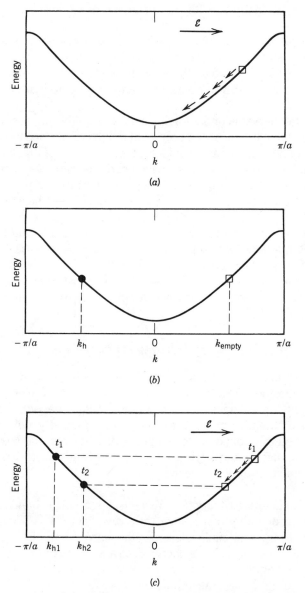

**FIGURE 9-3** An electron energy band for a one-dimensional crystal. The square marks a state that is initially empty in an otherwise filled band. (a) When an electric field is turned on in the direction shown, the states indicated by arrows successively become empty as electrons make transitions to the left. (b) The band is completely filled except for the state marked by the square. Except for the electron in the state marked by a circle, each electron can be paired with another so the sum of their crystal momenta vanishes. The total crystal momentum for the band and the crystal momentum of the hole are both $\hbar k_h$. (c) The empty state and unpaired electron for two times $t_2 > t_1$ when an electric field is in the direction shown. The change in $k_h$ is in the direction of the field.

sign in Eq. 9-19 is included with the reciprocal mass tensor since the force on the hole is $+e\mathcal{E}$.

We may think of any band either as a collection of negative electrons, one for each occupied state, or as a collection of positive holes, one for each empty state. Provided we are careful about determining crystal momenta, velocities, and effective masses, we will obtain the same results no matter which of these views we take.

For a crystalline semiconductor in thermal equilibrium, a small fraction of conduction band states, near the bottom of that band, are occupied by electrons and a small fraction of valence band states, near the top of that band, are empty. Since all conduction band electrons have nearly the same velocity and effective mass, contributions of that band to material properties are usually discussed in terms of electrons. On the other hand, valence band holes have nearly the same velocity and effective mass so contributions of that band are described in terms of holes. Furthermore, the curvature of a band near its minimum is positive, so most conduction band electrons have positive effective masses and their accelerations are roughly opposite to the direction of an applied field. The curvature of a band near its maximum is negative, so most valence band holes have positive effective masses and their accelerations are roughly in the same direction as an applied field.

## 9.2 THE BOLTZMANN TRANSPORT EQUATION

In addition to making field-induced transitions, electrons scatter from phonons, impurities, and other defects. These interactions result in electron transitions and accompanying changes in crystal momentum. The Boltzmann transport equation is a differential equation that describes changes produced in the electron distribution by applied fields, temperature gradients, and scattering. In this section we will discuss the origins of the equation and consider the solution for a solid in a constant uniform electric field.

*Relaxation Time Approximation.* We consider a single band and let $f(\mathbf{k}, t)$ represent the probability that a state with wave vector $\mathbf{k}$ is occupied at time $t$. If the solid is in thermal equilibrium, $f$ is the Fermi-Dirac distribution function, which depends on $\mathbf{k}$ through the energy $E(\mathbf{k})$ and is constant in time. If the solid is not in thermal equilibrium, $f$ may change with time because both the electric field and scattering cause transitions, so we write

$$\frac{\partial f}{\partial t} = \left[\frac{\partial f}{\partial t}\right]_{\text{field}} + \left[\frac{\partial f}{\partial t}\right]_{\text{scatt}}. \tag{9-20}$$

The first term is calculated as the limit

$$\left[\frac{\partial f}{\partial t}\right]_{\text{field}} = \lim_{\Delta t \to 0} \frac{f(\mathbf{k}, t + \Delta t) - f(\mathbf{k}, t)}{\Delta t}\bigg|_{\text{field}}, \tag{9-21}$$

where the numerator is the change in $f$ caused by the field in time $\Delta t$. Note that the same value of $\mathbf{k}$ appears in both terms. Equation 9-21 describes the change in occupation probability for a particular state. In time $\Delta t$ an electron in the state $\mathbf{k} + (e\mathcal{E}/\hbar)\,\Delta t$ moves to the state $\mathbf{k}$, so $f(\mathbf{k}, t + \Delta t) = f(\mathbf{k} + e\mathcal{E}\,\Delta t/\hbar, t) = f(\mathbf{k}, t) + e\mathcal{E} \cdot \nabla_k f\,\Delta t/\hbar$, in the limit as $\Delta t$ becomes small. Thus

$$\left[\frac{\partial f}{\partial t}\right]_{\text{field}} = +\frac{e}{\hbar}\mathcal{E} \cdot \nabla_k f. \tag{9-22}$$

The scattering term is taken to be

$$\left[\frac{\partial f}{\partial t}\right]_{\text{scatt}} = -\frac{f(\mathbf{k}, t) - f_0(\mathbf{k})}{\bar{t}}, \tag{9-23}$$

where $\bar{t}$ is a parameter called the relaxation time and $f_0(\mathbf{k})$ is the Fermi-Dirac distribution function. Equation 9-23 is an approximation. With the proper choice of $\bar{t}$, however, it leads to an accurate description of the electron distribution in many situations. Most importantly for our purposes, it describes the tendency of scattering to restore thermal equilibrium. If $f > f_0$, then $(\partial f/\partial t)_{\text{scatt}}$ is negative and $f$ is reduced by scattering. On the other hand, if $f < f_0$, then $(\partial f/\partial t)_{\text{scatt}}$ is positive and $f$ is increased.

Suppose the electrons have a nonequilibrium distribution $f(\mathbf{k}, 0)$ at time $t = 0$. If no electric field exists the distribution function satisfies $\partial f/\partial t = -(f - f_0)/\bar{t}$, a differential equation that has the solution

$$f(\mathbf{k}, t) = f_0(\mathbf{k}) + [f(\mathbf{k}, 0) - f_0(\mathbf{k})]e^{-t/\bar{t}}. \tag{9-24}$$

The distribution relaxes exponentially to the equilibrium distribution and, after an interval of time equal to a small multiple of $\bar{t}$, the electron system is essentially in thermal equilibrium.

When Eqs. 9-22 and 9-23 are substituted into Eq. 9-20, the result is

$$\frac{\partial f}{\partial t} = \frac{e}{h}\mathcal{E} \cdot \nabla_k f - \frac{f - f_0}{\bar{t}}. \tag{9-25}$$

This equation is an example of the Boltzmann transport equation, specialized to a situation for which only an electric field is applied.

We now obtain a solution appropriate when the applied electric field is small and the electron distribution is nearly the thermal equilibrium distribution. Take $f(\mathbf{k}, t) = f_0(\mathbf{k}) + f_1(\mathbf{k}, t)$, where $f_1$ is small, and assume $\mathcal{E} \cdot \nabla_k f_1$ is negligible. Then Eq. 9-25 becomes

$$\frac{\partial f_1}{\partial t} = \frac{e}{\hbar}\mathcal{E} \cdot \nabla_k f_0 - \frac{f_1}{\bar{t}} = e\mathcal{E} \cdot \mathbf{v}\frac{\partial f_0}{\partial E} - \frac{f_1}{\bar{t}}. \tag{9-26}$$

To obtain the second equality, the chain rule for differentiation was used to write $\nabla_k f_0 = (\partial f_0/\partial E)\,\nabla_k E$, then $\hbar\mathbf{v}$ was substituted for $\nabla_k E$. If the field was turned on at time $t = 0$, with the electron distribution in thermal equilibrium, then

$$f_1(\mathbf{k}, t) = e\mathcal{E} \cdot \mathbf{v}\bar{t}\frac{\partial f_0}{\partial E}[1 - e^{-t/\bar{t}}]. \tag{9-27}$$

For $t \gg \bar{t}$, the electron system reaches a steady state with

$$f(\mathbf{k}) = f_0(\mathbf{k}) + f_1(\mathbf{k}) = f_0(\mathbf{k}) + e\mathscr{E} \cdot \mathbf{v}\bar{t}\, \frac{\partial f_0}{\partial E}, \qquad (9\text{-}28)$$

where the argument $t$ has been omitted since the distribution is no longer time dependent.

Since $\partial f_0 / \partial E$ is negative, Eq. 9-28 predicts that states for which the electron velocity is directed opposite to the field have a slightly higher probability of being occupied than they do in thermal equilibrium. Similarly, states for which the electron velocity is in the same direction as the field have a slightly lower probability. Figure 9-4 depicts the situation.

The behavior of the average crystal momentum, defined by $\hbar\mathbf{k}_{\text{ave}} = \Sigma\hbar\mathbf{k}f(\mathbf{k}, t)/N$ is revealing. Multiply each term of Eq. 9-20 by $\hbar\mathbf{k}$ and sum over all states of the band. The left side becomes $d(\hbar\mathbf{k}_{\text{ave}})/dt$. Since $(d\hbar\mathbf{k}/dt)_{\text{field}} = -e\mathscr{E}$ for each electron, $(d\hbar\mathbf{k}_{\text{ave}}/dt)_{\text{field}}$ is also $-e\mathscr{E}$. If the relaxation time is

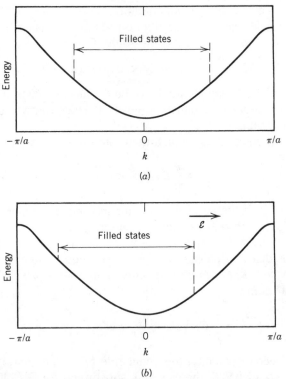

(a)

(b)

FIGURE 9-4   Occupied states of a partially filled electron energy band. (a) In the absence of an electric field the lowest energy states are filled. Both the total crystal momentum and the average electron velocity vanish. (b) When an electric field is turned on, the distribution shifts and the net crystal momentum and average electron velocity no longer vanish. Steady state is reached when field induced transitions, to the left, are balanced on average by scattering induced transitions, to the right. The shift is greatly exaggerated on the diagram.

independent of $\mathbf{k}$, $(d\hbar\mathbf{k}_{ave}/dt)_{scatt} = -\hbar\mathbf{k}_{ave}/\bar{t}$. So

$$\frac{d\hbar\mathbf{k}_{ave}}{dt} = -e\mathcal{E} - \frac{\hbar\mathbf{k}_{ave}}{\bar{t}}. \tag{9-29}$$

In the steady state

$$\hbar\mathbf{k}_{ave} = -e\mathcal{E}\bar{t}. \tag{9-30}$$

The steady state average crystal momentum is not zero, as it is when the electrons are in thermal equilibrium.

Equation 9-30 has an interesting interpretation. Imagine an electron that behaves like the average of all electrons in the band and suppose its crystal momentum is zero at time $t = 0$. It accelerates for a time interval $\bar{t}$, during which it acquires crystal momentum $-e\mathcal{E}\bar{t}$, then it is scattered back to the $\mathbf{k} = 0$ state and the sequence repeats. In this interpretation, $\bar{t}$ represents the average time between scattering events and hence is called the mean free time.

*Mobility.* In the steady state the average electron velocity is proportional to the electric field. The constant of proportionality is called the mobility. For many materials the average electron velocity is not parallel to the electric field and the mobility is represented by a tensor. If $v_{ave\ i}$ and $\mathcal{E}_j$ are Cartesian components, then elements $\mu_{ij}$ of the mobility tensor are defined by

$$v_{ave\ i} = -\sum_j \mu_{ij}\mathcal{E}_j. \tag{9-31}$$

The average velocity is often called the electron drift velocity.

To obtain an expression for $\mu_{ij}$, start with

$$v_{ave\ i} = \frac{1}{N}\sum_{states} v_i(\mathbf{k})f(\mathbf{k}), \tag{9-32}$$

where $N$ is the number of electrons in the band, and substitute for $f(\mathbf{k})$ from Eq. 9-28. Terms containing $f_0(\mathbf{k})$ sum to zero and terms containing $\partial f_0/\partial E$ yield

$$v_{ave\ i} = \frac{e}{N}\sum_{states} v_i\mathcal{E}\cdot\mathbf{v}\bar{t}\frac{\partial f_0}{\partial E}. \tag{9-33}$$

Comparison with Eq. 9-32 reveals that

$$\mu_{ij} = -\frac{e}{N}\sum_{states} \bar{t}v_i v_j \frac{\partial f_0}{\partial E}. \tag{9-34}$$

Since $\partial f_0/\partial E$ decreases rapidly as the difference between the energy and the chemical potential increases, those states that are closest in energy to the chemical potential dominate the sum in Eq. 9-34. For example, the average velocity of electrons in a core band is zero because $f_0 = 1$ for all states in the band and $\partial f_0/\partial E = 0$. For a metal, the mobility is large only for bands that contain the Fermi level. For a semiconductor only mobilities associated with conduction and valence bands are significant.

For cubic crystals the mobility is a scalar. For $i \neq j$, $v_i v_j$ and $-v_i v_j$ occur with equal weighting factors, so terms containing these factors cancel. Furthermore, weighting factors for $v_x^2$, $v_y^2$, and $v_z^2$ are all the same, so $\mu_{xx} = \mu_{yy} = \mu_{zz}$. We take the scalar mobility $\mu$ to be any one of the diagonal elements. Then

$$\mathbf{v}_{ave} = -\mu \mathcal{E}, \tag{9-35}$$

where

$$\mu = \tfrac{1}{3}(\mu_{xx} + \mu_{yy} + \mu_{zz}) = -\frac{e}{3N} \sum_{states} \bar{t} v^2 \frac{\partial f_0}{\partial E}. \tag{9-36}$$

Here $v$ is the magnitude of the velocity of an electron in a state with wave vector **k**. Since $\partial f_0/\partial E$ is negative, $\mu$ is positive and the direction of the average velocity is opposite that of the field. Equation 9-35 also holds for noncubic crystals if the band energies and relaxation time depend only on the magnitude of **k** and not on its direction.

*Holes.* A similar analysis can be carried out for holes. In the relaxation time approximation, the distribution function $f_h$ for holes obeys

$$\frac{\partial f_h}{\partial t} = -\frac{e}{\hbar} \mathcal{E} \cdot \nabla_{kh} f_h - \frac{f_h - f_{h0}}{t}, \tag{9-37}$$

where $f_{h0}$ is the equilibrium hole distribution function $[e^{-\beta(E-\eta)} - 1]^{-1}$. Equation 9-37 is like Eq. 9-25, except the hole crystal momentum replaces the electron crystal momentum and the sign of the charge is changed.

When steady state is reached in a weak electric field,

$$f_h(\mathbf{k}_h) = f_{h0}(\mathbf{k}_h) + e\mathcal{E} \cdot \mathbf{v}\bar{t} \frac{\partial f_{h0}}{\partial E}, \tag{9-38}$$

where $\nabla_{kh} f_{h0} = (\partial f_0/\partial E) \nabla_{kh} E = -(\partial f_{h0}/\partial E)\hbar \mathbf{v}$ was used. If there are $N$ holes in the band, their average velocity is

$$\mathbf{v}_{h\ ave} = \frac{1}{N} \sum_{states} \mathbf{v}(\mathbf{k}) f_h(\mathbf{k}) \tag{9-39}$$

and this is linear in the electric field. The hole mobility tensor, with elements $\mu_{ij}$, is defined by

$$v_{h\ ave\ i} = \sum_j \mu_{ij} \mathcal{E}_j. \tag{9-40}$$

This definition differs from that for electrons only in the sign.

Once Eq. 9-38 is used in Eq. 9-39 and a comparison is made with Eq. 9-40,

$$\mu_{ij} = \frac{e}{N} \sum_{states} v_i v_j \frac{\partial f_{h0}}{\partial E} \tag{9-41}$$

is obtained. If the crystal is cubic, off-diagonal elements of the mobility tensor vanish and diagonal elements are all equal. Then

$$\mathbf{v}_{\text{h ave}} = \mu \mathcal{E},$$ (9-42)

where

$$\mu = \frac{e}{3N} \sum_{\text{states}} v^2 \bar{t} \frac{\partial f_{h0}}{\partial E}.$$ (9-43)

Since $\partial f_{h0}/\partial E$ is positive, $\mu$ is positive and the average hole velocity is in the same direction as the applied field.

*Electron Mobilities in Metals and Semiconductors.* We consider a band for which the mobility is a scalar and start by converting Eq. 9-36 to an integral over energy:

$$\mu = -\frac{e}{3N} \int \bar{t} v^2 \frac{\partial f_0}{\partial E} \rho(E) \, dE,$$ (9-44)

where $\rho(E)$ is the density of states.

We first apply this expression to electrons in a metal with its Fermi level in a single isotropic band. Equation 9-44 has the same form as integrals investigated in Appendix D, with $h(E) = (e/3N)\bar{t}v^2\rho(E)$. To lowest order in the temperature we use the first term of Eq. D-7 and replace $\eta$ by $E_F$. Then

$$\mu = \frac{e}{3N} \bar{t} v_F^2 \rho(E_F),$$ (9-45)

where $v_F$ is the speed of an electron at the Fermi surface. If the relaxation time depends on energy, we use the value appropriate to the Fermi energy.

The net effect of an electric field is to transfer electrons from states just inside the Fermi surface, with velocities roughly in the direction of the field, to states just outside, with velocities roughly in the opposite direction. If the density of states is large near the Fermi surface a large number of electrons make transitions and the average velocity is high. It is also high if the velocity associated with individual states is great.

For the special case of parabolic band, $v_F^2 = 2E_F/m^*$, $\rho(E_F) = \sqrt{2}\tau_s(m^*)^{3/2}E_F^{1/2}/\pi^2\hbar^3$, and $E_F = (\hbar^2/2m^*)(3\pi^2 n)^{2/3}$, where $n$ is the electron concentration. Equation 9-45 becomes

$$\mu = \frac{e\bar{t}}{m^*}.$$ (9-46)

The mobility is proportional to the relaxation time and inversely proportional to the effective mass. A long relaxation time means that, on the average, the electrons accelerate for long times between scattering events and so achieve a high average velocity. A small effective mass means a large acceleration and a correspondingly high average velocity.

**EXAMPLE 9-3** In a certain copper sample the electron drift velocity is 2.16 m/s in an electric field of 500 V/m. Estimate (a) the electron mobility and (b) the relaxation time.

**SOLUTION** (a) The mobility is given by $\mu = v_{ave}/\mathcal{E} = 2.16/500 = 4.32 \times 10^{-3}$ m²/V · s. (b) The relaxation time is given by $\bar{t} = m^*\mu/e$. If $m^*$ is taken to be the free electron mass, then $\bar{t} = 9.11 \times 10^{-31} \times 4.32 \times 10^{-3}/1.60 \times 10^{-19} = 2.46 \times 10^{-14}$ s. ∎

Next consider the conduction band of a semiconductor, well above the chemical potential. The Fermi-Dirac distribution function can then be approximated by $f_0(E) = e^{-\beta\Delta E}$, where $\Delta E = E - \eta$, so $\partial f_0/\partial E = -\beta e^{-\beta\Delta E}$. If electron energies are given by $E(\mathbf{k}) = (\hbar^2/2m^*)k^2$, then the density of states is $AE^{1/2}$, where $A$ is a constant, and Eq. 9-44 becomes

$$\mu_e = \frac{2eA\beta}{3Nm^*} \int_0^\infty \bar{t}E^{3/2}e^{-\beta\Delta E}\, dE. \tag{9-47}$$

As we will see, the relaxation time is different for different states and must be considered a function of energy.

The number of electrons in the band is given by $N = \int f_0(E)\rho(E)\, dE$ and an integration by parts yields $N = (2A\beta/3)\int E^{3/2}e^{-\beta\Delta E}\, dE$. When this expression is used to replace $A/N$ in Eq. 9-47, factors containing the chemical potential cancel

**TABLE 9-1** Band Gap and Electron and Hole Mobilities for Selected Pure Semiconductors at 300 K

| Crystal | Band Gap (eV) | Electron Mobility (m²/V · s) | Hole Mobility (m²/V · s) |
|---|---|---|---|
| Germanium | 0.66 | 0.390 | 0.190 |
| Silicon | 1.12 | 0.150 | 0.045 |
| Aluminum antimonide | 1.58 | 0.020 | 0.042 |
| Gallium antimonide | 0.72 | 0.50 | 0.085 |
| Gallium arsenide | 1.42 | 0.850 | 0.040 |
| Gallium phosphide | 2.26 | 0.011 | 0.075 |
| Indium antimonide | 0.17 | 8.0 | 0.125 |
| Indium arsenide | 0.36 | 3.3 | 0.046 |
| Indium phosphide | 1.35 | 0.46 | 0.015 |
| Cadmium sulfide | 2.42 | 0.034 | 0.005 |
| Zinc sulfide | 3.68 | 0.0165 | 0.0005 |
| Lead sulfide | 0.41 | 0.060 | 0.070 |
| Lead telluride | 0.31 | 0.60 | 0.40 |

*Source: Handbook of Electronic Materials,* compiled by the Electronics Properties Information Center, Hughes Aircraft Company.

and the result is

$$\mu_e = \frac{e\langle \bar{t} \rangle}{m^*}, \tag{9-48}$$

where

$$\langle \bar{t} \rangle = \frac{\int \bar{t} E^{3/2} e^{-\beta E} \, dE}{\int E^{3/2} e^{-\beta E} \, dE}. \tag{9-49}$$

Both integrals are over the band. Equation 9-48 has the same form as Eq. 9-46 but an average relaxation time is used rather than the value corresponding to the Fermi level. The average is computed using the weighting factor $E^{3/2} e^{-\beta E}$.

To illustrate hole mobility, we consider a valence band with energy given by $E = -(\hbar^2/2m_h^*)k^2$, where $m_h^*$ is the hole effective mass and the zero of energy is taken to be at the top of the band. The density of states can be written $\rho(E) = A(-E)^{1/2}$. If the valence band is well below the chemical potential, then the equilibrium hole distribution function can be approximated by $f_{h0} = e^{\beta \Delta E}$, where $\Delta E = E - \eta$. The calculation is similar to that for electrons and the result is

$$\mu_h = \frac{e\langle \bar{t} \rangle}{m_h^*}, \tag{9-50}$$

where

$$\langle \bar{t} \rangle = \frac{\int \bar{t}(-E)^{3/2} e^{\beta E} \, dE}{\int (-E)^{3/2} e^{\beta E} \, dE}. \tag{9-51}$$

Both integrals are over valence band energies. The average relaxation time for holes in a valence band is not usually the same as that for electrons in a conduction band. Table 9-1 lists electron and hole mobilities for some semiconductors.

**EXAMPLE 9-4**  Electrons in the conduction band of silicon have an effective mass of $0.259m_0$ and a mobility of $0.1350$ m$^2$/V $\cdot$ s while holes in one of the valence bands have an effective mass of $0.537m_0$ and a mobility of $0.0480$ m$^2$/V $\cdot$ s. Here $m_0$ is the free electron mass. Find the average relaxation time for (a) electrons and (b) holes.

**SOLUTION**  Use Eqs. 9-48 and 9-50. (a) For electrons in the conduction band,

$$\langle \bar{t} \rangle = m_e^* \mu/e = 0.259 \times 9.11 \times 10^{-31} \times 0.1350/1.6 \times 10^{-19}$$

$$= 1.99 \times 10^{-13} \text{ s}.$$

(b) For holes in the valence band,

$$\langle \bar{t} \rangle = m_h^* \mu/e = 0.537 \times 9.11 \times 10^{-31} \times 0.0480/1.6 \times 10^{-19}$$

$$= 1.47 \times 10^{-13} \text{ s}. \qquad \blacksquare$$

## 9.3 ELECTRICAL CONDUCTION

*Current Density and Conductivity.* The current density for a collection of particles, each moving with velocity **v** and carrying charge $q$, is given by $\mathbf{J} = qn\mathbf{v}$, where $n$ is the particle concentration.[*] **J** is a vector in the direction of particle velocity for positive charges and in the opposite direction for negative charges.

To find the current density in a solid, take $f(\mathbf{k})/\tau_s$, where $\tau_s$ is the sample volume, to be the electron concentration associated with a state and sum the contributions of all electrons:

$$\mathbf{J} = -\frac{e}{\tau_s} \sum_{\text{states}} f(\mathbf{k})\mathbf{v}(\mathbf{k}). \tag{9-52}$$

For a crystal, the sum over states can be carried out one band at a time and the sum $\Sigma f\mathbf{v}$ over states of a single band can be identified with $N\mathbf{v}_{\text{ave}}$, where $N$ is the number of electrons in the band and $\mathbf{v}_{\text{ave}}$ is their average velocity. So

$$\mathbf{J} = -e \sum_{\text{bands}} \frac{N}{\tau_s} \mathbf{v}_{\text{ave}} = -e \sum_{\text{bands}} n\mathbf{v}_{\text{ave}}, \tag{9-53}$$

where $n = N/\tau_s$ is the contribution of the band to the electron concentration.

Equation 9-31 holds in the steady state if the field is weak, so

$$J_i = e \sum_{\text{bands}} \left[ n \sum_j \mu_{ij}\mathcal{E}_j \right]. \tag{9-54}$$

According to Eq. 9-54, the current density is proportional to the applied field. The relationship is usually written

$$J_i = \sum_j \sigma_{ij}\mathcal{E}_j, \tag{9-55}$$

where

$$\sigma_{ij} = e \sum_{\text{bands}} n\mu_{ij} \tag{9-56}$$

is an element of the conductivity tensor. A tensor must be used since the current density may not be in the same direction as the field. For amorphous materials and cubic crystals the conductivity tensor is diagonal, with the three diagonal elements equal, so

$$\mathbf{J} = \sigma\mathcal{E}, \tag{9-57}$$

where $\sigma$ is one of the diagonal elements. The current density and field are in the same direction for these materials. Equation 9-57 is often written $\mathcal{E} = \rho\mathbf{J}$, where $\rho = 1/\sigma$ is the electrical resistivity of the material.

---

[*]See, for example, J. R. Reitz, F. J. Milford, and R. W. Christy, *Foundations of Electromagnetic Theory* (Reading, MA: Addison-Wesley, 1979).

**EXAMPLE 9-5**   Energies in a certain band are given by $E(\mathbf{k}) = E_0 + (\hbar^2/2m^*)k^2$. (a) Suppose the band is occupied by a single electron, with crystal momentum $\hbar\mathbf{k}_1$. Find an expression for the current density. (b) Suppose the band is occupied by a second electron, with crystal momentum $-\hbar\mathbf{k}_1$. Show that the current density then vanishes. (c) Suppose an electric field $\mathcal{E}$ is turned on at time $t = 0$. If the initial condition is as described in part b, find an expression for the current density at a later time $t$. Ignore scattering.

**SOLUTION**   (a) The velocity of the electron is $\mathbf{v} = \hbar\mathbf{k}_1/m^*$ and the current density is $\mathbf{J} = -(e/\tau_s)\mathbf{v} = -(e\hbar/\tau_s m^*)\mathbf{k}_1$. (b) For the second electron $\mathbf{v} = -\hbar\mathbf{k}_1/m^*$, so the total current density is $\mathbf{J} = -(e\hbar/\tau_s m^*)\mathbf{k}_1 - (-e\hbar/\tau_s m^*)\mathbf{k}_1 = 0$. (c) After the field has been on for time $t$ the first electron has crystal momentum $\hbar\mathbf{k}_1 - e\mathcal{E}t$ and the second has crystal momentum $-\hbar\mathbf{k}_1 - e\mathcal{E}t$, so the total current density is $\mathbf{J} = -(e/\tau_s m^*)(\hbar\mathbf{k}_1 - e\mathcal{E}t) - (e/\tau_s m^*)(-\hbar\mathbf{k}_1 - e\mathcal{E}t) = (2e^2 t/\tau_s m^*)\mathcal{E}$. ■

*Metals.*   Consider a metal with its Fermi surface in a single band and take the electron energy to be given by $E(\mathbf{k}) = (\hbar^2/2m^*)k^2$, where $m^*$ is the effective mass. If the relaxation time $\bar{t}$ is also isotropic, then the mobility tensor is diagonal and its diagonal elements are given by $\mu = e\bar{t}/m^*$. To a good approximation we may assume this band makes the only significant contribution to the conductivity tensor. Then that tensor is diagonal and a diagonal element is given by

$$\sigma = \frac{e^2 n \bar{t}}{m^*}, \tag{9-58}$$

where $n$ is the concentration of electrons in the band.

**EXAMPLE 9-6**   The electrical resistivity of a certain copper sample is $1.77 \times 10^{-8}\ \Omega \cdot m$. Use the free electron approximation to estimate (a) the relaxation time and (b) the average speed of electrons in a field of 100 V/m. Copper is FCC with a cube edge of 3.61 Å and each atom contributes one electron to a nearly free electron band.

**SOLUTION**   (a) Since there are 4 copper atoms for each cubic unit cell, the concentration of electrons in the band is $n = 4/a^3 = 4/(3.61 \times 10^{-10})^3 = 8.50 \times 10^{28}$ electrons/m³. The conductivity $\sigma$ is the reciprocal of the resistivity $\rho$ so

$$\bar{t} = \frac{m^*}{ne^2\rho} = \frac{9.11 \times 10^{-31}}{8.50 \times 10^{28} \times (1.60 \times 10^{-19})^2 \times 1.77 \times 10^{-8}} = 2.37 \times 10^{-14}\ s,$$

where we have taken $m^*$ to be the mass of a free electron. (b) The average speed is

$$v_{ave} = \frac{e\bar{t}}{m^*}\mathcal{E} = \frac{1.6 \times 10^{-19} \times 2.37 \times 10^{-14}}{9.11 \times 10^{-31}} \times 100 = 0.416 \text{ m/s} .$$

This speed is much less than the speed of an electron at the Fermi surface (about $1.6 \times 10^6$ m/s for copper). ■

Table 9-2 gives the room temperature conductivities of some metals. Cadmium, magnesium, and zinc are hexagonal close packed, while indium is tetragonal. For these metals $\sigma_\parallel$ refers to fields that are parallel to the axis of highest symmetry and $\sigma_\perp$ refers to fields that are perpendicular to that axis.

*Semiconductors.* The current density may also be described in terms of holes, rather than electrons. Since $f(\mathbf{k}) = 1 - f_h(\mathbf{k})$, the contribution of a band to the current density can be written

$$\mathbf{J} = \frac{e}{\tau_s} \sum [1 - f_h(\mathbf{k})]\mathbf{v} = +\frac{e}{\tau_s} \sum f_h(\mathbf{k})\mathbf{v} , \qquad (9\text{-}59)$$

where the sums are over all states of the band. The last equality is valid since the sum of velocities associated with all states in a band vanishes. Note that holes contribute to the current density as particles with positive charge.

For a semiconductor, contributions of conduction band states are written in terms of electrons while contributions of valence band states are written in

**TABLE 9-2**  Room Temperature Electrical Conductivity of Selected Metals

| Crystal | $\sigma$ ($10^7 \ \Omega^{-1} \cdot m^{-1}$) | Crystal | $\sigma$ ($10^7 \ \Omega^{-1} \cdot m^{-1}$) |
|---|---|---|---|
| Aluminum | 4.12 | Magnesium | |
| Cadmium | | $\sigma_\parallel$ | 2.87 |
| $\sigma_\parallel$ | 1.28 | $\sigma_\perp$ | 2.39 |
| $\sigma_\perp$ | 1.59 | Nickel | 1.60 |
| Calcium | 3.25 | Platinum | 1.02 |
| Cesium | 0.56 | Potassium | 1.61 |
| Copper | 6.49 | Rubidium | 0.89 |
| Gold | 4.92 | Silver | 6.82 |
| Indium | | Sodium | 2.33 |
| $\sigma_\parallel$ | 1.27 | Tin | 0.77 |
| $\sigma_\perp$ | 1.23 | Tungsten | 2.07 |
| Iridium | 2.15 | Zinc | |
| Iron | 1.16 | $\sigma_\parallel$ | 1.79 |
| Lead | 0.52 | $\sigma_\perp$ | 1.86 |
| Lithium | 1.20 | Zirconium | 0.26 |

*Source:* K. H. Hellwege (Ed.), *Landolt-Bornstein Numerical Data and Functional Relationships in Science and Technology*, New Series, Group III, Vol. 15a (Berlin: Springer-Verlag, 1982).

terms of holes. Thus

$$J = -\frac{e}{\tau_s} \sum_{CB} f(\mathbf{k})\mathbf{v} + \frac{e}{\tau_s} \sum_{VB} f_h(\mathbf{k})\mathbf{v}, \tag{9-60}$$

where the first sum is over conduction band states and the second is over valence band states. In the steady state both the average electron and average hole velocities are proportional to the applied electric field and, for a semiconductor with a single conduction band and a single valence band, both isotropic,

$$J = e(n\mu_e + p\mu_h)\mathscr{E}. \tag{9-61}$$

Here $n$ is the concentration of conduction band electrons, $p$ is the concentration of valence band holes, $\mu_e$ is the electron mobility, and $\mu_h$ the hole mobility. The conductivity is given by

$$\sigma = e(n\mu_e + p\mu_h). \tag{9-62}$$

For an intrinsic semiconductor, of course, $n = p$ and $\sigma = en(\mu_e + \mu_h)$. For many tetrahedrally bonded semiconductors the sum in Eq. 9-62 must be extended to include more than one valence band.

## 9.4 THERMAL CONDUCTION

*Thermal Conductivity.* Consider two contiguous regions with different particle concentrations. In the course of their motions, particles cross the boundary in both directions, but because the concentrations are different in the two regions more leave the region of high concentration per unit time than enter it. Particles are said to diffuse from regions of high concentration to regions of low concentration. In this section we are interested in the energy carried by electrons and phonons as they diffuse in a temperature gradient. The role of the gradient is simply to maintain different particle concentrations in different regions of a solid.

The energy flux for a collection of particles, each carrying energy $E$ and moving with velocity $\mathbf{v}$, is given by $\mathbf{Q} = En\mathbf{v}$, where $n$ is the particle concentration. Compare this expression with the analogous expression for current density. We sum contributions from all normal modes and electron states, each with its own energy, particle concentration, and particle velocity. Phonons and electrons are treated separately.

*Phonons.* We divide the sample into macroscopically small regions, large enough that the vibration spectrum for each region is essentially the same as for the bulk material. Let $n(\mathbf{q}, \mathbf{r}, t)$ be the concentration of phonons near $\mathbf{r}$ in a mode with propagation vector $\mathbf{q}$. The phonon concentration changes with time because phonons diffuse into and out of the region and because scattering processes may change the number of phonons in any mode. That is

$$\frac{\partial n}{\partial t} = \left(\frac{\partial n}{\partial t}\right)_{diff} + \left(\frac{\partial n}{\partial t}\right)_{scatt}. \tag{9-63}$$

Since vibrational energy is transported with the group velocity of the wave, we take the phonon velocity to be the wave group velocity: $\mathbf{v} = \nabla_q \omega$. For long wavelength modes in an acoustical branch, $v$ is the speed of sound. If scattering does not take place, phonons in the neighborhood of $\mathbf{r} - \mathbf{v}\Delta t$ at time $t$ travel to the neighborhood of $\mathbf{r}$ in time $\Delta t$. This means $n(\mathbf{q}, \mathbf{r}, t + \Delta t) = n(\mathbf{q}, \mathbf{r} - \mathbf{v}\Delta t, t) = n(\mathbf{q}, \mathbf{r}, t) - \mathbf{v} \cdot \nabla n \, \Delta t$, to first order in $\Delta t$. So $(\partial n/\partial t)_{\text{diff}} = -\mathbf{v} \cdot \nabla n$. We use the relaxation time approximation for the second term of Eq. 9-63 and take $(\partial n/\partial t)_{\text{scatt}} = -(n - n_0)/\bar{t}$, where $\bar{t}$ is the relaxation time and $n_0$ is the equilibrium phonon concentration: $(1/\tau_s)(e^{\beta\hbar\omega} - 1)^{-1}$. Equation 9-63 becomes

$$\frac{\partial n}{\partial t} = -\mathbf{v} \cdot \nabla n - \frac{n - n_0}{\bar{t}}. \tag{9-64}$$

This is the Boltzmann transport equation for phonons.

We suppose the temperature gradient is small and seek the steady state solution to Eq. 9-64: place $\partial n/\partial t = 0$ and solve for $n$. Since $n$ does not differ much from $n_0$, replace $\nabla n$ with $\nabla n_0 = (\partial n_0/\partial T) \, \nabla T$, with the derivative evaluated for the average temperature of the sample. Thus

$$n(\mathbf{q}, \mathbf{r}) = n_0(\mathbf{q}, \mathbf{r}) - \bar{t} \frac{\partial n_0}{\partial T} \mathbf{v} \cdot \nabla T. \tag{9-65}$$

Since this is the steady state solution, $n$ does not depend on $t$.

Substitute Eq. 9-65 into $\mathbf{Q} = En\mathbf{v}$ to find the contribution of the mode to the energy flux, then sum over all modes to find the total flux. The term containing $n_0$ sums to zero. Each phonon carries energy $E = \hbar\omega$, so

$$\mathbf{Q} = \sum_{\text{modes}} \hbar\omega \mathbf{v}(\mathbf{q})n(\mathbf{q}, \mathbf{r}) = -\sum_{\text{modes}} \hbar\omega\bar{t} \frac{\partial n_0}{\partial T} \mathbf{v}\mathbf{v} \cdot \nabla T. \tag{9-66}$$

Equation 9-66 is usually written

$$Q_i = -\sum_j \kappa_{ij} \frac{\partial T}{\partial x_j} \tag{9-67}$$

where

$$\kappa_{ij} = \sum_{\text{modes}} \hbar\omega\bar{t} \frac{\partial n_0}{\partial T} v_i v_j. \tag{9-68}$$

$\kappa_{ij}$ is an element of the thermal conductivity tensor. For cubic and amorphous materials the tensor is diagonal and Eq. 9-67 becomes

$$\mathbf{Q} = -\kappa \, \nabla T, \tag{9-69}$$

where $\kappa$ is the scalar thermal conductivity, one of the diagonal elements of the tensor. Since $\kappa = \frac{1}{3}(\kappa_{xx} + \kappa_{yy} + \kappa_{zz})$, Eq. 9-68 yields

$$\kappa = \frac{1}{3} \sum_{\text{modes}} \hbar\omega\bar{t} \frac{\partial n_0}{\partial T} v^2. \tag{9-70}$$

Equation 9-69 is sometimes written $\nabla T = -W\mathbf{Q}$, where $W = 1/\kappa$ is the thermal resistivity of the sample.

Note that the right side of Eq. 9-65 consists of the first two terms of a power series expansion of $n_0(\mathbf{q}, \mathbf{r} - \mathbf{v}\bar{t})$. The phonon concentration at $\mathbf{r}$ is the thermal equilibrium concentration for the temperature at $\mathbf{r} - \mathbf{v}\bar{t}$, not for the temperature at $\mathbf{r}$. Imagine a group of phonons associated with propagation vector $\mathbf{q}$, in equilibrium at the temperature which prevails at $\mathbf{r} - \mathbf{v}\bar{t}$. They travel with velocity $\mathbf{v}$ down the temperature gradient for a time $\bar{t}$, the average time between scattering events. During this time the total energy of the group does not change. At $\mathbf{r}$, however, phonons are lost through scattering and the group reaches equilibrium at the temperature which prevails there.

The energy lost by the phonon group equals the product of their contribution to the heat capacity and the difference in the temperatures at $\mathbf{r} - \mathbf{v}\bar{t}$ and $\mathbf{r}$. The temperature difference is $\bar{t}\mathbf{v} \cdot \nabla T$ and the energy lost per unit volume is $c\bar{t}\mathbf{v} \cdot \nabla T$, where $c$ is the contribution per unit volume of the mode to the constant volume heat capacity. This also gives the net energy carried down the temperature gradient per unit volume. To find the total energy flux, it is multiplied by $\mathbf{v}$ and summed over modes. We expect

$$\kappa = \frac{1}{3\tau_s} \sum_{\text{modes}} c\bar{t}v^2 . \tag{9-71}$$

Since $c = \hbar\omega(\partial n_0/\partial T)$, Eqs. 9-70 and 9-71 agree.

**EXAMPLE 9-7** Use the Debye approximation to estimate the phonon contribution to the low-temperature thermal conductivity of a solid with primitive unit cell volume $\tau$, Debye temperature $T_D$, and average speed of sound $v$. Take the relaxation time to be the same for all modes.

**SOLUTION** Assume the vibrational spectrum is isotropic. Then the thermal conductivity is a scalar. If $g(\omega)$ is the density of modes, Eq. 9-70 becomes

$$\kappa = \frac{1}{3\tau_s} \frac{\bar{t}v^2}{k_B T^2} \int_0^{\omega_m} \frac{(\hbar\omega)^2 e^{\beta\hbar\omega}}{(e^{\beta\hbar\omega} - 1)^2} g(\omega)\, d\omega ,$$

where we have used $\partial n_0/\partial T = \hbar\omega e^{\beta\hbar\omega}/k_B T^2 (e^{\beta\hbar\omega} - 1)^2$ and taken $v$ to be independent of $\omega$. In the Debye approximation $g(\omega) = (9N\hbar^3/k_B^3 T_D^3)\omega^2$, where $N$ is the number of primitive unit cells in the sample. This density of modes automatically includes a sum over the three acoustical branches. Optical branches do not usually contribute significantly to the thermal conductivity at low temperatures and we neglect them. When $\tau_s/N$ is replaced by $\tau$, $\beta\hbar\omega$ by $x$, and the upper limit of the integral over $x$ by $\infty$, the expression for $\kappa$ becomes

$$\kappa = \frac{3\bar{t}v^2}{\tau} k_B \left(\frac{T}{T_D}\right)^3 \int_0^\infty \frac{x^4 e^x}{(e^x - 1)^2}\, dx .$$

The value of the integral is $4\pi^2/15$ so

$$\kappa = \frac{4\pi^4 \bar{t} v^2}{5\tau} k_B \left(\frac{T}{T_D}\right)^3.$$

■

*Electrons.* We suppose the crystal is divided into regions that are small on a macroscopic scale but are large enough that energy bands in each of them are essentially the same as for the crystal as a whole. First consider a single band and take $f(\mathbf{k}, \mathbf{r}, t)/\tau_s$ to be the concentration of electrons near $\mathbf{r}$ in a state with crystal momentum $\hbar\mathbf{k}$. In terms of $f$, the energy flux is given by

$$\mathbf{Q} = \frac{1}{\tau_s} \sum_{\text{states}} E(\mathbf{k})\mathbf{v}(\mathbf{k})f(\mathbf{k}, \mathbf{r}, t), \tag{9-72}$$

where $E(\mathbf{k})$ is the electron energy and $\mathbf{v}(\mathbf{k})$ is the electron velocity.

In addition to changes in $f$ caused by an electric field and by scattering, changes occur because electrons diffuse from regions of high concentration to regions of low concentration. The rate at which $f$ changes because diffusion occurs is given by $-\mathbf{v} \cdot \nabla f$, so the Boltzmann equation is

$$\frac{\partial f}{\partial t} = \frac{e}{h} \mathcal{E} \cdot \nabla_k f - \mathbf{v} \cdot \nabla f - \frac{f - f_0}{\bar{t}}, \tag{9-73}$$

where $f_0$ is the Fermi-Dirac distribution function. To solve Eq. 9-73, let $f = f_0 + f_1$ and suppose $f_1$ is sufficiently small that it may be neglected in the field and diffusion terms. In the steady state, $\partial f/\partial t = 0$ and

$$f_1 = \frac{\bar{t}e}{h} \mathcal{E} \cdot \nabla_k f_0 - \bar{t}\mathbf{v} \cdot \nabla f_0. \tag{9-74}$$

We are interested in the flow of energy when the current is zero and, to produce this situation, an electric field must be present. The field may be applied externally or it may be due to charge in the material but, in either event, we retain the field dependent term of Eq. 9-74.

Since $\nabla_k f_0 = (\partial f_0/\partial E) \nabla_k E = \hbar(\partial f_0/\partial E)\mathbf{v}$ and $\nabla f_0 = (df_0/dT) \nabla T$, Eq. 9-74 can be written

$$f_1 = \bar{t} \frac{\partial f_0}{\partial E} \mathbf{v} \cdot \mathcal{E} - \bar{t} \frac{\partial f_0}{\partial T} \mathbf{v} \cdot \nabla T. \tag{9-75}$$

We can factor some common quantities from the field and diffusion terms if we note that

$$\frac{\partial f_0}{\partial T} = \frac{\partial f_0}{\partial E} T \frac{d}{dT} \left(\frac{E - \eta}{T}\right), \tag{9-76}$$

a relationship that is obvious once the two derivatives, $\partial f_0/\partial E$ and $\partial f_0/\partial T$, are compared. Note that the temperature dependence of the chemical potential $\eta$

is taken into account. Equation 9-75 becomes

$$f_1 = \bar{t} \frac{\partial f_0}{\partial E} \mathbf{v} \cdot \left[ e\mathcal{E} + \nabla T \left( \frac{E}{T} + T \frac{d}{dT} \frac{\eta}{T} \right) \right]. \tag{9-77}$$

According to Eq. 9-52, the current density is

$$\mathbf{J} = -\frac{e}{\tau_s} \sum_{states} \bar{t} \frac{\partial f_0}{\partial E} \mathbf{v}\mathbf{v} \cdot \left[ e\mathcal{E} + \nabla T \left( \frac{E}{T} + T \frac{d}{dT} \frac{\eta}{T} \right) \right] \tag{9-78}$$

and according to Eq. 9-72, the energy flux is

$$\mathbf{Q} = \frac{1}{\tau_s} \sum_{states} \bar{t} \frac{\partial f_0}{\partial E} E\mathbf{v}\mathbf{v} \cdot \left[ e\mathcal{E} + \nabla T \left( \frac{E}{T} + T \frac{d}{dT} \frac{\eta}{T} \right) \right]. \tag{9-79}$$

The appearance of $\partial f_0/\partial E$ in the expressions for $\mathbf{J}$ and $\mathbf{Q}$ means that states with energy near the chemical potential dominate the sums. For a metal, only electrons near the Fermi surface are responsible for the transport of energy and charge while, for a semiconductor, only electrons near the bottom of the conduction band and holes near the top of the valence band are responsible.

$\mathbf{J}$ is set equal to zero in Eq. 9-78, then Eqs. 9-78 and 9-79 are solved simultaneously for $\mathbf{Q}$ and $\mathcal{E}$. Although the idea is simple, the procedure is complicated by the appearance of $\mathcal{E}$ and $\nabla T$ in scalar products with $\mathbf{v}$. Rather than treat a general situation, we consider the simple example of a metal for which the energy flux is due to electrons in a single isotropic band and take the temperature gradient to be in the $z$ direction, so $\nabla T = (dT/dz)\hat{z}$. Then, $\mathbf{J}$, $\mathcal{E}$, and $\mathbf{Q}$ are also in the $z$ direction.

Once the sums are converted to integrals over the band, Eqs. 9-78 and 9-79 become

$$\mathbf{J} = \Gamma_1 \mathcal{E} + \Gamma_2 \frac{dT}{dz} \tag{9-80}$$

and

$$\mathbf{Q} = -\Gamma_3 \mathcal{E} - \Gamma_4 \frac{dT}{dz}, \tag{9-81}$$

respectively. Here

$$\Gamma_1 = -\frac{e^2}{3\tau_s} \int \bar{t} v^2 \frac{\partial f_0}{\partial E} \rho(E) \, dE, \tag{9-82}$$

$$\Gamma_2 = -\frac{e}{3\tau_s} \int \bar{t} v^2 \left[ \frac{E}{T} + T \frac{d}{dT} \frac{\eta}{T} \right] \frac{\partial f_0}{\partial E} \rho(E) \, dE, \tag{9-83}$$

$$\Gamma_3 = -\frac{e}{3\tau_s} \int \bar{t} v^2 E \frac{\partial f_0}{\partial E} \rho(E) \, dE, \tag{9-84}$$

and

$$\Gamma_4 = -\frac{1}{3\tau_s} \int \bar{t}v^2 E \left[ \frac{E}{T} + T \frac{d}{dT} \frac{\eta}{T} \right] \frac{\partial f_0}{\partial E} \rho(E)\, dE \,. \tag{9-85}$$

J is set equal to zero and Eq. 9-80 is solved for $\mathcal{E}$.

$$\mathcal{E} = -\frac{\Gamma_2}{\Gamma_1} \frac{dT}{dz} \tag{9-86}$$

is the electric field that must exist if the current is to vanish. Once Eq. 9-86 is substituted into Eq. 9-81, that equation becomes

$$Q = -\left[ \Gamma_4 - \frac{\Gamma_2\Gamma_3}{\Gamma_1} \right] \frac{dT}{dz} \,. \tag{9-87}$$

Thus the thermal conductivity is given by

$$\kappa = \Gamma_4 - \frac{\Gamma_2\Gamma_3}{\Gamma_1} \,. \tag{9-88}$$

Each of the coefficients $\Gamma_1, \Gamma_2, \Gamma_3$, and $\Gamma_4$ can be evaluated easily for a parabolic band and uniform relaxation time. Replace $v^2$ by $2E/m^*$ and $\rho(E)$ by $[\sqrt{2}\tau_s(m^*)^{3/2}/\pi^2\hbar^3]E^{1/2}$. Then each of the integrals has a form like that studied in Appendix D. We must be careful to include terms proportional to $T$ since the lowest order terms cancel in Eq. 9-88. We give only the results:

$$\Gamma_1 = e^2 K E_F^{3/2} \,, \tag{9-89}$$

$$\Gamma_2 = e \frac{\pi^2}{3} K k_B^2 T E_F^{1/2} \,, \tag{9-90}$$

$$\Gamma_3 = e K E_F^{5/2} \,, \tag{9-91}$$

and

$$\Gamma_4 = \frac{2\pi^2}{3} K k_B^2 T E_F^{3/2} \,, \tag{9-92}$$

where

$$K = \frac{2\sqrt{2}(m^*)^{1/2}\bar{t}}{3\pi^2\hbar^3} \,. \tag{9-93}$$

When we substitute these expressions into Eq. 9-89 and replace $E_F$ by $(\hbar^2/2m^*)(3\pi^2n)^{2/3}$, where $n$ is the electron concentration, we find

$$\kappa = \frac{\pi^2 n\bar{t}}{3m^*} k_B^2 T \,. \tag{9-94}$$

If the relaxation time or effective mass depend on the energy, then values corresponding to the Fermi energy are used in Eq. 9-94. Table 9-3 gives the thermal conductivity of some metals at room temperature. For a hexagonal or tetragonal crystal $\kappa_{\parallel}$ and $\kappa_{\perp}$ give the thermal conductivity for a temperature gradient parallel and perpendicular, respectively, to the $c$ axis.

**TABLE 9-3**   Room Temperature Thermal Conductivities of Selected Metals

| Crystal | $\kappa$ $(W \cdot m^{-1} \cdot K^{-1})$ | Crystal | $\kappa$ $(W \cdot m^{-1} \cdot K^{-1})$ |
|---|---|---|---|
| Aluminum | 237 | Lead | 35.2 |
| Antimony | 24.3 | Lithium | 76.8 |
| Beryllium | 200 | Nickel | 90.5 |
| Bismuth | | Platinum | 71.4 |
| $\kappa_{\parallel}$ | 5.28 | Rubidium | 58.2 |
| $\kappa_{\perp}$ | 9.15 | Silver | 427 |
| Cadmium | | Sodium | 132 |
| $\kappa_{\parallel}$ | 83.0 | Tellurium | |
| $\kappa_{\perp}$ | 104 | $\kappa_{\parallel}$ | 3.96 |
| Copper | 398 | $\kappa_{\perp}$ | 2.08 |
| Gold | 315 | Tungsten | 178 |

*Source:* Y. S. Touloukian (series Ed.), *Thermophysical Properties of Matter* (New York: Plenum, various years).

**EXAMPLE 9-8**   Estimate the thermal conductivity of the copper sample described in Example 9-6. Assume the data is for 300 K and that the relaxation time is the same for electrical and thermal conductivity.

**SOLUTION**   Assume the free electron model is valid, with $m^* = 9.11 \times 10^{-31}$ kg. Then

$$\kappa = \frac{\pi^2 n \bar{t}}{3m^*} k_B^2 T = \frac{\pi^2 \times 8.5 \times 10^{28} \times 2.37 \times 10^{-14}}{3 \times 9.11 \times 10^{-31}} \times (1.38 \times 10^{-23})^2 \times 300$$

$$= 4.16 \times 10^2 \ W \cdot m^{-1} \cdot K^{-1} .$$

This is slightly greater than the experimental result, about $4.0 \times 10^2$ $W \cdot m^{-1} \cdot K^{-1}$. A phonon contribution must also be added, making the calculated total conductivity even larger. The error is in the assumption that the electron relaxation time is the same for electrical and thermal conduction. In fact, at room temperature the relation time for thermal conduction is about 0.9 that for electrical conduction and phonon flow accounts for about 10% of the energy flux in a temperature gradient.   ∎

## 9.5   SCATTERING

In the simplest model of a crystal, atomic equilibrium positions are arranged throughout all space with perfect periodicity and each atom is assumed to move under the influence of harmonic forces only. Electrons are assumed to occupy single particle states and to influence each other only through their average electrostatic interaction. Their wave functions distort as the atoms move, but the distortions are such that, at each instant of time, the wave functions obey the Schrödinger equation that holds for the atoms fixed at the positions they occupy at that instant. In such a crystal, the number of electrons in each state and the number of phonons in each vibrational mode remain constant in time. Relaxation times are infinite and charge or energy flux continues unabated once it is created.

Deviations from ideal crystallinity alter electron and phonon distributions and lead to the scattering terms in Boltzmann transport equations. They are crucial for the temperature dependence of electrical and thermal conductivities.

*Scattering by Phonons.* The potential energy function that describes interactions between atoms is not strictly quadradic in their displacements but contains anharmonic terms, as we have seen in connection with thermal expansion. Classically, expressions for atomic displacements have the form of traveling waves but the amplitudes are time dependent. Energy is transferred between harmonic normal modes or, in the language of quantum mechanics, phonon–phonon scattering occurs.

In the most likely phonon–phonon event, two phonons disappear and another appears. Energies and crystal momenta obey

$$\hbar\omega_1 + \hbar\omega_2 = \hbar\omega_3 \qquad (9\text{-}95)$$

and

$$\hbar\mathbf{q}_1 + \hbar\mathbf{q}_2 = \hbar\mathbf{q}_3 + \hbar\mathbf{G}, \qquad (9\text{-}96)$$

respectively, where subscripts 1 and 2 refer to the original phonons and 3 refers to the phonon that is created.

The reciprocal lattice vector $\mathbf{G}$ is chosen so $\mathbf{q}_1$, $\mathbf{q}_2$, and $\mathbf{q}_3$ are all in the Brillouin zone. If $\mathbf{q}_1$ and $\mathbf{q}_2$ are near the center of the zone, their sum is within the zone and $\mathbf{G} = 0$. Such events are called *normal* scattering processes. On the other hand, if $\mathbf{q}_1$ and $\mathbf{q}_2$ correspond to short wavelengths, their vector sum may be outside the zone and a reciprocal lattice vector must be added to $\mathbf{q}_3$ to bring it inside. Events for which $\mathbf{G} \neq 0$ are called *Umklapp* processes, from a German word meaning "flipping over." Normal and Umklapp processes are illustrated in Fig. 9-5. As we will see, Umklapp processes are extremely important for relaxation of the phonon system.

In addition to distorting electron wave functions, vibrating atoms induce electron transitions in what is called electron–phonon scattering. In the fundamental process a single electron is scattered from some initial state $\mathbf{k}$ to some final state $\mathbf{k}'$ and a vibrational mode with propagation vector $\mathbf{q}$ either gains or loses a phonon. Energies and crystal momenta obey

$$E(\mathbf{k}') = E(\mathbf{k}) \pm \hbar\omega(\mathbf{q}) \qquad (9\text{-}97)$$

and

$$\hbar\mathbf{k}' = \hbar\mathbf{k} \pm \hbar\mathbf{q} + \hbar\mathbf{G}, \qquad (9\text{-}98)$$

respectively. The upper sign is used if the mode loses a phonon and the lower sign is used if it gains a phonon. Propagation vectors $\mathbf{k}'$, $\mathbf{k}$, and $\mathbf{q}$ are all in the Brillouin zone. If $\mathbf{G} = 0$ the process is normal; otherwise it is an Umklapp process.

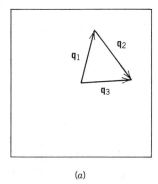

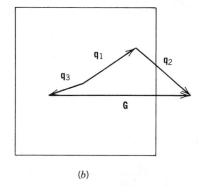

(a)                                      (b)

**FIGURE 9-5** Phonon-phonon scattering events. Phonons with propagation vectors $\mathbf{q}_1$ and $\mathbf{q}_2$ disappear and a phonon with propagation vector $\mathbf{q}_3$ appears. All three propagation vectors are in the Brillouin zone. (a) A normal event: $\mathbf{q}_1 + \mathbf{q}_2 = \mathbf{q}_3$. (b) An Umklapp event: $\mathbf{q}_1 + \mathbf{q}_2 = \mathbf{q}_3 + \mathbf{G}$, where $\mathbf{G}$ is a reciprocal lattice vector.

The number of scattering events per unit time is proportional to the number of phonons available with appropriate propagation constants and is therefore temperature dependent. The relaxation time is inversely proportional to the number of scattering events per unit time.

*Scattering by Defects.* All structural defects change atomic vibrations and electron wave functions. Compared to an atom in an ideal crystal, a point defect may have a different distribution of atoms around it, may couple with different force constants to neighboring atoms, may have a different mass, and may be responsible for a different electron potential energy function. Similar effects occur for boundaries and dislocations. As a result, defects provide mechanisms by which the number of phonons in a normal mode and the occupation probability of an electron state may change.

The influence of structural defects is conveniently discussed in terms of the mean free path $\ell$ of electrons or phonons. It is related to the relaxation time by $\ell = v\bar{t}$, where $v$ is the electron or phonon speed, and it gives the average distance traveled by a phonon or electron between scattering events. When point defect scattering dominates, we expect the mean free path for both phonons and electrons to be roughly on the order of the average distance between defects. When boundary scattering dominates, it is on the order of the sample dimensions. In either event, it is relatively insensitive to the temperature and, as a result, so is the relaxation time.

*Multiple Scattering Mechanisms.* Usually more than one scattering mechanism operates and, in the simplest model, different mechanisms do not influence each other. Scattering by phonons, for example, is not changed appreciably by the addition of point defects. To find $(\partial f / \partial t)_{\text{scatt}}$ we sum similar terms, each due to a different mechanism and each calculated or measured for the solid in the absence of other types of scatterers. In the relaxation time approximation, each scattering term is inversely proportional to a relaxation time which is characteristic of the scattering mechanism, so the overall relaxation time $\bar{t}$ is given by

$$\frac{1}{\bar{t}} = \sum_i \frac{1}{\bar{t}_i},$$ (9-99)

where $\bar{t}_i$ is the relaxation time associated with mechanism $i$ and the sum is over all mechanisms. If the relaxation time for one mechanism is much shorter than all others, scattering takes place predominately via that mechanism.

Thermal and electrical resistivities, not conductivities, are proportional to the reciprocal of the relaxation time so, in each case, the total resistivity may be written as a sum of contributions of the various scattering mechanisms. For example, the phonon contribution to the thermal resistivity $W$ of a cubic crystal might be

$$W = W_p + W_b + W_i + W_v,$$ (9-100)

where $W_p$ is due to phonon–phonon scattering, $W_b$ is due to phonon-boundary scattering, $W_i$ is due to phonon-impurity scattering, and $W_v$ is due to phonon-vacancy scattering. Of course, there might be other terms, corresponding to other scattering mechanisms. Once all the terms have been summed, the phonon contribution to the thermal conductivity is calculated as the reciprocal of $W$. An expression similar to Eq. 9-100 can be written for the electron contribution to the thermal resistivity. Electron and phonon contributions to the *conductivity* are summed to find the total conductivity.

*Thermal Conductivities of Insulators.* Figure 9-6 shows the thermal conductivity $\kappa$ as a function of the temperature $T$ for magnesium oxide, an insulator. At low temperatures $\kappa$ is proportional to $T^3$ while at high temperatures it is proportional

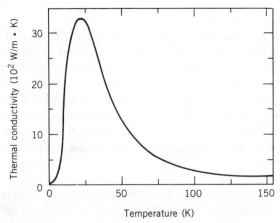

**FIGURE 9-6** Thermal conductivity of magnesium oxide as a function of temperature. At low temperature the curve is nearly proportional to $T^3$, while at high temperature it is proportional to $1/T$. (From Y. S. Touloukian, R. W. Powell, C. Y. Ho, and P. G. Klemens; *Thermal Conductivity of Nonmetallic Solids* (New York: Plenum; 1970); Used with permission.)

to $1/T$. To explain this behavior we must examine the temperature dependence of the relaxation time.

At high temperatures, relaxation is brought about chiefly by phonon–phonon scattering. Umklapp processes are more important than normal processes. Normal processes do indeed change mode occupation numbers and they may also change the energy flux, but these changes occur in such a way that the total phonon crystal momentum does not change, as Eq. 9-96 with $\mathbf{G} = 0$ shows. For many Umklapp events, on the other hand, $\mathbf{q}_3$ in Eq. 9-96 is much closer to the center of the zone than either $\mathbf{q}_1$ or $\mathbf{q}_2$ and the total crystal momentum is reduced toward its equilibrium value of zero.

The number of Umklapp events that occur per unit time is proportional to the number of short wavelength phonons present. Since the number of phonons associated with any mode is proportional to $T$ at high temperatures, we expect the high-temperature phonon-phonon relaxation time to be inversely proportional to $T$. This means the high-temperature thermal resistivity of an insulator is proportional to $T$. To understand this, refer to Eq. 9-71 and recall that the contribution of each mode to the heat capacity is independent of temperature at high temperature.

For modes with angular frequency $\omega \gg k_B T/\hbar$ the number of phonons is proportional to $e^{-\beta\hbar\omega}$. When the temperature is well below the Debye temperature this category includes all modes that participate in Umklapp processes, so we expect the phonon–phonon relaxation time at low temperature to be proportional to $e^{\alpha T_D/T}$, where $\alpha$ is a positive parameter that depends on the crystal structure and on the shape of the dispersion curves near the Brillouin zone boundary. As the temperature decreases, the number of Umklapp processes decreases dramatically and the phonon-phonon relaxation time increases, just as dramatically. Even Umklapp processes are ineffectual in reducing energy flux at low temperatures.

Eventually the phonon-phonon relaxation time becomes much longer than the relaxation time for defect or boundary scattering and these mechanisms dominate. We use the Debye approximation and a calculation similar to that of Example 9-7 to find the temperature dependence of the thermal conductivity at low temperatures. The calculation is slightly complicated because the relaxation time depends on normal mode frequency. It is proportional to $\omega^{-4}$ for point defect scattering and to $1/\omega$ for dislocation scattering, while it is independent of $\omega$ for boundary scattering.

If $\bar{t}$ is proportional to $\omega^{-n}$, then evaluation of Eq. 9-70 using the Debye approximation leads to a thermal resistivity which is proportional to $T^{n-3}$. Thus $W$ has the form

$$W = \frac{A}{T^3} + \frac{B}{T^2} + CT, \tag{9-101}$$

where the first term describes boundary scattering, the second describes dislocation scattering, and the third describes point defect scattering. $A$ depends on sample dimensions, $B$ is proportional to the number of dislocations, and $C$

is proportional to the number of point defects. At low temperatures vacancy and interstitial concentrations are small, so $C$ depends primarily on the impurity concentration. For any crystal at sufficiently low temperature, boundary scattering dominates and the conductivity is proportional to $T^3$. For some samples the exponents of $T$ in Eq. 9-101 vary somewhat from those given because the Debye approximation is not strictly valid.

*Thermal Conductivities of Metals.* Electrons are the primary carriers of energy in a metal. At high temperature, electron-phonon interactions provide the dominant relaxation mechanism, so the relaxation time is inversely proportional to the temperature, just as for phonon-phonon relaxation. According to Eq. 9-94, the electron contribution to the thermal conductivity is independent of temperature. As the temperature decreases the conductivity increases dramatically because the number of phonons decreases. To illustrate, the thermal conductivity of a copper sample is shown as a function of temperature in Fig. 9-7.

Both normal and Umklapp events contribute to electron relaxation. At low temperature, the total number of phonons is proportional to $T^3$, so we expect the low-temperature electron-phonon relaxation time to be proportional to $T^{-3}$. According to Eq. 9-94, the phonon scattering term in the electron contribution to the low-temperature thermal conductivity is proportional to $T^2$. This result, however, is valid only for metals with spherical Fermi surfaces. For most metals the term is proportional to $T^n$, where measured values of $n$ are usually between 2 and 3.

At low temperature, defect or boundary scattering dominates. Relevant relaxation times are nearly independent of temperature, unlike those for phonon-defect scattering. The difference arises because, in the case of electrons, we

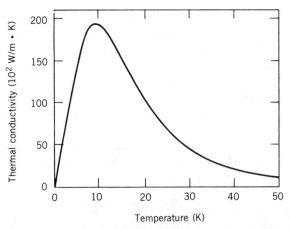

**FIGURE 9-7** Thermal conductivity of copper as a function of temperature. It approaches a constant at high temperature. Defect scattering dominates at low temperature and the conductivity is proportional to $T$. (From Y. S. Touloukian, R. W. Powell, C. Y. Ho, and P. G. Klemens, *Thermal Conductivity of Metallic Elements and Alloys* (New York: Plenum, 1970). Used with permission.)

evaluate the relaxation time at the Fermi energy and do not average over all electrons. The defect scattering term in the thermal resistivity is proportional to $1/T$.

According to the discussion above, the electron contribution to the thermal resistivity of a cubic metal at low temperature can be written

$$W_e = AT^n + \frac{B}{T}, \tag{9-102}$$

where $A$ and $B$ are constants. The first term arises from electron-phonon scattering while the second arises from electron-defect and electron-boundary scattering. For the copper sample of Fig. 9-7, the second term clearly dominates. The thermal conductivity is proportional to $T$.

For transition metals, with high electron densities of states, electron-electron scattering also serves to restore thermal equilibrium. The electron-electron contribution to the thermal resistivity is proportional to $T$ at low temperatures and a term of the form $CT$ must be added to Eq. 9-102.

The electron contribution to the thermal conductivity is calculated as the reciprocal of $W_e$. To this must be added the phonon contribution, typically on the order of 10% of the total. The phonon thermal resistivity is calculated in exactly the same way as for insulators, except it has an additional term, due to electron-phonon scattering. Scattering by electrons aids in phonon relaxation and, as a result, the phonon contribution to the conductivity is significantly less than it would be in the absence of such scattering.

*Electrical Conductivities of Metals.* Figure 9-8 shows the electrical resistivity of a copper sample as a function of temperature. To obtain qualitative understand-

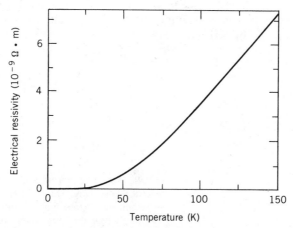

**FIGURE 9-8** Electrical resistivity of a metal as a function of temperature. For most of the temperature range shown electron-phonon scattering dominates and the resistivity is proportional to $T$. It does not vanish except perhaps at $T = 0$ K. Low-temperature values are too small to be seen on the graph.

ing, we consider a free electron metal, with electrical conductivity given by Eq. 9-58. At high temperatures, electron–phonon scattering dominates and we expect the relaxation time to be inversely proportional to the temperature. The electrical resistivity then varies as $T$. This temperature dependence prevails over most of the range shown in the figure.

At low temperatures the situation is more complicated. The number of phonons is proportional to $T^3$, but not all of them are equally effective in restoring the electron system to thermal equilibrium, with zero net crystal momentum. A scattering event that is effective is illustrated in Fig. 9-9. Roughly speaking, an electric field $\mathcal{E}$ to the right increases the occupation probability for states just outside the left side of the Fermi surface and, to restore equilibrium, electrons in these states must be scattered to states on the opposite side. For the relaxation time to be small, a large number of large angle scattering events must occur per unit time. Small-angle scattering occurs only rarely since, for these events, prospective final states are near initial states in reciprocal space and are already occupied with high probability.

Careful analysis for free electrons, using the Debye approximation, shows that the fraction of phonons that produce significant large-angle scattering is proportional to $(T/T_D)^2$, so the number of such phonons is proportional to $T^5$. This means the low-temperature electron-phonon relaxation time is proportional to $T^{-5}$ and, in the absence of other scattering mechanisms, the resistivity approaches zero as $T^5$. The temperature dependence is somewhat different for crystals with nonspherical Fermi surfaces. The low-temperature relaxation time for electron–phonon scattering is proportional to $T^{-n}$, where $n$ is typically between 3 and 5.

At sufficiently low temperature, defect and boundary scattering dominate and the relaxation time associated with these events is independent of temperature. In addition, electron-electron scattering, with a relaxation time proportional to $T^{-2}$, may also be influential, especially in transition metals. When the three mechanisms are taken into account, the low-temperature electrical resistivity of a cubic metal can be written

$$\rho(T) = \rho_0 + BT^n + CT^2, \tag{9-103}$$

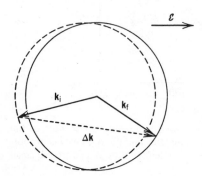

**FIGURE 9-9** A scattering event that contributes to electrical resistivity. The solid circle represents the Fermi sphere of a free electron metal in thermal equilibrium, while the dotted circle represents the sphere as it is displaced to the left by an electric field to the right. To reach steady state electrons must be scattered from occupied states on the left to unoccupied states on the right.

where the first term, called the residual resistivity, is due to defect scattering, the second is due to electron-phonon scattering, and the third is due to electron-electron scattering. Except for extremely high quality crystals, the electron-electron term is not usually observed.

The residual resistivity $\rho_0$ depends sensitively on the number of impurities present in the sample and, for low-impurity concentrations, on sample dimensions. To illustrate, Fig. 9-10 shows the resistivity of two copper samples at low temperatures. The upper curve is for a sample that has a somewhat higher impurity concentration than the lower and so has a higher residual resistivity.

Equation 9-103 is the basis for what is known as Matthiessen's rule: the resistivity of many metals is the sum of two terms, the first of which depends on the impurity concentration but not on the temperature and the second of which depends on the temperature but not on the impurity concentration. This rule is often used to find the phonon scattering contribution to the electrical resistivity. The low-temperature resistivity, in the temperature domain for which it is constant, is first measured. Then it is subtracted from the measured resistivity at other temperatures. For any particular material, the resistivity ratio $\rho(273 \text{ K})/\rho_0$ is often used as a measure of crystal quality. It is large if the sample has few defects.

Perhaps the most notable deviation from the temperature dependence described above occurs when magnetic impurities are added to a metal: iron or manganese impurities in copper, silver, or gold, for example. The magnetic impurity contribution to the electrical resistivity contains a term that is proportional to $-\ln(T)$ and to the concentration of magnetic impurities. This term increases as the temperature decreases, so the resistivity has a minimum at a temperature about absolute zero. This phenomenon is known as the Kondo effect.

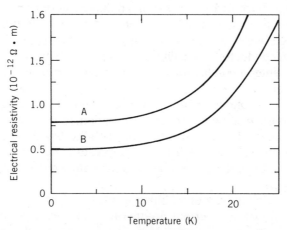

**FIGURE 9-10** Low-temperature electrical resistivities of two metal samples as functions of temperature. Impurity scattering dominates and each resistivity approaches a constant value as the temperature decreases. Sample A has a higher impurity concentration, so its resistivity reaches a higher limiting value than that of sample B. At high temperature both curves become similar to the curve of Fig. 9-8.

*Wiedemann-Franz Ratio.* If Eqs. 9-58 and 9-94 are used to calculate the ratio $\kappa/\sigma T$ for a free electron metal with the same relaxation time for both electrical and thermal conductivity, the result is $(\pi^2/3)(k_B/e)^2 = 2.443 \times 10^{-8}$ $J^2/K^2 \cdot C^2$. This value, known as the Lorentz number, is independent of the temperature. For most simple metals at high temperatures, the Wiedemann-Franz ratio $\kappa/\sigma T$ is very nearly the Lorentz number. For example, the value for gold at 373 K is $2.40 \times 10^{-8}$ $J^2/K^2 \cdot C^2$ while the value for copper at the same temperature is $2.33 \times 10^{-8}$ $J^2/K^2 \cdot C^2$.

Because electric fields and temperature gradients distort the electron distribution differently, relaxation times for electrical and thermal conduction may be different. Both large- and small-angle scattering contribute to the thermal resistivity but only the former contributes to the electrical resistivity. However, careful analysis shows that the two relaxation times are nearly the same if the change on scattering of the electron energy is small compared to $k_B T$. Since nearly all scattering is from states near the Fermi surface to other states near the Fermi surface, the change in energy is small. At high temperatures, $k_B T$ is large and the criterion is easily satisfied. At sufficiently low temperatures, $k_B T$ is less than the energy exchanged with a phonon in a typical scattering event and the criterion is not satisfied for electron-phonon scattering. When the temperature is so low that defect scattering predominates, the criterion is again satisfied because most defect scattering events are elastic. As the temperature decreases, the Wiedemann-Franz ratio typically dips below the Lorentz number, then increases again to that number.

*Semiconductors.* At low concentrations electrons and holes in a semiconductor do not significantly influence the thermal conductivity. Their contribution to the energy flux in a temperature gradient is typically less than the phonon contribution by a factor of $10^{-4}$ or less. In addition, electron-phonon scattering does not contribute significantly to phonon relaxation. Thus the thermal conductivity is much like that of an insulator and Eq. 9-68 applies. On the other hand, if the semiconductor is heavily doped and the temperature is high, the Fermi level may enter a conduction or valence band; the material then behaves more like a metal.

For a cubic semiconductor the electrical conductivity is given by $\sigma = e(n\mu_e + p\mu_h)$, where $n$ is the electron concentration, $\mu_e$ is the electron mobility, $p$ is the hole concentration, and $\mu_h$ is the hole mobility. Mobilities for an intrinsic semiconductor are far less sensitive to temperature than electron and hole concentrations, so the latter determine the temperature dependence of the electrical conductivity.

The electrical conductivity of an intrinsic semiconductor increases dramatically with temperature as electrons are thermally excited from valence to conduction bands. Since electron and hole concentrations are proportional to $e^{-\beta E_g/2}$, a plot of $\ln(\sigma)$ as a function of $\beta$ is nearly a straight line with a negative slope equal in magnitude to half the gap. An example is shown in Fig. 9-11. Plots of this kind are used to determine the gap experimentally.

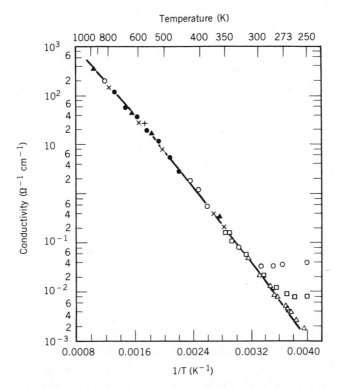

Temperature (K)

**FIGURE 9-11** The natural logarithm of the electrical conductivity of intrinsic germanium as a function of $1/T$. The curve is nearly a straight line with slope proportional to the energy gap between conduction and valence bands. (From F. J. Morin and J. P. Maita, *Phys. Rev.* **94**:1525, 1954. Used with permission.)

As shown in Fig. 8-14*b*, electron and hole concentrations in doped semiconductors may be nearly independent of temperature over a wide temperature range. In this range the temperature dependence of the conductivity is determined by the mobilities. As Eq. 9-47 implies, the energy dependence of the relaxation time is important and gives rise to a temperature dependence that is different from that of electron mobilities for metals. A simple model of electron-phonon scattering predicts $\bar{t}$ is proportional to $E^{-1/2}/T$ at high temperatures and, if the bands are parabolic and the Debye approximation is valid, the model leads to mobilities which are proportional to $T^{-3/2}$.

The situation for many semiconductors is complicated by the occurrence of several equivalent minima in the conduction band and several valence bands: scattering from one minimum to another or from one band to another can occur. In addition, bands are not parabolic and, at high temperatures, scattering by optical branch phonons occurs. As a result, mobilities are proportional to $T^{-n}$, where $n$ is usually larger than $\frac{3}{2}$. For example, the value of $n$ is 2.42 for electrons in silicon and 2.20 for holes.

## 9.6 REFERENCES

A. A. Abrikosov, *Introduction to the Theory of Normal Metals* (New York: Academic, 1972).

F. J. Blatt, *Physics of Electronic Conduction in Solids* (New York: McGraw-Hill, 1968).

E. Conwell, "Transport: The Boltzmann Equation" in *Handbook on Semiconductors* (T. S. Moss, Ed.), Vol. 1 (Amsterdam: North-Holland, 1982).

P. G. Klemens, "Thermal Conductivity and Lattice Vibrational Modes" in *Solid State Physics* (F. Seitz and D. Turnbull, Eds.), Vol. 7, p. 1 (New York: Academic, 1958).

G. A. Slack, "The Thermal Conductivity of Nonmetallic Crystals" in *Solid State Physics* (H. Ehrenreich, F. Seitz, and D. Turnbull, Eds.), Vol. 34, p. 1 (New York: Academic, 1979).

## PROBLEMS

1.  For a certain one-dimensional crystal, energies in a nearly free electron band are given by

$$E = \frac{\hbar^2 k_0^2}{2m} \left[ \frac{k^2}{k_0^2} + 0.61 \frac{k^4}{k_0^4} - 0.74 \frac{k^6}{k_0^6} \right],$$

where $k_0 = 5.4 \times 10^9 \text{ m}^{-1}$ is the propagation constant at the Brillouin zone boundary. Find the velocity of an electron with propagation constant (a) $k = 0$, (b) $k = \frac{1}{2}k_0$, (c) $k = -\frac{1}{2}k_0$, (d) $k = k_0$, and (e) $k = -k_0$.

2.  The one-dimensional crystal of Problem 1 is subjected to a 100-V/m electric field in the direction of positive $k$. Find the acceleration of an electron with propagation constant (a) $k = 0$, (b) $k = \frac{1}{2}k_0$, (c) $k = -\frac{1}{2}k_0$, (d) $k = k_0$, and (e) $k = -k_0$. Neglect scattering.

3.  Electron energies for a certain one-dimensional crystal are given by

$$E = \frac{\hbar^2 k^2}{2m} \left[ 1.00 + 4.71 \times 10^{-20} k^2 - 4.41 \times 10^{-39} k^4 \right],$$

in SI units. Suppose the band is empty except for an electron with $k = 3.10 \times 10^9 \text{ m}^{-1}$. (a) What is the velocity of the electron? (b) If a 150-V/m electric field is turned on in the direction of positive $k$, what is the propagation constant of the electron $5.00 \times 10^{-9}$ s later? Neglect scattering. (c) What is the velocity of the electron $5.00 \times 10^{-9}$ s after the field is turned on? (d) What is the change in the energy of the electron during the first $5.00 \times 10^{-9}$ s the field is on? Has it increased or decreased?

4.  Suppose the band of Problem 3 is filled with electrons except for a state with $k = 3.10 \times 10^9 \text{ m}^{-1}$, which is empty. A 150-V/m electric field is turned on in the direction of positive $k$. (a) What is the propagation constant of the state which is empty $5.00 \times 10^{-9}$ s later? (b) What is the change in the total crystal momentum of the electrons during the first $5.00 \times 10^{-9}$ s the field is on? (c) What is the change in total band energy during the first $5.00 \times 10^{-9}$ s the field is on? Has the energy increased or decreased?

5.  Near a minimum in the conduction band, energies for electrons in silicon

can be written

$$E = \frac{\hbar^2}{2m} [3.30k_x^2 + 7.75k_y^2 + 5.57k_z^2 + 6.24k_x k_z],$$

where the coordinate system has been placed so the minimum is at $\mathbf{k} = 0$. (a) Find the acceleration of an electron with $\mathbf{k} = 0$, in a 100-V/m electric field in the positive $x$ direction. (b) Find the acceleration of an electron with $\mathbf{k} = 0$, in a 100-V/m electric field in the positive $z$ direction. (c) In what directions can an electric field be applied so the acceleration of an electron with $\mathbf{k} = 0$ is directed opposite to the field?

6.  For the one-dimensional band of Problem 1, find: (a) the effective mass of an electron with $k = \frac{1}{4}k_0$, (b) the effective mass of a hole with $k_h = -\frac{1}{4}k_0$, (c) the acceleration of an electron with $k = \frac{1}{4}k_0$ in a 150-V/m electric field, and (d) the acceleration of a hole with $k_h = -\frac{1}{4}k_0$ in a 150-V/m electric field. In the last two parts, take the field to be in the direction of positive $k$.

7.  for the one-dimensional band of Problem 1, find: (a) the effective mass of an electron with $k = \frac{3}{4}k_0$, (b) the effective mass of a hole with $k_h = \frac{3}{4}k_0$, (c) the acceleration of an electron with $k = \frac{3}{4}k_0$ in a 150-V/m electric field, and (d) the acceleration of a hole with $k_h = \frac{3}{4}k_0$ in a 150-V/m electric field. In the last two parts, take the field to be in the direction of positive $k$.

8.  Estimate the change $f_1(\mathbf{k})$ induced in the steady state electron distribution function at the Fermi energy by a 150-V/m electric field. Take the temperature to be 300 K and assume the electrons are free, with a Fermi velocity of $5.0 \times 10^5$ m/s and a relaxation time of $1.0 \times 10^{-14}$ s. Neglect the slight difference between the Fermi energy and the chemical potential. In particular, calculate $f_1$ for electrons with velocity (a) in the direction of the field, (b) in the direction opposite to the field, and (c) normal to the field. (d) Calculate the work done on each of these electrons by the electric field during a time interval equal to the relaxation time.

9.  Consider a metal with a simple cubic structure (cube edge $a$) and suppose the Fermi level is in a single nearly free electron band with energies given by

$$E(\mathbf{k}) = \frac{\hbar^2}{2m} k^2 \left[ 1 - \frac{a^2}{2\pi^2} k^2 \right].$$

Each unit cell contributes one electron to the band. (a) Show that the density of states at the Fermi level is

$$\rho(E) = 102.2 \frac{Na^2 m}{\pi^2 \hbar^2},$$

where $N$ is the number of unit cells in the crystal. (b) Show that the electron mobility is $\mu = 0.0303 e\bar{t}/m$, where $\bar{t}$ is the relaxation time for electrons at the Fermi surface.

10. (a) Suppose the relaxation time $\bar{t}$ for electrons in the conduction band of a certain intrinsic semiconductor is independent of the temperature but is proportional to $E^{-2}$. Assume a parabolic band and find the temperature dependence of the average relaxation time $\langle \bar{t} \rangle$. (b) If, instead, the average relaxation time is found to be independent of temperature, what is the energy dependence of $\bar{t}$?

11. If a coordinate system is oriented with its $x$ and $y$ axes along edges of a square cell face of a tetragonal crystal, the conductivity tensor is diagonal and $\sigma_{xx} = \sigma_{yy}$. (a) Show that for any electric field in the $xy$ plane the current density is in the same direction as the field. (b) Show that for an electric field in the $xz$ plane, but not along either the $x$ or $z$ axis, the current density is not in the same direction as the field.

12. Use the two-level model of a semiconductor to estimate the electrical conductivity of intrinsic silicon at 300 K. Take the electron mobility to be 0.1350 m$^2$/V·s, the hole mobility to be 0.0480 m$^2$/V·s, and the gap to be 1.11 eV. For purposes of computing the number of states per unit volume in the bands, the electron effective mass is $1.08m_0$ and the hole effective mass is $0.81m_0$, where $m_0$ is the free electron mass.

13. Suppose the silicon sample of the previous problem is doped with $7.50 \times 10^{16}$ donors/m$^3$. Assume all donors are singly ionized and that doping does not change the mobilities. Estimate the electrical conductivity at 300 K.

14. Use the free electron model and the conductivity given in Table 9-2 to estimate the room temperature relaxation time for conduction electrons in sodium. Then estimate the electron mobility. Also estimate the speed of an electron at the Fermi surface and the mean distance such an electron travels between collisions. Sodium is body-centered cubic with a cube edge of 4.225 Å and each atom contributes one electron to the nearly free electron band. The Fermi energy is 3.22 eV above the bottom of the band.

15. When the coordinate system is appropriately positioned, the electron energy near a minimum of the conduction band of silicon is given by

$$E(\mathbf{k}) = \tfrac{1}{2}\hbar^2 \left[ \frac{k_z^2}{m_1} + \frac{k_x^2 + k_y^2}{m_2} \right],$$

where $1/m_1$ and $1/m_2$ are elements of the reciprocal effective mass tensor. Assume all energies in the band are much greater than the chemical potential and that the relaxation time is independent of $\mathbf{k}$. (a) Show that if the only electrons in the band are those near this minimum then the mobility tensor elements are $\mu_{zz} = e\bar{t}/m_1, \mu_{xx} = \mu_{yy} = e\bar{t}/m_2$, and all other elements vanish. (b) The band actually has six such minima, related to each other by cubic symmetry. Show that the mobility for electrons in the band is given by

$$\mu_{xx} = \mu_{yy} = \mu_{zz} = \frac{e\bar{t}}{3} \left( \frac{1}{m_1} + \frac{2}{m_2} \right),$$

with all other elements vanishing. (c) The quantity $m^*$, defined by $\mu_{xx} = e\bar{t}/m^*$, is called the mobility effective mass. For silicon $m_1 = 0.9163m_0$ and $m_2 = 0.1905m_0$, where $m_0$ is the free electron mass. Calculate $m^*$ and compare it with the density of states effective mass $m_d^* = (m_1m_2^2)^{1/3}$. See Problem 19 of Chapter 8. (d) The mobility of electrons in the conduction band of silicon is $0.1350 \text{ m}^2/\text{V} \cdot \text{s}$ at 300 K. Estimate the relaxation time for that temperature.

16. To understand why electrical conductivities of metals are greater than electrical conductivities of intrinsic semiconductors, compare the conductivity $\sigma_m$ of a cubic nearly free electron metal with the conductivity $\sigma_s$ of an intrinsic cubic semiconductor. Assume all bands are parabolic. Take all effective masses to be equal, all relaxation times to be equal, and the number of states per unit volume in all bands to be equal. Assume the free electron band of the metal is half occupied and show that the ratio of the conductivities is $\sigma_m/\sigma_s = \frac{1}{4}e^{\beta E_g/2}$, where $E_g$ is the semiconductor gap. Evaluate this ratio for $E_g = 1.0$ eV and $T = 300$ K. The ratio for actual metals and semiconductors may vary from that given by the above expression by several orders of magnitude.

17. A certain semiconductor has a single conduction band, with energies $E(\mathbf{k}) = E_c + (\hbar^2/2m^*)k^2$. Assume the Fermi level is in the gap, far from both bands, and neglect variations in the chemical potential with temperature. Take the relaxation time $\bar{t}$ to be independent of $\mathbf{k}$. (a) Show that the electron contribution to the current density when there is a temperature gradient along the $z$ axis is given by

$$J_z = \frac{en\bar{t}}{m^*}\left[e\mathcal{E}_z + \left(\frac{5}{2}k_B + \frac{E_c}{T}\right)\frac{dT}{dz}\right],$$

where $n$ is the concentration of electrons in the conduction band. (b) Show that the energy flux is

$$Q_z = -\frac{n\bar{t}}{m^*}\left[\left(E_c + \frac{5}{2}k_BT\right)e\mathcal{E} + \left(\frac{35}{4}k_B^2T + 5k_BE_c + \frac{E_c^2}{T}\right)\right]\frac{dT}{dz}.$$

(c) Assume the semiconductor is heavily doped $n$ type (but with the chemical potential in the gap), so holes do not contribute significantly to the current density and energy flux. Show the electron contribution to the thermal conductivity is then $\kappa = (5n\bar{t}/2m^*)k_B^2T$. (d) For an intrinsic semiconductor, explain why the sum of $\kappa$, given in part c, and a similar term for holes does not give the electron contribution to the thermal conductivity, even if the individual terms are correct for $n$- and $p$-type materials, respectively.

18. Consider the electrical conductivity $\sigma$ of a semiconductor as a function of dopant concentration. Assume impurity levels are in the gap, far from band edges, and show that $\sigma$ is a minimum if $n = n_i\sqrt{\mu_h/\mu_e}$, where $n_i$ is the intrinsic electron concentration, $\mu_e$ is the electron mobility, and $\mu_h$ is the

hole mobility. Show that $\sigma_{min} = 2en_i\sqrt{\mu_e\mu_h}$. Calculate the ratio of $\sigma_{min}$ to the intrinsic conductivity for gallium arsenide, with $\mu_e = 0.850$ m$^2$/V·s and $\mu_h = 0.0400$ m$^2$/V·s at room temperature.

19. Use the Debye model in the low-temperature limit to estimate the phonon contribution to the thermal conductivity of germanium at 300 K. Germanium has a diamond structure with cube edge $a = 5.658$ Å and its Debye temperature is 370 K. Take the phonon relaxation time to be $1.90 \times 10^{-12}$ s. (b) Use the result of Problem 17 to estimate the electron contribution to the thermal conductivity of germanium. Take the electron concentration to be $2.35 \times 10^{21}$ electrons/m$^3$, about 100 times the intrinsic concentration at 300 K. Take the relaxation time to be $8.65 \times 10^{-12}$ s and the electron effective mass to be $0.0393m_0$, where $m_0$ is the free electron mass.

20. Use the Debye approximation and the experimentally measured thermal conductivity of copper to estimate the relaxation time due to phonon scattering at room temperature. Assume the phonon contribution to the thermal conductivity is 10% of the total. Copper is face-centered cubic with a cube edge of 3.61 Å and its Debye temperature is 343 K.

21. The first term in Eq. 9-77 is often called the drift current density, while the sum of the second and third terms is called the diffusion current density. In an open-circuit measurement of the thermal conductivity, the electric field is such that these current densities have the same magnitude and opposite directions. Suppose a temperature gradient of 1.00 K/m is maintained in the positive $z$ direction along the copper sample of Example 9-6. Assume the data is for $T = 300$ K and that the electron concentration is $8.50 \times 10^{28}$ electrons/m$^3$. (a) Find the magnitude and direction of the diffusion current density. (b) Find the magnitude and direction of the electric field.

22. (a) For a simple cubic crystal with cube edge $a$, what is the magnitude of the longest propagation vector such that the initial phonons in an Umklapp scattering process cannot both have shorter propagation vectors? (b) What fraction of modes in acoustical branches have propagation vectors shorter than the length found in part a? (c) Assume an isotropic speed of sound with magnitude $5.50 \times 10^3$ m/s and take $a = 5.65$ Å. Compare the number of phonons in a mode with propagation vector at a Brillouin zone corner to the number in a mode with a propagation vector of the magnitude found in part a. Take the temperature to be 300 K, then repeat the calculation for 10 K.

23. Use the Debye approximation in the low-temperature limit to show that the thermal resistivity of a cubic insulator is proportional to $T^{3-n}$ if the relaxation time is proportional to $\omega^{-n}$.

24. Low-temperature electrical resistivity data for a certain gold crystal can be fitted to the function $\rho = \rho_0 + AT^n$, where $\rho_0 = 7.8 \times 10^{-4}$ μΩ·cm, $n = 3.99$, and $A = 5.07 \times 10^{-8}$ μΩ·cm/K$^n$. (a) At what temperature does the phonon scattering contribution match the defect scattering contribu-

tion? (b) Suppose the defect concentration is reduced by a factor of 10. At what temperature do the two contributions now match?

25. (a) Suppose the relaxation time for electrons in an isotropic parabolic conduction band of a semiconductor is given by $\bar{t} = AE^{-s}$, where $A$ is independent of energy. Show that the average relaxation time $\langle \bar{t} \rangle$, defined by Eq. 9-49, is given by

$$\langle \bar{t} \rangle = \frac{2A\Gamma(\frac{5}{2} - s)}{3\sqrt{\pi}(k_B T)^s},$$

where $\Gamma$ is the gamma function. (b) We can obtain plausible values for $A$ and $s$ if we assume the mean free path is inversely proportional to the temperature. Take $\ell = \gamma/T$, where $\gamma$ is a constant, use $\bar{t} = \ell/v$, and show that

$$\bar{t} = \frac{\gamma\sqrt{m^*}E^{-1/2}}{\sqrt{2}T}$$

for a parabolic band. (c) Take $s = \frac{1}{2}$ and $A = \gamma\sqrt{m^*}/\sqrt{2}T$ and show that $\langle \bar{t} \rangle$ is then proportional to $T^{-3/2}$.

# Chapter 10

## DIELECTRIC AND OPTICAL PROPERTIES

The experimental floor of the National Synchrotron Light Source at Brookhaven. Beams of intense x rays and ultraviolet light are led from the source to experimental apparatus set up throughout the room.

In this chapter we consider the application of static electric fields to insulators and oscillating electric fields to both insulators and metals. Positive and negative charges are pushed in opposite directions by a field and the distorted charge distribution gives rise to its own electric field. If a static field is applied to a metal, nearly free electrons move to screen it and if the circuit is not completed, the macroscopic field vanishes inside. Electrons in insulators, on the other hand, are tightly bound to atoms and are displaced only slightly by an applied field. An electric field exists in the interior.

The distorted charge distribution in an insulator can be approximated by a collection of electric dipoles. Dipoles result from shifts in the average positions of tightly bound electrons relative to their respective nuclei and, in ionic and partially ionic solids, from shifts in the relative positions of oppositely charged ions. Molecules in some molecular solids have permanent dipole moments and are more or less free to rotate. An applied field tends to align the moments and so induces a nonvanishing net moment.

If the applied field oscillates, as it does in an electromagnetic wave, charges in the interior oscillate in response and again change the field. Both reflection and refraction of electromagnetic waves are intimately associated with the oscillations of charges, as is the absorption of electromagnetic energy.

## 10.1  STATIC DIELECTRIC PROPERTIES

*Fundamental Concepts.*  The electric dipole moment of two charges $-q$ and $+q$ is given by $\mathbf{p} = q\mathbf{r}$, where $\mathbf{r}$ is the displacement of the positive charge from the negative. Each electron in an atom is paired with the nucleus to form a dipole and the individual moments are summed to find the total moment. If $n(\mathbf{r})$ is the electron concentration at $\mathbf{r}$, measured relative to the nucleus, then the total dipole moment is given by the volume integral

$$\mathbf{p} = -e \int \mathbf{r}n(\mathbf{r})\,d\tau, \tag{10-1}$$

over the atom. Since the average electron position is $\mathbf{r}_{ave} = (1/N)\int \mathbf{r}n(\mathbf{r})\,d\tau$, where $N$ is the number of electrons in the atom, Eq. 10-1 can be written $\mathbf{p} = -eN\mathbf{r}_{ave}$. For most atoms in solids the center of the electron distribution is at the nucleus and $\mathbf{p} = 0$. When an electric field is turned on, however, $\mathbf{r}_{ave}$ is no longer zero and the atom acquires a dipole moment. Figure 10-1 is a schematic representation of the shift in the electron distribution produced by an electric field.

For weak fields the average electron displacement from the nucleus is proportional to the field and the atomic dipole moment is given by

$$\mathbf{p} = \alpha\mathcal{E}_{loc}, \tag{10-2}$$

where $\mathcal{E}_{loc}$ is the field at the site of the atom and $\alpha$ is a constant of proportionality, called the polarizability of the atom. It is calculated by solving the Schrödinger equation with a term corresponding to the local field, then replacing $n$ in Eq. 10-1 with the sum $\Sigma\psi^*\psi$ over occupied states. Electronic polarizabilities of some atoms and ions are given in Table 10-1.

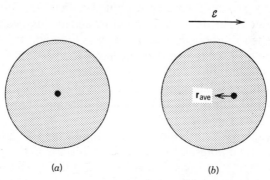

**FIGURE 10-1** A schematic representation of the electron distribution around a nucleus, represented by the solid circle. (a) The applied electric field is zero. The average electron position coincides with the nucleus and the electric dipole moment vanishes. (b) A field applied to the right displaces the distribution to the left and the nucleus to the right. The dipole moment is $\mathbf{p} = -eN\mathbf{r}_{ave}$, where $N$ is the number of electrons in the distribution.

Strictly speaking, Eq. 10-2 holds only for free atoms. Because neighboring atoms influence the shift in the electron distribution, the dipole moment of an atom in a solid may not be in the same direction as the local field and the polarizability must be represented by a tensor.

In an ionic solid the net dipole moment of any region containing a large number of ions is also proportional to the local field. Consider a simple one dimensional model consisting of two oppositely charged ions per unit cell, as shown in Fig. 10-2. Let $u_-$ and $u_+$ represent the displacements of the ions from their equilibrium positions and let $\gamma$ be the force constant associated with their mutual interaction. If no other interatomic forces are significant, the force on the negative ion is $-2\gamma(u_- - u_+) - Q\mathcal{E}_{loc}$ while the force on the positive ion is $-2\gamma(u_+ - u_-) + Q\mathcal{E}_{loc}$. At the new equilibrium positions the force on each ion vanishes and $u_+ - u_- = Q\mathcal{E}_{loc}/2\gamma$. The dipole moment of the pair is

$$p = Qa + Q(u_+ - u_-) = Qa + \frac{Q^2}{2\gamma}\mathcal{E}_{loc}, \qquad (10\text{-}3)$$

**TABLE 10-1** Electronic Polarizabilities of Selected Atoms and Ions (expressed as $\alpha/4\pi\epsilon_0$ in $10^{-30}$ m³)

| H | 0.66 | Ne | 0.390 | K$^+$ | 1.136 |
|---|---|---|---|---|---|
| H$^-$ | 10.0 | Na | 27 | Ca$^{2+}$ | 0.47 |
| He | 0.201 | Na$^+$ | 0.312 | Br$^-$ | 4.276 |
| Li | 12 | Mg$^{2+}$ | 0.094 | Kr | 2.46 |
| Li$^+$ | 0.029 | Cl$^-$ | 3.063 | Rb$^+$ | 1.758 |
| Be$^{2+}$ | 0.008 | Ar | 1.62 | Xe | 3.99 |
| O$^{2-}$ | 3.88 | K | 34 | Cs$^+$ | 3.015 |
| F$^-$ | 0.867 | | | | |

*Sources:* A. M. Portis, *Electromagnetic Fields* (New York: Wiley, 1978); L. Pauling, *Proc. R. Soc. (London) Ser. A* **114:** 181, 1927; J. Tessman, A. Kahn, and W. Shockley, *Phys. Rev.* **92:** 890, 1953.

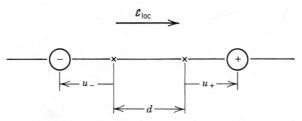

**FIGURE 10-2** Two oppositely charged ions in an electric field. They are displaced in opposite directions and their dipole moment is increased over its equilibrium value. Equilibrium sites are marked by ×'s.

where $a$ is the equilibrium separation in the absence of a field. To find the net dipole moment of a region, Eq. 10-3 is summed over all ion pairs in the region. The first term sums to zero and the remaining terms are all proportional to the local field. Distortions of the electron distribution about each ion also contribute to the net moment but are not included in Eq. 10-3.

The electric field produced by a dipole with moment $\mathbf{p}$ is given by

$$\mathcal{E}(\mathbf{r}) = \frac{3(\mathbf{p} \cdot \mathbf{r})\mathbf{r} - r^2\mathbf{p}}{4\pi\epsilon_0 r^5}, \tag{10-4}$$

where $\mathbf{r}$ is the displacement vector from the dipole to the field point. The total field in a solid is the vector sum of the applied field and the fields due to the individual dipoles. We must be careful to distinguish between the total field and the local field, which induces a moment on any given atom. All dipoles contribute to the total field, while the local field does not include the field of the dipole on which it acts. The field that is measured is the macroscopic field, an average of the total field over a volume that is macroscopically small but nevertheless contains many atoms. The averaging process smooths variations from point to point between neighboring atomic positions, so the macroscopic field is typically smaller in magnitude than the local field.

*Polarization.* Averaging can be carried out conveniently in terms of the polarization field $\mathbf{P}(\mathbf{r})$, the dipole moment per unit volume at the point $\mathbf{r}$. $\mathbf{P}(\mathbf{r})$ is calculated as the ratio of dipole moment to volume of a small region around $\mathbf{r}$, in the limit as the region becomes small. Polarization is a macroscopic quantity: the limiting volume used for its calculation is large on an atomic scale but small macroscopically. For a crystal we may take it to be the net dipole moment of a unit cell divided by the cell volume. If the concentration of atoms in a given region is $n$ and each atom has dipole moment $\mathbf{p}$, then their contribution to the polarization is $\mathbf{P} = n\mathbf{p}$.

The net dipole moment of a region with volume $d\tau'$ located at $\mathbf{r}'$ is $\mathbf{P}(\mathbf{r}') \, d\tau'$, so the macroscopic electric field produced at $\mathbf{r}$ by dipoles of a solid is given by

$$\mathcal{E}(\mathbf{r}) = \int \frac{3\mathbf{P}(\mathbf{r}') \cdot (\mathbf{r} - \mathbf{r}')(\mathbf{r} - \mathbf{r}') - \mathbf{P}(\mathbf{r}')|\mathbf{r} - \mathbf{r}'|^2}{4\pi\epsilon_0|\mathbf{r} - \mathbf{r}'|^5} \, d\tau'. \tag{10-5}$$

As shown in most electromagnetic theory texts,* Eq. 10-5 is mathematically identical to

$$\mathscr{E}(\mathbf{r}) = \frac{1}{4\pi\epsilon_0} \int \frac{-\nabla' \cdot \mathbf{P}(\mathbf{r}')(\mathbf{r} - \mathbf{r}')}{|\mathbf{r} - \mathbf{r}'|^3} \, d\tau'$$

$$+ \frac{1}{4\pi\epsilon_0} \int \frac{\mathbf{P}(\mathbf{r}') \cdot \hat{\mathbf{n}}(\mathbf{r} - \mathbf{r}')}{|\mathbf{r} - \mathbf{r}'|^3} \, dS', \tag{10-6}$$

where the first integral is over the volume of the polarized material and the second is over its surface. The unit vector $\hat{\mathbf{n}}$ is the outward normal to the surface and $\epsilon_0$ is the permittivity of free space, $8.854 \times 10^{-12}$ F/m. $\mathscr{E}(\mathbf{r})$ is precisely the field produced by a volume distribution of charge with charge density $-\nabla \cdot \mathbf{P}$ and a surface distribution of charge with surface density $\mathbf{P} \cdot \hat{\mathbf{n}}$. Most samples we consider are uniformly polarized, so $\nabla \cdot \mathbf{P} = 0$ and the only contributions come from polarization charge $\mathbf{P} \cdot \hat{\mathbf{n}}$ on the surface.

For most solids in weak electric fields, components of the polarization at any point are proportional to the macroscopic field at that point. Such solids are called linear dielectrics and for them

$$P_i(\mathbf{r}) = \sum_j \chi_{ij}\epsilon_0\mathscr{E}_j(\mathbf{r}), \tag{10-7}$$

where $P_i$ is component $i$ of the polarization and $\mathscr{E}_j$ is component $j$ of the electric field. The constants of proportionality $\chi_{ij}$ are elements of the electric susceptibility tensor. A tensor is required since the polarization and electric fields may not be in the same direction. As we will see, Eq. 10-7 results from the linear relationship between the dipole moment of an atom or ion pair and the local field. It does not hold for solids discussed in Section 10.2, which may be permanently polarized in the absence of an applied field.

For most cubic and amorphous materials the susceptibility tensor is diagonal with all three diagonal elements identical. For these materials the polarization and the field are in the same direction and

$$\mathbf{P} = \chi\epsilon_0\mathscr{E}, \tag{10-8}$$

where $\chi$ is any one of the diagonal elements of the susceptibility tensor. To avoid mathematical complexity we will usually discuss solids for which Eq. 10-8 is valid.

Quantities other than the susceptibility are sometimes used to characterize dielectric solids. Elements of the permittivity tensor are defined by $\epsilon_{ij} = \epsilon_0(\delta_{ij} + \chi_{ij})$ and elements of the dielectric constant are defined by $K_{ij} = \epsilon_{ij}/\epsilon_0 \, (= \delta_{ij} + \chi_{ij})$, where $\delta_{ij}$ is the Kronecker delta. If the susceptibility is a scalar these definitions reduce to $\epsilon = \epsilon_0(1 + \chi)$ and $K = \epsilon/\epsilon_0$, respectively.

Susceptibilities, permittivities, and dielectric constants are listed in many handbooks. A sample list of dielectric constants is given in Table 10-2. For many

---

*See, for example, Section 4-2 of J. R. Reitz, F. J. Milford, and R. W. Christy, *Foundations of Electromagnetic Theory* (Reading, MA: Addison-Wesley, 1979).

**TABLE 10-2**  Static Dielectric Constants of Selected Pure Solids at about 20 K

| Solid | K | Solid | K |
|---|---|---|---|
| Aluminum oxide | 9.5 | Lead telluride | 800 |
| Ammonium chloride | 7.22 | Lithium fluoride | 9.27 |
| Barium fluoride | 7.34 | Lithium niobate | 80 |
| Cadmium iodide | 68 | Nickel monoxide | 12.6 |
| Cadmium telluride | 11.00 | Potassium bromide | 4.7 |
| Calcium fluoride | 6.81 | Potassium chloride | 4.68 |
| Cesium bromide | 6.60 | Potassium iodide | 5.1 |
| Cesium chloride | 6.83 | Selenium | |
| Cesium iodide | 6.49 | (amorphous) | 6.0 |
| Cobalt oxide | 10.6 | Sodium bromide | 5.99 |
| Europium oxide | 24 | Sodium chloride | 5.8 |
| Gallium arsenide | 13.08 | Sodium iodide | 4.94 |
| Germanium | 15.3 | Thallium iodide | 20.4 |
| Lead fluoride | | Titanium dioxide | 44.7 |
| (cubic) | 29.3 | Zinc oxide | 16.2 |

Sources: *Digest of Literature on Dielectrics,* published annually by the National Academy of Sciences; I. S. Zheludev, *Physics of Crystalline Dielectrics* (New York: Plenum, 1971).

binary insulators both electronic and ionic contributions to the dielectric constant are significant. For example, the static dielectric constant of sodium chloride is 5.7, of which 2.2 is due to electronic polarization and 3.5 is due to ionic polarization. On the other hand, the extremely large dielectric constants of titanium dioxide and many other oxides arise almost exclusively from the large electronic polarizability of oxygen.

The following example illustrates some of the ideas discussed above.

**EXAMPLE 10-1**  Consider a parallel plate capacitor with surface charge density $+\sigma$ on one plate and $-\sigma$ on the other. The region between the plates is filled with a uniform isotropic dielectric with susceptibility $\chi$, as shown in Fig. 10-3. Find expressions for the macroscopic electric and polarization fields in the dielectric. Neglect fringing.

**SOLUTION**  The electric field due surface charge density $\sigma$ on a plane is $\sigma/2\epsilon_0$. In the interior of a capacitor the total field due to the two plates is $\sigma/\epsilon_0$ and points from the positive toward the negative plate. The polarization

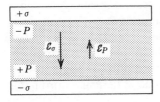

**FIGURE 10-3.**  A parallel plate capacitor containing a dielectric. The upper plate has surface charge density $+\sigma$ while the upper surface of dielectric has surface polarization charge density $-P$. The lower plate and dielectric surfaces have surface charge densities $-\sigma$ and $+P$, respectively. The field $\mathcal{E}_\sigma$ of the plate charge and the field $\mathcal{E}_P$ of the polarization charge are indicated by arrows.

is uniform and points from the negative toward the positive plate, so the polarization surface charge density is $\mathbf{P} \cdot \hat{\mathbf{n}} = -P$ at the positive plate and $\mathbf{P} \cdot \hat{\mathbf{n}} = +P$ at the negative plate. The electric field associated with polarization is $P/\epsilon_0$ and points from the negative toward the positive plate. Thus the total field in the interior is $\mathcal{E} = (\sigma - P)/\epsilon_0$. Since $P = \chi\epsilon_0\mathcal{E}$, $\mathcal{E} = (\sigma - \chi\epsilon_0\mathcal{E})/\epsilon_0$ or $\mathcal{E} = \sigma/\epsilon_0(1 + \chi) = \sigma/K\epsilon_0$. The polarization is $P = \chi\epsilon_0\mathcal{E} = (K - 1)\sigma/K$.

The electric field is reduced by the factor $K$ from what it would be in the absence of a dielectric. As $K$ increases the polarization approaches $\sigma$ and the field it produces approaches $-\sigma/\epsilon_0$. The total field vanishes in the limit of large $K$. ∎

The capacitance of a dielectric filled capacitor is $KC_0$, where $C_0$ is the capacitance in the absence of a dielectric. Thus dielectric constants can be determined experimentally by measuring capacitance. If the sample is crystalline, several measurements must be made, with the crystal aligned differently each time. High accuracy can be obtained by using a capacitance bridge with a low-frequency alternating source. As we will see, the dielectric constant is nearly independent of frequency for low frequencies, so this method gives the static dielectric constant.

*The Local Field.* Electronic and ionic polarizabilities are the fundamental microscopic parameters that determine the susceptibility and dielectric constant of a solid. To find the relationship we must first derive an expression for the local field, which appears in Eqs. 10-2 and 10-3. This field is the sum of the applied field and the fields due to all dipoles except the one being considered.

Consider only electronic polarization and suppose the sample is a homogeneous linear dielectric with $n$ atoms per unit volume in a uniform applied field. To find the local field at an atom, imagine a sphere of radius $R$ with the atom at its center, as depicted in Fig. 10-4. If $R$ is sufficiently large, the contribution of material outside the sphere to the local field at the center is nearly the macroscopic field it produces. Since the material is uniformly polarized, the macroscopic field is produced by polarization charge on the surfaces of the sample and sphere. On the other hand, we treat atoms inside the sphere as discrete dipoles and sum the fields they produce. In summary, the local field is given by

$$\mathcal{E}_{\text{loc}} = \mathcal{E}_a + \mathcal{E}_1 + \mathcal{E}_2 + \mathcal{E}_3, \tag{10-9}$$

where $\mathcal{E}_a$ is the applied field, $\mathcal{E}_1$ is due to polarization charge on the sample surface, $\mathcal{E}_2$ is due to polarization charge on the sphere surface, and $\mathcal{E}_3$ is due to dipoles within the sphere. $\mathcal{E}_a + \mathcal{E}_1$ is the macroscopic field $\mathcal{E}$.

$\mathcal{E}_2$, given by the second integral of Eq. 10-6, can be evaluated easily. Place a coordinate system with its origin at the sphere center and its $z$ axis in the direction of the polarization. The outward normal to the surface points toward the center of the sphere so, in terms of the usual spherical coordinates,

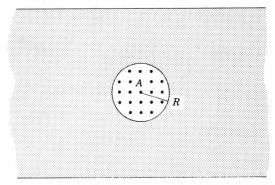

**FIGURE 10-4** A dipole is at $A$ in the interior of a solid sample. To find the local field at $A$, imagine a sphere of radius $R$, centered on $A$. The fields of dipoles inside the sphere are computed individually while the field of dipoles outside is approximated by the macroscopic field they produce.

$\mathbf{P} \cdot \hat{\mathbf{n}} = -P \cos \theta$, $\mathbf{r} = 0$, $\mathbf{r}' = R(\hat{\mathbf{x}} \sin \theta \cos \phi + \hat{\mathbf{y}} \sin \theta \sin \phi + \hat{\mathbf{z}} \cos \theta)$, and $dS = R^2 \sin \theta \, d\theta \, d\phi$. The $x$ and $y$ components of $\mathcal{E}$ vanish, while the $z$ component is

$$\mathcal{E}_2 = \frac{1}{4\pi\epsilon_0} \int_{\theta=0}^{\pi} \int_{\phi=0}^{2\pi} \frac{P \cos \theta}{R^3} R \cos \theta \, R^2 \sin \theta \, d\theta \, d\phi = \frac{P}{3\epsilon_0}. \quad (10\text{-}10)$$

In vector form, $\mathcal{E}_2 = \mathbf{P}/3\epsilon_0$, so Eq. 10-9 becomes

$$\mathcal{E}_{\text{loc}} = \mathcal{E} + \frac{\mathbf{P}}{3\epsilon_0} + \mathcal{E}_3, \quad (10\text{-}11)$$

where the macroscopic field $\mathcal{E}$ was substituted for $\mathcal{E}_a + \mathcal{E}_1$.

The burden of the calculation now rests with $\mathcal{E}_3$. If the crystal has one atom per primitive unit cell, all atoms experience the same local field and have the same dipole moment $\mathbf{p}$, so

$$\mathcal{E}_3 = \sum{}' \frac{3(\mathbf{p} \cdot \mathbf{r}_i)\mathbf{r}_i - r_i^2 \mathbf{p}}{4\pi\epsilon_0 r_i^5}, \quad (10\text{-}12)$$

where $\mathbf{r}_i$ is the position of atom $i$ relative to the sphere center and the sum is over all atoms in the sphere except the one at its center.

In principle, Eq. 10-12 is used to substitute for $\mathcal{E}_3$ in Eq. 10-11, $\mathbf{p}$ is replaced by $\alpha \mathcal{E}_{\text{loc}}$, and the result is solved for $\mathcal{E}_{\text{loc}}$. If, however, the environment of the atom at the origin has cubic symmetry, then $\mathcal{E}_3 = 0$ (see Problem 3) and the calculation simplifies considerably. We consider such a situation. Set $\mathcal{E}_3 = 0$ in Eq. 10-11 and replace $\mathbf{P}$ by $\chi\epsilon_0 \mathcal{E}$ to find

$$\mathcal{E}_{\text{loc}} = \left(1 + \frac{\chi}{3}\right) \mathcal{E}. \quad (10\text{-}13)$$

Since $\chi$ is positive the local field is larger than the macroscopic field.

Since $\mathbf{P} = n\alpha\mathcal{E}_{\text{loc}}$ from a microscopic viewpoint and $\mathbf{P} = \chi\epsilon_0\mathcal{E}$ from a macroscopic viewpoinot, $n\alpha\mathcal{E}_{\text{loc}} = \chi\epsilon_0\mathcal{E}$. When Eq. 10-13 is used,

$$\chi = \frac{n\alpha/\epsilon_0}{1 - n\alpha/3\epsilon_0} \tag{10-14}$$

is obtained. Equation 10-14 is known as the Clausius-Mossotti relationship between $\chi$ and $\alpha$. Local field effects account for a denominator that is different from 1. If they are neglected the polarization is $n\alpha\mathcal{E}$, where $\mathcal{E}$ is the macroscopic field, and $\chi = n\alpha/\epsilon_0$, much smaller than indicated by Eq. 10-14.

Since $\chi = K - 1$, Eq. 10-14 can be written in the form

$$\frac{K - 1}{K + 2} = \frac{n\alpha}{3\epsilon_0}. \tag{10-15}$$

For a crystal with more than one atom in its primitive basis the generalization of Eq. 10-15 is

$$\frac{K - 1}{K + 2} = \frac{1}{3\epsilon_0} \sum_i n_i\alpha_i , \tag{10-16}$$

where the sum is over atoms of the basis. For Eq. 10-16 to be valid, $\mathcal{E}_3$ must vanish for each atom of a unit cell.

**EXAMPLE 10-2**    In Example 10-1 the region between the capacitor plates is filled by amorphous selenium with a dielectric constant of 6.0 and a concentration of $3.67 \times 10^{28}$ atoms/m$^3$. (a) Estimate the polarizability of a selenium atom. (b) Estimate the local field at a selenium atom if the charge on the plates produces a field of 1500 V/m. (c) Estimate the dipole moment of a selenium atom in the field of part b. (d) What would the dielectric constant be if the local field were the same as the macroscopic field?

**SOLUTION**    (a) From Eq. 10-15,

$$\alpha = \frac{3\epsilon_0}{n} \frac{K - 1}{K + 2} = \frac{3 \times 8.85 \times 10^{-12}}{3.67 \times 10^{28}} \frac{5.0}{8.0} = 4.5 \times 10^{-40} \text{ F} \cdot \text{m}^2.$$

(b) The macroscopic field is $\mathcal{E} = \mathcal{E}_a/K = 1500/6 = 250$ V/m and the local field is

$$\mathcal{E}_{\text{loc}} = \left(1 + \frac{K - 1}{3}\right)\mathcal{E} = \left(1 + \frac{5.0}{3}\right)250 = 670 \text{ V/m},$$

more than twice as great. (c) The dipole moment is $p = \alpha\mathcal{E}_{\text{loc}} = 4.5 \times 10^{-40} \times 670 = 3.0 \times 10^{-37}$ C $\cdot$ m. (d) $K = 1 + \chi = 1 + n\alpha/\epsilon_0 = 1 + 3.67 \times 10^{28} \times 4.5 \times 10^{-40}/8.85 \times 10^{-12} = 2.87$, where we have used $\chi = n\alpha/\epsilon_0$ rather than Eq. 10-14. ∎

*Dipolar Solids.*    Consider a solid composed of molecules with permanent dipole moments. An electric field exerts a torque $\mathbf{p} \times \mathcal{E}$ on a dipole and tends to align it with the field. The field does work on the dipole as it rotates and the energy of interaction is given by $U = -\mathbf{p} \cdot \mathcal{E} = -p\mathcal{E} \cos \theta$, where $\theta$ is the angle between the field and the moment. Alignment is opposed by thermal motion. If a dipole moment is free to rotate, the probability it makes an angle $\theta$ with the field is given by $Ae^{-\beta p\mathcal{E} \cos\theta}$, where $A$ is a constant. Since the angular distribution of dipoles must be normalized, the reciprocal of $A$ is the integral of $e^{-\beta p\mathcal{E}\cos\theta}$ over solid angle.

Components of the average moment in directions perpendicular to the field vanish. To find the component along the field, multiply $Ae^{-\beta p\mathcal{E}\cos\theta}$ by $p \cos \theta$ and integrate over solid angle. An element of solid angle is $\sin \theta \, d\theta \, d\phi$, where $\theta$ and $\phi$ are the usual spherical coordinates, so

$$P_{ave} = \frac{\int e^{-\beta p\mathcal{E}\cos\theta} \, p \cos \theta \sin \theta \, d\theta}{\int e^{-\beta p\mathcal{E}\cos\theta} \sin \theta \, d\theta} = p\left[ \coth(\beta p\mathcal{E}) - \frac{1}{\beta p\mathcal{E}} \right], \quad (10\text{-}17)$$

where the integration limits are 0 and $\pi$. Integrals over $\phi$ cancel from the numerator and denominator. The magnitude of the polarization is given by

$$P = np_{ave} = np\left[ \coth(\beta p\mathcal{E}) - \frac{1}{\beta p\mathcal{E}} \right], \quad (10\text{-}18)$$

where $n$ is the number of molecules per unit volume. **P** is in the direction of the field.

The Langevin function $L(x) = \coth(x) - 1/x$ is plotted in Fig. 10-5. In terms of this function $P = npL(\beta p\mathcal{E})$. For low temperatures and high fields $\beta p\mathcal{E} \gg 1$

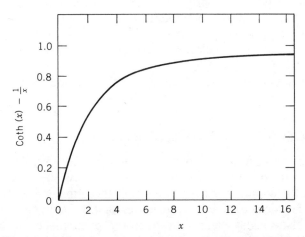

**FIGURE 10-5** The Langevin function $L(x) = \coth(x) - 1/x$. Near $x = 0$ the function is nearly linear with slope 1/3 but it approaches 1 in the limit as $x$ becomes large. The polarization of material composed of freely rotating dipolar molecules is proportional to $L$. For weak fields or high temperatures the material is a linear dielectric, but for strong fields or low temperatures the polarization saturates.

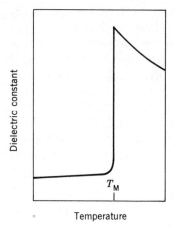

Dielectric constant

$T_M$

Temperature

**FIGURE 10-6** The dielectric constant of a polar substance as a function of temperature. Above the melting point $T_M$ molecules are free to rotate. The dielectric constant is large and proportional to $1/T$. Below the melting point, molecular rotation is hindered by intermolecular interactions.

and $L$ is nearly 1. All dipoles are aligned with the field and the polarization has its maximum value, $np$. It is said to be saturated. For moderate fields at room temperature, $\beta p \mathcal{E} \ll 1$ and the Langevin function is very nearly linear. A power series expansion shows that $L(\beta p \mathcal{E}) \rightarrow \frac{1}{3}\beta p \mathcal{E}$ as $\beta p \mathcal{E}$ becomes small and in this limit $P = \frac{1}{3}n\beta p^2 \mathcal{E}$. The solid is now a linear dielectric and if local field effects can be neglected the dipolar contribution to the susceptibility is given by

$$\chi = \frac{n\beta p^2}{3\epsilon_0}. \tag{10-19}$$

In this limit the susceptibility varies as $1/T$. At high temperature few dipoles are aligned with the field and $\chi$ is small.

Equation 10-18 is more appropriate for gases and liquids than for solids. Internal interactions in a solid limit dipole orientations to a few directions, with high potential energy barriers between. The effect is illustrated in Fig. 10-6. Above the melting point the molecules can rotate freely so the dielectric constant is large and proportional to $1/T$. Below the melting point rotation is hindered so the dielectric constant is much smaller and its temperature dependence is not as strong.

## 10.2 FERROELECTRICS AND PIEZOELECTRICS

*Ferroelectric Solids.* A small number of solids, called pyroelectrics, may be polarized even in the absence of an external field. As the name implies, the magnitude of the spontaneous polarization in these materials depends on the temperature. Ferroelectrics are pyroelectrics that, in addition to spontaneous polarization, also display electric hysteresis, a phenomenon to be explained later.* Some ferroelectrics and their properties are listed in Table 10-3.

---

*The word ferroelectric is used because these solids are electrical counterparts to ferromagnetic materials, which display spontaneous magnetization, not because they contain iron.

At high temperatures, most ferroelectrics behave like the dielectrics discussed in the last section. They are polarized only in an external field and their polarizations vanish when the field is turned off. These high-temperature phases are labeled paraelectric. Some ferroelectrics have no paraelectric phase, the temperature required to reach such a phase presumably being above the melting point. The transition between a ferroelectric and paraelectric phase occurs at a well-defined temperature and is usually accompanied by a change in atomic structure.

Crystalline barium titanate is an important example. Above 393 K it is paraelectric and has the perovskite structure, with the unit cell shown in Fig. 10-7a. The lattice is simple cubic, with $Ba^{2+}$ ions at cube corners, $O^{2-}$ ions at face centers, and $Ti^{4+}$ ions at body centers. Below 393 K it has three ferroelectric phases. Between 278 and 393 K the lattice is tetragonal, formed by a slight elongation along one of the cubic fourfold axes. See Fig. 10-7b. Between 180 and 273 K the lattice is orthorhombic and below 180 K it is trigonal. Changes in the lattice are extremely small. In going from cubic to tetragonal, for example, titanium ions are displaced by about 0.13 Å, compared to a cube edge slightly greater than 4 Å.

Below 393 K, spontaneous polarization in barium titanate arises from a coupling between ionic and electronic dipole moments. A displacement of a titanium ion during the course of its vibratory motion creates an ionic dipole moment in the direction of the displacement and the associated electric field causes some oxygen ions to become polarized in the opposite direction, as shown in Fig. 10-7c. If the titanium displacement is along a $\langle 100 \rangle$ direction, the field produced by oxygen ions tends to increase titanium displacements and titanium equilibrium sites move away from cube centers. Both the local field at oxygen

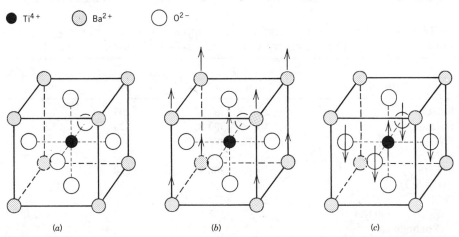

FIGURE 10-7 Unit cells of barium titanate. (a) The cubic cell of the paraelectric state. (b) The tetragonal cell of the highest temperature ferroelectric state. Ba and Ti ions are displaced relative to O ions as indicated by arrows and the cell is slightly elongated along the vertical axis. (c) Arrows indicate the direction of the local field at Ti and O ions for the ferroelectric state.

sites and the oxygen dipole moments increase and the process continues. Polarization does not increase indefinitely because anharmonic elastic forces limit the titanium displacement. The end result is a change in structure and a net dipole moment.

For each ferroelectric phase there is a direction of easy polarization and the net dipole moment of a unit cell is aligned with this direction. For the tetragonal phase it is along a $\langle 100 \rangle$ direction,* for the orthorhombic phase it is along a $\langle 110 \rangle$ direction, and for the trigonal phase it is along a $\langle 111 \rangle$ direction.

For most paraelectric phases near the transition temperature the dielectric constant as a function of temperature has the form

$$K = \frac{C}{T - T_c}, \qquad (10\text{-}20)$$

where $C$ and $T_c$ are parameters that vary from solid to solid. Equation 10-20 is called the Curie-Weiss law for ferroelectrics, a name taken from a similar equation that is valid for ferromagnetic materials. $C$ is called the Curie constant and $T_c$ is called the Curie temperature of the solid. For many ferroelectrics $T_c$ nearly coincides with the transition temperature but for others it is a few degrees less. Some values for $C$ and $T_c$ are given in Table 10-3. The validity of the law for a given material is checked by plotting $1/K$ as a function of $T$. If a straight line results, the law holds and values for $C$ and $T_c$ can be found from the slope and intercept.

*Ferroelectric Domains and Hysteresis.* When the temperature of a ferroelectric is lowered so the material passes from the paraelectric to a ferroelectric phase in the absence of an applied field, the net dipole moment is nearly zero even though each unit cell is spontaneously polarized. As depicted schematically in

**TABLE 10-3** Spontaneous Polarization $P_s$ (at temperature in parentheses), Curie Point $T_c$, and Curie Constant $C$ for Selected Ferroelectrics

| Material | $P_s$ ($\mu$C/cm$^2$) | | $T_c$ (K) | $C$ (K) |
|---|---|---|---|---|
| BaTiO$_3$ | 26 | (300 K) | 293 | $1.6 \times 10^5$ |
| PbTa$_2$O$_6$ | 10 | (300 K) | 533 | $1.5 \times 10^5$ |
| KNO$_3$ | 6.3 | (397 k) | 397 | 4300 |
| NaNO$_3$ | 6.4 | (416 K) | 433 | 5000 |
| NH$_4$HSO$_4$ | 0.8 | (200 k) | 270 | — |
| (NH$_4$)$_2$SO$_4$ | 6.4 | (const) | −50 | — |
| SbSI | 25 | (273 K) | 295 | — |
| NaKC$_4$H$_4$O$_6$ · 4H$_2$O (Rochelle salt) | 0.25 | (278 K) | 24 | 2240 |
| C(NH$_2$)$_3$Al(SO$_4$)$_2$ · 6H$_2$O | 0.35 | (300 K) | — | — |
| (NH$_2$CH$_2$COOH)$_3$H$_2$BeF$_4$ | 3.2 | (300 K) | 70 | 2350 |

*Source:* J. C. Burfoot, *Ferroelectrics* (Princeton, NJ: Van Nostrand, 1967).

---

*Directions are specified here as if the unit cell were cubic.

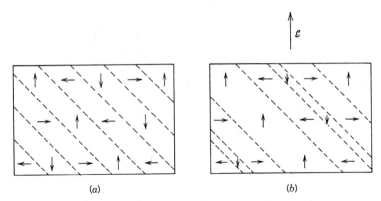

**FIGURE 10-8**  Schematic representation of ferroelectric domains in a barium titanate sample. Arrows show the directions of spontaneous polarization. (a) No external field is applied and the net polarization is zero. (b) An applied field causes domains with polarization in the direction of the field to increase in size and other domains to decrease. New domains with polarization along the field may also be created.

Fig. 10-8a, the sample can be divided into domains of macroscopic size such that the polarization is uniform in each domain but is in different directions in neighboring domains. Domain walls, which separate one domain from its neighbors, are typically a few unit cells thick. Outer domain surfaces can be observed using polarized light or by sprinkling charged powder on the sample surface.

The dipole moment of a domain with polarization **P** and volume $V$ is **P**$V$ and the total dipole moment of the sample vanishes if domain volumes are such that $\Sigma \mathbf{P}_i V_i = 0$, where the sum is over all domains. If an electric field is applied, domains with polarization roughly in the same direction as the field tend to grow in volume while domains with polarization in other directions shrink, as depicted in Fig. 10-8b. As a result, the sample acquires a net dipole moment. The moment remains when the external field is removed.

The net polarization of a ferroelectric depends not only on the strength of the applied field but also on previous polarization conditions. This phenomenon, called electric hysteresis, is illustrated by Fig. 10-9, a plot of the net polarization as a function of applied field strength. Points to the right of the origin represent, for example, a field applied in the positive $z$ direction, while points to the left represent a field applied in the negative $z$ direction. Similarly, points above the origin represent polarization in the positive $z$ direction and points below represent polarization in the negative $z$ direction.

Assume the material is initially in an unpolarized ferroelectric phase and the applied field is zero. This condition is represented by the origin of the plot. As the field is increased in the positive $z$ direction, domains with polarization in that direction grow while others shrink and the net polarization increases, its value being given by points on the curve $OA$. Once $A$ is reached the sample is essentially a single domain and further increases in $P$ are slight. If the field is now decreased the polarization follows $BAC$ and is nearly constant as long as the applied field remains in the positive $z$ direction. At $C$ the field reverses direction and the polarization follows $CDEF$. Domains with polarization in the

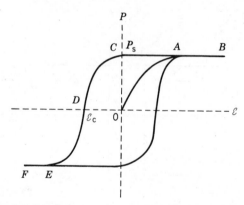

**FIGURE 10-9** Polarization as a function of an applied electric field for an ideal ferroelectric, showing hysteresis. $P_s$ is the saturation polarization and $\mathcal{E}_c$ is the coercive force. When an increasing electric field is applied to a ferroelectric with a net polarization that is initially zero, the polarization increases along $OAB$. When the field is decreased the polarization follows $BACDEF$. The electric field reverses at $C$ but the polarization does not reverse until $D$.

new direction of the field are created and these grow as the magnitude of the field increases. At first most of the sample volume is polarized in the old field direction: the net polarization does not reverse direction until $D$ is reached.

Hysteresis curves for real samples usually depart from the ideal form shown in the figure. The chief reason is that defects in the sample inhibit changes in domain sizes. Typically, $BAC$ is not horizontal and the loop sides are less steep than shown.

A hysteresis loop is characterized by the spontaneous polarization $P_s$, the value of the polarization at $C$, and by the coercive force $\mathcal{E}_c$, the magnitude of the field at $D$. Spontaneous polarizations for some ferroelectrics are reported in Table 10-3.

*Piezoelectric Solids.* Ionic equilibrium positions change when stress is applied to a sample and, in some materials, polarization results. Stress-induced polarization is known as the piezoelectric effect and materials for which it occurs are called piezoelectrics. These materials also exhibit the inverse effect: an applied electric field causes the sample to change size or shape. All ferroelectrics are piezoelectrics, many while in paraelectric phases. Some piezoelectrics, such as quartz, are not ferroelectric.

Figure 10-10 shows how strain-induced polarization can come about in a nonferroelectric. The unit cell in the figure has four ions, a single $A^{3-}$ ion at the center of the pattern with three $B^+$ ions around it. When the sample is not strained the three dipole moments cancel and the net polarization vanishes. When the sample is strained uniaxially, the ions move to the configuration shown in ($b$) and the net dipole moment does not vanish. Clearly the geometry of the cell is important for piezoelectricity. Unit cells of piezoelectric crystals cannot have an inversion center, for example.

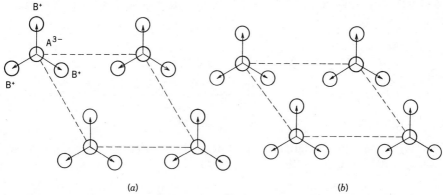

**FIGURE 10-10** A model of a piezoelectric solid. One negative and three positive ions form three dipole moments. (a) In the unstrained state the moments cancel. (b) When strained, the sample has a net dipole moment, in this case along the axis of uniaxial strain.

For a piezoelectric, polarization and strain are both proportional to the electric field and to the applied stress. Consider a sample subjected to uniaxial stress $T$ and electric field $\mathcal{E}$, in the same direction. We suppose this produces polarization $P$ and uniaxial strain $e$ along the same axis. The polarization is given by

$$P = \chi \epsilon_0 \mathcal{E} + dT \qquad (10\text{-}21)$$

and the strain by

$$e = sT + d\mathcal{E}, \qquad (10\text{-}22)$$

where $d$ is called the polarization strain constant and $s$ is an elastic compliance constant, the reciprocal of an elastic constant. Note that the same constant $d$ appears in both equations. For quartz $d = 2.3 \times 10^{-12}$ m/V while for barium titanate $d = 3 \times 10^{-10}$ m/V when the applied stress is in the [100] direction.

**EXAMPLE 10-3** Consider a 0.30-mm-thick wafer of single-crystal quartz with its faces parallel to cubic unit cell faces. How much should the wafer be compressed to obtain a 350-mV potential difference between its faces? The dielectric constant of quartz is 4.5 and the appropriate compliance constant is $s = 0.941 \times 10^{-11}$ m²/N. Assume the electric field and polarization are uniform and normal to the faces.

**SOLUTION** Since the fractional compression is so small, we use the original thickness to calculate the desired electric field. It is $350 \times 10^{-3}/0.30 \times 10^{-3} = 1170$ V/m. The field is generated by polarization charge on the faces of the wafer and is given by $\mathcal{E} = -P/\epsilon_0$. Substitute $P = -\epsilon_0 \mathcal{E}$ into

Eq. 10-21 and solve for $T$: $T = -(1 + \chi)\epsilon_0 \mathcal{E}/d = -K\epsilon_0 \mathcal{E}/d$. Substitute for $T$ in Eq. 10-22 and solve for $e$:

$$e = \frac{-sK\epsilon_0 + d^2}{d}\mathcal{E}$$

$$= \frac{-0.941 \times 10^{-11} \times 4.5 \times 8.85 \times 10^{-12} + (2.3 \times 10^{-12})^2}{2.3 \times 10^{-12}} \times 1170$$

$$= -1.9 \times 10^{-7}.$$

This is the fractional change in thickness. The wafer must be compressed by $1.9 \times 10^{-7} \times 0.30 \times 10^{-3} = 5.6 \times 10^{-11}$ m. ∎

Both the direct and inverse piezoelectric effect are widely used in strain gauges, transducers, and other devices. A piezoelectric strain gauge is bonded to a sample, then the potential difference across it is used to determine the extent of sample strain. An ac potential difference applied to a piezoelectric transducer, bonded to a sample, causes it to oscillate mechanically and produce sound waves in the sample.

## 10.3   ELECTROMAGNETIC WAVES IN SOLIDS

*Electromagnetic Waves.*   We are chiefly interested in ultraviolet frequencies and below, for which wavelengths are much longer than interatomic distances and diffraction effects are not important. For these waves a macroscopic description is adequate.

Maxwell's equations for the electromagnetic field in a nonmagnetic dielectric medium with free charge density $\rho$ and free current density $\mathbf{J}$ are

$$\nabla \cdot \mathcal{E} = \frac{\rho}{\epsilon_0} - \frac{\nabla \cdot \mathbf{P}}{\epsilon_0}, \tag{10-23}$$

$$\nabla \cdot \mathbf{B} = 0, \tag{10-24}$$

$$\nabla \times \mathcal{E} = -\frac{\partial \mathbf{B}}{\partial t}, \tag{10-25}$$

and

$$\nabla \times \mathbf{B} = \mu_0 \mathbf{J} + \mu_0\epsilon_0 \frac{\partial \mathcal{E}}{\partial t} + \mu_0 \frac{\partial \mathbf{P}}{\partial t}, \tag{10-26}$$

where $\mu_0$ is the permeability of free space, $4\pi \times 10^{-7}$ H/m. Note that $-\nabla \cdot \mathbf{P}$ and $\partial \mathbf{P}/\partial t$ enter as a charge and current densities, respectively. If the charges of an oscillating dipole are in motion, they contribute what is called a polarization current.

Suppose a sinusoidal plane wave with angular frequency $\omega$ and propagation vector $\mathbf{s}$ travels in a linear dielectric with scalar permittivity $\epsilon$. Assume no free charge or current are present, so $\rho$ and $\mathbf{J}$ are both zero. If complex exponentials are used to represent the electric and magnetic induction fields,

$$\mathcal{E}(\mathbf{r}, t) = \mathcal{E}_0 e^{i(\mathbf{s} \cdot \mathbf{r} - \omega t)} \tag{10-27}$$

and

$$\mathbf{B}(r, t) = \mathbf{B}_0 e^{i(\mathbf{s} \cdot \mathbf{r} - \omega t)}, \tag{10-28}$$

respectively. Physical fields are represented by the real parts of these expressions. When Eqs. 10-27, 10-28, and the relationship $\mathbf{P} = \chi \epsilon_0 \mathcal{E}$ are substituted into Maxwell's equations the results are

$$\mathbf{s} \cdot \mathcal{E}_0 = 0, \tag{10-29}$$

$$\mathbf{s} \cdot \mathbf{B}_0 = 0, \tag{10-30}$$

$$\mathbf{s} \times \mathcal{E}_0 = \omega \mathbf{B}_0, \tag{10-31}$$

and

$$\mathbf{s} \times \mathbf{B}_0 = -\omega \mu_0 \epsilon \mathcal{E}_0, \tag{10-32}$$

where $\epsilon = (1 \times \chi)\epsilon_0$. The first two equations indicate that both $\mathcal{E}_0$ and $\mathbf{B}_0$ are normal to the direction of propagation whereas the third and fourth indicate that $\mathcal{E}_0$ and $\mathbf{B}_0$ are perpendicular to each other.

Suppose the wave travels in the positive $z$ direction with its electric field along the $x$ axis and its magnetic field along the $y$ axis. Since $\mathbf{s}$ is in the $z$ direction, the first two of Maxwell's equations are identically satisfied. The third and fourth become $s\mathcal{E}_0 = \omega B_0$ and $sB_0 = \omega \mu_0 \epsilon \mathcal{E}_0$, respectively. They can be solved readily for $s$ as a function of $\omega$ and for the ratio of the field amplitudes:

$$s = \omega \sqrt{\mu_0 \epsilon}. \tag{10-33}$$

and

$$\frac{B_0}{\mathcal{E}_0} = \sqrt{\mu_0 \epsilon}. \tag{10-34}$$

$P = \chi \epsilon_0 \mathcal{E}$ is valid even if the electric field and polarization are not in phase, but the susceptibility $\chi$ is a complex number. If $\chi = |\chi| e^{i\phi}$, then $P = \chi \epsilon_0 \mathcal{E}_0 e^{i(sz - \omega t)} = |\chi| \epsilon_0 \mathcal{E}_0 e^{i(sz - \omega t + \phi)}$, where $\phi$ is the phase of $P$ relative to that of $\mathcal{E}$. Since $\epsilon = 1 + \chi$, the permittivity is also complex and, as Eq. 10-33 shows, so is the propagation constant.

The amplitude of a wave with a complex propagation constant decreases exponentially with distance into the solid. Let $s'$ and $s''$ be the real and imaginary parts of $s$, respectively, and substitute $s = s' + is''$ into Eq. 10-27 to obtain

$$\mathcal{E} = \mathcal{E}_0 e^{-s''z} e^{i(s'z - \omega t)}. \tag{10-35}$$

Over a distance equal to $1/s''$ the wave amplitude decreases by the factor $1/e$ ($\approx 0.37$). If $1/s''$ is small compared to sample dimensions, little radiation is transmitted and the sample is opaque. On the other hand, the sample is transparent if $1/s''$ is large. The same sample may be transparent to waves of one frequency and opaque to waves of another.

Equation 10-33 can be used to find $s'$ and $s''$ in terms of the real and imagninary parts of $\epsilon$. Substitute $s = s' + is''$ and $\epsilon = \epsilon' + i\epsilon''$ into that equation, then equate the real parts of the two sides and do the same for the imaginary parts. The two resulting equations yield

$$s'^2 = \tfrac{1}{2}\mu_0\epsilon'\omega^2\left[1 \pm \left(1 + \frac{\epsilon''^2}{\epsilon'^2}\right)^{1/2}\right]$$  (10-36)

and

$$s''^2 = \tfrac{1}{2}\mu_0\epsilon'\omega^2\left[\pm\left(1 + \frac{\epsilon''^2}{\epsilon'^2}\right)^{1/2}\right].$$  (10-37)

Signs are chosen so both $s'$ and $s''$ are real. If $\epsilon$ is real and positive, the propagation constant is real and has the value $s = \sqrt{\mu_0\epsilon}\omega$. Fields propagate in such a sample with undiminished amplitude. If $\epsilon$ is real and negative then $s' = 0$ and $s'' = \sqrt{-\mu_0\epsilon}\omega$. Fields do not propagate but rather oscillate with amplitudes that diminish with distance into the sample. If $\epsilon$ is complex then $s$ is also complex and fields propagate but are attenuated.

We now show the connection between the imaginary part of the permittivity and the rate at which electromagnetic energy is dissipated in a solid. The calculation is carried out in terms of the Poynting vector $\mathbf{\Pi} = (1/\mu_0)\mathcal{E} \times \mathbf{B}$, a vector in the direction of wave propagation with magnitude equal to the rate per unit area at which energy crosses a plane normal to the direction of propagation. For the wave we are discussing, the magnitude of the Poynting vector is $\Pi = \mathcal{E}B/\mu_0$.

Consider two planes, each of area $A$ and both normal to the direction of propagation, one at $z$ and the other at $z + \Delta z$. The rate at which energy is removed from the wave in the region between is $A[\Pi(z) - \Pi(z + \Delta z)]$, or $-A(d\Pi/dz)\,\Delta z$ in the limit of small $\Delta z$. The rate per unit volume is $-d\Pi/dz$.

The real parts of Eqs. 10-27 and 10-28 must be used to calculate $\Pi$. Since the real part of $\mathcal{E}$ is $\mathcal{E}_0 e^{-s''z}\cos(s'z - \omega t)$ and the real part of $B$ ($=s\mathcal{E}/\omega$) is $(\mathcal{E}_0/\omega)e^{-s''z}[s' \cos(s'z - \omega t) - s'' \sin(s'z - \omega t)]$, the magnitude of the Poynting vector is $(\mathcal{E}_0^2/\mu_0\omega)e^{-2s''z}[s' \cos^2(s'z - \omega t) - s'' \sin(s'z - \omega t)\cos(s'z - \omega t)]$. For some values of $t$ this is positive while for others it is negative, indicating that energy flows back and forth between solid and field. Only energy that flows to the solid and is not returned is dissipated, so we use the average of the Poynting vector over a cycle to calculate the rate of dissipation. Since the average of $\cos^2(s'z - \omega t)$ is $\tfrac{1}{2}$ and the average of $\cos(s'z - \omega t)\sin(s'z - \omega t)$ is 0, the average of $\Pi$ is $\tfrac{1}{2}(s'/\mu_0\omega)\mathcal{E}_0^2 e^{-2s''z}$ and the rate of energy dissipation per unit volume is $(-d\Pi/dz)_{ave} = (s's''/\mu_0\omega)\mathcal{E}_0^2 e^{-2s''z}$.

The product of Eqs. 10-36 and 10-37 yields $s's'' = \tfrac{1}{2}\mu_0|\epsilon''|\omega^2$, so the rate of energy dissipation per unit volume is also given by

$$-\left[\frac{d\Pi}{dz}\right]_{ave} = \tfrac{1}{2}\omega|\epsilon''|\mathcal{E}_0^2 e^{-2s''z},$$  (10-38)

a quantity that is proportional to $\epsilon''$. Losses occur if $\epsilon$ is complex and, as the discussion following Eq. 10-37 shows, the wave amplitude is attenuated. If $\epsilon$ is real and negative, attenuation occurs without energy loss. The permittivity is often written $|\epsilon|e^{i\delta}$, where $\tan \delta = \epsilon''/\epsilon'$. The phase angle $\delta$ is called the loss angle and is tabulated in some handbooks.

*Optical Constants.* Recall that the propagation constant is related to the permittivity by $s = \omega\sqrt{\mu_0\epsilon}$, or if $c = 1/\sqrt{\mu_0\epsilon_0}$ is used, by $s = (\omega/c)\sqrt{\epsilon/\epsilon_0}$. The relationship is often written

$$s = \frac{\omega}{c}(n + i\kappa),  \tag{10-39}$$

where $n$ and $\kappa$ are the real and imaginary parts, respectively, of $\sqrt{\epsilon/\epsilon_0}$. The first of these is called the index of refraction and the second is called the extinction coefficient. Sometimes the combination $n + i\kappa$ is referred to as the complex index of refraction.

Since $\epsilon/\epsilon_0 = (n + i\kappa)^2 = (n^2 - \kappa^2) + 2in\kappa$,

$$\frac{\epsilon'}{\epsilon_0} = n^2 - \kappa^2  \tag{10-40}$$

and

$$\frac{\epsilon''}{\epsilon_0} = 2n\kappa .  \tag{10-41}$$

These equations yield

$$2\epsilon_0 n^2 = \epsilon' \pm (\epsilon'^2 + \epsilon''^2)^{1/2}  \tag{10-42}$$

and

$$2\epsilon_0 \kappa^2 = -\epsilon' \pm (\epsilon'^2 + \epsilon''^2)^{1/2},  \tag{10-43}$$

where the signs are chosen to make $n$ and $\kappa$ real. If $\epsilon' = \epsilon_0$ and $\epsilon'' = 0$, as for propagation in empty space, then $n = 1$ and $\kappa = 0$. For propagation in a solid, $n$ is larger than 1 and $\kappa$ is larger than 0. We will encounter a situation for which $\epsilon'$ is negative and $\epsilon''$ vanishes. Then the lower signs must be used and $n = 0$ while $\kappa = \sqrt{-\epsilon'/\epsilon_0}$.

An electromagnetic wave propagates with phase velocity $v_p = \omega/s' = c/n$. Because an electric field polarizes a solid, the phase velocity is reduced from its value for a vacuum by the factor $1/n$. Since the speed of a wave changes as it passes into the sample from vacuum, the wave is refracted at the sample surface. Snell's law is used to measure the index of refraction if the transmitted wave can be traced.

The imaginary part of the propagation constant can be measured directly if the solid is sufficiently transparent. A slab of thickness $z$, say, is illuminated with monochromatic light. If $I_0$ is the intensity that enters the solid and $I$ is the intensity just before exiting on the other side, then $I = I_0 e^{-2s''z}$ and $s'' = (1/2z) \ln(I_0/I)$.

Incident and transmitted intensities are measured and adjusted to account for reflection at both front and back surfaces. The extinction coefficient is given by $\kappa = (c/2z\omega) \ln(I_0/I)$. The attenuation coefficient $\alpha = 2s''$ is often used instead of the extinction coefficient. Its reciprocal, called the skin depth, is a measure of the distance a wave penetrates into the sample.

The reflectivity of a sample is another optical constant of interest. Consider radiation incident normally on the surface of a sample, as depicted in Fig. 10-11, and place the $z$ axis along the normal, with $z = 0$ at the boundary. The incident wave has propagation constant $s_1 = \omega/c$ and travels in the positive $z$ direction. The transmitted wave travels in the same direction but, because it is in the sample, its propagation constant is $s_2 = (n + i\kappa)s_1$. The reflected wave travels in the negative $z$ direction and so has propagation constant $-s_1$.

Continuity of the electric field at $z = 0$ leads to

$$\mathscr{E}_{inc} + \mathscr{E}_{refl} = \mathscr{E}_{trans} \tag{10-44}$$

while continuity of the magnetic field leads to

$$\frac{s_1}{\omega} \mathscr{E}_{inc} - \frac{s_1}{\omega} \mathscr{E}_{refl} = \frac{s_2}{\omega} \mathscr{E}_{trans} , \tag{10-45}$$

where Eqs. 10-33 and 10-34 were used. Equation 10-44 is used to eliminate $\mathscr{E}_{trans}$ from Eq. 10-45 and obtain $\mathscr{E}_{refl}/\mathscr{E}_{inc} = -(s_2 - s_1)/(s_2 + s_1) = -(n + i\kappa - 1)/(n + i\kappa + 1)$. The reflectivity $R(\omega)$ of the sample is the ratio of the reflected intensity to the incident intensity:

$$R(\omega) = \frac{|\mathscr{E}_{refl}|^2}{|\mathscr{E}_{inc}|^2} = \frac{(n - 1)^2 + \kappa^2}{(n + 1)^2 + \kappa^2}. \tag{10-46}$$

Reflectivity values range from 0 to 1. $R(\omega) = 0$ and no reflection occurs if $n = 1$ and $\kappa = 0$. The reflectivity approaches 1 as either $n$ or $\kappa$ becomes large.

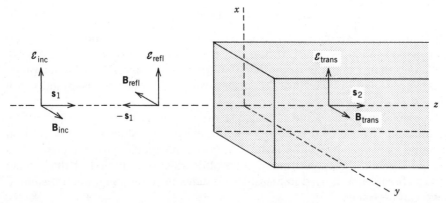

**FIGURE 10-11** Reflection and transmission of an electromagnetic wave at the surface of a solid. The surface is in the $xy$ plane and the wave is incident from the left along the $z$ axis. Electric fields $\mathscr{E}_{inc}$, $\mathscr{E}_{refl}$, and $\mathscr{E}_{trans}$ for the incident, reflected, and transmitted waves, respectively, are shown for some instant of time. Magnetic induction fields and propagation vectors are also shown.

**EXAMPLE 10-4**   Radiation with angular frequency $\omega = 7.2 \times 10^{12}$ rad/s is incident on a 0.020-mm-thick sample of cadmium sulfide, with index of refraction 11.7 and extinction coefficient 8.5. (a) What is the speed of the wave in the sample? (b) What is the wavelength of the wave before and after it enters the sample? (c) As a fraction of the intensity entering the sample what is the intensity just before exiting? (d) What are the real and imaginary parts of the susceptibility at this frequency?

**SOLUTION**   (a) The phase velocity is $v_p = c/n = 3.0 \times 10^8/11.7 = 2.6 \times 10^7$ m/s. (b) For the wave in vacuum $\lambda = 2\pi/s' = 2\pi c/\omega = 2\pi \times 3.0 \times 10^8/7.2 \times 10^{12} = 2.6 \times 10^{-4}$ m. For the wave in the solid $\lambda = 2\pi c/\omega n = 2\pi \times 3.0 \times 10^8/7.2 \times 10^{12} \times 11.7 = 2.2 \times 10^{-5}$ m. (c) $I/I_0 = e^{-2s''z} = e^{-2(\omega/c)\kappa z} = \exp[-2(7.2 \times 10^{12}/3.0 \times 10^8) \times 8.5 \times 2.0 \times 10^{-5}] = 2.8 \times 10^{-4}$. (d) $\epsilon'/\epsilon_0 = n^2 - \kappa^2 = 11.7^2 - 8.5^2 = 64$. $\epsilon''/\epsilon_0 = 2n\kappa = 2 \times 11.7 \times 8.5 = 200$.   ∎

## 10.4   FREQUENCY-DEPENDENT POLARIZABILITIES

Figure 10-12a schematically depicts various contributions to the real part of the permittivity of a typical insulator or semiconductor as functions of frequency. Similarly, (b) depicts contributions to absorption. Data for a real solid may show much more structure than these plots. On the other hand, some of the structure shown may not actually be present for a particular solid. Not all solids contain dipolar molecules or ions, for example, and free carrier contributions are significant only for semiconductors (and metals).

Valence and conduction band electrons typically contribute at ultraviolet frequencies and below, while core electrons contribute at ultraviolet frequencies and above. Dipolar molecules, when they exist in a solid, usually contribute only at microwave frequencies and below, while ions typically contribute up to an infrared frequency. The remainder of the chapter is devoted to understanding graphs such as these.

*An Oscillating Dipole.*   We start with a single dipole in an oscillating local electric field and examine the frequency dependence of its polarizability. Suppose the negative particle has mass $m_-$ and is at $\mathbf{r}_-(t)$ while the positive particle has mass $m_+$ and is at $\mathbf{r}_+(t)$. Suppose further that, for small displacements from equilibrium, the force of one particle on the other is linear in the relative displacement and can be characterized by the force constant $\gamma$. If $Q$ is the magnitude of either charge, then the equations of motion for the particles are

$$m_- \frac{d^2\mathbf{r}_-}{dt^2} = -\gamma(\mathbf{r}_- - \mathbf{r}_+) - Q\mathcal{E}_{loc}. \tag{10-47}$$

and

$$m_+ \frac{d^2\mathbf{r}_+}{dt^2} = -\gamma(\mathbf{r}_+ - \mathbf{r}_-) - Q\mathcal{E}_{loc}. \tag{10-48}$$

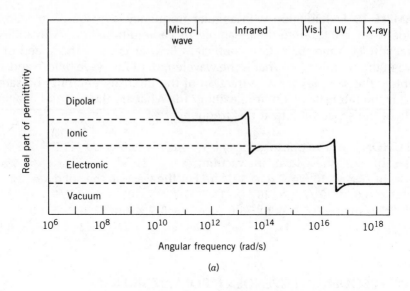

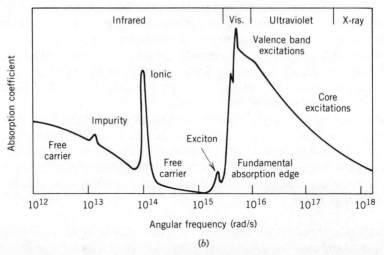

**FIGURE 10-12**   (a) Schematic diagram showing approximately the frequency ranges in which dipolar, electronic, and ionic contributions to the real part of the permittivity are significant. For all solids electrons contribute at frequencies in the ultraviolet and below. Ions in ionic and partially ionic solids contribute at frequencies in the infrared and below. Permanent dipoles in dipolar materials contribute at microwave frequencies and below. (b) Absorption of radiation in a hypothetical semiconductor. Below the visible region electrons in conduction bands and holes in valence bands (nearly free carriers) absorb radiation. Impurities may also absorb. The phonon system of an ionic or partially ionic solid absorbs at infrared frequencies. Interband absorption occurs near the visible region and above: valence band electrons are excited at the fundamental absorption edge and just above, while core electrons are excited at higher frequencies. For an insulator the free carrier contribution is small and the fundamental absorption edge is at a higher frequency than for a semiconductor.

The dipole moment is given by $\mathbf{p}(t) = Q(\mathbf{r}_+ - \mathbf{r}_-)$. To find the differential equation it satisfies, multiply Eq. 10-47 by $Qm_+$ and Eq. 10-48 by $Qm_-$, then subtract the first from the second. The result is

$$\frac{d^2\mathbf{p}}{dt^2} + \frac{\gamma}{M}\mathbf{p} = \frac{Q^2}{M}\mathcal{E}_{loc}, \qquad (10\text{-}49)$$

where $M = m_+m_-/(m_+ + m_-)$ is the reduced mass of the system.

We can include energy dissipation by adding a force term proportional to $d\mathbf{p}/dt$. Such a term might come about through interactions of the oscillating dipole with phonons or defects. We also note that $\gamma = M\omega_0^2$, where $\omega_0$ is the natural angular frequency of the dipole. When these changes are made, Eq. 10-49 becomes

$$\frac{d^2\mathbf{p}}{dt^2} + \rho\frac{d\mathbf{p}}{dt} + \omega_0^2\mathbf{p} = \frac{Q^2}{M}\mathcal{E}_{loc}, \qquad (10\text{-}50)$$

where $\rho$ is a constant with units of reciprocal time.

To gain some understanding of the role played by the dissipation term, suppose the local field is removed and the dipole moment is initially $\mathbf{p}_0$. Equation 10-50 then has the solution

$$\mathbf{p} = \mathbf{p}_0 e^{-\rho t/2} e^{-i\omega t}, \qquad (10\text{-}51)$$

where $\omega^2 = \omega_0^2 - \frac{1}{4}\rho^2$. If $\omega^2 > 0$ the dipole oscillates but the amplitude of oscillation decreases exponentially with time. The quantity $2/\rho$ is a relaxation time that provides a measure of the time required for oscillations to damp out. If relaxation is produced by phonon collisions, we expect $\rho$ to increase with temperature.

Now return to a dipole in an oscillating local field. Take $\mathcal{E}_{loc} = \mathcal{E}_0 e^{-i\omega t}$ and $\mathbf{p} = \alpha\mathcal{E}_{loc}$, then solve Eq. 10-50 for $\alpha$:

$$\alpha(\omega) = \frac{Q^2/M}{\omega_0^2 - \omega^2 - i\omega\rho} = \frac{Q^2}{M}\frac{\omega_0^2 - \omega^2 + i\omega\rho}{(\omega_0^2 - \omega^2)^2 + \omega^2\rho^2}. \qquad (10\text{-}52)$$

The last form follows from the first when both numerator and denominator are multiplied by $\omega_0^2 - \omega^2 + i\omega\rho$. Because energy dissipation introduces a phase difference between the local field and dipole moment, $\alpha$ is complex. Expressions for the real part $\alpha'$ and the imaginary part $\alpha''$ can be obtained from Eq. 10-52 by inspection.

As Fig. 10-13a shows, $\alpha'$ has a sharp positive peak when the dipole is in resonance with the local field. If $\rho = 0$ the resonance angular frequency is $\omega_0$ but for $\rho \neq 0$ it is shifted somewhat. The maximum value of $\alpha'$, given by $Q^2/M\omega^2\rho$, is limited by dissipation.

The real part of the polarizability rises gently from its static value until $\omega$ is close to the resonance frequency, then it increases sharply. For frequencies less than the resonance frequency $\alpha'$ is positive, indicating that the local field and dipole are in phase if $\rho = 0$. The frequency range for which $\alpha'$ is nearly constant

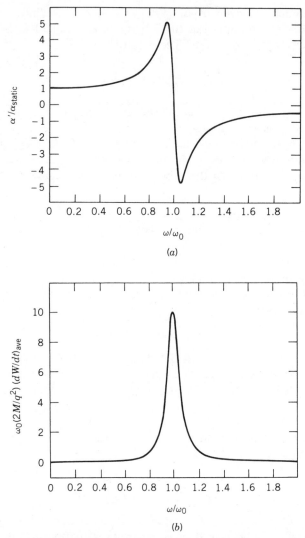

**FIGURE 10-13** (a) The real part of the polarizability ratio $\alpha/\alpha_{static}$ for a single dipole with natural frequency $\omega_0$ and relaxation constant $\rho = 0.1\omega_0$, as a function of $\omega/\omega_0$. Resonance occurs at the positive peak, slightly below the natural frequency. At low frequencies the ratio approaches 1; at high frequencies it is negative and approaches 0. (b) The dissipation function $\omega_0(2M/Q^2)(dW/dt)_{ave}$ for a single dipole. The rate of energy dissipation is a maximum for $\omega = \omega_0$.

or increases slowly is called the region of normal dispersion. The region of strong frequency dependence is called the region of anomalous dispersion.

Beyond resonance $\alpha'$ becomes negative, indicating that the local field and dipole are 180° out of phase if $\rho = 0$. Finally, $\alpha'$ tends toward zero in proportion to $1/\omega^2$ as the frequency is increased. At high frequencies the particles cannot follow the field and the amplitude of the dipole moment is diminished.

Energy is continually transferred between the local field and dipole, sometimes in one direction and sometimes in the other. The rate $dW/dt$ at which the field does work on the dipole is given by the scalar product of the real parts of $d\mathbf{p}/dt$ ($= \alpha d\mathcal{E}_{loc}/dt$) and $\mathcal{E}_{loc}$. Only energy that is not returned to the field is dissipated, so the dissipation rate is calculated by averaging $dW/dt$ over a cycle. The result is

$$\left[\frac{dW}{dt}\right]_{ave} = \frac{1}{2}\,\alpha''\omega\mathcal{E}_0^2 = \frac{1}{2}\frac{Q^2}{M}\frac{\omega^2\rho}{(\omega_0^2 - \omega^2)^2 + \omega^2\rho^2}\mathcal{E}_0^2. \tag{10-53}$$

As Fig. 10-13$b$ shows, absorption occurs chiefly in the region of anomalous dispersion and maximum absorption occurs for $\omega \approx \omega_0$.

**EXAMPLE 10-5**   For an ion pair in NaCl, $\omega_0 = 3.1 \times 10^{13}$ rad/s and $M = 2.32 \times 10^{-26}$ kg. Take $\rho = 5.65 \times 10^{11}$ s$^{-1}$ and calculate the real part of the polarizability for $\omega = 0$, $\omega_0/10$, $0.99\omega_0$, $1.01\omega_0$, and $10\omega_0$. Also calculate the rate of energy dissipation when the local field has an amplitude of $1.5 \times 10^4$ V/m and an angular frequency of $\omega_0$.

**SOLUTION**   If $x = \omega/\omega_0$ and $u = \rho/\omega_0$ ($= 5.65 \times 10^{11}/3.1 \times 10^{13} = 1.82 \times 10^{-2}$), then the real part of $\alpha$ is given by

$$\alpha' = \frac{Q^2}{M\omega_0^2}\frac{1 - x^2}{(1 - x^2)^2 + u^2x^2} = \frac{(1.6 \times 10^{-19})^2}{2.32 \times 10^{-26} \times (3.1 \times 10^{13})^2}\frac{1 - x^2}{(1 - x^2) + u^2x^2}$$

$$= 1.15 \times 10^{-39}\frac{1 - x^2}{(1 - x^2)^2 + u^2x^2}.$$

For $x = 0$, $\alpha'$ is $1.15 \times 10^{-39}$ F $\cdot$ m$^2$; for $x = 0.1$, it is $1.16 \times 10^{-39}$ F $\cdot$ m$^2$; for $x = 0.99$, it is $3.18 \times 10^{-38}$ F $\cdot$ m$^2$; for $x = 1.01$, it is $-3.17 \times 10^{-38}$ F $\cdot$ m$^2$; and for $x = 10$, it is $-1.16 \times 10^{-41}$ F $\cdot$ m$^2$. For $\omega = \omega_0$, Eq. 10-53 reduces to

$$\left[\frac{dW}{dt}\right]_{ave} = \frac{1}{2}\frac{Q^2}{M\rho}\mathcal{E}_0^2 = \frac{1}{2}\frac{(1.60 \times 10^{-19})^2 \times (1.5 \times 10^4)^2}{2.32 \times 10^{-26} \times 1.82 \times 10^{-2}}$$

$$= 6.8 \times 10^{-3}\text{ J/m}^3 \cdot \text{s}. \qquad \blacksquare$$

## 10.5   ELECTRONIC POLARIZABILITY

*Quantum Interpretation.*   In an oscillating local field the electron system oscillates relative to the positive nuclei. Classically, $\mathbf{r}_-$ in Eq. 10-47 is interpreted as the center of the electron charge distribution in the region around a nucleus at $\mathbf{r}_+$. To obtain an accurate picture, however, we should carry out a quantum mechanical calculation of the polarizability. Electron wave functions change under the influence of the local field and, as a result, so do dipole moments. Resonance corresponds to an electron transition from one state to another and occurs when $\hbar\omega$ matches the difference in energy of two levels of the electron system. An equation similar to Eq. 10-52 is obtained for a single resonance but, in the quantum description, the frequency $\omega_0$ must be interpreted as $\Delta E/\hbar$, where $\Delta E$ is the separation of two levels.

To complete the quantum picture, we suppose electromagnetic radiation of frequency $\omega$ consists of photons, each with energy $\hbar\omega$. When the photon energy matches the energy difference of two electron levels a photon may be absorbed and, if one is, an electron undergoes a transition to a higher state. In the absence of dissipation the electron returns to its initial state with the emission of a photon. Dissipation occurs if some or all of the energy originally absorbed by the electron is given up to the phonon system or to impurities.

Many electron resonances contribute to the polarizability and we must sum their contributions. Since the wave functions associated with different transitions are different, resonances do not contribute equally. In addition, a state at the lower level must be occupied and a state at the upper level must be unoccupied. As a result, a factor called the oscillator strength and denoted by $f$ multiplies each term of the sum. The final expression for the polarizability is

$$\alpha(\omega) = \frac{e^2}{m} \sum_i f_i \frac{\omega_i^2 - \omega^2 + i\omega\rho_i}{(\omega_i^2 - \omega^2)^2 + \omega^2\rho_i^2}, \tag{10-54}$$

where the sum is over all resonances. The oscillator strength is defined so the factor in front contains the charge and mass of a single electron. In the remainder of this section we treat transitions from one band to another. Intraband transitions are discussed in the following section.

*Interband Transitions.*  For an insulator or semiconductor the interband resonance region begins when conditions are right for an electron to absorb a photon and jump the gap between the valence and conduction bands. The corresponding frequency is known as the fundamental absorption edge. For insulators the gap may be on the order of 10 eV wide, corresponding to an angular frequency on the order of $10^{16}$ rad/s, in the ultraviolet. For semiconductors gaps are from just below 1 eV to several eV. For many, including Si, Ge, and GaAs, the fundamental absorption edge is in the infrared. For some, such as GaP, CdS, and SiC, the fundamental absorption edge corresponds to visible light.

Crystal momentum as well as energy is conserved in an absorption process. If $\hbar\mathbf{k}_f$ is the crystal momentum of the final electron state, $\hbar\mathbf{k}_i$ is the crystal momentum of the initial electron state, and $\mathbf{s}$ is the propagation vector of the electromagnetic wave, then the oscillator strength vanishes unless $\mathbf{k}_f = \mathbf{k}_i + \mathbf{s}$. Since $\mathbf{s}$ for frequencies of interest is much shorter than any Brillouin zone dimension, we may neglect it and take $\mathbf{k}_f = \mathbf{k}_i$. A vertical line connects the initial and final states on a reduced zone band diagram and the transitions are called vertical transitions.

To find the lowest frequency interband resonance we must search the gap for the smallest energy difference with both states at the same $\mathbf{k}$. If the conduction band minimum occurs at the same $\mathbf{k}$ as the valence band maximum then $\omega_0 = E_g/\hbar$. If it does not then $\omega_0$ is somewhat greater. The two possibilities are illustrated in Fig. 10-14.

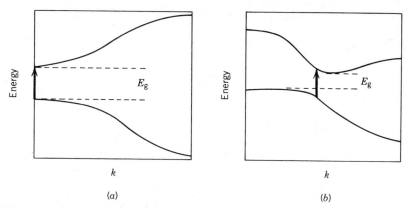

**FIGURE 10-14** Valence and conduction bands of two semiconductors. In (a) the minimum of the conduction band occurs at the same point in the Brillouin zone as the maximum of the valence band and the fundamental absorption edge is at angular frequency $\omega = E_g/\hbar$. GaAs, GaSb, InP, ZnS, and CdS are examples. In (b) the conduction band minimum and valence band maximum occur at different points and the fundamental absorption edge occurs at a frequency greater than $E_g/\hbar$. Si, Ge, AlSb, GaP, PbS are examples.

Many closely spaced resonances occur just beyond the fundamental absorption edge as valence band electrons are excited through greater energy differences by photons associated with higher frequency waves. In this region the absorption curve is punctuated by peaks. For example, Fig. 10-15 shows several in the absorption spectrum of silicon for $\omega$ between $10^{14}$ and $10^{15}$ rad/s. Density of states considerations show that peaks exist at frequencies such that $\nabla_k E_f(\mathbf{k}) = \nabla_k E_i(\mathbf{k})$, where $E_i$ is the initial electron energy and $E_f$ is the final electron energy. Frequencies associated with peaks are measured and used to verify electron band calculations.

At high frequencies, beyond the mid to high ultraviolet region, oscillator strengths are small and valence electron resonances no longer influence optical properties.

*Core Electrons.* Core electrons contribute to the polarizability at high frequencies. These electrons have energies that are ten to a thousand or more electron volts below the conduction band so, their resonance region extends from the far ultraviolet well into the x-ray region. Oscillator strengths are small, however; hence, so are core contributions to the polarizability. As a result, the index of refraction is nearly 1. The speed of an x-ray wave in a solid is nearly $c$.

Absorption does occur, however, and its study has aided greatly in understanding core states in solids. Absorption is also important for crystallography since it limits the penetration of x rays into a sample.

Synchrotron radiation, produced by high-speed electrons traveling in circular orbits, has proved extremely useful for studying absorption in the ultraviolet and x-ray regions. With the advent of particle storage rings and electron accel-

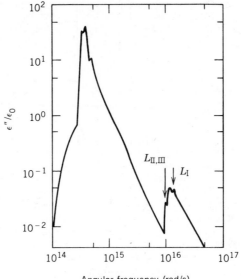

Angular frequency (rad/s)

**FIGURE 10-15** Optical absorption by silicon. Peaks on the left are due to the excitation of valence electrons; peaks on the right are due to the excitation of core electrons. Core transitions shown are from $L$ shell states. $K$ shell electrons are excited at frequencies beyond the right edge of the graph. (From F. C. Brown, "Ultraviolet Spectroscopy of Solids with the Use of Synchrotron Radiation" in *Solid State Physics* (H. Ehrenreich, F. Seitz, and D. Turnbull, Eds.), Vol. 29, p. 1; New York: Academic, 1974). Used with permission. Data are from H. R. Philipp and H. Ehrenreich; *Phys. Rev.* **129**: 1550, 1963, and C. Gahwiller and F. C. Brown, *Phys. Rev. B* **2**: 1918, 1970.)

erators such radiation has become fairly easy to obtain, although absorption experiments must, of course, be carried out at accelerator sites. Synchrotron radiation is favored over more conventional sources, such as hydrogen or deuterium discharge tubes, because an intense beam is obtained with a continuous distribution of frequencies from the near ultraviolet to the near x-ray regions.

Figure 10-15 shows the imaginary part of the dielectric constant of silicon, obtained from synchrotron radiation reflectivity measurements. The large peak to the left is due to excitation of valence electrons while the smaller peak to the right is due to excitation of core electrons. Note that the peak for core electron excitation is less than that for valence electron excitation by a factor of about $10^{-3}$.

Electrons may be ejected from the solid by high-energy photons. Their energies are measured and analyzed to obtain information about electron states. X-ray photoemission spectroscopy (XPS) is used to study core states and ultraviolet photoemission spectroscopy (UPS) is used to study valence band states. Photoemission techniques are also used for the study of surface states.

*Frequencies below the Fundamental Absorption Edge.* The discussion above implies that the polarizability has little structure for frequencies below the fundamental absorption edge but this is not true. If a crystal has several valence bands, for example, transitions from one to another may occur. Since the gap is small, the excitation frequency is low. Usually only a relatively few states are vacant, so these contributions are small. More important contributions come from other processes. We describe some here.

*Excitons.* When an electron jumps from a valence band state to a conduction band state it leaves behind a hole. The electron and hole are oppositely charged and may bind together to form an exciton. Figure 10-16 illustrates the process and the exciton contribution to an absorption spectrum. Binding is most probable if the transition is such that the electron and hole velocities match.

In some materials, such as $Cu_2O$, excitons are somewhat like hydrogen atoms, with the hole playing the role of a proton. Energy levels may be approximated by

$$E_n = E_c - \frac{Me^4}{2\hbar\epsilon^2 n^2},\qquad (10\text{-}55)$$

where $M = m_e m_h/(m_e + m_h)$ is the reduced mass of the electron-hole pair, $\epsilon$ is the static susceptibility of the solid, $E_c$ is the conduction band energy, and $n$ is an integer. Because $\epsilon > \epsilon_0$ these levels are much more shallow than hydrogen atomic levels. In different materials the lowest exciton level may be from a few millielectron volts to about 1 eV below the conduction band. As shown in Fig. 10-16, the gap contains a series of levels just below the conduction band. As a result, absorption spectra show a series of peaks just below the fundamental absorption edge.

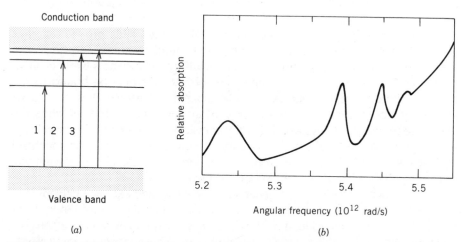

(a)                                                    (b)

**FIGURE 10-16**   (a) Exciton energy levels (not to scale). The lowest level is the ground state of the exciton; other levels correspond to excited states. Arrows indicate possible transitions. The longest corresponds to the fundamental absorption edge. These excitations might produce an absorption spectrum like that shown in (b).

*Phonon-Assisted Transitions.* A phonon-assisted transition occurs when a phonon is created simultaneously with the absorption of a photon. If the phonon frequency is $\omega_p$, then the resonance condition becomes $\Delta E/\hbar = \omega - \omega_p$. Since $\hbar\omega_p$ is always much less than the gap, $\Delta E/\hbar \approx \omega$. The phonon, however, substantially changes the electron crystal momentum. If $\mathbf{q}$ is the phonon wave vector then conservation of crystal momentum requires $\mathbf{k}_f = \mathbf{k}_i - \mathbf{q} + \mathbf{s}$ or, if $\mathbf{s}$ is ignored, $\mathbf{k}_f = \mathbf{k}_i - \mathbf{q}$.

Phonon-assisted transitions are particularly important when the valence band maximum and conduction band minimum are not at the same $\mathbf{k}$. Transitions between these states are possible only with phonon assistance and are called indirect transitions. They lead to absorption peaks at frequencies below the fundamental absorption edge. Since indirect processes require simultaneous absorption of a photon and creation of a phonon, absorption peaks are much lower than peaks for direct transitions.

Other phonon-assisted transitions are possible. In a solid at high temperature, with a large number of phonons, phonons may be absorbed as well as created. Conservation of crystal momentum then becomes $\mathbf{k}_f = \mathbf{k}_i + \mathbf{q}$. In addition, multiphonon processes may occur. In these events more than one phonon is created or absorbed or a phonon with one wave vector is created while one with another wave vector is absorbed. Absorption peaks for such processes are, of course, lower than those for single-phonon events.

Raman and Brillouin processes involve the direct interaction of electromagnetic radiation with atomic vibrations. Absorption or scattering of a photon is accompanied by the creation or loss of a phonon. The process is called a Brillouin process if the phonon is in an acoustical branch and a Raman process if it is in an optical branch. Energies of the incident and emitted photons are measured and the results are used to infer the phonon spectrum of the solid. This technique has proved to be highly successful in the study of surface phonon states.

*Color Centers.* Vacancies in ionic and partially ionic solids provide another absorption mechanism for frequencies below the fundamental absorption edge. If a negative ion is missing from its site the resulting vacancy acts like a positive charge and, if the solid contains excess electrons, one may be bound to the vacancy. There are several energy levels associated with each of these vacancies and resonance occurs when the energy of an incident photon matches the difference in energy of any two of them. Such vacancies are called color centers because in many cases absorption occurs in the visible region of the spectrum and results in the coloring of an otherwise colorless crystal. NaCl, for example, becomes yellow when color centers are formed.

Note that two ingredients are necessary: vacancies and excess electrons. Vacancies exist in all solids but excess electrons are normally not available. They are produced by bombarding a solid with x rays, $\gamma$ rays, neutrons, or electrons. They are also produced if a sample is heated in an atmosphere containing atoms that are ionized on entering the solid.

An electron is more loosely bound to a vacancy than an electron in an analogous atomic state, chiefly because the electron interacts with neighboring ions.

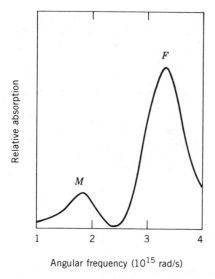

FIGURE 10-17 Typical absorption spectrum for $F$ and $M$ color centers in an ionic solid. An $F$ center consists of an electron trapped at a vacancy while an $M$ center is composed of two adjacent $F$ centers.

Energy level differences are about 2 eV in CsCl, RbCl, and KBr and about 5 eV in LiF. Thus resonance frequencies range from $3 \times 10^{15}$ rad/s to $8 \times 10^{15}$ rad/s, from the red-yellow portion of the visible spectrum well into the ultraviolet. Figure 10-17 shows a color center absorption spectrum.

Color centers consisting of vacancy aggregates may also form and, if they do, add much structure to the absorption spectrum of a sample. A single vacancy and its associated electron is called an $F$ center, after the German word for color, Farbe. Two adjacent $F$ centers are known as an $M$ center while three adjacent centers are called an $R$ center. Presumably a color center could be formed by a hole bound to a positive ion vacancy but no such center has been observed. $V_K$ centers, consisting of holes bound to a pair of adjacent negative ions, account for absorption in the same frequency range as $F$ centers. No vacancies are involved.

*Impurities.* Donor and acceptor impurities, with energy levels in the gap between valence and conduction bands, also contribute to the electronic polarizability below the fundamental absorption edge. An electron in an impurity state may be driven to the conduction band by absorption of a photon. Similarly, an electron from a valence band may be driven to an impurity level. In both cases the photon frequency is clearly less than that of the fundamental absorption edge.

## 10.6 FREE CARRIER EFFECTS

In this section we will be concerned with intraband transitions made by electrons in partially filled bands of semiconductors and metals. These electrons make what are called free carrier contributions to optical properties, although they are not strictly free.

*Electromagnetic Waves in the Presence of Free Carriers.* The intraband response of free carriers to an oscillating electric field can be described in terms of the electrical conductivity $\sigma$. For simplicity we consider an amorphous solid or cubic crystal, with a scalar conductivity: the current density $\mathbf{J}$ and field $\mathcal{E}$ are related by $\mathbf{J} = \sigma\mathcal{E}$. Since the field is oscillatory so is the current density. The conductivity is a function of frequency and is complex if the current density and field are not in phase.

Consider a plane electromagnetic wave traveling in the positive $z$ direction, with its electric field given by $\mathcal{E} = \mathcal{E}_0\hat{\mathbf{x}}e^{i(sz-\omega t)}$ and its magnetic field by $\mathbf{B} = B_0\hat{\mathbf{y}}e^{i(sz-\omega t)}$. Take the polarization to be $\mathbf{P} = \chi\epsilon_0\mathcal{E}$. Then Eqs. 10-25 and 10-26 become

$$s\mathcal{E}_0 = \omega B_0 \tag{10-56}$$

and

$$sB_0 = (i\mu_0\sigma + \omega\mu_0\epsilon)\mathcal{E}_0, \tag{10-57}$$

respectively. $B_0 = (s/\omega)\mathcal{E}_0$, from Eq. 10-56, is substituted into Eq. 10-57 to obtain

$$s^2 = \omega^2\mu_0\left(\epsilon + i\frac{\sigma}{\omega}\right). \tag{10-58}$$

This equation replaces Eq. 10-33 when the medium is conducting. According to Eq. 10-58 we may think of

$$\epsilon_{\text{eff}} = \epsilon + i\frac{\sigma}{\omega} \tag{10-59}$$

as the effective permittivity of the material. It replaces $\epsilon$ in calculations of the index of refraction, the extinction coefficient, and the reflectivity.

Free carriers oscillate in the field and, in combination with positive ions, form oscillating dipoles. They contribute $i\sigma/\omega\epsilon_0$ to the susceptibility, as the following derivation shows. The equation of motion for a single electron in an electric field $\mathcal{E}$ is

$$m\frac{d^2\mathbf{r}}{dt^2} = -e\mathcal{E} \tag{10-60}$$

and, if the electric field has the form $\mathcal{E}_0e^{-i\omega t}$, then $\mathbf{r} = (e/m\omega^2)\mathcal{E}$. The dipole moment is $\mathbf{p} = -e\mathbf{r} = -(e^2/m\omega^2)\mathcal{E}$. If there are $n$ free electrons per unit volume, their polarization is $\mathbf{P} = -(ne^2/m\omega^2)\mathcal{E}$ and their contribution to the susceptibility is $\chi = -ne^2/m\omega^2\epsilon_0$. On the other hand, the electron velocity is $\mathbf{v} = d\mathbf{r}/dt = -i\omega\mathbf{r} = -i(e/m\omega)\mathcal{E}$ and the current density is $\mathbf{J} = -ne\mathbf{v} = i(ne^2/m\omega)\mathcal{E}$, so the conductivity is $\sigma = ine^2/m\omega$. Thus $\chi = i\sigma/\omega\epsilon_0$, in agreement with Eq. 10-59.

*Optical Conductivity.* An expression for the frequency dependent conductivity can be obtained from the Boltzmann transport equation. Consider electrons in a single band and use the relaxation time approximation to write

$$\frac{\partial f}{\partial t} = \frac{e}{\hbar} \mathcal{E} \cdot \nabla_k f - \mathbf{v} \cdot \nabla f - \frac{f - f_0}{\bar{t}},$$ (10-61)

where $f(\mathbf{k}, \mathbf{r}, t)/\tau_s$ gives the concentration at $\mathbf{r}$ of electrons in a state with crystal momentum $\hbar\mathbf{k}$, $f_0$ is the Fermi-Dirac distribution function, and $\bar{t}$ is the relaxation time. Although we assume the temperature gradient vanishes, we must include the second term on the right because the electric field depends on $\mathbf{r}$. In principle the magnetic field also exerts a force, but it is small and we neglect it here.

Let $\mathcal{E} = \mathcal{E}_0 e^{i(sz-\omega t)}$ be a weak field along the $x$ axis and assume $f = f_0 + f_{10} e^{i(sz-\omega t)}$, where the second term is small compared to the first. Substitute this form into Eq. 10-61 and neglect the product $f_{10}\mathcal{E}$. Also make the substitution $\nabla_k f = (\partial f_0/\partial E)\nabla_k E = (\partial f_0/\partial E)\hbar\mathbf{v}$, then solve for $f_{10}$. The result is

$$f_{10} = \frac{e\mathcal{E}_0 v_x \bar{t}}{1 - i\omega\bar{t} + isv_z\bar{t}} \frac{\partial f_0}{\partial E}.$$ (10-62)

The ratio $sv_z\bar{t}/\omega\bar{t}$ is small compared to 1 so the last term in the denominator can be neglected. Then the $x$ component of the current density is

$$J = -\frac{e}{\tau_s} \sum v_x f_{10} e^{i(sz-\omega t)} = -\frac{e^2\mathcal{E}}{\tau_s} \sum \frac{\bar{t}v_x^2}{1 - i\omega\bar{t}} \frac{\partial f_0}{\partial E},$$ (10-63)

where the sum is over all states of the band. The factor that multiplies $\mathcal{E}$ is the conductivity:

$$\sigma(\omega) = -\frac{e^2}{\tau_s} \sum \frac{\bar{t}v_x^2}{1 - i\omega\bar{t}} \frac{\partial f_0}{\partial E}.$$ (10-64)

If the relaxation time is independent of $\mathbf{k}$, $1/(1 - i\omega\bar{t})$ can be factored from the sum and Eq. 10-64 can be written

$$\sigma(\omega) = \frac{\sigma_0}{1 - i\omega\bar{t}} = \frac{\sigma_0(1 + i\omega\bar{t})}{1 + \omega^2\bar{t}^2},$$ (10-65)

where $\sigma_0$ is the conductivity for $\omega = 0$. The real and imaginary parts of $\sigma$ are displayed explicitly in the second expression. Equation 10-65 remains a reasonable approximation even if the relaxation time depends on $\mathbf{k}$, provided $\bar{t}$ is replaced by an appropriate average.

For $\omega\bar{t}$ large, the real part of $\sigma$ is proportional to $1/\omega^2\bar{t}^2$ and the imaginary part is proportional to $1/\omega\bar{t}$. Both are small and we conclude free carriers are important for optical properties only if the angular frequency is on the order of $1/\bar{t}$ or less. At room temperature $\bar{t}$ is typically around $10^{-12}$ and the upper limit of free carrier influence is in the microwave or infrared portion of the spectrum. Over the frequency range for which free carriers are important, $\epsilon'$ is essentially the static permeability and $\epsilon''$ vanishes for most materials. An exception may occur if the solid is a partially ionic semiconductor. Ionic resonances, discussed in the next section, contribute to $\epsilon$ at infrared frequencies.

*Absorption.* The electric field accelerates carriers and they transfer energy to phonons and defects in collision events. We investigate the rate at which the electric field does work on free carriers and show that the rate of energy dissipation is proportional to the real part of the conductivity.

An electric field does work on an individual electron at the rate $-e\mathbf{v} \cdot \mathcal{E}$, so the rate at which energy is transferred per unit volume to a collection of electrons is $dW/dt = \mathbf{J} \cdot \mathcal{E}$. The real parts of $\mathbf{J}$ ($= \sigma\mathcal{E}$) and $\mathcal{E}$ must be used to compute $dW/dt$. We suppress the dependence of the field on $z$ and take the real part of the local field in some region of the sample to be $\mathcal{E}_0 \cos(\omega t)$. If $\sigma = \sigma' + i\sigma''$, the real part of the current density is $\mathcal{E}_0[\sigma'\cos(\omega t) + \sigma''\sin(\omega t)]$ and $dW/dt = \mathcal{E}_0^2\cos(\omega t)[\sigma'\cos(\omega t) + \sigma''\sin(\omega t)]$.

Energy passes back and forth between the field and the electrons. The rate at which it is dissipated per unit volume is the average of $dW/dt$ over a cycle, or $\frac{1}{2}\sigma'\mathcal{E}_0^2$. Once $\sigma' = \sigma_0/(1 + \omega^2\bar{t}^2)$ is used, this can be written

$$\left[\frac{dW}{dt}\right]_{\text{ave}} = \frac{1}{2}\frac{\sigma_0\mathcal{E}_0^2}{1 + \omega^2\bar{t}^2}. \tag{10-66}$$

Figure 10-18 shows $(2/\sigma_0\mathcal{E}_0^2)(dW/dt)_{\text{ave}}$ as a function of $\omega\bar{t}$. It has its greatest value for $\omega = 0$, then diminishes toward 0 as $\omega$ increases.

*The Effective Permittivity.* We study free carrier contributions to optical properties in terms of the effective permittivity $\epsilon_{\text{eff}} = \epsilon + (i\sigma/\omega)$, with real part $\epsilon'_{\text{eff}} = \epsilon' - \sigma_0\bar{t}/(1 + \omega^2\bar{t}^2)$ and imaginary part $\epsilon''_{\text{eff}} = \sigma_0/\omega(1 + \omega^2\bar{t}^2)$, where we have taken $\epsilon''$ to be zero. Assume the band is parabolic and use $\sigma_0 = ne^2\bar{t}/m^*$, where $m^*$ is the effective mass, to write

$$\epsilon'_{\text{eff}} = \epsilon' - \frac{ne^2\bar{t}^2}{m^*(1 + \omega^2\bar{t}^2)} = \epsilon'\left[1 - \frac{\omega_p^2\bar{t}^2}{1 + \omega^2\bar{t}^2}\right] \tag{10-67}$$

and

$$\epsilon''_{\text{eff}} = \frac{ne^2\bar{t}}{m^*\omega(1 + \omega^2\bar{t}^2)} = \epsilon'' + \frac{\epsilon'\omega_p^2\bar{t}^2}{\omega\bar{t}(1 + \omega^2\bar{t}^2)}, \tag{10-68}$$

where $\omega_p = \sqrt{ne^2/m^*\epsilon'}$ is called the plasma angular frequency. This is an important parameter for the determination of free carrier contributions to optical properties. As we will see later, it is also the natural oscillation frequency of the free carrier system.

For intrinsic semiconductors at room temperature $n$ is typically between $10^{15}$ carriers/m$^3$ and $10^{20}$ carriers/m$^3$, corresponding to plasma frequencies from about $10^9$ rad/s, in the radio wave region, to about $10^{12}$ rad/s, in the microwave region. Plasma frequencies are less at low temperature. They are well below the fundamental absorption edge for intrinsic and lightly doped semiconductors at all temperatures of interest.

For a typical intrinsic or lightly doped semiconductor, $\omega_p\bar{t}$ is less than 1 and $\epsilon'_{\text{eff}}$ is positive for all frequencies. Figure 10-19 shows the ratio $\epsilon'_{\text{eff}}/\epsilon'$ as a function of $\omega$ for $\omega_p\bar{t} = 0.5$. For frequencies below the plasma frequency free carriers act

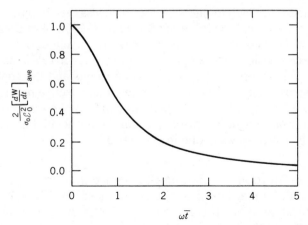

**FIGURE 10-18** Free carrier absorption: $(2/\sigma_0\mathscr{E}_0^2)(dW/dt)_{ave}$ as a function $\omega\bar{t}$, where $\omega$ is the angular frequency and $\bar{t}$ is the relaxation time. Absorption is high for angular frequencies below $1/\bar{t}$, but tends toward 0 as the frequency becomes large. For most semiconductors and insulators at room temperature $1/\bar{t}$ is on the order of $10^{12}$ rad/s, well below interband transitions.

to decrease the effective permeability. Equation 10-67 gives $\epsilon'_{eff} = \epsilon'(1 - \omega_p^2\bar{t}^2)$ for $\omega\bar{t} \ll 1$. At high frequencies, $\epsilon'_{eff}$ is nearly $\epsilon'$.

Equations 10-42 and 10-43 can be used to find the index of refraction $n$ and extinction coefficient $\kappa$. At low frequencies. $n = \kappa = [\epsilon'\omega_p^2\bar{t}/2\epsilon_0\omega]^{1/2}$. Since $n$ and $\kappa$ are both much larger than 1 the reflectivity is nearly 1. Waves that do enter the sample are severely attenuated. At high frequencies, $n = \sqrt{\epsilon'/\epsilon_0}$ and $\kappa = (\omega_p^2/2\omega^2\bar{t})\sqrt{\epsilon'/\epsilon_0}$. Free carriers do not significantly influence the high-frequency index of refraction and attenuation is slight.

For most compound semiconductors the limiting forms given above for $\omega\bar{t} \gg 1$ are not usually seen experimentally since the frequency dependence of $\epsilon$

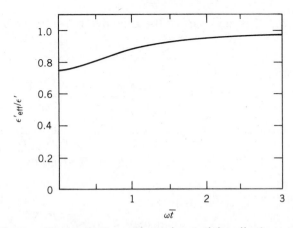

**FIGURE 10-19** Free carrier contribution to the real part of the effective permittivity, plotted for $\omega_p\bar{t} = 0.5$. Note that $\epsilon'_{eff}$ is less than $\epsilon'$ but tends to $\epsilon'$ as the frequency becomes large.

becomes important at high frequencies. Nevertheless the attenuation coefficient for infrared radiation can often be expressed as $C\omega^{-s}$, where $C$ and $s$ are empirically determined constants. The exponent $s$ is usually between 2.5 and 4.5.

We have been considering a single band. However, carriers in all valence and conduction bands contribute to the optical constants and the contributions of the various bands must be summed to find the total effective permittivity. Equations 10-67 and 10-68 hold for each band, with values of $\sigma_0$ and $\bar{t}$ dependent on the band. Contributions of valence bands can be formulated in terms of holes rather than electrons, as was done in Chapter 9.

**EXAMPLE 10-6**  At room temperature the concentration of electrons in the conduction band of an intrinsic germanium sample is $2.5 \times 10^{19}$ electrons/m³, the static dielectric constant is 15.3, and the relaxation time is $2.0 \times 10^{-12}$ s for both electrons and holes. Take the electron effective mass to be $0.55m_0$ and the hole effective mass to be $0.37m_0$, where $m_0$ is the free electron mass. Calculate the index of refraction and the skin depth for frequencies of $2.0 \times 10^{14}$ rad/s and $2.0 \times 10^{10}$ rad/s. These frequencies are well below the fundamental absorption edge.

**SOLUTION**  The electron plasma angular frequency is

$$\omega_{pe} = \left[\frac{ne^2}{m_e^*\epsilon'}\right]^{1/2} = \left[\frac{2.5 \times 10^{19} \times (1.6 \times 10^{-19})^2}{0.55 \times 9.11 \times 10^{-31} \times 15.3 \times 8.85 \times 10^{-12}}\right]^{1/2}$$

$$= 9.71 \times 10^{10} \text{ rad/s}$$

and $\omega_{pe}\bar{t} = 0.194$. The hole plasma angular frequency is

$$\omega_{ph} = \left[\frac{ne^2}{m_h^*\epsilon'}\right]^{1/2} = \left[\frac{2.5 \times 10^{19} \times (1.6 \times 10^{-19})^2}{0.37 \times 9.11 \times 10^{-31} \times 15.3 \times 8.85 \times 10^{-12}}\right]^{1/2}$$

$$= 1.18 \times 10^{11} \text{ rad/s}$$

and $\omega_{ph}\bar{t} = 0.237$. The real part of the effective permittivity is given by

$$\epsilon'_{eff} = \epsilon'\left[1 - \frac{\omega_{pe}^2\bar{t}^2 + \omega_{ph}^2\bar{t}^2}{1 + \omega^2\bar{t}^2}\right]$$

$$= 15.3\epsilon_0\left[1 - \frac{(0.194)^2 + (0.237)^2}{1 + (2.0 \times 10^{-12})^2\omega^2}\right],$$

where the electron and hole contributions have been summed. For $\omega = 2.0 \times 10^{14}$ rad/s, $\epsilon'_{eff} = 15.3\epsilon_0$ and for $\omega = 2.0 \times 10^{10}$ rad/s, $\epsilon'_{eff} = 13.9\epsilon_0$. The imaginary part of the effective permittivity is given by

$$\epsilon''_{eff} = \epsilon'\frac{\omega_{pe}^2\bar{t}^2 + \omega_{ph}^2\bar{t}^2}{\omega\bar{t}(1 + \omega^2\bar{t}^2)}$$

$$= 15.3\epsilon_0\frac{(0.194)^2 + (0.237)^2}{\omega \times 2.0 \times 10^{-12}[1 + (2.0 \times 10^{-12})^2\omega^2]}.$$

It is $2.24 \times 10^{-8}\epsilon_0$ for $\omega = 2.0 \times 10^{14}$ rad/s and $35.8\epsilon_0$ for $\omega = 2.0 \times 10^{10}$ rad/s. When Eq. 10-42 is used to calculate the index of refraction, the result is 3.91 for the higher frequency and 5.11 for the lower. When Eq. 10-43 is used to calculate the extinction coefficient, the result is $5.66 \times 10^{-9}$ for the higher frequency and 3.50 for the lower. The skin depth, given by $c/2\omega\kappa$, is 132 m for the higher frequency and $2.14 \times 10^{-3}$ m for the lower. ■

*Optical Properties of Metals.*   The general principles outlined in previous sections are also valid for metals. As might be expected, free carrier effects are much more important in metals than in intrinsic semiconductors and insulators. Look first at the imaginary part of the effective permittivity for aluminum, plotted in Fig. 10-20. The general trend of the curve is determined by intraband contributions but several peaks associated with interband transitions are discernable.

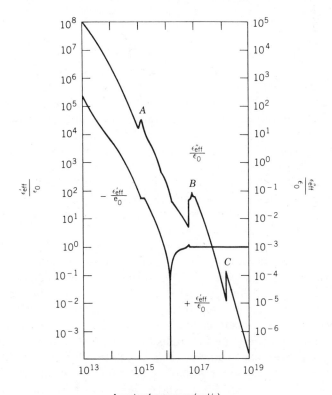

Angular frequency (rad/s)

**FIGURE 10-20**   The real and imaginary parts of the effective permittivity of aluminum as functions of frequency. The plasma frequency is $\omega_p = 1.4 \times 10^{16}$ rad/s. The negative of $\epsilon'_{eff}$ is plotted for $\omega < \omega_p$. *A* marks the·onset of interband transitions, *B* marks *L*-shell core transitions, and *C* marks *K*-shell core transitions. Absorption below *A* is due to free carriers. (From D. Y. Smith, E. Shiles, and M. Inokuti, "The Optical Properties of Metallic Aluminum" in *Handbook of Optical Constants of Solids* (E. D. Palik, Ed.) (New York: Academic, 1985). Data are from E. Shiles, T. Sasaki, M. Inokuti, and D. Y. Smith, *Phys. Rev. B* **22**: 1612, 1980. Used with permission.)

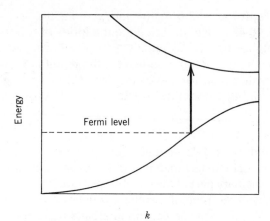

$k$

**FIGURE 10-21** The onset of interband transitions for a nearly free electron metal. The arrow connects filled and empty states with the same value of **k**, in different bands.

Figure 10-21 illustrates the lowest frequency interband transition for a nearly free electron metal. Fundamental absorption for aluminum begins at an angular frequency of about $1.0 \times 10^{15}$ rad/s, in the near infrared, and evidence for it on the graph extends to somewhat beyond $10^{16}$ rad/s. Core absorption begins at an angular frequency of about $9.0 \times 10^{15}$ rad/s, in the near ultraviolet. Note that absorption is quite low in the visible region and above.

This curve is typical of $\epsilon''$ for simple metals. Fundamental absorption begins at about $3.3 \times 10^{15}$ rad/s for sodium and at about $2.3 \times 10^{15}$ rad/s for potassium. Semimetals and transition metals are somewhat different. For these materials, two or more bands that overlap in energy are partially filled. At many points in the Brillouin zone occupied states in one band are close in energy to empty states in another. Frequencies for the onset of interband transitions are typically an order of magnitude lower than for simple metals and absorption curves usually display some structure over much of the infrared region.

We now turn to the real part of $\epsilon_{\text{eff}}$. Because metals have high concentrations of nearly free electrons their plasma frequencies are high. For example, if the electron concentration is $10^{29}$ electrons/$m^3$ then $\omega_p = 1.8 \times 10^{16}$ rad/s, in the near ultraviolet. Since relaxation times are around $10^{-14}$ s, $\omega_p \bar{t} > 1$. As a consequence metals are nearly perfect reflectors for incident radiation with a frequency below the plasma frequency. This conclusion follows directly from Eq. 10-67, which predicts that the real part of the effective permittivity is negative for $\omega^2 < \left( \omega_p^2 - 1/\bar{t}^2 \right) \approx \omega_p^2$. A transverse wave with frequency less than the plasma frequency does not propagate in a metal. If $\epsilon_{\text{eff}}''$ can be neglected then $n = 0$ and $R = 1$.

Figure 10-20 shows the real part of $\epsilon_{\text{eff}}$ for aluminum. Since it is negative for frequencies below the plasma frequency $-\epsilon_{\text{eff}}'$ is plotted in that range, while $+\epsilon_{\text{eff}}'$ is plotted for $\omega > \omega_p$. The combination of a plasma frequency in the ultraviolet and weak absorption above the visible means reflection and transmission properties of a simple metals exhibit a dramatic change. A metal is an excellent reflector for frequencies up to its plasma frequency but then it suddenly becomes transparent.

The same phenomenon occurs for highly doped semiconductors, with high carrier concentrations, and for highly pure semiconductors at extremely low temperatures, with long relaxation times. In the first case doping increases the plasma frequency, perhaps to the near infrared or visible region. In the second case plasma frequencies decrease to perhaps well into the microwave region but the increase in $\bar{t}$ compensates for this.

*Plasma Waves.* Longitudinal waves are possible in a medium containing a high electron concentration. Electrons are displaced along the line of propagation and the electric field created supplies a restoring force. The natural frequency of oscillation is the plasma frequency.

A simple one-dimensional model can be used to find the frequency. We suppose the material initially has a uniform electron concentration $n$ and, for neutrality, an identical concentration of positive charge. Suppose a thin layer of electrons, parallel to the $xy$ plane, moves a distance $z$ in the positive $z$ direction. Layers of positive and negative charge are created and the electric field in the evacuated region has magnitude $\mathcal{E} = enz/\epsilon$, where $\epsilon$ is the real part of the permittivity. The field pulls an electron toward its original position with a force given by $-e^2nz/\epsilon$ and the equation of motion for an electron is

$$m \frac{d^2z}{dt^2} = -\frac{e^2n}{\epsilon} z ,$$

(10-69)

identical to the equation of motion for a harmonic oscillator with natural angular frequency $\omega = (e^2n/\epsilon m)^{1/2}$.

A more complete analysis shows that a distortion in the electron concentration propagates through the material as a wave. The concentration has the form $n + Ae^{i(sz-\omega t)}$. For electrons with an extremely long relaxation time in a parabolic band, the frequency $\omega$ and propagation constant $s$ of a plasma wave are related by

$$\omega^2 = \omega_p^2 + \tfrac{3}{5}s^2v_F^2 ,$$

(10-70)

where $v_F$ is the Fermi velocity, $\sqrt{2E_F/m^*}$. Since $sv_F$ is much smaller than $\omega_p$, all plasma waves have frequencies near the plasma frequency, regardless of wavelength.

Plasma waves are excited by incident radiation at the plasma frequency. For a wave at normal incidence plasma waves travel along the surface of the solid, parallel to the incident electric field. At other angles of incidence, waves that penetrate the sample are excited. Perhaps the most common method used to excite plasma waves is by means of fast electrons injected into a sample. Oscillations are caused by the coulomb interaction between the entering electron and electrons of the solid.

## 10.7 IONIC POLARIZABILITY

*Ionic Resonances.*   For ionic and partially ionic solids another resonance region, associated with ionic polarizability, occurs below the fundamental absorption edge. Because the forces of an electric field on oppositely charged ions are in opposite directions, the field couples to atomic vibrational modes. Resonance occurs when the frequency of an electromagnetic wave nearly matches the frequency of such a mode. In particular, the strongest coupling is to transverse optical modes near $\mathbf{q} = 0$ in the Brillouin zone. Recall that for these modes oppositely charged ions move in opposite directions while their center of mass remains stationary.

For most ionic solids optical modes have frequencies in the infrared portion of the electromagnetic spectrum. An electromagnetic wave in resonance with one of these modes has a propagation constant that is much less than a Brillouin zone dimension. If $\omega$ is about $10^{13}$ rad/s, for example, the real part of the propagation constant, given approximately by $\omega/c$, is about $10^4$ m$^{-1}$. On the other hand, a Brillouin zone dimension may be $10^9$ m$^{-1}$ or more.

Since the wavelength, about $10^{-6}$ m, is much greater than unit cell dimensions, we neglect variations in atomic displacements from cell to cell and assume equivalent ions in different cells have identical motions. We may use Eq. 10-52 directly if we interpret $M$ as the reduced mass of the two ions in a pair and $\omega_0$ as a $\mathbf{q} = 0$ transverse optical frequency. For this discussion we label the angular frequency $\omega_T$. $Q$ is less than the charge on an isolated ion because electron distributions around neighboring ions overlap somewhat. For NaCl, $Q = 0.74e$; for GaAs, $Q = 0.51e$; and for ZnS, $Q = 0.96e$.

To obtain the total polarizability of an ion pair we add the electronic polarizabilities, $\alpha_+$ for the positive ion and $\alpha_-$ for the negative ion, to the ionic polarizability. The total for an ion pair is

$$\alpha(\omega) = \alpha_+ + \alpha_- + \frac{Q^2}{M}\frac{\omega_T^2 - \omega^2 + i\omega\rho}{(\omega_T^2 - \omega^2) + \omega^2\rho^2}. \tag{10-71}$$

For simplicity we neglect the difference between the local and macroscopic fields and take the susceptibility to be

$$\epsilon(\omega) = (1 + N\alpha)\epsilon_0$$

$$= (1 + N\alpha_+ + N\alpha_-)\epsilon_0 + \frac{NQ^2\epsilon_0}{M}\frac{\omega_T^2 - \omega^2 + i\omega\rho}{(\omega_T^2 - \omega^2)^2 + \omega^2\rho^2}, \tag{10-72}$$

where $N$ is the number of ion pairs per unit volume. The real part is often written

$$\epsilon(\omega) = \epsilon(\infty) + [\epsilon(0) - \epsilon(\infty)]\frac{(\omega_T^2 - \omega^2)\omega_T^2}{(\omega_T^2 - \omega^2)^2 + \omega^2\rho^2}, \tag{10-73}$$

where $\epsilon(0)$ is the permittivity for a frequency below the region of ionic resonance and $\epsilon(\infty)$ is the permittivity for a frequency well above the ionic resonance, for which the second term is negligibly small. If free carrier contributions can be neglected, as they can for an insulator, $\epsilon(0)$ is the static permittivity. A frequency in the visible region, below interband electron transitions, is used to measure $\epsilon(\infty)$. We may assume $\epsilon(0)$ and $\epsilon(\infty)$ are both constant in the ionic resonance region.

The real part of the polarizability is similar to the function plotted in Fig. 10-13a, with the peak occurring at about $\omega_T$. The imaginary part is similar to the function plotted in Fig. 10-13b. Energy dissipation comes about chiefly through phonon-phonon interactions and is temperature dependent. Table 10-4 gives polarizability parameters for some ionic crystals and Fig. 10-22 shows the index of refraction and extinction coefficient for NaCl.

*Dispersion Relations.* The vibration spectrum changes dramatically near $q = 0$ when an electromagnetic wave is present. Take the electronic polarizabilities to be independent of frequency and neglect dissipation. Then $\epsilon$ is real and

$$s^2 = \frac{\omega^2 \epsilon}{c^2 \epsilon_0} = \frac{\omega^2}{c^2 \epsilon_0} \left[ \epsilon(\infty) + \frac{\epsilon(0) - \epsilon(\infty)}{\omega_T^2 - \omega^2} \omega_T^2 \right]. \tag{10-74}$$

This equation is solved for $\omega^2$ as a function of $s$, with the result

$$\omega^2 = \frac{1}{2} \left[ \frac{\epsilon_0}{\epsilon(\infty)} c^2 s^2 + \frac{\epsilon(0)}{\epsilon(\infty)} \omega_T^2 \right]$$

$$\pm \frac{1}{2} \left\{ \left[ \frac{\epsilon_0}{\epsilon(\infty)} c^2 s^2 + \frac{\epsilon(0)}{\epsilon(\infty)} \omega_T^2 \right]^2 - \frac{4\epsilon_0}{\epsilon(\infty)} \omega_T^2 c^2 s^2 \right\}^{1/2}. \tag{10-75}$$

As shown in Fig. 10-23, the dispersion curve has two branches, corresponding to the two possible signs for the root in Eq. 10-75. In the resonance region each solution corresponds to a situation in which an electromagnetic wave and a vibrational wave travel together with the same frequency and wavelength. For

**TABLE 10-4** Ionic Permittivity Parameters for Selected Ionic Crystals

| Crystal | $\epsilon(0)/\epsilon_0$ | $\epsilon(\infty)/\epsilon_0$ | $\omega_T$ ($10^{13}$ rad/s) | $Q/e$ |
|---------|----------|----------|----------|-------|
| LiF | 9.3 | 1.92 | 5.8 | 0.83 |
| NaF | 6.0 | 1.74 | 4.6 | 0.94 |
| NaCl | 5.8 | 2.25 | 3.1 | 0.76 |
| NaBr | 6.0 | 2.62 | 2.5 | 0.85 |
| NaI | 4.9 | 2.91 | 2.2 | 0.71 |
| KCl | 4.7 | 2.13 | 2.7 | 0.80 |
| KI | 5.1 | 2.69 | 1.8 | 0.69 |
| RbCl | 5.0 | 2.19 | 2.2 | 0.86 |
| CsCl | 6.8 | 2.60 | 1.8 | 0.88 |
| TlCl | 32 | 5.10 | 1.6 | 1.11 |
| CuCl | 10 | 3.57 | 3.6 | 1.10 |
| CuBr | 8 | 4.08 | 3.3 | 1.0 |

*Sources:* B. Szigeti, *Trans. Faraday Soc.* **45**:155, 1949; I. S. Zheludev, *Physics of Crystalline Dielectrics* (New York: Plenum, 1971).

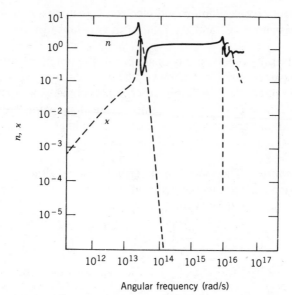

**FIGURE 10-22** Index of refraction $n$ (solid curve) and extinction coefficient $\kappa$ (dotted curve) for NaCl. Ionic resonance occurs at about $3 \times 10^{13}$ rad/s and the fundamental absorption edge is at about $10^{26}$ rad/s. (From J. E. Eldridge and E. D. Palik, "Sodium Chloride" in *Handbook of Optical Constants of Solids* (E. D. Palik, Ed.) (New York: Academic, 1985). Used with permission.)

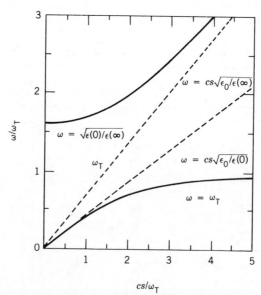

**FIGURE 10-23** Dispersion curves in the Reststrahlen region of an ionic crystal with $\epsilon(0)/\epsilon_0 = 5.62$ and $\epsilon(\infty)/\epsilon(0) = 2.25$, the values for NaCl. The $q = 0$ optical phonon frequency is $\omega_T$. The plot covers only a small portion of the Brillouin zone.

frequencies well above or well below resonance, however, each wave is either purely electromagnetic or purely vibrational in character.

For long wavelengths, near the left side of the graph, the lower branch approaches the straight line $\omega = \sqrt{\epsilon_0/\epsilon(0)}cs$. The wave is electromagnetic and travels in a medium with dielectric constant $\epsilon(0)/\epsilon_0$. The upper curve, on the other hand, corresponds to an atomic vibration with angular frequency given by

$$\omega^2 = \frac{\epsilon(0)}{\epsilon(\infty)}\,\omega_T^2 . \tag{10-76}$$

Because an electromagnetic wave is present the frequency is changed from the usual optical phonon frequency by the factor $\sqrt{\epsilon(0)/\epsilon(\infty)}$.

For short wavelengths, the lower curve bends to coincide with the line $\omega = \omega_T$, the dispersion curve for a pure atomic vibration at the optical phonon frequency. A more precise calculation takes into account forces between ions in different cells and produces a dispersion curve like the one shown in Fig. 6-7 for the optical branch. In the short wavelength limit the upper curve corresponds to an electromagnetic wave traveling in a medium with dielectric constant $\epsilon(\infty)/\epsilon_0$. Since the ions no longer follow the oscillating field only electronic polarization is important.

No wave propagates with an angular frequency in the range from $\omega_T$ to $\sqrt{\epsilon(0)/\epsilon(\infty)}\omega_T$. According to Eq. 10-73, $\epsilon$ is negative, so the reflectivity is 1 for these frequencies. If dissipation is included in the model, the reflectivity is slightly less than 1. Nevertheless, ionic solids exhibit high reflectivity for angular frequencies in what is known as the Reststrahlen band, between $\omega_T$ and $\sqrt{\epsilon(0)/\epsilon_0}\omega_T$. The reflectivity of NaCl is shown in Fig. 10-24.

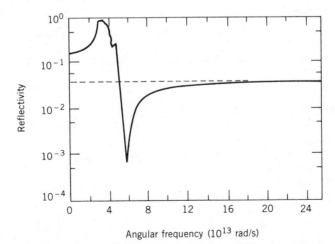

Angular frequency ($10^{13}$ rad/s)

**FIGURE 10-24** Reflectivity of crystalline NaCl as a function of angular frequency. The region of high reflectivity lies between $\omega_T$ and $\sqrt{\epsilon(0)/\epsilon(\infty)}\omega_T$. It is not sharply delineated because energy dissipation takes place. A dotted line indicates the reflectivity for visible light. (From D. Y. Smith and C. A. Manogue, *J. Opt. Soc. Am.* **71**: 935, 1981. Used with permission.)

Tails of dissipation curves associated with ionic resonances are quite small in the visible region. On the other hand, electronic absorption may not begin until the frequency reaches the ultraviolet. Transmission is high in a fairly wide band, reaching from the near infrared to the near ultraviolet and including the visible. Most ionic insulators are highly transparent in the visible.

**EXAMPLE 10-7** For NaCl calculate: (a) the limits of the frequency gap for which no waves propagate; (b) the speed of an electromagnetic wave with frequency below the infrared resonance; and (c) the speed of an electromagnetic wave with frequency in the visible region.

**SOLUTION** Relevant data are given in Table 10-4. For NaCl $\omega_T = 3.1 \times 10^{13}$ rad/s, $\epsilon(0)/\epsilon_0 = 5.8$, and $\epsilon(\infty)/\epsilon_0 = 2.25$. (a) The bottom of the gap is at $\omega = \omega_T = 3.1 \times 10^{13}$ rad/s while the top of the gap is at $\omega = [\epsilon(0)/\epsilon(\infty)]^{1/2}\omega_T = [5.8/2.25]^{1/2} \times 3.1 \times 10^{13} = 5.0 \times 10^{13}$ rad/s. (b) The wave speed for frequencies below resonance is $v_p = \sqrt{\epsilon_0/\epsilon(0)}c = \sqrt{1/5.8} \times 3.0 \times 10^8 = 1.2 \times 10^8$ m/s. (c) Just above resonance the wave speed is $v_p = \sqrt{\epsilon_0/\epsilon(\infty)}c = \sqrt{1/2.25} \times 3.0 \times 10^8 = 2.0 \times 10^8$ m/s. ∎

## 10.8   REFERENCES

### Electromagnetic Theory

A. M. Portis, *Electromagnetic Fields: Sources and Media* (New York: Wiley, 1978).
J. R. Reitz, F. J. Milford, and R. W. Christy, *Foundations of Electromagnetic Theory* (Reading, MA: Addison-Wesley, 1979).

### Static Dielectric Properties

A. Nussbaum, *Electronic and Magnetic Behavior of Materials* (Englewood Cliffs, NJ: Prentice-Hall, 1967).
I. S. Zheludev, *Physics of Crystalline Dielectrics,* Vol. 2 New York: Plenum, 1971).

### Ferroelectrics

J. C. Burfoot, *Ferroelectrics* (Princeton, NJ: Van Nostrand, 1967).
J. Grindlay, *An Introduction to the Phenomenological Theory of Ferroelectricity* (Elmsford, NY: Pergamon, 1970).
F. Iona and G. Shirane, *Ferroelectric Crystals* (New York: McMillan, 1962).
W. Kanzig, "Ferroelectrics and Antiferroelectrics" in *Solid State Physics* edited by (F. Seitz and D. Turnbull, Eds.), Vol. 4, p. 1 (New York: Academic, 1957).
H. D. Megaw, *Ferroelectricity in Crystals* (London: Methuen, 1957).

### Optical Properties

A. Bienenstock and H. Winick, "Synchrotron Radiation Research—An Overview." *Phys. Today* **36**(6):48, June 1983.
F. C. Brown, "Ultraviolet Spectroscopy of Solids with the Use of Synchrotron Ra-

diation" in *Solid State Physics* H. Ehrenreich, F. Seitz, and D. Turnbull, Eds.), Vol. 29, p. 1 (New York, Academic, 1974).

P. N. Butcher, "AC Conductivity" in *Handbook on Semiconductors* (W. Paul, Ed.), Vol. 1 (Amsterdam: North-Holland, 1982).

J. N. Hodgson, *Optical Absorption and Dispersion in Solids* (London: Chapman & Hall, 1970).

J. C. Phillips, "The Fundamental Optical Spectra of Solids" in *Solid State Physics* (F. Seitz and D. Turnbull, Eds.), Vol. 18, p. 55 (New York: Academic, 1966).

F. Stern, "Elementary Theory of the Optical Properties of Solids" in *Solid State Physics* (F. Seitz and D. Turnbull, Eds.), Vol. 15, p. 299 (New York: Academic, 1963).

J. P. Wolfe, "Thermodynamics of Excitons in Semiconductors." *Phys. Today* **35**(3): 46, March 1982.

## PROBLEMS

1. A certain parallel plate capacitor has a plate separation of 1.00 mm. It is charged to 1000 V, then the source of emf is removed and a 0.90-mm-thick sample is placed between the plates, with the same space on either side. The potential difference is found to be 675 V. What is the dielectric constant of the sample?

2. The force constant for two neighboring atoms in NaCl is about 36 N/m. Their equilibrium separation is 2.82 Å. (a) Take the magnitude of the charge on each ion to be $e$ and find the dipole moment of an ion pair when they are at their equilibrium sites. (b) Find the change in their separation produced by a 1500 V/m local electric field. (c) Find the change in the dipole moment. (d) Estimate the static ionic polarizability.

3. Consider a crystal with a simple cubic structure (cube edge $a$) and suppose each atom has the same dipole moment $\mathbf{p}$. (a) Show that the electric field produced at the site of any atom due to all atoms a distance $a$ away vanishes. (b) Repeat for the field of all atoms a distance $\sqrt{2}a$ away. (c) Repeat for the field of all atoms a distance $\sqrt{3}a$ away.

4. Consider a crystal with a simple tetragonal structure (square edge $a$ and height $c$) and suppose each atom has dipole moment $\mathbf{p}$. (a) Show that the electric field produced at the site of any atom by all atoms a distance $a$ away is $(1/2\pi\epsilon_0)(\mathbf{p} - 3p_z\hat{\mathbf{z}})/a^3$, where the $z$ axis is along the tetrad. (b) Find an expression for the field produced by all atoms a distance $c$ away. (c) Show that the resultant of the fields found in parts a and b vanishes if $c = a$.

5. The static electronic polarizabilities of $Na^+$ and $Cl^-$ ions are $3.47 \times 10^{-41}$ $C^2 \cdot m/N$ and $3.41 \times 10^{-40}$ $C^2 \cdot m/N$, respectively, while the static ionic polarizability of a sodium chlorine ion pair is $3.56 \times 10^{-40}$ $C^2 \cdot m/N$. (a) Use the Clausius-Mossotti relationship to estimate the static dielectric constant of sodium chloride. NaCl is FCC with a cube edge of 5.64 Å. (b) If a 1500-V/m electric field is applied perpendicularly to the face of a slab,

what is the local field at an ion pair? What are the macroscopic electric and polarization fields in the sample?

6. Consider a crystal with a tetragonal structure (square edge $a$ and cell height $c$, with $c > a$). In a static electric field $\mathcal{E}$ each atom has the same dipole moment $\mathbf{p}$. To estimate the local field at an atom use a sphere with radius slightly larger than $c$. Consider 2 atoms a distance $c$ from the center and 4 atoms a distance $a$ from the center. Place the $z$ axis along the tetrad and the other coordinate axes along edges of the square base. Use $\mathbf{p} = \alpha\mathcal{E}_{loc}$ and $\mathbf{P} = \mathbf{p}/a^2c$ to show the $x$ component of the local field is $\mathcal{E}_x/A$, the $y$ component is $\mathcal{E}_y/A$, and the $z$ component is $\mathcal{E}_z/B$, where $A = 1 - (\alpha/3\epsilon_0 a^2 c) - (\alpha/2\pi\epsilon_0)(1/a^3 - 1/c^3)$ and $B = 1 - (\alpha/3\epsilon_0 a^2 c) + (\alpha/\pi\epsilon_0)(1/a^3 - 1/c^3)$. (b) Find expressions for the elements of the susceptibility tensor. (c) Compare $\chi_{xx}$ and $\chi_{zz}$ for $\alpha = 10^{-40}$ C$^2 \cdot$ m/N, $a = 5.0$ Å, and $c = 9.0$ Å. (d) How can a uniaxial stress be applied to increase the difference?

7. For a solid that obeys the Clausius-Mossotti relationship show that the temperature coefficient of the dielectric constant, defined as $(1/K)(dK/dT)$, is $-(K - 1)(K + 2)\Delta_\tau/3K$, where $\Delta_\tau$ is the coefficient of volume expansion, defined by $\Delta_\tau = (1/\tau_s)(d\tau_s/dT)$. Assume the atomic concentration changes with temperature but polarizabilities do not.

8. Suppose the concentration of dipolar molecules is $1.6 \times 10^{28}$ molecules/m$^3$ and that each molecule has a permanent dipole moment of $3.5 \times 10^{-26}$ C$\cdot$m. Assume the Langevin formulation is valid. (a) Compute the saturation polarization. (b) What is the polarization at 300 K in an electric field of $2.5 \times 10^4$ V/m? (c) Neglect local field effects and compute the susceptibility at 300 K.

9. Suppose a crystal is composed of identical dipolar molecules at equivalent sites and that each dipole moment $\mathbf{p}$ is either in the positive $z$ direction or in the negative $z$ direction. Show that the thermodynamic average dipole moment is $p_{ave} = p\,\tanh(\beta p\mathcal{E}_z)$, where $\beta = 1/k_B T$. Show that for weak fields the susceptibility is $np^2\beta/\epsilon_0$, where $n$ is the molecular concentration. Except for a numerical factor on the order of 1, this is the same as the result for molecules that can rotate freely.

10. A 1000-V/m field is applied to a slab of quartz, 0.500 mm thick. The field is normal to the slab faces, which are (100) crystal planes. (a) If no stress is applied to the slab, what is the fractional change in its thickness? (b) What uniaxial stress must be applied to the faces so the thickness of the slab does not change? Use data from Example 10-3.

11. 500-nm light is incident normally on a sample with index of refraction $n = 1.653$ and extinction coefficient $\kappa = 2.35 \times 10^{-2}$. (a) What is the speed of the wave in the sample? (b) What is the wavelength of the wave in the sample? (c) Over what distance is the wave intensity diminished by half? (d) What is the reflectivity of the sample?

12. A sample has a reflectivity of 0.250 for light with an angular frequency of $2.56 \times 10^{15}$ rad/s, incident normally. In the sample, the intensity decreases by half in a distance of 5.00 mm. For this frequency find (a) the extinction coefficient, (b) the index of refraction, (c) the real part of the permittivity, and (d) the imaginary part of the permittivity.

13. Suppose a binary ionic solid consists of $N$ ion pairs per unit volume and each pair has reduced mass $M$, natural angular frequency $\omega_0$, and relaxation constant $\rho$ ($\ll \omega_0$). The magnitude of the charge on each ion is $Q$. Neglect electronic polarization and take the permittivity to be $\epsilon = \epsilon_0 + N\alpha$, where $\alpha$ is the ionic polarizability. (a) For $\omega \ll \omega_0$ show that the index of refraction is given by $n^2 = 1 + NQ^2/M\epsilon_0\omega_0^2$ and the extinction coefficient approaches 0 in proportion to $\omega$. (b) Derive an expression for the low-frequency reflectivity. Evaluate it using the parameters given in Example 10-5 for NaCl. Take $Q$ to be $1.60 \times 10^{-19}$ C and the cube edge to be 5.63 Å.

14. For the solid of problem 13 take $\omega \gg \omega_0$ and $\rho \ll \omega_0$. Show that the index of refraction $n$ is less than 1 but approaches 1 in the limit as $\omega$ becomes large. Show that the extinction coefficient and reflectivity both approach 0 in the same limit. In reality, several resonances may contribute to the index of refraction and extinction coefficient.

15. Consider the solid of Problem 13. (a) Show that the resonance frequency $\omega_R$ is given by $\omega_R^2 = \omega_0^2 - \rho\omega_0$. (b) Take $\rho \ll \omega_0$ and show that the real and imaginary parts of the polarizability are $\alpha' = \alpha'' = Q^2/2M\omega_0\rho$ at resonance. (c) Show that at resonance the index of refraction $n$ is given by $n^2 = (1 + \sqrt{2})NQ^2/4M\epsilon_0\omega_0\rho$ and the extinction coefficient is given by $\kappa^2 = (\sqrt{2} - 1)NQ^2/4M\epsilon_0\omega_0\rho$. (d) Evaluate $n$ and $\kappa$ using the parameters given in Example 10-5. Take the cube edge to be 5.62 Å, the charge to be $1.6 \times 10^{-19}$ C, and the relaxation constant $\rho$ to be $0.01\omega_0$. (e) Use your results to estimate the reflectivity of NaCl at resonance.

16. Figure 10-25 shows the index of refraction $n$ and extinction coefficient $\kappa$ for gallium arsenide. Identify the regions of Reststrahlen, fundamental absorption, and core absorption. Use the graph to estimate the optical phonon frequency, the fundamental absorption edge, and the lowest core excitation frequency. Calculate the band gap and the lowest core excitation energy.

17. Suppose that the permittivity $\epsilon$ is real and constant in the region of free carrier absorption and that the free carrier relaxation time is long so $\omega\bar{t} \gg 1$ for all frequencies of interest. Show that $\epsilon_{\text{eff}} = \epsilon(1 - \omega_p^2/\omega^2)$ and find expressions for the index of refraction, extinction coefficient, and reflectivity as functions of angular frequency. Consider $\omega > \omega_p$ and $\omega < \omega_p$ separately.

18. Consider the nearly free electrons in a solid for which the permittivity has the constant real value $5.0\epsilon_0$ at low frequencies. This is the contribution of all sources other than free electrons. (a) Suppose $\omega_p\bar{t} = 0.70$, where $\omega_p$ is the plasma angular frequency and $\bar{t}$ is the relaxation time. Evaluate the

index of refraction, extinction coefficient, and reflectivity for $\omega = \omega_p/2$ and for $\omega = 2\omega_p$. (b) Do the same for $\omega_p \bar{t} = 1.4$.

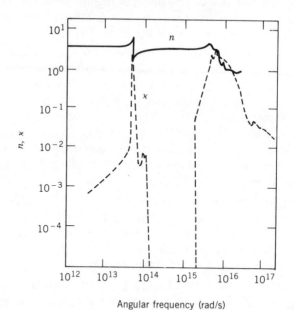

Angular frequency (rad/s)

**FIGURE 10-25** The index of refraction $n$ (solid line) and extinction coefficient $\kappa$ (dotted line) of gallium arsenide. (From E. D. Palik, "Gallium Arsenide" in *Handbook of Optical Constants of Solids* (E. D. Palik, Ed.) (New York: Academic, 1985). Used with permission.)

# Chapter 11

## MAGNETIC PROPERTIES

Ferromagnetic domains where they meet the surface of an iron sample. The photograph was taken with polarized light and uses the dependence of reflectance on magnetic field.

Electrons and nuclei in solids produce magnetic fields, both because they are moving charges and because they have intrinsic magnetic dipole moments. Fields produced by nuclei are typically much smaller than those produced by electrons, so we will be concerned chiefly with the latter.

Electrons in some materials produce a macroscopic field only when an external field is applied. For paramagnets the induced field is in the same direction as the applied field, while for diamagnets it is in the opposite direction. For other materials, called ferromagnets and ferrimagnets, a macroscopic field exists even in the absence of an applied field. Most of this chapter deals with the fundamental question: How do electron motions give rise to the wide range of magnetic properties observed?

## 11.1 FUNDAMENTAL CONCEPTS

*Sources of Magnetic Fields.* The magnetic induction field $\mathbf{B}(\mathbf{r})$ produced at $\mathbf{r}$ by a steady current density $\mathbf{J}(\mathbf{r}')$ is given by the Biot-Savart law:

$$\mathbf{B}(\mathbf{r}) = \frac{\mu_0}{4\pi} \int \frac{\mathbf{J}(\mathbf{r}') \times (\mathbf{r} - \mathbf{r}')}{|\mathbf{r} - \mathbf{r}'|^3} \, d\tau', \tag{11-1}$$

a volume integral with the primed coordinates as integration variables. At points far from the current distribution the field is given by

$$\mathbf{B}(\mathbf{r}) = \frac{\mu_0}{4\pi} \left[ \frac{3(\boldsymbol{\mu} \cdot \mathbf{r})\mathbf{r} - \boldsymbol{\mu} r^2}{r^5} \right]. \tag{11-2}$$

where $\boldsymbol{\mu}$ is the dipole moment of the distribution, defined by

$$\boldsymbol{\mu} = \frac{1}{2} \int \mathbf{r}' \times \mathbf{J}(\mathbf{r}') \, d\tau'. \tag{11-3}$$

If the current distribution is in the form of a filament loop carrying current $I$ and lying in a plane, the magnitude of the dipole moment is $IA$, where $A$ is the area of the loop. Its direction is normal to the plane and is given by a right-hand rule; when the fingers of the right hand curl around the loop in the sense of the current, the thumb points in the direction of $\boldsymbol{\mu}$.

Figure 11-1 shows an electron traveling with uniform angular speed $\omega$ around a circle of radius $R$. The period of the motion is $2\pi/\omega$, so the current is $e\omega/2\pi$ and the magnitude of the dipole moment is $\mu = (e\omega/2\pi)\pi R^2 = \frac{1}{2}e\omega R^2$. The direction of the moment is shown in the figure.

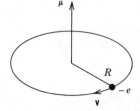

FIGURE 11-1 An electron traversing a circular orbit. Its magnetic dipole moment $\boldsymbol{\mu}$, averaged over a cycle, is normal to the plane of the orbit, in the direction shown. The dipole moment of an electron traveling in the opposite direction is opposite that shown.

The dipole moment of a circulating charge is proportional to its angular momentum. An electron in uniform circular motion has angular momentum with magnitude $L = m\omega R^2$, so $\omega R^2 = L/m$ and $\mu = (e/2m)L$. For a negative charge $\mu$ and $\mathbf{L}$ are in opposite directions and

$$\mu = -\frac{e}{2m}\mathbf{L}. \tag{11-4}$$

The ratio $\mu/L$ is called the gyromagnetic ratio. It is $e/2m = 8.78 \times 10^{10}$ C/kg for a circulating electron.

A magnetic dipole precesses about the direction of a constant magnetic field. The field exerts a torque $\mu \times \mathbf{B}$ on a dipole, so the angular momentum obeys $d\mathbf{L}/dt = \mu \times \mathbf{B}$. Substitute $\mathbf{L} = -(2m/e)\mu$ to obtain

$$\frac{d\mu}{dt} = -\frac{e}{2m}\mu \times \mathbf{B} \tag{11-5}$$

for an electron. If $\mathbf{B}$ is in the $z$ direction, Eq. 11-5 has the solution $\mu_x = A \cos(\omega_L t)$, $\mu_y = A \sin(\omega_L t)$, $\mu_z =$ constant, where $\omega_L$ is the Larmor angular frequency, given by $\omega_L = eB/2m$. The magnitude of $\mu$ remains constant, while the projection of $\mu$ on the $xy$ plane moves with angular speed $\omega_L$ around a circle.

The energy of a dipole in a field $\mathbf{B}$ is given by

$$E = -\mu \cdot \mathbf{B}. \tag{11-6}$$

$E$ represents the work that must be done by an external agent to reorient a moment that is initially perpendicular to the field. A uniform field does not exert a net force on a dipole but the force of an inhomogeneous field is given by

$$\mathbf{F} = (\mu \cdot \nabla)\mathbf{B}. \tag{11-7}$$

An electron also has an intrinsic magnetic dipole moment, associated with its spin angular momentum $\mathbf{S}$. It is

$$\mu = -g_s \frac{e}{2m}\mathbf{S}, \tag{11-8}$$

where $g_s = 2.0023$. For most magnetic phenomena $g_s$ may be approximated by 2.00 and the spin dipole moment may be taken to be $-(e/m)\mathbf{S}$. The spin gyromagnetic ratio is nearly $e/m$. Like an orbital moment, a spin moment produces a magnetic field and precesses in an applied field. Its energy in an applied field is given by Eq. 11-6 and it experiences a force given by Eq. 11-7.

*Magnetization and the Magnetic Field.*   The magnetization $\mathbf{M}(\mathbf{r})$ at any point $\mathbf{r}$ in a sample is defined as the magnetic dipole moment per unit volume of a macroscopically small region around $\mathbf{r}$. For a crystal the magnetization is the total dipole moment of a single unit cell, divided by the cell volume.

The magnetic induction field of a magnetized sample is exactly the same as that of an identical unmagnetized region with current density $\nabla \times \mathbf{M}$ in its interior.* Magnetization current enters Maxwell's equations in exactly the same way as macroscopic current, which arises from the transport of charge over macroscopic distances. For static conditions

$$\nabla \cdot \mathbf{B} = 0 \tag{11-9}$$

and

$$\nabla \times \mathbf{B} = \mu_0 \mathbf{J} + \mu_0 \nabla \times \mathbf{M}, \tag{11-10}$$

where $\mathbf{J}$ is the macroscopic current density.

At sample boundaries the magnetization may be replaced by a surface current density $\mathbf{M} \times \hat{\mathbf{n}}$, where $\hat{\mathbf{n}}$ is the unit outward normal to the surface. If $d\ell$ is an infinitesimal line element on the surface then the magnetization current through it is $(\mathbf{M} \times \hat{\mathbf{n}}) \cdot d\ell$. We consider situations for which the magnetization is uniform, so the volume current vanishes, but we must usually take a surface current into account.

**EXAMPLE 11-1** Consider a sample in the shape of a long cylinder with uniform magnetization **M** parallel to its axis. Use Ampere's law to find the magnetic induction field in the interior, far from the ends.

**SOLUTION** The geometry is shown in Fig. 11-2. For **M** in the direction shown, the sample acts like a long solenoid with current out of the diagram on upper portions of the surface and into the diagram on lower portions. The field is nearly zero outside the sample and is along the axis inside. According to Ampere's law $\oint \mathbf{B} \cdot d\ell = \mu_0 I$ for any closed loop. Here the integral is a line integral around the loop and $I$ is the current through the

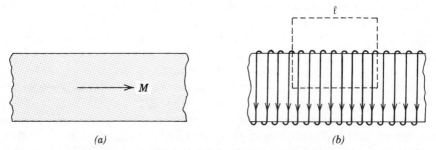

*(a)*          *(b)*

**FIGURE 11-2** The magnetic induction field inside a uniformly magnetized cylinder is the same as that inside a solenoid with surface current density $\mathbf{M} \times \hat{\mathbf{n}}$, shown in *(b)*. The current per unit length of solenoid is $M$ and is out of the page at the top of the solenoid. A dotted line shows the loop used to evaluate the integral in Ampere's law. The current through the loop is $M\ell$ where $\ell$ is its lentgth, so $\mathbf{B} = \mu_0\mathbf{M}$.

---

*See, for example, Chapter 9 of J. R. Reitz, F. J. Milford, and R. W. Christy, *Foundations of Electromagnetic Theory* (Reading, MA: Addison-Wesley, 1979).

loop. For the dotted loop in the figure the integral is $B\ell$, where $\ell$ is the length of the lower loop edge. The current though the loop is $M\ell$ so $B = \mu_0 M$. **B** and **M** are in the same direction. ∎

*Magnetic Susceptibility and Permittivity.* We now specialize to materials for which the magnetization vanishes unless a field is applied. If the field is sufficiently weak the magnetization at any point is proportional to **B** at that point. For historical reasons the relationship is described in terms of the magnetic field, defined by $\mathbf{H}(\mathbf{r}) = (1/\mu_0)\mathbf{B}(\mathbf{r}) - \mathbf{M}(\mathbf{r})$, rather than in terms of the magnetic induction field **B**. It is

$$\mathbf{M}(r) = \chi \mathbf{H}(r), \tag{11-11}$$

where the constant of proportionality $\chi$ is called the magnetic susceptibility of the sample. Equation 11-11 is analogous to Eq. 10-8 for polarization in the presence of an electric field. For some materials the magnetization is not parallel to the magnetic field and the susceptibility must be written as a tensor. We confine our discussion to solids with scalar susceptibilities.

To find the relationship between **M** and **B**, substitute $\mathbf{H} = \mathbf{M}/\chi$ into $\mathbf{H} = (1/\mu_0)\mathbf{B} - \mathbf{M}$ and solve for **M**. The result is

$$\mathbf{M} = \frac{\chi \mathbf{B}}{\mu_0(1 + \chi)}. \tag{11-12}$$

The quantity $\mu_0(1 + \chi)$ is called the permeability of the sample and is usually denoted by $\mu$. However, we will continue to use $\mu$ to denote a dipole moment and use $\mu_0(1 + \chi)$ for the permeability. For materials we consider, $|\chi| \ll 1$ and $\mathbf{M} = \chi \mathbf{B}/\mu_0$.

Unlike electric susceptibilities, magnetic susceptibilities can be either positive or negative. A diamagnetic solid has a negative magnetic susceptibility: **M** and **H** are in opposite directions and the permeability is less than $\mu_0$. A paramagnetic solid has a positive susceptibility: **M** and **H** are in the same direction and the permeability is greater than $\mu_0$.

**EXAMPLE 11-2** Suppose the sample of Example 11-1 has magnetic susceptibility $\chi$ and is placed in a uniform applied field $\mathbf{B}_a$, along its axis. Find the magnitude of the magnetization **M**, the magnetic induction field **B**, and the magnetic field **H** at points inside.

**SOLUTION** Suppose the material is paramagnetic so the magnetization is in the same direction as the applied field. According to the results of Example 11-1, the magnetization produces an induction field $\mu_0 \mathbf{M}$ in the same direction as the applied field, so $B = B_a + \mu_0 M$. Now $M = \chi B/\mu_0(1 + \chi) = \chi(B_a + \mu_0 M)/\mu_0(1 + \chi)$ or $M = \chi B_a/\mu_0$. The induction field is $B = B_a + \mu_0 M = (1 + \chi)B_a$ and the magnetic field is $H = (1/\mu_0)B - M = B_a/\mu_0$.

The results for $M$, $B$, and $H$ also hold if the material is diamagnetic. The susceptibility is then negative. $H$ is the same as before but $B$ is less than $B_a$. ∎

Susceptibility is determined experimentally by measuring the force on the sample when it is in an inhomogeneous applied field. To find an expression for the force, replace $\mu$ in Eq. 11-7 by $\mathbf{M} \, d\tau$ and integrate over the sample volume:

$$\mathbf{F} = \int (\mathbf{M} \cdot \nabla)\mathbf{B} \, d\tau. \tag{11-13}$$

In a Gouy balance, a long thin specimen is placed in a field and the component of the force along its axis is measured. If the $z$ axis is parallel to the long dimension of the sample, Eq. 11-13 yields $F_z = [\chi/2\mu_0(1 + \chi)]A(B_2^2 - B_1^2)$, where $A$ is the sample cross section, $B_1$ is the induction field at one end, and $B_2$ is the induction field at the other end. $F_z$, $A$, $B_1$, and $B_2$ are measured, then $\chi$ is calculated. In practice, one end of the sample is usually placed in a region of zero field.

In a Faraday or Curie balance an extremely small sample is placed in an inhomogeneous field. The force is given by $F_z = (\chi/2\mu)\tau_s \, dB^2/dz$, where $\tau_s$ is the sample volume. $F_z$ is measured and the result is used to determine $\chi$.

Inductance measurements are also used to determine $\chi$. The sample is used as a transformer core linking two circuits and the current in one is changed. The emf induced in the second circuit is proportional to the permeability $\mu_0(1 + \chi)$ of the sample. Since $\chi$ is typically on the order of $10^{-5}$, great precision is required.

## 11.2 DIAMAGNETISM AND PARAMAGNETISM

Susceptibilities of the chemical elements are plotted in Fig. 11-3. Elements in transition and rare earth series, with partially filled $d$ or $f$ shells, are strongly paramagnetic. Each of these atoms has a net angular momentum and nonvanishing dipole moment. In the absence of an applied field the moments are randomly distributed in angle but, as we will see, they tend to align with an applied field. Alkali metals are weakly paramagnetic. Individual atoms do not have dipole moments, but intrinsic moments of nearly free electrons tend to align with an applied field. The noble metals and most, but not all nonmetals are diamagnetic. In an applied field the atoms acquire dipole moments that are directed opposite the field.

We first discuss a classical model of diamagnetism, then show how a quantum mechanical calculation of $\chi$ is carried out for both diamagnetic and paramagnetic materials. In this chapter we concentrate on contributions of core electrons. Free electrons are discussed in the next chapter.

*Diamagnetic Response.* An electric field accompanies a magnetic field as it is turned on. Classically, the electric field changes the angular velocity of a circulating electron and, consequently, changes its orbital dipole moment. The change $\Delta\mu$ is directed opposite to the magnetic field and leads to diamagnetism.

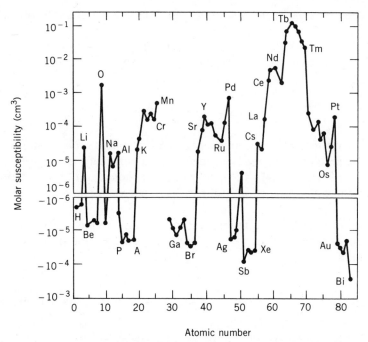

**FIGURE 11-3** Molar susceptibilities (in Gaussian units) of the chemical elements, plotted as a function of atomic number. Diamagnetic elements are plotted in the lower portion, paramagnetic in the upper. The gap is at iron, cobalt, and nickel, which are spontaneously magnetized. (From R. M. Bozorth, T. R. McGuire, and R. P. Hudson, "Magnetic Properties of Materials" in *American Institute of Physics Handbook* (D. W. Gray, Ed.) (New York: McGraw-Hill, 1967). Used with permission.)

Consider an electron executing uniform motion around a circle of radius $R$ in the $xy$ plane, as illustrated in Fig. 11-4, and suppose a magnetic field is turned on in the positive $z$ direction. The induced electric field is tangent to the orbit, as shown. Its magnitude can be obtained from Faraday's law: $\oint \mathcal{E} \cdot d\ell = -d\Phi_B/dt$, where the line integral is around the orbit and $\Phi_B$ is the magnetic flux through the orbit. The value of the integral is $2\pi R\mathcal{E}$ so $\mathcal{E} = (1/2\pi R)(d\Phi_B/dt)$. $\mathcal{E}$ is in the clockwise direction when viewed from the positive $z$ axis.

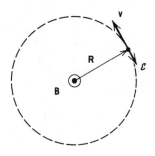

**FIGURE 11-4** An electron traversing a circular orbit in an increasing uniform magnetic field, out of the page. Electric field lines are circles and one of them coincides with the orbit. The torque on the electron is $-e\mathbf{r} \times \mathcal{E}$ and is out of the page. The change produced in the electron dipole moment by the torque is directed opposite the magnetic field; into the page for the situation shown.

The change in angular momentum is the time integral of the torque $eR\mathcal{E}\hat{z}$ so $\Delta L = (e/2\pi)\Delta\Phi_B\hat{z} = \frac{1}{2}eBR^2\hat{z}$, where we have written $\Phi_B = \pi R^2 B$ for the flux through the orbit when the field has reached its final value. The magnetic moment is $\mu = -(e/2m)(L_0 + \Delta L) = -(e/2m)L_0 - (e^2/4m)R^2B$, so the change induced by the field is

$$\Delta\mu = -\frac{e^2R^2}{4m}\,\mathbf{B}\,, \tag{11-14}$$

a quantity with magnitude proportional to $B$ and direction opposite to that of $\mathbf{B}$. The same result is obtained for an electron moving in the clockwise direction. Equation 11-14 is also valid if the field is not perpendicular to the orbit, provided $R$ is taken to be the projection of the orbit radius on a plane perpendicular to the field.

Unlike an orbital moment, the magnitude of a spin moment is fixed and does not change as a field is turned on. Spin moments do not contribute to diamagnetism.

*Quantum Mechanical Formulation.*    Classically, the total dipole moment of an electron is

$$\mu = -\frac{e}{2m}\,(\mathbf{L} + 2\mathbf{S}) - \frac{e^2R^2}{4m}\,\mathbf{B}\,. \tag{11-15}$$

The quantum mechanical prescription for calculating the dipole moment of an electron with wave function $\Psi$ closely follows Eq. 11-15. Each component of $\mathbf{L}$ and $\mathbf{S}$ is replaced by the corresponding average value for the electron state and $R^2$ is replaced by the average of the square of the projection of the electron position vector on a plane perpendicular to $\mathbf{B}$. If $\mathbf{B}$ is in the $z$ direction, for example, the $z$ component of $\mu$ is

$$\mu_z = -\frac{e}{2m}\,\langle L_z + 2S_z\rangle - \frac{e^2}{4m}\,\langle x^2 + y^2\rangle B\,. \tag{11-16}$$

An operator is associated with each component of angular momentum: $L_{z\,op}\Psi = -\hbar(x\,\partial\Psi/\partial y - y\,\partial\Psi/\partial x)$, for example. Then the average value is given by

$$\langle L_z\rangle = \int \Psi^*L_{z\,op}\Psi\,d\tau\,, \tag{11-17}$$

where the integral is over all space. For the $z$ component of spin $S_{z\,op}\Psi = \pm\frac{1}{2}\hbar\Psi$, where the positive sign is used for a spin-up electron and the negative sign is used for a spin-down electron. Thus $\langle S_z\rangle$ is $+\frac{1}{2}\hbar$ for some electrons and $-\frac{1}{2}\hbar$ for others. Finally,

$$\langle x^2 + y^2\rangle = \int \Psi^*(x^2 + y^2)\Psi\,d\tau\,. \tag{11-18}$$

The wave functions used to compute $\langle L_z \rangle$, $\langle S_z \rangle$, and $\langle x^2 + y^2 \rangle$ are solutions to a Schrödinger equation that includes the interaction of an electron with the magnetic field. A spin-orbit interaction term must also be included. In the rest frame of an electron, its environment is in motion and produces a magnetic field. The field is proportional to the orbital angular momentum of the electron and exerts a torque on the electron spin. In the laboratory rest frame both orbital and spin angular momenta exert torques on each other. Some consequences will be described later.

The magnetization at $\mathbf{r}$ is found by summing the dipole moments of all electrons in a macroscopically small region around $\mathbf{r}$, then dividing by the volume of the region. In many cases we can associate dipole moments with individual atoms. If all atoms are identical and each has dipole moment $\mathbf{\mu}$, then $\mathbf{M(r)} = n(\mathbf{r})\mathbf{\mu}$, where $n$ is the atomic concentration.

*Core Diamagnetism.* We first consider an inert gas solid, with all electrons in filled core shells. Then the sum $\Sigma \langle L_z + 2S_z \rangle$ over all electrons vanishes and the solid is diamagnetic.

Atomic orbitals can be used to estimate $\langle x^2 + y^2 \rangle$. If the probability density $\Psi^*\Psi$ for a state is spherically symmetric $\langle x^2 \rangle = \langle y^2 \rangle = \langle z^2 \rangle$ and $\langle x^2 + y^2 \rangle = \frac{2}{3}\langle r^2 \rangle$, where $r$ measures the distance from the nucleus. Although $\Psi^*\Psi$ may not be spherically symmetric for an individual state, the sum over all states in a shell is. If an atom contains $Z$ electrons, all in closed shells, and $\langle r^2 \rangle$ is their mean square distance from the nucleus, then

$$\mu_z = -\frac{Ze^2}{6m} \langle r^3 \rangle B \, , \tag{11-19}$$

for the atom as a whole.

The induction field that appears in Eq. 11-19 is the local field at the site of the atom, not the applied or macroscopic field. In principle, we should obtain an expression for the local field in terms of the macroscopic field. The derivation proceeds as in Section 10.1 and the result is $\mathbf{B}_{loc} = \mathbf{B} + \frac{1}{3}\mu_0\mathbf{M}$, where $\mathbf{B}$ is the macroscopic field. In practice, $\mu_0 M \ll B$, so the local field is very nearly the same as the macroscopic field and local field effects can be neglected. All atoms are identical, so $M = n\mu_z = -(nZe^2/6m)\langle r^2 \rangle B$ and

$$\chi = \frac{\mu_0 M}{B} = -\frac{\mu_0 nZe^2}{6m} \langle r^2 \rangle . \tag{11-20}$$

Most diamagnetic susceptibilities are nearly independent of temperature; slight dependence arises from changes in atomic concentrations that accompany thermal expansion.

Equation 11-20 may be used to compute the contribution of core electrons to the susceptibility of any solid. $Z$ is the number of core electrons in an atom and $\langle r^2 \rangle$ is their mean square distance from the nucleus. The equation also gives the contributions of closed shell ions to the susceptibility of an ionic solid. For

**TABLE 11-1** Experimentally Determined Molar Susceptibilities of Selected Closed Shell Atoms and Ions (expressed as $\chi_{molar}/4\pi$ in $10^{-12}$ m³/mol)

| Helium | −1.9 | Lithium (+) | −0.7 |
|---|---|---|---|
| Neon | −7.6 | Sodium (+) | −6.1 |
| Argon | −19 | Potassium (+) | −14.6 |
| Krypton | −29 | Rubidium (+) | −22.0 |
| Xenon | −44 | Cesium (+) | −35.1 |
| Fluorine (−) | −9.4 | Magnesium (2+) | −4.3 |
| Chlorine (−) | −24.2 | Calcium (2+) | −10.7 |
| Bromine (−) | −34.5 | Strontium (2+) | −18.0 |
| Iodine (−) | −50.6 | Barium (2+) | −29.0 |

*Source:* W. R. Myers, *Rev. Mod. Phys.* **24**:1, 1952.

example, it may be used to calculate the contributions of $Na^+$ or $Cl^-$ ions in NaCl. The susceptibility is the sum of terms, one for each ion type.

Most covalent and mixed covalent-ionic solids are diamagnetic, but Eq. 11-20 is valid only for the core contribution. Because electrons outside the core have wave functions that extend into interstitial regions and are far from spherically symmetric, it is not valid for them. Some semiconductors have small paramagnetic susceptibilities, a result of conduction electron contributions.

Some experimental values are listed in Table 11-1. The molar susceptibility is given by $\chi_{molar} = N_A\chi/n$, where $N_A$ is Avogadro's number. To obtain $\chi$ in Gaussian units, used in many handbooks, divide the SI value by $4\pi$.

**EXAMPLE 11-3** Estimate the susceptibility of solid argon. Argon has atomic number 18 and at 4 K its concentration is $2.66 \times 10^{28}$ atoms/m³. Take the root mean square distance of an electron from the nearest nucleus to be 0.62 Å. Also calculate the magnetization of solid argon in a 2.0 T induction field.

**SOLUTION** Substitute into Eq. 11-20 to obtain

$$\chi = -\frac{4\pi \times 10^{-7} \times 2.66 \times 10^{28} \times 18 \times (1.60 \times 10^{-19})^2 \times (0.62 \times 10^{-10})^2}{6 \times 9.11 \times 10^{-31}}$$

$$= -1.08 \times 10^{-5}.$$

The magnitude of the magnetization is given by

$$M = \frac{|\chi|B}{\mu_0} = \frac{1.08 \times 10^{-5} \times 2.0}{4\pi \times 10^{-7}} = 17.2 \text{ A/m}.$$

This value justifies the approximation $\mu_0 M \ll B$, used above. ∎

*Core Paramagnetism.* If $\langle L_z \rangle$ and $\langle S_z \rangle$ do not both vanish for an atom, the atom has a permanent magnetic dipole moment and is paramagnetic. Perhaps the easiest examples to understand are salts formed by rare earth or transition metal ions in combination with closed shell ions from the right side of the periodic table. $FeF_2$ and $GdCl_3$ are examples. Magnetic ions in salts are far enough apart that orbitals associated with partially filled shells do not overlap appreciably and, to a good approximation, each magnetic ion has a localized magnetic moment.

Suppose an ion has total orbital angular momentum **L**, total spin angular momentum **S**, and total angular momentum $\mathbf{J} = \mathbf{L} + \mathbf{S}$. A naive interpretation of Eq. 11-16 might lead us to expect the moment to be proportional to $-(\mathbf{L} + 2\mathbf{S})$, but it is not. As a result of spin-orbit interactions, both spin and orbital angular momenta precess about the direction of **J** and components of **L** and **S** along directions perpendicular to **J** average to zero. Spin-orbit interactions do not alter the magnitudes of **L** and **S**, so $\langle \mathbf{L} + 2\mathbf{S} \rangle = (\mathbf{L} + 2\mathbf{S}) \cdot \mathbf{J} \mathbf{J}/J^2$ and we may write

$$\boldsymbol{\mu} = -g\mu_B \frac{\mathbf{J}}{\hbar}, \tag{11-21}$$

where $g$ is the Landé $g$ factor, given by $(\mathbf{L} + 2\mathbf{S}) \cdot \mathbf{J}/J^2$, and $\mu_B$ is the Bohr magneton, given by $e\hbar/2m = 9.27 \times 10^{-24}$ J/T. Atomic angular momenta are on the order of $\hbar$ so dipole moments are on the order of $\mu_B$.

An expression for $g$ can be found in terms of $J^2$, $L^2$, and $S^2$. Since $\mathbf{J} = \mathbf{L} + \mathbf{S}$, $g = (L^2 + 2S^2 + 3\mathbf{L} \cdot \mathbf{S})/J^2$. Now $J^2 = L^2 + S^2 + 2\mathbf{L} \cdot \mathbf{S}$, so $\mathbf{L} \cdot \mathbf{S} = \frac{1}{2}(J^2 - L^2 - S^2)$ and $g = (3J^2 + S^2 - L^2)/2J^2$. This is usually written

$$g = 1 + \frac{J^2 + S^2 - L^2}{2J^2}. \tag{11-22}$$

The magnitudes of **L**, **S**, and **J** are quantized: $L^2$ has one of the values $L'(L' + 1)\hbar^2$, $S^2$ has one of the values $S'(S' + 1)\hbar^2$, and $J^2$ has one of the values $J'(J' + 1)\hbar^2$. Here $L'$ is a positive integer, while $S'$ and $J'$ are positive integers or positive half integers. The $z$ components of each of the angular momenta are also quantized. $J_z$, for example, is given by $M_J\hbar$, where $M_J$ can have any one of the values $-J'$, $-J' + 1, \ldots, +J'$. Similar statements hold for $S_z$ and $L_z$. States of a given shell, occupied by a given number of electrons, are characterized by the values of $L'$, $S'$, $J'$, and $M_J$. As described in many texts,* possible values can be found by considering individual electron angular momenta.

Electron-electron and spin-orbit interactions produce energy differences between states with different values of $J'$. For rare earth and transition metal ions, except europium and samarium, excited states are separated from the ground state by large energy differences and so are essentially unoccupied. We need not consider them. Hund's rules provide a way of determining $J'$, $L'$, and $S'$ for the ground state. These rules were originally developed empirically from spectroscopic data but they are substantiated by detailed model calculations.

Rule 1 states that each electron, up to half the number of states in the shell,

---

*See, for example, Chapter 8 of R. B. Leighton, *Principles of Modern Physics* (New York: McGraw-Hill, 1959).

contributes $+\frac{1}{2}$ to $S'$. Electrons beyond this number each contribute $-\frac{1}{2}$. Thus $S'$ has the largest possible value consistent with the Pauli exclusion principle. An $f$ shell can hold up to 14 electrons. If there are 7, then each contributes $+\frac{1}{2}$ to $S'$ and $S' = \frac{7}{2}$. If there are 8, then 7 contribute $+\frac{1}{2}$ each and 1 contributes $-\frac{1}{2}$ so $S' = 3$.

Each electron in a $d$ shell contributes either $-2, -1, 0, +1,$ or $+2$ to $L'$; each electron in an $f$ shell contributes either $-3, -2, -1, 0, +1, +2,$ or $+3$. Two electrons with the same spin cannot make the same contribution, however. Rule 2 states that $L'$ is the maximum value possible consistent with rule 1. If the shell is less than half full and the maximum contribution of an individual electron is $\ell$, then $L' = \ell + (\ell - 1) + (\ell - 2) + \cdots$, where the sum is over all electrons in the shell. If the shell is exactly half full, $L' = 0$ and, if the shell is more than half full, $L' = \ell + (\ell - 1) + (\ell - 2) + \cdots$, where the sum is over electrons in excess of those that half fill the shell.

Finally, $J'$ is given by rule 3: $J'$ is $|L' - S'|$ if the shell is less than half full and is $L' + S'$ if the shell is more than half full. $J' = S'$ for a shell that is exactly half full. In the ground state **L** and **S** are in opposite directions for a less than half full shell and in the same direction for a more than half full shell. Rule 3 originates in the spin-orbit interaction.

**EXAMPLE 11-4**   Find the Landé $g$ factor for the ground state of a praseo-dymium ion, with 2 $f$ electrons, and for the ground state of an erbium ion, with 11 $f$ electrons.

**SOLUTION**   Both electrons in praseodymium have spin $+\frac{1}{2}$ so $S' = 1$. The largest value of $L'$ is obtained if one electron contributes 3 and the other 2, so $L' = 5$. Note that they cannot both contribute 3 because their contri-butions to $S'$ are the same. Finally, the shell is less than half full so $J' = |L' - S'| = 5 - 1 = 4$. According to Eq. 11-22,

$$g = 1 + \frac{4 \times 5 + 1 \times 2 - 5 \times 6}{2 \times 4 \times 5} = 0.55.$$

Seven electrons in erbium contribute $+\frac{1}{2}$ to $S'$ while four contribute $-\frac{1}{2}$, so $S' = \frac{3}{2}$. Two electrons contribute 3 to $L'$, two contribute 2, two contribute 1, and two contribute 0. The other three contribute $-1, -2,$ and $-3$, re-spectively, so $L' = 6$. The shell is more than half full so $J' = L' + S' = \frac{15}{2}$. Equation 11-22 yields $g = 1.2$   ∎

Consider a salt in which all magnetic ions are identical and have the same value of $J'$, the value appropriate to the ground state. In the absence of a magnetic field every allowed value of $J_z$ is equally likely, so the average value of an ionic dipole moment is zero. Suppose, however, a magnetic field exists in the positive $z$ direction. Then states with different values of $J_z$ have different energies and different probabilities of occupation. The $z$ component of a dipole moment is $\mu_z = -g\mu_B(J_z/\hbar) = -g\mu_B M_J$ and its energy is $E = -\mu_z B = +g\mu_B M_J B$.

To illustrate, energy levels for an ion with $J' = 2$ are diagrammed in Fig. 11-5. An ion is most likely to be in the lowest energy state, that with $M_J = -J'$ and $\mu_z = +g\mu_B J'$. Its dipole moment is in nearly the same direction as the field. On the other hand, the ion is least likely to be in the state with $M_J = +J'$ and $\mu_z = -g\mu_B J'$. In short, more ions have moments with positive $z$ components than have moments with negative $z$ components.

The probability that an ion is in a state with $J_z = M_J\hbar$ is proportional to $e^{-\beta E} = e^{-\beta g\mu_B M_J B}$, so the $z$ component of the average dipole moment is given by

$$\langle \mu_z \rangle = \frac{\Sigma - g\mu_B M_J e^{-\beta\mu_B M_J B}}{\Sigma e^{-\beta g\mu_B M_J B}}, \tag{11-23}$$

where both sums run from $M_J = -J'$ to $M_J = +J'$. Equation 11-23 can be written in closed form: $\langle \mu_z \rangle = g\mu_B J' B_{J'}(\beta g\mu_B J' B)$, where $B_{J'}$ is a Brillouin function, defined by

$$B_{J'}(x) = \frac{2J' + 1}{2J'} \coth\left[\frac{(2J' + 1)x}{2J'}\right] - \frac{1}{2J'} \coth\left[\frac{x}{2J'}\right]. \tag{11-24}$$

The magnetization is given by $n\langle \mu_z \rangle$, where $n$ is the concentration of magnetic ions. Thus

$$M = ng\mu_B J' B_{J'}(\beta g\mu_B J' B). \tag{11-25}$$

If states with other values of $J'$ are nearby in energy, both the numerator and denominator of Eq. 11-23 must be summed over $J'$.

Several Brillouin functions are plotted as functions of $x = g\mu_B J' B$ in Fig. 11-6. If $g\mu_B J' B \gg k_B T$ ($x \gg 1$ in the figure), level separations are much greater than the average thermal energy and nearly all ions are in the lowest state. All dipoles are then aligned with the field and the magnetization is said to be saturated. $B_{J'}$ is nearly 1 and $M = ng\mu_B J'$.

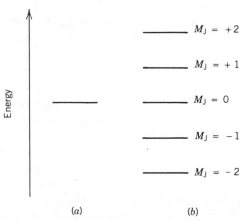

(a)　　　　　　　　(b)

**FIGURE 11-5** Energy levels for an ion with $J' = 2$. (a) No magnetic field is applied. (b) In a magnetic field $B\hat{z}$ adjacent levels are separated by $g\mu_B B$. More ions are in the lowest level than any other. Their dipole moments are aligned with the field.

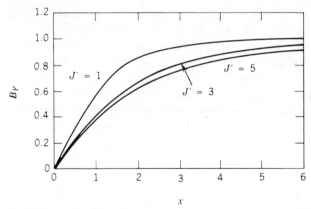

**FIGURE 11-6** The Brillouin functions $B_{J'}(x)$ for $J' = 1$, 3, and 5. As $x$ becomes large each function tends toward 1. For $x$ small, $B_{J'}$ tends toward $(J' + 1)x/3J'$, a linear function of $x$.

If $g\mu_B J'B \ll k_B T$ ($x \ll 1$ in the figure), an ion has nearly the same probability of being in any of the states and the magnetization is small. We use a power series to evaluate the Brillouin function in the limit of small $x$. The expansion of $\coth(u)$ is $(1/u) + (u/3) - (u^2/45) + \cdots$ and the leading term of $B_{J'}(x)$ is $(J' + 1)x/3J'$, so the magnetization is

$$M = n\beta g^2 \mu_B^2 \frac{J'(J' + 1)}{3} B. \tag{11-26}$$

In weak fields the magnetization is proportional to the induction field and the sample is a linear paramagnetic. Of course, diamagnetic contributions must also be included, but these are typically small compared to paramagnetic contributions.

Equation 11-26 is valid for most paramagnetic salts in laboratory fields at room temperature. Strictly speaking, the induction field $B$ in this equation is the local field at a magnetic ion. Here we neglect local field corrections and assume $\chi \ll 1$. Then

$$\chi = \frac{\mu_0 M}{B} = \frac{C}{T}, \tag{11-27}$$

where $C = \mu_0 ng^2 \mu_B^2 J'(J' + 1)/3k_B$. This is the Curie law for the susceptibility of a paramagnetic salt. It can be verfied experimentally by plotting $1/\chi$ as a function of $T$, as in Fig. 11-7.

The Curie constant can be written $C = \mu_0 np^2 \mu_B^2/3k_B$, where $p$ is called the effective number of Bohr magnetons per ion and is given by $p = g\sqrt{J'(J' + 1)}$. Once $C$ is obtained from experimental data, it can be used to estimate $p$. Some values are given in Table 11-2. For rare earth ions these compare favorably with $g\sqrt{J'(J' + 1)}$ and so substantiate the theory presented above.

Agreement is not obtained, however, for transition metal ions. Unlike $f$ wave functions, $d$ wave functions extend to interstitial regions and, as a consequence,

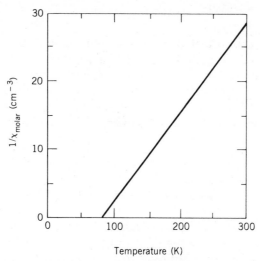

FIGURE 11-7 The reciprocal of the molar susceptibility (in Gaussian units) as a function of temperature for EuO. The line represents theoretical values for noninteracting ions in states with $L' = 0$ and $S' = \frac{7}{2}$, while the dots represent experimental values. The linearity of the plot attests to the validity of the Curie law. (From D. H. Martin, *Magnetism in Solids* (Cambridge, MA: MIT Press, 1967). Used with permission.)

TABLE 11-2 Effective Number of Bohr Magnetons $p$ for Rare Earth and Transition Metal Ions (Experimentally Determined)

| Ion | Number of Electrons in Shell | $L'$ | $S'$ | $J'$ | $P$ |
|---|---|---|---|---|---|
| Cerium(3+) | 1 | 3 | $\frac{1}{2}$ | $\frac{5}{2}$ | 2.39 |
| Praseodymium(3+) | 2 | 5 | 1 | 4 | 3.60 |
| Neodymium(3+) | 3 | 6 | $\frac{3}{2}$ | $\frac{9}{2}$ | 3.62 |
| Promethium(3+) | 4 | 6 | 2 | 4 | — |
| Samarium(3+) | 5 | 5 | $\frac{5}{2}$ | $\frac{5}{2}$ | 1.54 |
| Europium(3+) | 6 | 3 | 3 | 0 | 3.61 |
| Gadolinium(3+) | 7 | 0 | $\frac{7}{2}$ | $\frac{7}{2}$ | 8.2 |
| Terbium(3+) | 8 | 3 | 3 | 6 | 9.6 |
| Dysprosium(3+) | 9 | 5 | $\frac{5}{2}$ | $\frac{15}{2}$ | 10.5 |
| Holmium(3+) | 10 | 6 | 2 | 8 | 10.5 |
| Erbium(3+) | 11 | 6 | $\frac{3}{2}$ | $\frac{15}{2}$ | 9.5 |
| Thulium(3+) | 12 | 5 | 1 | 6 | 7.2 |
| Ytterbium(3+) | 13 | 3 | $\frac{1}{2}$ | $\frac{7}{2}$ | 4.4 |
| Vanadium(2+) | 3 | 3 | $\frac{3}{2}$ | $\frac{3}{2}$ | 3.8 |
| Chromium(2+) | 4 | 2 | 2 | 0 | 4.9 |
| Manganese(2+) | 5 | 0 | $\frac{5}{2}$ | $\frac{5}{2}$ | 5.9 |
| Iron(2+) | 6 | 2 | 2 | 4 | 5.4 |
| Cobalt(2+) | 7 | 3 | $\frac{3}{2}$ | $\frac{9}{2}$ | 4.8 |
| Nickel(2+) | 8 | 3 | 1 | 4 | 3.2 |
| Copper(2+) | 9 | 2 | $\frac{1}{2}$ | $\frac{5}{2}$ | 1.9 |

*Source: American Institute of Physics Handbook* (D. W. Gray, Ed.) (New York: McGraw-Hill, 1963).

are more closely approximated by linear combinations of atomic $d$ orbitals than by single orbitals. $\langle L_z \rangle$ is quite small for the appropriate linear combinations and orbital angular momentum is said to be quenched. Equations 11-26 and 11-27 are valid, but $J' \approx S'$ and $g \approx 2$. The effective number of Bohr magnetons per ion is then $2\sqrt{S'(S' + 1)}$, a result that agrees quite well with experimental data. The spin quantum number $S'$ is still given by Hund's first rule.

**EXAMPLE 11-5**  Compute the effective number of Bohr magnetons for a nickel ion. Assume first that orbital angular momentum is not quenched, then that it is. Nickel has 8 electrons in its $3d$ shell.

**SOLUTION**  Since a $d$ shell can hold up to 10 electrons, 5 electrons in nickel contribute $+\frac{1}{2}$ to $S'$ and three contribute $-\frac{1}{2}$, so $S' = 1$. If quenching is ignored, 2 electrons contribute $+2$ to $L'$, 2 contribute $+1$, 2 contribute 0, 1 contributes $-1$, and 1 contributes $-2$, so $L' = 3$. The shell is more than half full so $J' = L' + S' = 4$ and $g = 1.25$. The effective number of Bohr magnetons is $g\sqrt{J'(J' + 1)} = 1.25 \times \sqrt{20} = 5.59$. If, on the other hand, $L' = 0$ then $J' = S' = 1$ and $g = 2$, so $p = 2 \times \sqrt{2} = 2.83$. Actually quenching is not quite complete in nickel: the experimental value for $p$ is 3.2. ∎

## 11.3  SPONTANEOUS MAGNETIZATION AND FERROMAGNETISM

Spontaneous magnetization occurs in some materials composed of atoms with unfilled shells. Atomic dipole moments then exhibit long-range order, which may take one of the forms depicted schematically in fig. 11-8. In a ferromagnet, moments tend to be aligned; in an antiferromagnet or a ferrimagnet, atoms with moments in one direction are interspersed in a systematic way with atoms having moments in the opposite direction. Magnitudes of oppositely directed moments are not the same in a ferrimagnet, so the net magnetization does not vanish. Oppositely directed moments in an antiferromagnet have the same magnitude, so the net magnetization vanishes.

Iron, nickel, and cobalt are ferromagnetic except at high temperature. Gadolinium and terbium are ferromagnetic below room temperature and other rare earths are ferromagnetic at extremely low temperature. At intermediate temperatures all rare earths except gadolinium are antiferromagnetic. At high temperatures transition and rare earth metals are all paramagnetic.

*Spontaneous Magnetization.*  Spontaneous magnetization and long-range magnetic order can be understood in terms of a strong local field, called the Weiss effective field, at the site of each dipole. Later we will discuss the source of this field and explain why fields are strong in some solids and weak in others. For now we assume the local induction field at any atom can be written $\mathbf{B}_{loc} = \mathbf{B}_a + \mu_0 \gamma \mathbf{M}$, where $\mathbf{B}_a$ is an applied induction field, $\mathbf{M}$ is the magnetization, and $\gamma$ is a constant. For simplicity we ignore the field due to magnetization current on sample surfaces.

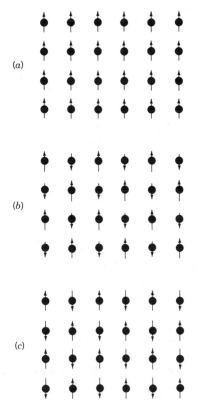

**FIGURE 11-8** Schematic representation of three types of magnetic ordering in solids. (a) Ferromagnetic: all spins are aligned. (b) Ferrimagnetic: spins on different sublattices are in different directions. The net magnetization does not vanish. (c) Antiferromagnetic: the same type structure as a ferrimagnet but the net magnetization vanishes.

Consider a collection of $n$ identical atoms per unit volume and suppose each atom has angular momentum $J$. Take all fields to be in the $z$ direction. If $B$ is replaced by $B_a + \mu_0 \gamma M$, Eq. 11-25 becomes

$$M = M_s B_{J'}[\beta g \mu_B J'(B_a + \mu_0 \gamma M)], \qquad (11\text{-}28)$$

where $M_s$ is the saturation magnetization, given by $n g \mu_B J'$. To find the spontaneous magnetization we set $B_a = 0$ and solve Eq. 11-28 for $M$. A solution cannot be obtained algebraically so we resort to a graphical technique. If $x = \mu_0 \beta g \mu_B J' \gamma M$, Eq. 11-28 becomes $x/\mu_0 \beta g \mu_B J' \gamma M_s = B_{J'}(x)$. Curve $a$ of Fig 11-9 shows $B_{J'}(x)$ while curve $b$ shows $x/\mu_0 \beta g \mu_B J' \gamma M_s$, both plotted as functions of $x$. The intersection of these two curves, at the point marked $A$, represents the solution to Eq. 11-28.

At low temperature the slope of curve $b$ is small and the intersection occurs at a point for which the magnetization is nearly saturated. As the temperature increases so does the slope of curve $b$, indicating that the magnetization decreases. Curve $c$ corresponds to a temperature for which the slope of the straight line is greater than the limiting slope of the Brillouin function. Now the only solution to Eq. 11-28 is $M = 0$, indicating that spontaneous magnetization cannot occur at this temperature. The sample is then paramagnetic. Figure 11-10 illustrates the temperature dependence of spontaneous magnetization.

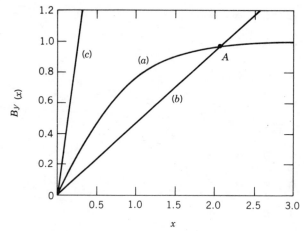

**FIGURE 11-9** Curve (a) is the Brillouin function $B_{J'}(x)$ for $J' = \frac{1}{2}$. Curves (b) and (c) represent $x/\mu_0\beta g\mu_B J'\gamma M_s$ for different temperatures. Solutions to Eq. 11-28 with $B_a = 0$ are given by the intersection of (a) with a straight line. At low temperature a line such as (b) is appropriate. It intersects (a) at A and spontaneous magnetization occurs. At high temperature a line such as (c) is appropriate. It intersects (a) only at $x = 0$ and spontaneous magnetization does not occur.

An expression for the Curie temperature $T_c$, which marks the boundary between ferromagnetic and paramagnetic behavior, can be found by equating the limiting slope of the Brillouin function, given by $(J' + 1)/3J'$, to the slope of curve $b$, given by $1/\mu_0\beta g\mu_B J'\gamma M_s$, then solving for $T$. It is

$$T_c = \frac{\mu_0 ng^2\mu_B^2\gamma}{3k_B}J'(J' + 1),\qquad(11\text{-}29)$$

after $M_s$ is replaced by $ng\mu_B J'$.

We now investigate the paramagnetic region. If an external field $B_a$ is applied,

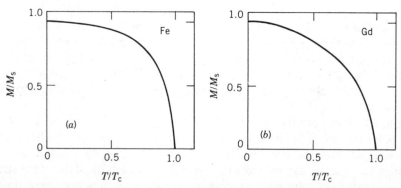

**FIGURE 11-10** Spontaneous magnetization as a function of temperature for single domain samples of (a) iron and (b) gadolinium. The magnetization is saturated at $T = 0$ K and vanishes at $T = T_c$ and above. (From D. H. Martin, *Magnetism in Solids* (Cambridge, MA: MIT Press, 1967). Used with permission.)

curve $a$ of Fig. 11-9 shifts to the left by $\beta g \mu_B J' B_a$ and intersects curve $b$ no matter what the values of $T$ and $B_a$. For typical laboratory fields applied at temperatures above $T_c$, the intersection is near the origin, where the Brillouin function is essentially linear.

For small fields and high temperatures, Eq. 11-28 becomes

$$M = \tfrac{1}{3} n \beta g^2 \mu_B^2 J'(J' + 1)(B_a + \mu_0 \gamma M), \qquad (11\text{-}30)$$

so

$$M = \frac{C}{\mu_0(T - C\gamma)} B_a, \qquad (11\text{-}31)$$

where $C$ is the Curie constant $\mu_0 n g^2 \mu_B^2 J'(J' + 1)/3k_B$. Since $C\gamma = T_c$ and $\chi = \mu_0 M/B_a$, the susceptibility is given by

$$\chi = \frac{C}{T - T_c}. \qquad (11\text{-}32)$$

This is the Curie-Weiss law for a ferromagnet in the paramagnetic region.

Curie-Weiss theory is qualitatively correct, at least for temperatures well above the Curie temperature. As Fig. 11-11 shows for nickel, $1/\chi$ as a function of $T$ is a straight line for high temperatures. Near the Curie temperature, however, the susceptibility is found to increase faster than $1/(T - T_c)$ as $T$ approaches $T_c$ from above.

Passage through the Curie temperature results in a significant change in magnetic order. Below $T_c$ most dipole moments are aligned, while above they are not. A large entropy change is associated with the change in order and, since $C_\tau = T(\partial S/\partial T)_\tau$, the heat capacity increases sharply around the Curie temperature. To illustrate, the heat capacity of nickel is shown in Fig. 11-12. Magnetic contributions to the heat capacity extend to temperatures above the Curie temperature and provide evidence of short-range magnetic order in the paramagnetic region.

Table 11-3 gives the experimentally determined Curie temperature and effective number of Bohr magnetons for ferromagnetic elements. For ferromagnets,

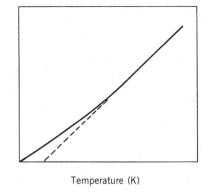

Temperature (K)

**FIGURE 11-11** Reciprocal of the magnetic susceptibility as a function of temperature for a ferromagnet above its Curie temperature. The dotted line is a continuation of the high-temperature curve.

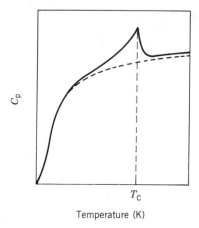

$C_p$

$T_c$

Temperature (K)

**FIGURE 11-12** Heat capacity at constant pressure for a ferromagnet. The peak occurs at the Curie temperature. The dotted curve shows the specific heat with the magnetic contribution subtracted.

$p$ is determined by the maximum $z$ component of an atomic dipole moment rather than by the magnitude of the moment: $p = gJ'$. For transition metal elements, $p$ is not strictly consistent with the discussion given above. If, for example, orbital angular momentum is completely quenched, $g = 2$ and $p$ should be an integer. Even if incomplete quenching is taken into account, experimental values do not correspond to integer numbers of electrons per atom. The discrepancy can be explained in terms of energy band theory. Before we consider the energy bands of a ferromagnet, however, we discuss the origin of the local field.

*Heisenberg Exhange Theory.* The magnetic field produced by atomic dipoles cannot account for the Weiss local field in a ferromagnet. The Curie constant for iron, with $g = 2, J' = 1$, and $n = 8.5 \times 10^{28}$ atoms/m³, is 1.77 K. $T_c = 1043$ K according to Table 11-3, so $\gamma = T_c/C = 1043/1.77 = 588$. $\gamma$ is much larger than $\frac{1}{3}$, the value it would have if the field were due to magnetization.

The dominant part of the local field is not a true magnetic field and does not, for example, enter Maxwell's equations. Like a true magnetic field, however, it

**TABLE 11-3** Saturation Magnetization $M_s$, Curie Temperature $T_c$, and Effective Number of Bohr Magnetons $p$ for Ferromagnetic Elements

| Material | $M_s$ (10⁶ A/m) | $T_c$ (K) | $p$ |
|---|---|---|---|
| Iron | 1.75 | 1043 | 2.219 |
| Cobalt | 1.45 | 1404 | 1.715 |
| Nickel | 0.512 | 631 | 0.604 |
| Gadolinium | 2.00 | 289 | 7.12 |
| Terbium | 1.44 | 230 | 4.95 |
| Dysprosium | 2.01 | 85 | 6.84 |
| Holmium | 2.55 | 20 | 8.54 |

*Sources: American Institute of Physics Handbook* (D. W. Gray, Ed.) (New York: McGraw-Hill, 1963).

exerts torques on spins and changes the energy levels of atoms. As we will see, its origin and strength can be understood in terms of the Pauli exclusion principle and electrostatic interactions between charges in solids.

In its fundamental form, the exclusion principle requires the wave function for a collection of electrons to be antisymmetric in the interchange of particle coordinates and spins. For illustrative purposes we consider a system composed of only two electrons. Let $\Psi(\mathbf{r}_1, \mathbf{s}_1; \mathbf{r}_2, \mathbf{s}_2)$ be a two-electron wave function. One electron has position vector $\mathbf{r}_1$ and spin $\mathbf{s}_1$, while the other has position vector $\mathbf{r}_2$ and spin $\mathbf{s}_2$. The exclusion principle requires $\Psi(\mathbf{r}_2, \mathbf{s}_2; \mathbf{r}_1, \mathbf{s}_1) = -\Psi(\mathbf{r}_1, \mathbf{s}_1; \mathbf{r}_2, \mathbf{s}_2)$.

To a good approximation the wave function is the product of space and spin parts: $\Psi(\mathbf{r}_1, \mathbf{s}_2; \mathbf{r}_2, \mathbf{s}_2) = f(\mathbf{r}_1, \mathbf{r}_2)g(\mathbf{s}_1, \mathbf{s}_2)$. The spin part is antisymmetric if the two spins are antiparallel and symmetric if they are parallel. In the first case the total spin is 0 and, because there is only one possible value for the $z$ component, the state is called a singlet state. The space part must be symmetric. In the second case the total spin is $\hbar$ and, because the $z$ component may be either $-\hbar$, 0, or $+\hbar$, the state is called a triplet state. The space part must be antisymmetric.

If $\psi_A$ and $\psi_B$ are single particle functions, $\Psi_s = N_s[\psi_A(\mathbf{r}_1)\psi_B(\mathbf{r}_2) + \psi_B(\mathbf{r}_1)\psi_A(\mathbf{r}_2)]$ is symmetric in the coordinates and can be used to represent a singlet state while $\Psi_t = N_t[\psi_A(\mathbf{r}_1)\psi_B(\mathbf{r}_2) - \psi_B(\mathbf{r}_1)\psi_A(\mathbf{r}_2)]$ is antisymmetric and can be used to represent a triplet state. $N_s$ and $N_t$ are normalization constants, chosen so $\int \Psi^*\Psi \, d\tau_1 d\tau_2 = 1$ in each case. Since $N_s \approx N_t$, we denote both by the same constant $N$.

Suppose the potential energy function $U(\mathbf{r}_1, \mathbf{r}_2)$ describes interactions of the electrons with ions of the material as well as with each other. It is symmetric: $U(\mathbf{r}_2, \mathbf{r}_1) = U(\mathbf{r}_1, \mathbf{r}_2)$. We estimate the potential energy for either the singlet or triplet state by evaluating integrals of the form $\langle U \rangle = \int \Psi^*U\Psi \, d\tau_1 d\tau_2$. For the singlet state

$$\langle U \rangle_s = 2|N|^2 \int \psi_A^*(\mathbf{r}_1)\psi_B^*(\mathbf{r}_2)U(\mathbf{r}_1, \mathbf{r}_2)\psi_A(\mathbf{r}_1)\psi_B(\mathbf{r}_2) \, d\tau_1 d\tau_2$$
$$+ 2|N|^2 \int \psi_A^*(\mathbf{r}_1)\psi_B^*(\mathbf{r}_2)U(\mathbf{r}_1, \mathbf{r}_2)\psi_B(\mathbf{r}_1)\psi_A(\mathbf{r}_2) \, d\tau_1 d\tau_2. \quad (11\text{-}33)$$

The expression for the triplet state is the same except the second term is subtracted from the first. The states differ in energy by

$$\langle U \rangle_s - \langle U \rangle_t = 4|N|^2 \int \psi_A^*(\mathbf{r}_1)\psi_B^*(\mathbf{r}_2)U(\mathbf{r}_1, \mathbf{r}_2)\psi_B(\mathbf{r}_1)\psi_A(\mathbf{r}_2) \, d\tau_1 d\tau_2. \quad (11\text{-}34)$$

The integral that appears in Eq. 11-34 is called an exhange integral since the arguments of the single particle functions on the left side of the integrand are exchanged with each other to produce the right side. For example, on the left side $\mathbf{r}_1$ is the argument of $\psi_A^*$, while on the right side it is the argument of $\psi_B$. $\langle U \rangle_s - \langle U \rangle_t$ is called an exchange energy.

Both singlet and triplet state potential energy can be written in the form

$$\langle U \rangle = U_0 - \frac{J_e}{\hbar^2} \mathbf{s}_1 \cdot \mathbf{s}_2, \quad (11\text{-}35)$$

where $J_e$ is called an exchange coefficient. Since $s^2 = |\mathbf{s}_1 + \mathbf{s}_2|^2 = s_1^2 + s_2^2 + 2\mathbf{s}_1 \cdot \mathbf{s}_2$ and $s_1^2 = s_2^2 = \frac{3}{4}\hbar^2$, $\mathbf{s}_1 \cdot \mathbf{s}_2 = \frac{1}{2}(s^2 - \frac{3}{2}\hbar^2)$. In the singlet state $s^2 = 0$ so $\mathbf{s}_1 \cdot \mathbf{s}_2 = -\frac{3}{4}\hbar^2$. In the triplet state $s^2 = 2\hbar^2$ so $\mathbf{s}_1 \cdot \mathbf{s}_2 = +\frac{1}{4}\hbar^2$. Equation 11-35 predicts $\langle U \rangle_s - \langle U \rangle_t = -(J_e/\hbar^2)(-\frac{3}{4} - \frac{1}{4})\hbar^2 = J_e.$*

The sign of $J_e$ is important. If it is positive parallel spins have lower energy than antiparallel spins. The sign depends on the potential energy function and on the nature of the single-particle wave functions. To understand how different signs might come about, write $U(\mathbf{r}_1, \mathbf{r}_2) = U_i(\mathbf{r}_1) + U_i(\mathbf{r}_2) + U_{ee}(\mathbf{r}_1, \mathbf{r}_2)$, where the first two terms give the potential energy of electron-ion interactions while the last gives the potential energy of electron-electron interactions: $U_{ee} = (e^2/4\pi\epsilon_0)(1/|\mathbf{r}_1 - \mathbf{r}_2|)$.

The electron-electron term contributes

$$[\langle U \rangle_s - \langle U \rangle_t]_{ee} = 4|N|^2 \int \psi_A^*(\mathbf{r}_1)\psi_B^*(\mathbf{r}_2)U_{ee}\psi_B(\mathbf{r}_1)\psi_A(\mathbf{r}_2) \, d\tau_1 d\tau_2 \quad (11\text{-}36)$$

to $J_e$. The integral is the energy of a charge distribution with charge density $-e\psi_A^*(\mathbf{r})\psi_B(\mathbf{r})$ and is positive. If it were the only contribution, most electron spins would be parallel and the system would be ferromagnetic.

The contribution of the electron-ion terms are identical and their sum is given by

$$[\langle U \rangle_s - \langle U \rangle_t]_i = 8|N|^2 \int \psi_A^*(\mathbf{r}_2)\psi_B(\mathbf{r}_2) \, d\tau_2 \int \psi_B^*(\mathbf{r}_1)U_i(\mathbf{r}_1)\psi_A(\mathbf{r}_1) \, d\tau_1 . \quad (11\text{-}37)$$

If $\psi_A$ and $\psi_B$ represent two different atomic orbitals centered on the same atom, then $\int \psi_A^*(\mathbf{r}_2)\psi_B(\mathbf{r}_2) \, d\tau_2 = 0$ and $J_e$ is given by Eq. 11-36. Alignment of the spins minimizes the energy. This is the basis of Hund's first rule.

More importantly for the study of magnetic structures, if the two single-particle functions are centered on different atoms, the right side of Eq. 11-37 is not zero. In fact, since $U_i$ results from the attraction of ions for an electron, it is negative. The second integral in Eq. 11-37 is an overlap integral, of the type considered in Chapter 5 as part of the discussion of bonding. It is large if the atoms are close and small if they are far apart. The sign of $J_e$ depends on the relative magnitudes of the two contributions. For small interatomic separations, $J_e$ is negative and antiparallel spin states have lower energy than parallel spin states. For large interatomic distances the reverse is true.

Equation 11-35 is often generalized by writing

$$E = -\frac{J_e}{\hbar^2} \mathbf{S}_i \cdot \mathbf{S}_j , \quad (11\text{-}38)$$

for the exhange energy of two atoms with spins $\mathbf{S}_i$ and $\mathbf{S}_j$, respectively. $J_e$ has a more complicated form than for two electrons but, as indicated by Fig. 11-13, it is negative for small atomic separations and positive for large. The change in sign explains why some transition metals are ferromagnetic and some are not. Manganese, for example, is not ferromagnetic but some of its compounds, such

---

*Many authors write $\langle U \rangle = U_0 - (2J_e/\hbar^2)\mathbf{s}_1 \cdot \mathbf{s}_2$, where $J_e = (\langle U \rangle_s - \langle U \rangle_t)/2$.

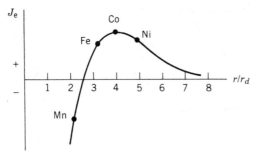

**FIGURE 11-13** Heisenberg exchange coefficient $J_e$ as a function of interatomic distance $r$ for transition metals. Here $r_d$ is the average distance of 4d electrons from a nucleus. Iron, cobalt, and nickel have positive coefficients and are ferromagnetic. Manganese has a negative coefficient and is antiferromagnetic.

as MnSb and MnAs, are. Manganese atoms are farther apart in compounds than in pure manganese.

To round out the Heisenberg theory, we formulate the Weiss local field in terms of the exhange coefficient. According to Eq. 11-38, the exchange energy of atom $i$ is $-(1/\hbar^2)(\Sigma J_e \mathbf{S}_j) \cdot \mathbf{S}_i$, where the sum is over all atoms except atom $i$. If the orbital contribution is quenched, the dipole moment of atom $i$ is $-(g\mu_B/\hbar)\mathbf{S}_i$ and its exchange energy can be written $(1/g\mu_B\hbar)(\Sigma J_e \mathbf{S}_j) \cdot \boldsymbol{\mu}$. Thus the exchange interaction can be replaced by the interaction between a spin dipole moment and an effective induction field given by $-(1/g\mu_B\hbar) \Sigma J_e \mathbf{S}_j$. The local field at the site of atom $i$ is

$$\mathbf{B}_{loc} = \mathbf{B}_a - \frac{1}{g\mu_B\hbar} \sum_j J_e \mathbf{S}_j = \mathbf{B}_a - \frac{zJ_e\mathbf{S}}{g\mu_B\hbar}, \tag{11-39}$$

where $\mathbf{B}_a$ is the applied field and the sum is over all atoms except atom $i$. The second form is valid if all atoms have the same spin $\mathbf{S}$ and $J_e$ vanishes except for the $z$ nearest neighbors of atom $i$.

Since $\mathbf{M} = n\boldsymbol{\mu} = -(ng\mu_B/\hbar)\mathbf{S}$, the effective exchange field is $(zJ_e/ng^2\mu_B^2)\mathbf{M}$ and the constant $\gamma$ that appears in Eq. 11-28 is given by

$$\gamma = \frac{zJ_e}{\mu_0 ng^2\mu_B^2}. \tag{11-40}$$

Equation 11-29 can be used to find an expression for the Curie temperature in terms of the exchange coefficient.

**EXAMPLE 11-6** Use the experimental value of the Curie temperature to estimate the exchange coefficient $J_e$ for iron. The concentration of iron atoms is $8.5 \times 10^{28}$ atoms/m³ and each iron atom has 12 nearest neighbors. Take $g = 2$.

**SOLUTION**  We have already found $\gamma = 588$ for iron. According to Eq. 11-40,

$$J_e = \frac{\mu_0 n g^2 \mu_B^2 \gamma}{z} = \frac{4\pi \times 10^{-7} \times 8.5 \times 10^{28} \times 4 \times (9.28 \times 10^{-24})^2 \times 588}{12}$$

$$= 1.8 \times 10^{-21} J = 11 \text{ meV}.$$

Electrostatic interactions easily account for this value. ∎

Overlap is small for $f$ shell wave functions in rare earth metals and the associated exchange field is weak. Nevertheless, exchange interactions between $f$ electrons and nearly free electrons gives rise to ferromagnetism in these materials at low temperatures. If $f$ electrons around one atom are spin down, say, exchange interactions lower energies of spin-up free electrons in the vicinity. These electrons move to other atoms where exchange interactions tend to lower energies of spin-down $f$ electrons. This process is known as indirect exchange.

In some compounds a nonmagnetic atom resides between two magnetic atoms. Since the magnetic atoms are far apart, direct exchange does not produce ferromagnetism. Electrons on nonmagnetic atoms, however, act as intermediaries. Superexchange, as this process is called, is responsible for ferromagnetism in oxides and other compounds of the transition metal elements.

*Electron Bands in Ferromagnets.*    The theory presented above must be modified to account for transition metal ferromagnetism. As indicated in Table 11-3, the effective number of Bohr magnetons is 2.219 for iron and 0.604 for nickel. These numbers are quite different from those predicted by $p = gS'$. Since wave functions for $d$ electrons in transition metals are not localized, electron band theory must be used.

Figure 11-14 is a band diagram for ferromagnetic nickel. Exchange effects have been included in the calculation of energy levels. Solid lines in the diagram refer to one spin direction, spin down say, and dotted lines refer to the opposite spin direction. Each spin-up curve is nearly parallel to a spin-down curve but is displaced from it by an exchange energy.

Nearly parabolic curves represent nearly free electron bands, while nearly flat curves represent $d$ bands. The Fermi level cuts two of the spin-up $d$ bands as well as both the spin-up and spin-down free electron bands. As a result, these bands are partially filled. Free electron bands contain 0.54 electrons/atom and $d$ bands contain 9.46 electrons/atom. Below the Curie temperature spin-down $d$ states are completely filled, while spin up $d$ states have $9.46 - 5 = 4.46$ electrons/atom. We expect the effective number of Bohr magnetons to equal the number of unpaired electrons, 0.54 per atom for nickel. Incomplete quenching of orbital angular momentum accounts for the discrepancy with the experimental value.

*Ferromagnetic Domains.*    As depicted in Fig. 11-15, a ferromagnetic sample is divided into regions, called domains, such that the direction of magnetization

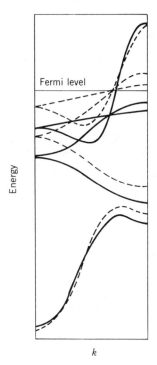

**FIGURE 11-14** Electron energy bands for ferromagnetic nickel. The propagation vector **k** is parallel to a cube edge. Solid curves represent states with dipole moments parallel to the magnetization, while dotted curves represent states with moments antiparallel to the magnetization. More electrons have moments aligned with the magnetization than have moments in the opposite direction. (From J. W. D. Connolly, *Phys. Rev.* **159:** 415, 1967. Used with permission.)

is different in neighboring domains. Domains may be observed where they intersect a sample surface by painting the surface with a colloidal suspension of small iron particles. The particles tend to collect in regions of high magnetization, between domain boundaries. This technique is useful for qualitatively studying changes that occur in the domain pattern when an applied field is varied.

All ferromagnetic crystals have directions of easy magnetization. For example, magnetization in iron is predominantly parallel to cube edges and magnetization vectors in neighboring domains usually make angles of 90 or 180° with each

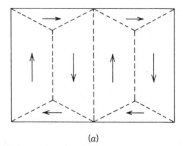
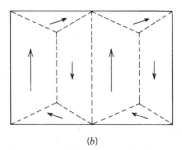

(a)                                        (b)

**FIGURE 11-15** Schematic representation of domains in a ferromagnetic crystal. Magnetization in each domain is indicated by an arrow. (a) The net magnetization is zero. (b) The net magnetization points toward the top of the diagram. Domain dimensions have changed and the magnetization has rotated in some domains.

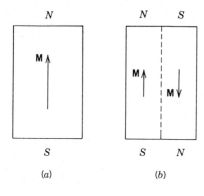

**FIGURE 11-16** (a) A single domain, magnetized in the direction indicated by the arrow. (b) Two neighboring domains, magnetized in opposite directions. The domains are equivalent to bar magnets with poles as marked. In going from (a) to (b) magnetic energy decreases but exchange and anisotropy energies increase.

other. A direction-dependent contribution to the energy arises because spatial parts of electron wave functions are somewhat spin dependent. Overlap integrals have slightly different values for spins that are oriented differently relative to the lattice.

Domain formation results in a considerable decrease in magnetic field energy. Consider a sample with a single domain, as in Fig. 11-16a. Most spins are aligned and together they produce a strong magnetic field. The field energy, given by $(\mu_0/4\pi) \int B^2 d\tau$, is large. It is reduced significantly, however, if the domain structure changes to that shown in (b). The situation is like that of two bar magnets: their energy is much less if opposite poles rather than like poles are adjacent to each other.

Domain formation is opposed by an increase in the energy of spins in the transition region between domains. Spins on opposite sides of a domain boundary are not parallel, so the exchange energy is higher in Fig. 11-16b than in (a). Exchange energy is minimized if the transition from one orientation to another takes place gradually over several hundred unit cells, in what is known as a domain or Bloch wall. Each spin in a wall is rotated only slightly from its neighbor. For example, the exchange energy associated with a line of 300 spins, each making an angle of $180°/300 = 0.6°$ with its neighbors, is much less than the energy of two spins making an angle of $180°$ with each other.

The foregoing discussion implies that the energy of a domain wall approaches zero as the thickness of the wall becomes large and the angle between adjacent spins becomes small. However, spins in a Bloch wall are not along directions of easy magnetization, so the energy of anisotropy increases with wall thickness. Typically, thicknesses are limited to a few hundred unit cells.

*Hysteresis.* As a relatively small field is applied, domain boundaries move so that domains with magnetization roughly in the same direction as the field become larger and domains with magnetization roughly in the opposite direction become smaller. The idea is illustrated in Fig. 11-15b. Boundary motion in defect-free samples is reversible. As the applies field is decreased, domains with magnetization in the field direction decrease in volume. Boundary motion, however, is impeded by defects, so reversibility is not exact for real samples.

Relatively high fields rotate the magnetization in some domains so it is more

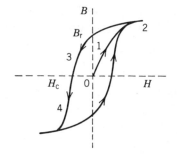

**FIGURE 11-17** Hysteresis curve for a ferromagnet. When an increasing magnetic field is applied to an initially unmagnetized sample, the induction field follows $0 \rightarrow 1 \rightarrow 2$. When the field is decreased it follows $2 \rightarrow 3 \rightarrow 4$. $H_c$ is the coercive field, $B_r$ is the remanence.

nearly aligned with the field. Domain rotation is also illustrated in Fig. 11-15$b$. High fields are required to rotate a magnetization vector past the high-energy barrier between two directions of easy magnetization. As a result, changes in magnetization due to domain rotation are not reversible. The spins cannot easily rotate back past the energy barrier when the applied field is decreased.

Irreversibility gives rise to hysteresis, a phenomenon that can be displayed by plotting either $M$ or $B = \mu_0(H + M)$ as a function of $H$. Such a plot is shown in Fig. 11-17. Positive values of $H$ and $B$ represent fields in one direction, while negative values represent fields in the opposite direction.

At the origin of the plot, the average magnetization is zero but, as the field is increased in the positive direction, $B$ follows the curve marked 0, 1, 2. In the linear portion of the curve, magnetization results chiefly from domain boundary motion and if the field is decreased from one of these values, $B$ follows the same curve back to the origin. If, on the other hand, the field is increased to 2 and then decreased, $B$ follows the curve marked 2, 3, 4. Domain rotation has now occurred.

The saturation induction field $B_s$ is the limiting value of $B$ as $H$ becomes large. The saturation magnetization can be calculated once $B_s$ is measured. The coercive force $H_c$ is the reverse field required to bring $B$ to zero and the remanence $B_r$ is the value of $B$ when $H$ is zero. Values of these quantities are often listed in handbooks.

Knowledge of ferromagnetic domains is important for the design of permanent magnets, electromagnets, and transformer cores, for example. In magnetic bubble memories designed for computers, a large number of closely spaced domains are positioned in a regular fashion in a ferromagnetic sample. Information is coded in terms of magnetization directions in the domains.

Ferromagnets need not be crystalline. Amorphous ferromagnets exist and, in fact, are of great technological importance. Since they do not have easy or hard directions of magnetization, these materials have low coercivities and their hysteresis curves are narrow.

## 11.4 FERRIMAGNETISM AND ANTIFERROMAGNETISM

Ions in most ferrimagnets and antiferromagnets are positioned on two sublattices such that the spins on each sublattice tend to be aligned with each other but spins on different sublattices tend to be in opposite directions. The material is an antiferromagnet if the net magnetic moment vanishes and a ferrimagnet if it does not. Materials with more than two sublattices exist, but we will not consider them.

*Ferrimagnetism.* Qualitatively, ferrimagnets are quite similar to ferromagnets. They are magnetized spontaneously at temperatures below their Curie temperatures and are paramagnetic for temperatures above. Domains are formed and hysteresis occurs in ferrimagnetic phases. Spontaneous magnetizations below $T_c$ and susceptibilities above, however, are different functions of temperature for ferromagnets and ferrimagnets.

Ferrites, which form an important group of ferrimagnets, all have the chemical composition $Fe_2MO_4$, where $M$ is a divalent metal ion, commonly copper, lead, magnesium, manganese, cobalt, nickel, or iron. Iron enters as a triply ionized atom, while the metal atom is doubly ionized. Magnetite, the magnetic component of lodestone, has the chemical formula $Fe_3O_4$ and contains both $Fe^{3+}$ and $Fe^{2+}$ ions.

A ferrite unit cell, shown in Fig. 11-18, contains tetrahedral sites, each with four neighboring oxygen atoms, and octahedral sites, each with six neighboring oxygen atoms. In the "normal" configuration tetrahedral sites are occupied by $M^{2+}$ ions and octahedral sites are occupied by $Fe^{3+}$ ions. In what is called the inverse ferrite configuration, tetrahedral sites are occupied by $Fe^{3+}$ ions, while octahedral sites are occupied by both $M^{2+}$ and $Fe^{3+}$ ions. Many intermediate configurations also occur. In any event, at low temperatures ions at tetrahedral

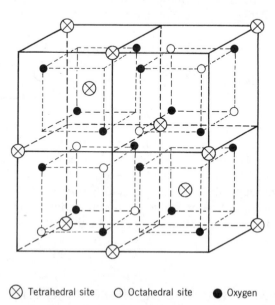

⊗ Tetrahedral site      ○ Octahedral site      ● Oxygen

**FIGURE 11-18** One-half a ferrite primitive cubic unit cell. The other half sits behind the one shown and is its mirror image. Spins of ions at tetrahedral sites tend to be in the same direction, parallel to a cube edge. Spins of ions at octahedral sites tend to be in the opposite direction.

sites have spins aligned along one of the cube edges while ions at octahedral sites have spins aligned in the opposite direction. Table 11-4 gives data for some ferrites.

Ions with spins that tend to align with each other can be assigned to the same sublattice. The two sublattices are interlaced so the nearest magnetic neighbors to an ion on one sublattice are on the other sublattice.

To produce ferrimagnetism, the exchange energy for two magnetic nearest neighbors must be negative. The energy is then a minimum when spins on different sublattices are antiparallel. In many cases the exchange energy for ions on the same sublattice is also negative but these interactions are weaker than nearest-neighbor interactions and do not greatly influence spin directions. Spins of the same sublattice tend to be aligned primarily because of exchange interactions with spins of the other sublattice, not because of interactions with spins of their own sublattice.

To analyze the magnetization of a ferrimagnet, suppose spins of sublattice A are predominantly in the positive $z$ direction and spins of sublattice B are predominantly in the negative $z$ direction. Let $M_A$ and $M_B$ represent the $z$ components of the magnetizations of the individual sublattices. Assume only nearest-neighbor exchange interactions are important and take the Weiss effective field to be $B_a - \mu_0 \gamma M_B$ at an A ion and $B_a - \mu_0 \gamma M_A$ at a B ion. Here $B_a$ is the applied field and $\gamma$ is a positive constant. The ferrimagnetic exchange field enters with a minus sign.

An equation similar to Eq. 11-28 can be written for each sublattice:

$$M_A = M_{sA} B_{J_A'}[\beta g_A \mu_B J_A'(B_a - \mu_0 \gamma M_B)] \tag{11-41}$$

and

$$M_B = M_{sB} B_{J_B'}[\beta g_B \mu_B J_B'(B_a - \mu_0 \gamma M_A)] \tag{11-42}$$

for the A and B sublattices, respectively. We have allowed for different saturation magnetizations, $g$ factors and values of $J'$. The saturation magnetization for A

**TABLE 11-4**  Curie Temperatures and Spontaneous Magnetizations of Selected Ferrites at $T = 0$ K

| Solid | $T_c$ (K) | $M_s$ ($10^5$ A/m) |
|---|---|---|
| $Fe_3O_4$ | 858 | 5.10 |
| $MnFe_2O_4$ | 573 | 5.60 |
| $NiFe_2O_4$ | 858 | 3.00 |
| $MgFe_2O_4$ | 713 | 1.40 |
| $BaFe_{12}O_{19}$ | 733 | 5.30 |
| $Ba_2Co_2Fe_{11}O_{22}$ | 613 | 2.04 . |
| $BaFe_{18}O_{27}$ | 728 | 5.20 |

*Source:* D. H. Martin, *Magnetism in Solids* (Cambridge, MA: MIT Press, 1967).

ions is given by $M_{sA} = n_A g_A \mu_B J_A'$, where $n_A$ is the concentration of A ions. A similar expression holds for B ions.

Equations 11-41 and 11-42 are solved simultaneously for $M_A$ and $M_B$, then the total magnetization is computed using $M = M_A + M_B$. We first consider spontaneous magnetization, with $B_a = 0$. Since the sign of a Brillouin function is the same as the sign of its argument, the equations clearly indicate that $M_A$ and $M_B$ are in opposite directions. Consider the low-temperature limit and suppose $M_A$ is positive. Then, according to Eq. 11-42, $M_B = -M_{sB}$ and, according to Eq. 11-41, $M_A = +M_{sA}$. The total magnetization is

$$M = M_A + M_B = n_A g_A \mu_B J_A' - n_B g_B \mu_B J_B' . \tag{11-43}$$

Figure 11-19 shows the spontaneous magnetic moment per unit mass density of magnesium, iron, and manganese ferrites as functions of temperature. Both $M_A$ and $M_B$ decrease as the temperature increases.

*Paramagnetic Behavior.* Spontaneous magnetization vanishes for a ferrimagnet at temperatures above its Curie temperature $T_c$ and the sample becomes paramagnetic. Curie temperatures are evident on the curves of Fig. 11-19.

To investigate paramagnetism, we retain the applied field in Eqs. 11-41 and 11-42 and use the limiting form of the Brillouin functions: $B_{J'}(x) = (J' + 1)x/3J'$. Then those equations become

$$M_A = \frac{C_A}{\mu_0 T} (B_a - \mu_0 \gamma M_B) \tag{11-44}$$

and

$$M_B = \frac{C_B}{\mu_0 T} (B_a - \mu_0 \gamma M_A) , \tag{11-45}$$

respectively. Here $C_A$ and $C_B$ are the Curie constants for the two sublattices, respectively. $C_A$, for example, is $\mu_0 n_A g_A^2 \mu_B^2 J_A'(J_A' + 1)/3k_B$. Equations 11-44 and

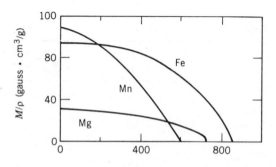

Temperature (K)

**FIGURE 11-19** Spontaneous magnetization of manganese, iron, and magnesium ferrites as functions of temperature. (From D. H. Martin, *Magnetism in Solids* (Cambridge, MA: MIT Press, 1967). Used with permission.)

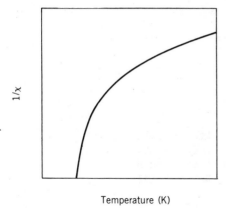

FIGURE 11-20 The reciprocal of the susceptibility as a function of temperature of a ferrite
above its Curie temperature.

11-45 predict spontaneous magnetization for only one temperature, the Curie
temperature $T_c$, given by $\gamma\sqrt{C_A C_B}$. To derive this expression, put $B_a = 0$ in both
equations and solve for $T$.

Equations 11-44 and 11-45 are solved for $M_A$ and $M_B$ and the results are summed
to obtain the total magnetization:

$$M = M_A + M_B = \frac{(C_A + C_B)T - 2C_A C_B \gamma}{\mu_0(T^2 - T_c^2)} B_a. \qquad (11\text{-}46)$$

The susceptibility is

$$\chi = \frac{\mu_0 M}{B_a} = \frac{(C_A + C_B)T - 2C_A C_B \gamma}{T^2 - T_c^2}. \qquad (11\text{-}47)$$

Figure 11-20 shows $1/\chi$ as a function of temperature for manganese ferrite.

*Antiferromagnetism.* Although the net magnetization of an antiferromagnet is
zero, the spins are highly ordered at low temperature and a large magnetization
is associated with each sublattice. Magnetic order is opposed by thermal agi-
tation and the magnetization of each sublattice decreases with increasing temperature. At what is known as the Néel temperature, magnetic order disappears
and the solid becomes paramagnetic. Néel temperatures for some antiferromagnets are listed in Table 11-5.

Fluorides of the iron group of transition metals are antiferromagnetic with
two sublattices. The unit cell, shown for $MnF_2$ in Fig. 11-21, is body-centered
tetragonal. At $T = 0$ K, all magnetic moments are parallel to the $c$ axis, with half
the Mn spins in one direction and the other half in the opposite direction. In
the diagram, spins at cell corners point up and spins at body centers point down.

Antiferromagnetism is easy to analyze for the simplest situation, that of two
interpenetrating sublattices with interactions between nearest magnetic neighbors only. Equations 11-41 and 11-42, changed to reflect the identical nature of
the atoms, become

$$M_A = M_s B_J'[\beta g \mu_B J'(B_a - \mu_0 \gamma M_B)] \qquad (11\text{-}48)$$

**TABLE 11-5** Néel Temperatures of Selected Antiferromagnetic Solids

| Solid | $T_N$ (K) | Solid | $T_N$ (K) |
|-------|-----------|-------|-----------|
| Ce | 12.5 | CuO | 230 |
| Cr | 308 | $FeBr_2$ | 11 |
| Dy | 179 | $FeCl_2$ | 24 |
| Er | 80 | $FeF_2$ | 79 |
| Ho | 132 | $FeF_3$ | 394 |
| Tm | 51 | FeO | 198 |
| Mn | 100 | FeS | 613 |
| Nd | 20 | FeTe | 70 |
| Tb | 230 | $KNiF_3$ | 275 |
| $CoCl_2$ | 25 | $MnCl_2$ | 2 |
| $CoF_2$ | 38 | $MnF_2$ | 68 |
| $CoF_3$ | 460 | MnO | 116 |
| $CrCl_2$ | 40 | $MnO_2$ | 90 |
| CrN | 273 | MnS | 160 |
| CrSb | 720 | $NdFeO_3$ | 760 |
| $CuBr_2$ | 193 | $NiCl_2$ | 50 |
| $CuCl_2$ | 70 | NiO | 525 |

*Source: American Institute of Physics Handbook* (D. E. Gray, Ed.) (New York: McGraw-Hill, 1963).

and

$$M_B = M_s B_J'[\beta g \mu_B J'(B_a - \mu_0 \gamma M_A)] , \qquad (11\text{-}49)$$

respectively. Now $M_s = \frac{1}{2} n g \mu_B J'$, where $n$ is the concentration of magnetic atoms. The factor 2 appears because half the atoms are on each sublattice. If $B_a = 0$, solutions to these equations obey $M_A = -M_B$ and the total magnetization vanishes, regardless of the temperature.

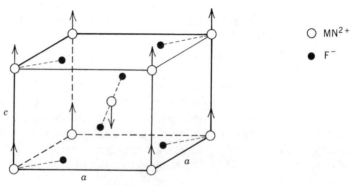

○ MN²⁺

● F⁻

**FIGURE 11-21** The tetragonal primitive unit cell of antiferromagnetic $MnF_2$. Arrows indicate spin directions. The net magnetization of the cell is zero.

When $B_a = 0$ and $M_B = -M_A$ are substituted into Eq. 11-48 the result is

$$M_A = M_s B_{J'}(\mu_0 \beta g \mu_B J' \gamma M_A), \qquad (11\text{-}50)$$

an equation that is similar to Eq. 11-28 (with $B_a = 0$) for the magnetization of a ferromagnet. As functions of temperature, $M_A$ and $M_B$ are like the magnetizations shown in Fig. 11-10.

For high temperature, Eqs. 11-48 and 11-49 become

$$M_A = \frac{C}{2\mu_0 T} (B_a - \mu_0 \gamma M_B) \qquad (11\text{-}51)$$

and

$$M_B = \frac{C}{2\mu_0 T} (B_a - \mu_0 \gamma M_A), \qquad (11\text{-}52)$$

respectively. $C$ is again the Curie constant. The Néel temperature is the only temperature for which these equations have a solution when $B_a = 0$. It is given by $T_N = \frac{1}{2}\gamma C$.

The total magnetization for $T > T_N$ can be found by adding the two equations, setting $M = M_A + M_B$, and then solving for $M$. The result is

$$M = \frac{C}{\mu_0(T + T_N)} B_a, \qquad (11\text{-}53)$$

so the susceptibility is

$$\chi = \frac{\mu_0 M}{B_a} = \frac{C}{T + T_N}. \qquad (11\text{-}54)$$

Like $1/\chi$ for a ferromagnet in its paramagnetic phase, $1/\chi$ is linear in the temperature. However, the Néel temperature, unlike the Curie temperature, enters the denominator with a plus sign. Experimental results substantiate the qualitative form of Eq. 11-54. Agreement is improved, however, if exchange interactions between ions of the same sublattice are taken into account.

**EXAMPLE 11-7** The susceptibility of $MnF_2$ is 1.02 at its Néel temperature (68 K). Assume Eq. 11-54 is valid and find values for the constants $C$ and $\gamma$.

**SOLUTION** According to Eq. 11-54, $\chi(T_N) = C/2T_N$ so $C = 2T_N\chi(T_N) = 2 \times 68 \times 1.02 = 139$ K. Since $T_N = \frac{1}{2}\gamma C$, $\gamma = 2T_N/C = 2 \times 68/139 = 0.98$. The exchange coefficient, in addition to being negative, is considerably smaller in magnitude than the coefficient for iron. ∎

*Neutron Scattering.* Because the amplitude of a neutron wave scattered from an atom depends on the relative orientations of the atomic and neutron spins, neutron scattering is used to study magnetic ordering. If the incident beam is polarized so neutron spins are all in the same direction, atomic scattering factors

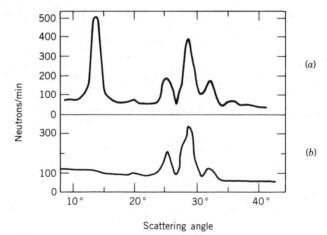

Scattering angle

**FIGURE 11-22** Neutron scattering powder patterns for $MnF_2$: (a) in the antiferromagnetic state and (b) in the paramagnetic state. The (100) peak disappears at high temperature. (From R. A. Erickson, *Phys. Rev.* **90:** 779, 1953. Used with permission.)

are different for atoms with spins in different directions. Diffraction patterns can be used to identify ferromagnetic, ferrimagnetic, antiferromagnetic, and paramagnetic ordering. Neutron scattering has been used to discover many antiferromagnets and to determine Néel temperatures.

Consider antiferromagnetic $MnF_2$, with the unit cell shown in Fig. 11-21. Suppose (100) reflections are investigated by orienting an incident beam of polarized monoenergetic neutrons in such a way that waves scattered from atoms at cell corners interfere constructively. In the paramagnetic phase all manganese atoms are equivalent. No diffraction peak is observed since waves scattered from atoms at body centers interfere destructively with those scattered from corner atoms. Below the Néel temperature, however, the spins of these two sets of atoms are not in the same direction. The atoms are not equivalent, complete destructive interference does not occur, and a peak is observed. Figure 11-22 shows typical sets of peaks for antiferromagnetic and paramagnetic $MnF_2$.

Neutron diffraction can be used to investigate spin directions. Suppose the spins of incident neutrons are along the direction of propagation and (001) reflections are observed. If the atomic spins are perpendicular to (001) planes, as they are in antiferromagnetic $MnF_2$, waves are scattered identically from atoms of both sublattices and no peak is observed. A weak (001) peak is observed for $NiF_2$, indicating that the atomic spins are not precisely perpendicular to (001) planes.

## 11.5 SPIN WAVES

Consider a line of $N$ magnetic atoms, each with spin $S$ and each interacting with its nearest neighbors via the Heisenberg exchange interaction. The system is in its ground state when all spins are aligned, along the $z$ axis say. We might expect the first excited state to consist of one atom with $S_z = S - \hbar$ and $N - 1$ atoms with $S_z = S$. It does not. The effective field produced by the unaligned spin is not along the $z$ axis. Neighboring spins precess around the effective field direction and soon move away from the $z$ axis. We must seek a collective description of the spins, much like the collective description of atomic vibrations.

Start with the equations of motion. The effective field at atom $p$ is given by $\mathbf{B}_{\text{eff}} = -(J_e/g\mu_B\hbar)(\mathbf{S}_{p-1} + \mathbf{S}_{p+1})$ and the torque acting on spin $\mathbf{S}_p$ is given by $\boldsymbol{\mu} \times \mathbf{B}_{\text{eff}} = -(g\mu_B/\hbar)\mathbf{S} \times \mathbf{B}_{\text{eff}}$, so

$$\frac{d\mathbf{S}_p}{dt} = \frac{J_e}{\hbar^2} \mathbf{S}_p \times (\mathbf{S}_{p-1} + \mathbf{S}_{p+1}). \tag{11-55}$$

Assume all spins are nearly in the positive $z$ direction, so the $x$ and $y$ components are small and their products can be neglected. Then Eq. 11-55 is satisfied by

$$S_{px} = A \cos(kpa - \omega t), \tag{11-56}$$

$$S_{py} = A \sin(kpa - \omega t), \tag{11-57}$$

and

$$S_{pz} = \sqrt{S^2 - A^2}, \tag{11-58}$$

where $a$ is the separation of adjacent spins. Both $S_{px}$ and $S_{py}$ have the form of traveling waves with amplitude $A$, angular frequency $\omega$, and propagation constant $k$. The waves are called spin waves. If periodic boundary conditions are applied, $k$ has one of values $2\pi\hbar/Na$, where $\hbar$ is an integer in the range $-\frac{1}{2}N \le \hbar < +\frac{1}{2}N$. Substitution of Eqs. 11-56, 11-57, and 11-58 into Eq. 11-55 yields the spin wave dispersion relationship

$$\omega = \frac{4J_eS_z}{\hbar^2} \sin^2(\tfrac{1}{2}ka). \tag{11-59}$$

The algebra is similar to that used in Chapter 6 to derive phonon dispersion relationships.

A spin wave is illustrated in Fig. 11-23. Each spin precesses about the $z$ axis with angular frequency $\omega$. The phase of the precessional motion advances from spin to spin, with adjacent spins having a phase difference of $ka$.

An expression for the spin wave energy can be derived easily. According to Eq. 11-38, the total exchange energy is given by

$$E = -\frac{J_e}{2\hbar^2} \sum_{p=1}^{N} \mathbf{S}_p \cdot (\mathbf{S}_{p-1} + \mathbf{S}_{p+1}), \tag{11-60}$$

where the factor $\frac{1}{2}$ is included to avoid double counting the interactions. When $\mathbf{S}_p = A \cos(kpa - \omega t)\hat{\mathbf{x}} + A \sin(kpa - \omega t)\hat{\mathbf{y}} + S_z\hat{\mathbf{z}}$ is used, Eq. 11-60 becomes

$$E = -\frac{NJ}{\hbar^2} [S^2 - 2A^2 \sin^2(\tfrac{1}{2}ka)]. \tag{11-61}$$

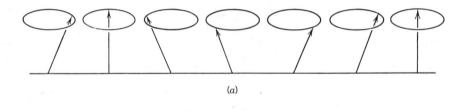

(a)

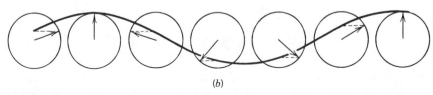

(b)

**FIGURE 11-23** (a) Side and (b) top views of atomic dipole moment orientations in a traveling spin wave. All moments have the same magnitude and z component but different phases in their precessional motions. If the propagation constant is k and the atomic separation is a, the phase advances by ka from one atom to the next.

Finally, the dispersion relationship is used to write

$$E = -\frac{NJ}{\hbar^2} S^2 + \frac{N\omega}{2S_z} A^2 = -\frac{NJ_e}{\hbar^2} S^2 + \frac{N\omega}{2S} A^2, \qquad (11\text{-}62)$$

where $S_z$ was approximated by $S$ in the denominator of the second term. The first term represents the energy when all spins are aligned, while the second represents the spin wave energy.

*Magnons.* According to quantum mechanical theory, the energy of a spin wave is quantized. In particular, a wave with angular frequency $\omega$ may have any one of the energies $\hbar\omega(n_M + \frac{1}{2})$ where $n_M$ is a positive integer or zero. A quantum of spin wave energy is called a magnon and $n_M$ is the number of magnons associated with the wave.

To make the connection with the classical theory, suppose the second term in Eq. 11-62 represents $n_M$ magnons and equate it to $n_M\hbar\omega$, then solve for $A^2$:

$$A^2 = \frac{2n_M\hbar}{N} S. \qquad (11\text{-}63)$$

The z component of each spin is given by

$$S_z = \sqrt{S^2 - A^2} = S - \frac{n_M\hbar}{N}, \qquad (11\text{-}64)$$

to lowest order in $n_M/N$.

According to Eq. 11-64, the z component of the total spin is $NS_z = NS - n_M\hbar$. For each magnon the z component of the spin on each atom decreases by $\hbar/N$ and the z component of the total decreases by $\hbar$. All magnons contribute equally to the spin angular momentum, regardless of spin wave frequency.

*Thermodynamics of Magnons.* One of the most important predictions of spin wave theory is the low-temperature behavior of the magnetization of a ferromagnet. Experimentally $M \rightarrow M_s - AT^{2/3}$ as $T$ becomes small. We outline a derivation of the spin wave theory prediction.

In thermodynamic equilibrium, magnons obey a distribution law that is identical to that for phonons. In particular, the thermodynamic average number of magnons associated with a wave of frequency $\omega$ is

$$\langle n_M \rangle = \frac{1}{e^{\beta \hbar \omega} - 1}, \tag{11-65}$$

where $\beta = 1/k_B T$. Thermodynamic equilibrium is attained through interchange of energy with the phonon system and with defects.

The total number of magnons associated with all spin waves is given by

$$\langle n_M \rangle_{total} = \int_0^{\omega_{max}} \frac{D(\omega)}{e^{\beta \hbar \omega} - 1} \, d\omega, \tag{11-66}$$

where $D(\omega) \, d\omega$ is the density of spin wave modes. Only low-frequency waves are excited at low temperature and, for them, $\omega$ is proportional to $k^2$. The density of modes is proportional to $\omega^{1/2}$ and we write

$$\langle n_M \rangle_{total} = A \int_0^{\omega_{max}} \frac{\omega^{1/2}}{e^{\beta \hbar \omega} - 1} \, d\omega = A \frac{(k_B T)^{3/2}}{\hbar^{3/2}} \int_0^{\beta \hbar \omega_{max}} \frac{x^{1/2} \, dx}{e^x - 1}, \tag{11-67}$$

where $A$ is a constant of proportionality and $x = \beta \hbar \omega$ is used as the variable of integration. For low temperatures the upper limit is large and may be replaced by $\infty$. The integral is then independent of $T$ and $\langle n_M \rangle_{total}$ is proportional to $T^{3/2}$. Since the magnetization is proportional to $NS - \langle n_M \rangle_{total}$, the theory correctly predicts its temperature dependence. As shown in Problem 17, Curie-Weiss theory predicts a different temperature dependence.

## 11.6  MAGNETIC RESONANCE PHENOMENA

Magnetic resonance experiments are used to determine Landé $g$ factors and to investigate local magnetic and exchange fields. Resonance measurements, along with theoretical arguments, are used to deduce the local environments of magnetic atoms.

*Magnetic Resonance.* The idea is simple. Suppose a static induction field $B_0 \hat{z}$ is applied to a paramagnetic solid containing identical magnetic ions. Each moment precesses around the direction of $\mathbf{B}_0$ with angular frequency $\omega_0 = (g \mu_B / \hbar) B_0$. Figure 11-24a shows a group of dipole moments with identical $z$ components. They are shown after steady state has been reached and they have a random distribution of phases. If all moments are drawn with their tails at the same point, as in $b$, they are distributed uniformly on the surface of a cone. As a result, components of the magnetization perpendicular to $\mathbf{B}$ vanish.

In a typical magnetic resonance experiment, a second magnetic field, in the

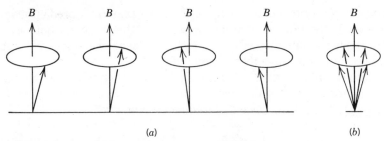

**FIGURE 11-24** (a) A line of magnetic atoms with their dipole moments precessing about a static magnetic induction field **B**. (b) Moment vectors drawn from the same origin. In steady state the vectors are distributed uniformly on the surface of a cone and the net magnetization is parallel to **B**. If different atoms have spins with different $z$ components, the diagram should show several concentric cones, one for each value of $\mu_z$.

$xy$ plane and rotating in the direction of dipole precession, is used to disturb the dipole distribution. Two phenomena occur simultaneously. The dipoles tend to precess around the direction of the combined field and they tend to relax toward the distribution shown in Fig. 11-24b. The rotating field is in resonance with the dipoles when the field has an angular speed that matches the angular frequency of dipole precession about the static field.

*Relaxation Times.* Classical mechanics can be used to obtain most of the fundamental relationships we require. If relaxation is neglected, the equation of motion for a dipole moment is $d\boldsymbol{\mu}/dt = -\gamma\boldsymbol{\mu} \times \mathbf{B}$, where $\gamma$ is $g\mu_B/\hbar$ and $\mathbf{B}$ is the sum of the static and rotating fields. The magnetization obeys $d\mathbf{M}/dt = -\gamma\mathbf{M} \times \mathbf{B}$ or, when a relaxation term is added,

$$\frac{d\mathbf{M}}{dt} = -\gamma\mathbf{M} \times \mathbf{B} + \left(\frac{d\mathbf{M}}{dt}\right)_{\text{relax}}. \tag{11-68}$$

Relaxation of the magnetization can be described in terms of two relaxation times, one for the $z$ component and the other for $x$ and $y$ components. The $z$ component approaches $M_0$ ($= \chi B_0/\mu_0$), so $(dM_z/dt)_{\text{relax}} = -(M_z - M_0)/\bar{t}_1$. The other components tend toward 0, so $(dM_x/dt)_{\text{relax}} = -M_x/\bar{t}_2$ and $(dM_y/dt)_{\text{relax}} = -M_y/\bar{t}_2$. The two relaxation times $\bar{t}_1$ and $\bar{t}_2$ are different because they are associated with different processes.

Relaxation of the $z$ component is primarily a result of spin-phonon interactions* and involves an exchange of energy between the dipole and phonon systems. Phonon interactions change the angle between a dipole moment and the $z$ axis. They also change the $x$ and $y$ components of the magnetization. However, they do not necessarily cause these components to vanish. Another process, known as spin-spin relaxation, spreads the moments uniformly over the surface of a cone. Each dipole is subjected to small fluctuating fields due to

---

*In discussions of resonance phenomena the word spin is used to denote angular momentum, even if a portion is orbital in nature.

neighbors, so the dipoles precess at slightly different frequencies and the precession frequencies fluctuate. Even if several dipoles start with nearly the same orientation they do not remain parallel as time goes on. Eventually their projections on the $xy$ plane become randomly oriented and both $M_x$ and $M_y$ vanish. The dipoles exchange energy among themselves but not with the environment. As a result, spin-spin relaxation is much faster than spin-phonon relaxation.

Once relaxation terms are included, the Cartesian components of Eq. 11-68 become

$$\frac{dM_z}{dt} = -\gamma[M_xB_y - M_yB_x] - \frac{M_z - M_0}{\bar{t}_1}, \tag{11-69}$$

$$\frac{dM_x}{dt} = -\gamma[M_yB_z - M_zB_y] - \frac{M_x}{t_2}, \tag{11-70}$$

and

$$\frac{dM_y}{dt} = -\gamma[M_zB_x - M_xB_z] - \frac{M_y}{t_2}. \tag{11-71}$$

These equations are known as the Bloch equations for the magnetization.

*Magnetization and the Absorption of Power.*   Take the rotating field to be $B_1[\cos(\omega t)\hat{x} + \sin(\omega t)\hat{y}]$. If $\omega$ is positive the field rotates about the $z$ axis in the direction of dipole precession. Substitute $\mathbf{B} = B_1\cos(\omega t)\hat{x} + B_1\sin(\omega t)\hat{y} + B_0\hat{z}$ into Eqs. 11-69, 11-70, and 11-71, then solve for $\mathbf{M}$. In the steady state, $M_x = M_1\cos(\omega t) + M_2\sin(\omega t)$, $M_y = M_1\sin(\omega t) - M_2\cos(\omega t)$, and $M_z =$ constant. This trial solution describes a magnetization of constant magnitude that rotates with the field around the $z$ axis. Expressions for $M_1$, $M_2$, and $M_z$ are obtained from Eqs. 11-69, 11-70, and 11-71. They are

$$M_1 = \frac{\gamma \bar{t}_2^2 M_0(\omega_0 - \omega)}{1 + (\omega - \omega_0)^2\bar{t}_2^2 + \gamma^2\bar{t}_1\bar{t}_2B_1^2} B_1, \tag{11-72}$$

$$M_2 = \frac{\gamma \bar{t}_2 M_0}{1 + (\omega - \omega_0)^2\bar{t}_2^2 + \gamma^2\bar{t}_1\bar{t}_2B_1^2} B_1, \tag{11-73}$$

and

$$M_z = M_0 - \frac{\gamma^2\bar{t}_1\bar{t}_2 M_0}{1 + (\omega - \omega_0)^2\bar{t}_2^2 + \gamma^2\bar{t}_1\bar{t}_2B_1^2} B_1^2, \tag{11-74}$$

where $\omega_0 = \gamma B_0$.

The energy per unit volume of the dipole system is $-\mathbf{M} \cdot \mathbf{B}$. In the steady state it is constant. Nevertheless, energy is supplied by the source of the field and absorbed by phonons. The power absorbed per unit volume of solid is given by $P = \mathbf{B} \cdot (d\mathbf{M}/dt)_{\text{relax}}$ or, when expressions for the components of $(d\mathbf{M}/dt)_{\text{relax}}$ are used, by

$$P = -\frac{B_0}{\bar{t}_1}(M_z - M_0) - \frac{B_1}{\bar{t}_2}M_1. \tag{11-75}$$

Finally, Eqs. 11-72 and 11-74 are used to obtain

$$P = \frac{\omega \gamma \bar{t}_2 M_0 B_1^2}{1 + (\omega - \omega_0)^2 \bar{t}_2^2 + \gamma^2 \bar{t}_1 \bar{t}_2 B_1^2} . \tag{11-76}$$

In Fig. 11-25a, $\omega_0$ is varied and the absorbed power $P$ is plotted as a function of $(\omega - \omega_0)\bar{t}_2$. It is a maximum when $\omega_0 = \omega$ or, what is the same, when $B_0 = \hbar\omega/g\mu_B$.

*Resonance Experiments.* In a resonance experiment the applied static field is produced by a large electromagnet and the rotating field is the magnetic component of a circularly polarized electromagnetic wave. Radiation intensity is measured before and after the wave interacts with the sample, then the power absorbed is computed.

For laboratory fields, resonance frequencies are in the microwave region of

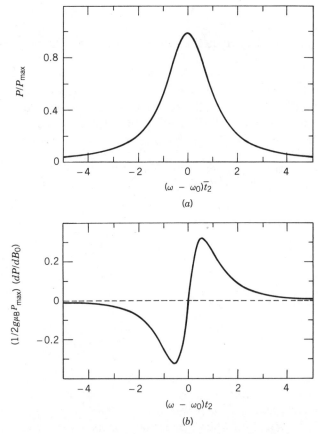

(a)

(b)

**FIGURE 11-25** Magnetic resonance curves. (a) The power absorbed as a function of $(\omega - \omega_0)\bar{t}_2$, where $\omega_0 = (g\mu_B/\hbar)B_0$. (b) The derivative of the power absorbed. In each case $\omega$ is the angular frequency of the applied radiation field and $\bar{t}_2$ is the spin-spin relaxation time. At resonance the power absorbed is a maximum and its derivative is zero.

the electromagnetic spectrum. Suppose a 0.3-T static field is applied to a dipole system with $g = 1$. The resonance angular frequency is $(g\mu_B/\hbar)B_0 = (1 \times 9.27 \times 10^{-24}/1.05 \times 10^{-34}) \times 0.3 = 2.6 \times 10^{10}$ rad/s, corresponding to a wavelength of 7.2 cm.

In most experiments the microwave frequency is fixed and the static field is varied until resonance is reached. $B_0$ can be varied easily by changing the current in the coils of the electromagnet. On the other hand, the frequency of the radiation field is proscribed by the geometry of the microwave cavity and waveguide used, so it is much more difficult to change.

The power absorbed is a small fraction of the input power and so is hard to detect. Usually a small oscillating induction field is added to the large static field. Then $P$ oscillates and the amplitude of its oscillation is detected. Data are usually displayed by plotting $dP/dB_0$ as a function of $B_0$, as in Fig. 11-25$b$. The value of $B_0$ corresponding to resonance is at the zero of the function.

Resonance spectrometers have a wide range of characteristics. Typically the static field may be varied from 0 to about 1 T, although some have an upper limit beyond 2 T. It is modulated at a frequency ranging from a few hundred hertz to a few hundred kilohertz, in different spectrometers. The rotating field typically has a frequency of about $10^{10}$ Hz and an amplitude between $10^{-10}$ and $10^{-5}$ T.

If the applied static field is much greater than the internally produced field, its value is nearly the same as $B_0$ and can be used to find the $g$ factor for electron states: $g = \hbar\omega/\mu_B B_0$. The value of $g$ can be used, for example, to determine the extent to which orbital angular momentum is quenched. Resonance experiments can also be performed on magnetically ordered materials and the data can be used to determine the exchange field.

Many samples produce a series of resonance lines, corresponding to states above the ground state. One of the most important uses of electron resonance experiments is to study the low-lying energy level structure of magnetic atoms. The sequence of $g$ factors obtained for a sample is often sufficient to identify the atoms it contains. Since the area under a resonance curve is proportional to the number of atoms that contribute to the curve, both the type and concentration of magnetic atoms can be determined experimentally. The sequence of low-lying levels in a ferromagnetic sample can often be used to infer the local environment of an atom.

*Nuclear Magnetic Resonance (NMR).*   Resonance experiments can also be performed with nuclear dipole moments. A proton has a magnetic moment of $2.79\,\mu_N$, in the direction of its spin, and a neutron has a magnetic moment of $1.91\,\mu_N$, opposite the direction of its spin. Here $\mu_N$ is the nuclear magneton, given by $e\hbar/2M_p = 5.05 \times 10^{-27}$ J/T, where $M_p$ is the mass of a proton. The nuclear magneton replaces the Bohr magneton in expressions of this section. Since it is three orders of magnitude smaller than the Bohr magneton, proton resonance frequencies are smaller than electron resonance frequencies. For $g = 1$ and $B_0 = 0.3$ T, the proton resonance angular frequency is $1.4 \times 10^7$ rad/s, corresponding to a wavelength of 130 m, in the radio region of the electromagnetic spectrum.

Magnetic fields produced by electrons contribute to the total field at a nucleus and are instrumental in determining the nuclear resonance frequency. As a consequence, the resonance frequency shifts slightly if electron distributions in the vicinity of nuclei are changed. The chemical shift, as it is called, is used to study the electron distributions around nuclei in molecules and solids.

NMR has become an important analytical tool in diagnostic medicine. The sample is a patient and the apparatus is tuned to the resonance line of a proton in hydrogen. The power absorbed indicates the concentration of hydrogen in the observed portion of the body. This is sufficient to distinguish various tissues from each other and to identify places where hydrogen concentration is abnormal, as at a tumor or thickened heart value. In many instances the procedure has replaced exploratory surgery.

## 11.7 REFERENCES

**General**

L. F. Bates, *Modern Magnetism* (London/New York: Cambridge Univ. Press, 1961).

S. Chikazumi, *Physics of Magnetism* (New York: Wiley, 1964).

B. D. Cullity, *Introduction to Magnetic Materials* (Reading, MA: Addison-Wesley, 1972).

D. H. Martin, *Magnetism in Solids* (Cambridge, MA: MIT Press, 1967).

D. C. Mattis, *Theory of Magnetism* (New York: Harper & Row, 1965).

A. H. Morrish, *Physical Principles of Magnetism* (New York: Wiley, 1966).

G. T. Rado and H. Suhl (Eds.), *Magnetism* (New York: Academic, 1963–); Multivolume set.

D. Wagner, *Introduction to the Theory of Magnetism* (Elmsford, NY: Pergamon, 1972).

R. M. White, *Quantum Theory of Magnetism* (New York: McGraw-Hill, 1970).

**Magnetically Ordered Materials**

B. R. Cooper, "Magnetic Properties of Rare Earth Metals" in *Solid State Physics* (F. Seitz, D. Turnbull, and H. Ehrenreich, Eds.), Vol. 21, p. 393, (New York: Academic, 1968).

D. J. Crack and R. S. Tebble, *Ferromagnetism and Ferromagnetic Domains* (Amsterdam: North-Holland, 1966).

E. Della Torre and A. H. Bobeck, *Magnetic Bubbles* (Amsterdam: North-Holland, 1974).

H. J. Jones, *Magnetic Interactions in Solids* (London/New York: Oxford Univ. Press, 1973).

C. Kittel and J. K. Galt, "Ferromagnetic Domain Theory" in *Solid State Physics* (F. Seitz and D. Turnbull, Eds.), Vol. 3, p. 437 (New York: Academic, 1956).

C. Kittel, "Indirect Exchange Interactions in Metals" in *Solid State Physics* (F. Seitz, D. Turnbull, and H. Ehrenreich, Eds.), Vol. 22, p. 1 (New York Academic, 1968).

J. S. Smart, *Effective Field Theories of Magnetism* (Philadelphia: Saunders, 1966).

J. H. Van Vleck, "A Survey of the Theory of Ferromagnetism" in *Rev. Mod. Phys.* **17:**27, 1945.

## Magnetic Resonance

A. Abragam, *The Principles of Nuclear Magnetism* (London/New York: Oxford Univ. Press, 1961).

M. Bersohn and J. C. Baird, *An Introduction to Electron Paramagnetic Resonance* (New York: Benjamin, 1966).

W. Low, *Paramagnetic Resonance in Solids* (New York: Academic, 1960).

G. E. Pake, *Paramagnetic Resonance* (New York: Benjamin, 1962).

C. P. Poole, Jr., *Electron Spin Resonance* (New York: Interscience, 1967).

J. Talpe, *Theory of Experiments in Paramagnetic Resonance* (Elmsford, NY: Pergamon, 1971).

J. E. Wertz and J. R. Bolton, *Electron Spin Resonance* (New York: McGraw-Hill, 1972).

## PROBLEMS

1. Consider a magnetic dipole $\boldsymbol{\mu}$ in a constant magnetic induction field $\mathbf{B}$, along the $z$ axis. (a) Verify that $\mu_x = A \cos(\omega_L t + \phi)$ and $\mu_y = A \sin(\omega_L t + \phi)$ satisfy $d\boldsymbol{\mu}/dt = -(e/2m)\boldsymbol{\mu} \times \mathbf{B}$ if $\omega_L = eB/2m$. (b) What is the sense of rotation of $\boldsymbol{\mu}$? (c) Show that $\boldsymbol{\mu}$ has constant magnitude. (d) Find expressions for the components of $\boldsymbol{\mu}$ if $\mathbf{B}$ is in the negative $z$ direction.

2. The normalized electron wave function for the ground state of a hydrogen atom is

$$\psi(\mathbf{r}) = \frac{1}{\sqrt{\pi}} \frac{1}{a_0^{3/2}} e^{-r/a_0},$$

where $a_0$ is the Bohr radius, $5.29 \times 10^{-11}$ m. (a) Calculate the molar diamagnetic susceptibility and compare with the accepted value of $2.97 \times 10^{-11}$ m³/mol. (b) What is the magnetic dipole moment of a hydrogen atom in an induction field of 0.5 T?

3. The diamagnetic core contribution to the molar susceptibility of sodium is $-6.1 \times 10^{-12}$ m³/mole. Estimate the average distance of a core electron from the nearest nucleus. Also estimate the dipole moment of a sodium core in a magnetic induction field of 0.50 T.

4. (a) Use Hund's rules to find $L'$, $S'$, and $J'$ for the ground state of a vanadium ion, with three electrons in its $3d$ shell. Assume orbital angular momentum is not quenched. What is the value of the Landé $g$ factor? What is the magnitude of the magnetic dipole moment? (b) If orbital angular momentum is completely quenched what are $L'$, $S'$, $J'$, and $g$? What is the magnitude of the dipole moment?

5. (a) Use Hand's rules to calculate the Landé $g$ factor for all possible $d$ shell occupation numbers (1 through 10). Assume the atoms are in their ground states and orbital angular momentum is not quenched. (b) For which occupation numbers does the magnetic dipole moment vanish? What atoms have these occupation numbers? (c) For which occupation number does the magnetic dipole moment have its greatest value? What atom has this

occupation number? (d) What are the answers to (b) and (c) if orbital angular momentum is completely quenched?

6. Suppose each magnetic ion in a certain paramagnetic salt has total angular momentum $\hbar$ and Landé $g$ factor 2. For a 0.70-T magnetic induction field, calculate the fraction of atoms with $J_z = +\hbar$, with $J_z = 0$, and with $J_z = -\hbar$ at $T = 300$ K. What is the average atomic dipole moment?

7. Equation 11-25 can be derived quite easily from Eq. 11-23. (a) Let $W = \Sigma\, e^{-\beta g \mu_B M_J B}$ and show that $B\langle\mu_z\rangle = d(\ln W)/d\beta$. (d) Show that

$$W = \frac{\sinh[\beta g \mu_B(J' + \frac{1}{2})B]}{\sinh[\frac{1}{2}\beta g \mu_B B]}.$$

(c) Finally, combine the results of (a) and (b) to show that

$$\langle\mu_z\rangle = g\mu_B(J' + \tfrac{1}{2})\coth[\beta g \mu_B(J' + \tfrac{1}{2})B] - \tfrac{1}{2}g\mu_B\coth[\tfrac{1}{2}\beta g \mu_B B].$$

8. Consider a paramagnetic salt in an induction field $B\hat{z}$. If each magnetic ion has angular momentum quantum number $J'$, the Helmholtz free energy $F$ per magnetic ion is given by

$$e^{-\beta F} = \sum_{M_J = -J'}^{J'} e^{-\beta g \mu_B M_J B}.$$

(a) Consider $F$ to be a function of $B$ and $\beta$ and show that $\partial(\beta F)/\partial\beta = -\langle\mu_z\rangle B$. This is the average energy $E$ per magnetic ion. (b) According to thermodynamics, $F = E - TS$, where $S$ is the entropy per magnetic ion and $E = -\langle\mu_z\rangle B$. Show that $S = k_B\beta^2(\partial F/\partial\beta)$. Then show that

$$S = k_B\ln \Sigma\, e^{-\beta g \mu_B M_J B} + k_B\beta\frac{\Sigma\, g\mu_B M_J B e^{-\beta g \mu_B M_J B}}{\Sigma\, e^{-\beta g \mu_B M_J B}}.$$

(c) The magnetic contribution to the heat capacity per ion, at constant volume and field, is $C = T(\partial S/\partial T)_{\tau, B}$. Show that in the weak field limit $(g\mu_B B \ll k_B T)$,

$$C = \frac{k_B}{3}\left[\frac{g\mu_B B}{k_B T}\right]^2 J'(J' + 1).$$

9. At room temperature oxygen is a paramagnetic gas with a molar susceptibility of $4.33 \times 10^{-8}$ m$^3$/mol. Estimate the effective number of Bohr magnetons per atom and show it is consistent with two electrons in the same $s$ shell.

10. The saturation magnetization of iron is $1.75 \times 10^6$ A/m. Show that this corresponds to 2.22 Bohr magnetons per atom. The concentration of iron atoms is $8.50 \times 10^{28}$ m$^{-3}$.

11. (a) Estimate the value of the effective field constant $\gamma$ for nickel. Take $n = 9.14 \times 10^{28}$ m$^{-3}$, $g = 2$, and $J' = S' = 0.3$. (b) Estimate the Curie constant $C$. (c) Why does nickel have a lower Curie temperature than iron? Consider the concentrations of atoms, effective fields, and atomic dipole moments.

12.  Take $J' = S' = \frac{1}{2}$ and show that Eq. 11-28 reduces to

$$\frac{T}{T_c} x = \tanh x,$$

in the absence of an applied field. Here $x = T_c M/T M_s$ and $T_c$ is the Curie temperature. Make a plot of $\tanh x$ as a function of $x$ and use it to evaluate $M/M_s$ for $T/T_c = 0.1, 0.5,$ and $0.9$.

13.  Two ferromagnets are identical in crystal structure and unit cell dimensions. The spins of their atoms are identical but the exchange coefficient $J_e$ for one is twice that for the other. Compare their effective field constants $\gamma$, Curie constants $C$, saturation magnetizations $M_s$, and Curie temperatures $T_c$.

14.  Two ferromagnets are identical in crystal structure and unit cell dimensions. Their exchange coefficients $J_e$ are the same but their atomic spins are different. For one $S' = 2$, while for the other $S' = 1$. Compare their effective field constants $\gamma$, Curie constants $C$, saturation magnetizations $M_s$, and Curie temperatures $T_c$.

15.  Show that the exchange coefficient $J_e$ is given by

$$J_e = \frac{3 k_B T_c}{z S'(S' + 1)}$$

for a ferromagnet with Curie temperature $T_c$. Each atom has $z$ identical nearest neighbors and each has spin $S'$. Orbital angular momentum is quenched. Use this expression to calculate $J_e$ for nickel. It has a face-centered cubic structure and a Curie temperature of 631 K.

16.  Estimate the change in magnetic field energy that occurs when a domain is formed. Consider a single domain of length $\ell$, width $w$, and thickness $t$, magnetized along its length, as shown in Fig. 11-16a. Its field can be approximated by that of a pole $P_1 = +Mwt$ at the center of one end and a pole $P_2 = -Mwt$ at center of the other. The interaction energy of two poles is given by $(\mu_0/4\pi)P_1 P_2/r$, where $r$ is their separation. Suppose the domain divides in two as in Fig. 11-16b. Each new domain has width $\frac{1}{2}w$. (a) What is the change in magnetic field energy? (b) Evaluate the change in energy if $\ell = 0.50$ cm, $w = 0.0020$ cm, $t = 0.0020$ cm, and $M = 5.0 \times 10^5$ A/m. (c) Estimate the increase in exchange energy. Assume the spins change abruptly from $S = \frac{1}{2}\hbar$ on one side of the boundary to $-\frac{1}{2}\hbar$ on the other. Take the exchange energy coefficient to be $J_e = 0.250$ meV and suppose atoms along the boundary are separated by $6.0 \times 10^{-10}$ m.

17.  Show that $\coth(x)$ can be approximated by $1 + 2e^{-2x}$ for large $x$ and use this result to show that Eq. 11-28 becomes

$$M = M_s \left[ 1 - \frac{1}{J'} e^{-\mu_0 \beta g \mu_B \gamma M} \right]$$

for low temperatures. Write $M = M_s - \Delta M$ and replace $M$ by $M_s$ in the exponent, then show that

$$\Delta M = \frac{M_s}{J'} e^{-uT_c/T},$$

where $u = 3/(J' + 1)$. This gives the Curie-Weiss prediction for the low-temperature behavior of a ferromagnet. It does not agree with experimental results but discrepancies are resolved by spin wave theory.

18. Take the spin wave dispersion relationship to be $\omega = Ak^2$ and show that the spin wave energy is given by

$$E = \frac{\tau_s}{4\pi^2} \frac{(k_BT)^{5/2}}{(A\hbar)^{3/2}} \int_0^\infty \frac{x^{3/2} \, dx}{e^x - 1}$$

for a three-dimensional ferromagnet at low temperatures. Use this result to show that the spin wave contribution to the low-temperature heat capacity is proportional to $T^{3/2}$.

19. Suppose next nearest neighbors of an atom in an antiferromagnetic contribute to the local field. Take the local fields for the two sublattices to be $B_A = -\mu_0(\gamma_1 M_B + \gamma_2 M_A)$ and $B_B = -\mu_0(\gamma_1 M_A + \gamma_2 M_B)$, respectively. (a) Show that $T_N = \frac{1}{2}C(\gamma_1 - \gamma_2)$ and $\chi = C/(T + \theta)$, where $\theta = \frac{1}{2}C(\gamma_1 + \gamma_2)$. (b) Estimate $\gamma_1$ and $\gamma_2$ for MnO, an antiferromagnet with the structure of NaCl. The cube edge is $4.445 \times 10^{-10}$ m and the dipole moment of each manganese ion is $5.0\mu_B$. $T_N = 116$ K and $\theta = 610$ K.

20. Neutron scattering is used to identify a simple cubic antiferromagnetic, with cube edge $a$. Suppose alternate atoms along $\langle 100 \rangle$ lines have spins in opposite directions and so have different atomic form factors. All atoms on any (100) plane have spins in the same direction. (a) Take the magnetic unit cell to be a cube with edge $2a$ and calculate the structure factor for each of the following peaks (indexed on the magnetic cell): (100), (110), (111), (200), (210), (220), (211), and (311). (b) Which peaks vanish for temperatures above the Néel temperature?

21. A certain electron magnetic resonance line occurs for $B_0 = 0.35$ T in a 9.5-GHz radiation field. (a) What is the value of $g$? (b) Assume two dipole moments, one in a field $B_0$ and the other in a field $B_0 + \delta B_0$, start aligned. How much later are they $\pi$ out of phase in their precessional motion? Take $\delta B_0 = 5.0 \times 10^{-5}$ T.

22. In one type resonance experiment the sample is placed in a magnetic induction field $\mathbf{B}_0$, along the $z$ axis. Its magnetization is $\mathbf{M}_0$, in the same direction. A transverse rotating field with amplitude $B_1$ is turned on and remains on for a time $t = \pi/2\gamma B_1$. The field rotates with angular velocity $\gamma B_0$ in the same sense as the magnetization. Suppose $t_1$ is much shorter than both $\bar{t}_1$ and $\bar{t}_2$, so relaxation effects can be neglected. (a) Show that $M_z$ then obeys $(d^2M_z/dt^2) = -\gamma^2B_1^2M_z$. (b) Show that $M_z = 0$ when the rotating field is turned off. (c) What is the value of $M_x^2 + M_y^2$ when the field is turned off? (d) Now take relaxation processes into account and sketch the $z$ component of the magnetization as a function of time after the field is turned off.

# Chapter 12

# FREE ELECTRONS AND MAGNETISM

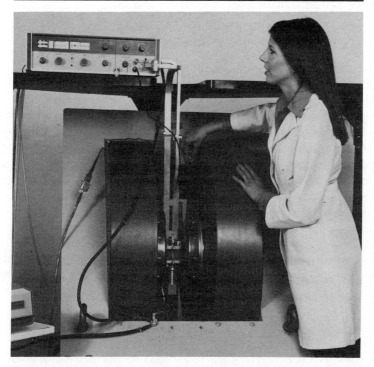

An electron spin resonance spectrometer. The sample is between the pole pieces of the large electromagnet; electromagnetic radiation is generated by the source above the magnet and fed to the sample through a wave guide.

- 12.1 FREE ELECTRONS IN A MAGNETIC FIELD

- 12.2 FREE ELECTRON DIAMAGNETISM

- 12.3 FREE ELECTRON PARAMAGNETISM

- 12.4 CHARGE TRANSPORT IN MAGNETIC FIELDS

Nearly free electrons contribute to the magnetic properties of metals and semiconductors. The application of a static magnetic field to a nearly free electron system has three important consequences. First, the propagation vector of each electron changes with time. It moves along a constant energy surface in reciprocal space. Second, the electron states and energy spectrum change. Constant energy surfaces in the presence of a field are different from surfaces in the absence of a field. Third, electron spins tend to align with the field. Electron orbital motions make a diamagnetic contribution to the susceptibility, while spins make a paramagnetic contribution.

## 12.1 FREE ELECTRONS IN A MAGNETIC FIELD

*Cyclotron Frequency.* We first consider the time dependence of an electron propagation vector. Classically, a free electron moving with speed $v$ in a plane perpendicular to a magnetic induction field **B** executes uniform circular motion with angular speed $eB/m$. The field exerts a centripetal force with magnitude $evB$, so Newton's second law becomes $evB = mv^2/R$, where $R$ is the radius of the orbit. The electron speed and orbit radius are related by $v = (eB/m)R$, so the angular frequency of rotation is $\omega_c = v/R = eB/m$. The cyclotron angular frequency, as $\omega_c$ is called, is twice the Larmor angular frequency. The component of the velocity along the field is constant and if it does not vanish, the electron moves in a spiral.

The momentum **p** of an electron in a magnetic field obeys $d\mathbf{p}/dt = -e\mathbf{v} \times \mathbf{B}$. Its magnitude and the component along the field are both constant, but its tip precesses with angular frequency $\omega_c$ about the direction of the field.

In the quantum description **p** is replaced by $\hbar\mathbf{k}$ and **v** is replaced by $(1/\hbar) \nabla_k E$, so

$$\frac{d\mathbf{k}}{dt} = -\frac{1}{\hbar^2} e\nabla_k E \times \mathbf{B}. \tag{12-1}$$

This equation is analogous to Eq. 9-12 for the rate of change of **k** in an electric field. The electric force has been replaced by the magnetic force.

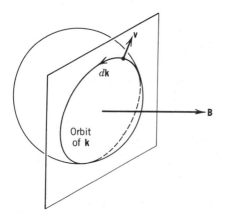

FIGURE 12-1 Electron transitions in a constant magnetic field. The sphere is a constant energy surface for free electrons. The propagation vector of an electron moves around the circle formed by the intersection of the sphere with a plane perpendicular to the magnetic field. The diagram does not take into account changes in constant energy surfaces produced by the field.

As the electron makes transitions in the field its energy does not change. $\nabla_k E$ is normal to a constant energy surface and, according to Eq. 12-1, the change in propagation vector is perpendicular to both $\nabla_k E$ and **B**. Thus the tip of **k** moves around the intersection of a constant energy surface and a plane perpendicular to **B**, as indicated in Fig. 12-1.

The direction of motion is determined by the directions of $\nabla_k E$ and **B**. Figure 12-2a shows a constant energy surface near the bottom of a band. Energy increases from the center of the zone toward the boundary and $\nabla_k E$ points outward. If **B** is out of the page, transitions are made in the counterclockwise direction. On the other hand, b shows a surface near the top of a band. Energy now increases from the zone boundary toward the center, $\nabla_k E$ points inward, and transitions are made in the clockwise direction. Orbits such as this are called hole orbits.

A third type, called an open orbit, is shown in Fig. 12-2c. The orbit intersects the zone boundary at two points that are separated by a reciprocal lattice vector. When the electron gets to $A$ it starts over again at $A'$. For the most part, we will deal with closed orbits.

We can easily derive a general expression for the time an electron takes to make transitions around a closed orbit. Refer to Fig. 12-3. Let $\mathbf{k}_\parallel$ be an infinitesimal

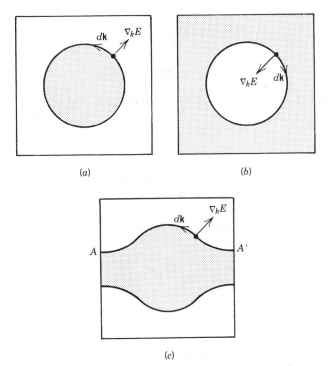

(a)  (b)

(c)

**FIGURE 12-2** A magnetic field directed out of the page causes an electron to jump to a succession of states on the same constant energy surface, as indicated by an arrow. The direction of increasing energy is indicated by $\nabla_k E$ and states with energy lower than that of the electron are indicated by shaded regions. (a) An electron orbit, (b) a hole orbit, and (c) an open orbit.

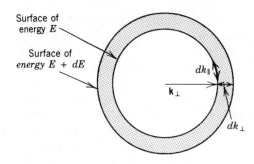

Surface of energy $E$

Surface of energy $E + dE$

$dk_\parallel$

$k_\perp$

$dk_\perp$

**FIGURE 12-3** Two orbits in reciprocal space, associated with energies $E$ and $E + dE$. The magnetic field is normal to the page. The time for a propagation vector to traverse an orbit is given by $(\hbar^2/eB)/|dS/dE|$, where $dS$ is the shaded area.

segment of the orbit and let $\mathbf{k}_\perp$ be the component of $\mathbf{k}$ perpendicular to both the orbit and to $\mathbf{B}$. Then Eq. 12-1 becomes $dk_\parallel/dt = (eB/\hbar^2)|\partial E/\partial k_\perp|$, or $dt = (\hbar^2/eB)(1/|\partial E/\partial k_\perp|)\, dk_\parallel$. The period $T$ of the motion is the line integral around the orbit:

$$T = \frac{\hbar^2}{eB} \oint \frac{dk_\parallel}{|\partial E/\partial k_\perp|}. \tag{12-2}$$

The integral can be written $\oint |dk_\perp/dE| dk_\parallel = (d/dE) \oint k_\perp dk_\parallel = dS/dE$, where $S$ is the area of the orbit. Thus

$$T = \frac{\hbar^2}{eB} \left| \frac{dS}{dE} \right|. \tag{12-3}$$

The component of $\mathbf{k}$ parallel to $\mathbf{B}$ is held constant when the derivative is evaluated.

The angular frequency associated with the orbit of $\mathbf{k}$ is the cyclotron angular frequency:

$$\omega_c = \frac{2\pi}{T} = \frac{2\pi eB}{\hbar |dS/dE|}. \tag{12-4}$$

Equation 12-4 is more complicated than the classical expression $\omega_c = eB/m$ if the orbit in reciprocal space is not circular. For nearly free electrons with scalar effective mass $m^*$, however, $S = \pi k_\perp^2 = \pi[(2m^*E/\hbar^2) - k_z^2]$, where $k_z$ is the component of $k$ along the field. So $dS/dE = 2\pi m^*/\hbar^2$ and

$$\omega_c = \frac{eB}{m^*}. \tag{12-5}$$

For typical laboratory fields the cyclotron angular frequency is on the order of $10^{10}$ rad/s. If $B = 0.5$ T and $m^* = 9.1 \times 10^{-31}$ kg, for example, $\omega_c = 8.8 \times 10^{10}$ rad/s.

Even if the band is not parabolic, Eq. 12-5 is used to define what is called the cyclotron effective mass, with magnitude given by

$$m^* = \frac{eB}{\omega_c} = \frac{eBT}{2\pi} = \frac{\hbar^2}{2\pi} \left| \frac{dS}{dE} \right|. \tag{12-6}$$

It may differ from the band effective mass.

*Electron Energies in a Magnetic Field.* For a field in the $z$ direction, both the $x$ and $y$ coordinates of an electron oscillate with angular frequency $\omega_c$. Quantum mechanics tells us that the energy of a simple harmonic oscilator is quantized in units of $\hbar\omega$ and we expect the same to be true for a circulating electron. In fact, the energy of a nearly free electron in a magnetic field $B\hat{z}$ is given by

$$E = \frac{\hbar^2 k_z^2}{2m^*} + \hbar\omega_c(p + \tfrac{1}{2}),\qquad(12\text{-}7)$$

where $p$ is a positive integer or zero.* To allow for the influence of the potential energy function the electron mass $m$ has been replaced by an effective mass $m^*$. The first term represents the contribution of motion along the field to the kinetic energy while the second is the oscillator energy. Allowed values of the energy, known as Landau levels, are designated by $k_z$ and $p$. They are plotted in Fig. 12-4.

All states with the same value of $p$ lie on a cylinder in reciprocal space. A series of cylinders is shown in Fig. 12-5. If we think of an electron in a magnetic field as making transitions from one free electron state to another then the tip of its **k** vector moves with angular velocity $\omega_c$ around a circular orbit on the surface of one cylinder. In this sense, a Landau state is represented by a circle in reciprocal space rather than by a point. Since the energy depends on $k_z$ as well as on $p$, a Landau cylinder is not a constant energy surface. Each of the curves in Fig. 12-4 shows how the energy varies along the length of a cylinder. The energy is constant on each circle around the surface of a cylinder.

For many solids, of course, the electron energy is not parabolic in $k$. The ideas discussed above are valid for them, but the Landau cylinders are not circular. Instead, their cross sections mimic cross sections of constant energy surfaces in the absence of a field. For simplicity we deal with circular cylinders.

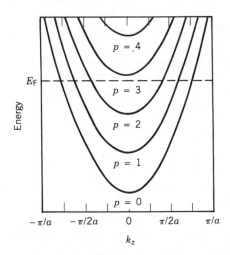

**FIGURE 12-4** Low-lying energy levels for a free electron system in a magnetic field.

---

*For a derivation of this result from the Schrödinger equation, see, for example, R. E. Peierls, *Quantum Theory of Solids* (London/New York: Oxford Univ. Press, 1954).

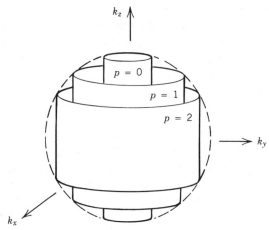

**FIGURE 12-5** Landau cylinders in reciprocal space. Each cylinder extends to infinity in both directions along the $k_z$ axis, but only portions inside a zero-field constant energy surface are shown. Electron energy varies with $k_z$ on each cylinder. A circle around any cylinder is a line of constant energy.

According to Eq. 12-7 the radius $k_\perp$ of a Landau cylinder is given by $(\hbar^2/2m^*)k_\perp^2 = \hbar\omega_c(p + \frac{1}{2})$ or $k_\perp^2 = (2m^*/\hbar)\omega_c(p + \frac{1}{2})$. Cylinder radii are directly proportional to the induction field and the number of cylinders through any fixed region of reciprocal space decreases with increasing field.

Landau levels are highly degenerate. Each circle around the surface of a cylinder represents many states. In fact, the number of states represented by a length $\Delta k_z$ of any given cylinder is exactly the same as the number of states represented by the volume between adjacent cylinders of the same length, in the absence of a field. It is the product of the volume enclosed by adjacent cylinders and the number of states per unit volume of reciprocal space, in the absence of a field.

The volume enclosed by segments of two adjacent Landau cylinders is $\pi\Delta(k_\perp^2) \Delta k_z = (2\pi m^*/\hbar)\omega_c \Delta k_z$. The number of states per unit volume of reciprocal space is $\tau_s/4\pi^3$, where $\tau_s$ is the sample volume and spin has been taken into account. Thus the number of states represented by a segment of a Landau cylinder with length $\Delta k_z$ is $(\tau_s m^*\omega_c/2\pi^2\hbar) \Delta k_z$. It is independent of $p$ and is proportional to $B$.

In effect, a field moves all states with zero-field energy between $(\hbar^2/2m)k_z^2 + \hbar\omega_c p$ and $(\hbar^2/2m)k_z^2 + \hbar\omega_c(p + 1)$ to the same level, $(\hbar^2/2m)k_z^2 + \hbar\omega_c(p + \frac{1}{2})$. We may think of points representing states in reciprocal space moving to the nearest cylinder, as illustrated in Fig. 12-6.

Spin-up and spin-down electrons have different energies in a magnetic field and the spin energy should be included in Eq. 12-7. There are actually two sets of Landau levels, one for each spin direction. Each has half the number of states given above. We defer discussion of free electron spin until Section 12.3.

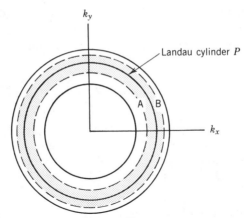

**FIGURE 12-6** A plane in reciprocal space normal to an applied magnetic field. Solid lines are circumferences of Landau cylinders. Cylinder $p$ has cross section $(2m^*/\hbar)\omega_c(p + \frac{1}{2})$. Cylinders A and B have cross sections $(2m^*/\hbar)\omega_c p$ and $(2m^*/\hbar)\omega_c(p + 1)$, respectively. The number of states per unit length of the Landau cylinder equals the number of states per unit length in the region between A and B, in the absence of a field. When the field is turned on all states originally between A and B move to the Landau cylinder.

**EXAMPLE 12-1** Suppose a solid has $5.1 \times 10^{28}$ free electrons/m³. Estimate the number of Landau cylinders that pierce the Fermi sphere in a magnetic induction field of 0.50 T.

**SOLUTION** According to Eq. 8-28, the radius of the Fermi sphere is

$$k_F = (3\pi^2 n)^{1/3} = (3\pi^2 \times 5.1 \times 10^{28})^{1/3} = 1.15 \times 10^{10}\ \text{m}^{-1}.$$

Take $k_F$ to be the radius of a Landau cylinder and solve $(\hbar^2/2m)k_F^2 = \hbar\omega_c(p + \frac{1}{2})$ for $p$:

$$p = \frac{\hbar k_F^2}{2m\omega_c} - \frac{1}{2}.$$

The cyclotron angular frequency is $\omega_c = eB/m = 1.6 \times 10^{-19} \times 0.5/9.1 \times 10^{-31} = 8.8 \times 10^{10}$ rad/s, so

$$p = \frac{1.05 \times 10^{-34}(1.15 \times 10^{10})^2}{2 \times 9.11 \times 10^{-31} \times 8.8 \times 10^{10}} - \frac{1}{2} = 8.7 \times 10^4.$$

In the absence of a field $5.1 \times 10^{28}$ states are uniformly distributed inside the Fermi sphere. In a 0.50-T induction field the same number of states are distributed on $8.7 \times 10^4$ cylinders. The reorganization of states can be seen only on a fine scale diagram. It is, however, important for magnetic properties. ∎

Landau levels are filled in accordance with Fermi-Dirac statistics. All states associated with low-lying levels are occupied, while all states associated with high-lying levels are empty. Occupation probabilities for states near the Fermi level are between 0 and 1 and vary with temperature.

The total energy of the electron system in a magnetic field is given by

$$E_T = \frac{\tau_s m^* \omega_c}{2\pi^2 \hbar} \sum_{p=0}^{\infty} \int_{-\infty}^{\infty} \frac{E(k_z, p)}{e^{\beta(E-\eta)} + 1} \, dk_z, \tag{12-8}$$

where $E$ is given by Eq. 12-7. The factor in front is the number of states per unit length of a cylinder. The chemical potential $\eta$ also depends on the field. It is determined by the condition that the total number of electrons $N$ is given by

$$N = \frac{\tau_s m^* \omega_c}{2\pi^2 \hbar} \sum_{p=0}^{\infty} \int_{-\infty}^{\infty} \frac{dk_z}{e^{\beta(E-\eta)} + 1}. \tag{12-9}$$

The chemical potential is a complicated function of the induction field. We can obtain a qualitative understanding, however, by calculating the Fermi energy ($\eta$ at $T = 0$ K). Refer to Fig. 12-4. The Fermi level is adjusted so the number of states with energy less than $E_F$ equals the number of electrons. If $E_F > \hbar \omega_c(p + \frac{1}{2})$ all states on cylinder $p$ are occupied from $k_z = -k_{z\,max}$ to $k_z = +k_{z\,max}$, where $k_{z\,max} = (2m^*/\hbar^2)^{1/2}[E_F - \hbar\omega_c(p + \frac{1}{2})]^{1/2}$. Thus $(\tau_s \hbar \omega_c/2\pi^2)(2m^*/\hbar^2)^{3/2}[E_F - \hbar\omega_c(p + \frac{1}{2})]^{1/2}$ states on cylinder $p$ are occupied and

$$N = \frac{\tau_s \hbar \omega_c}{2\pi^2} \left(\frac{2m^*}{\hbar^2}\right)^{3/2} \sum_{p=0}^{p_{max}} [E_F - \hbar\omega_c(p + \frac{1}{2})]^{1/2}, \tag{12-10}$$

where $p_{max}$ is the largest value of $p$ for which the quantity in the brackets is positive. States with $p > p_{max}$ are unoccupied at $T = 0$ K.

Numerical methods are used to solve Eq. 12-10 for $E_F$. Results are plotted in Fig. 12-7a. For $\hbar \omega_c \ll k_B T$ the distribution of states in reciprocal space is nearly the same as in the absence of a field and the Fermi energy is nearly independent of the field. At high fields, however, $E_F$ displays considerable structure.

As the field increases $E_F$ changes in such a way that the right side of Eq. 12-10 remains constant. Two competing effects occur. The number of states associated with each Landau level increases. As a result, the factor that multiplies the sum in Eq. 12-10 increases and $E_F$ tends to decrease in response. On the other hand, the energy of each Landau level increases. The term $\hbar\omega_c(p + \frac{1}{2})$ in the brackets of Eq. 12-10 increases and $E_F$ tends to increase in response. The closer $E_F$ is to $\hbar\omega_c(p + \frac{1}{2})$, the greater the influence of that level on $E_F$.

Suppose the Fermi level is just below the lowest $p = 3$ level, as in Fig. 12-4. It is then far above the lowest $p = 2$ level and it drops as $B$ increases. At the same time the energies of all states increase. As $E_F$ nears the lowest $p = 2$ level it begins to rise, but not as fast as the level itself, so the level crosses the Fermi level for some value of $B$. The Fermi level is then just below the lowest $p = 2$ level and the cycle begins again. Because the separation of Landau levels increases with the field, swings in the Fermi energy are greater at high fields than at low fields.

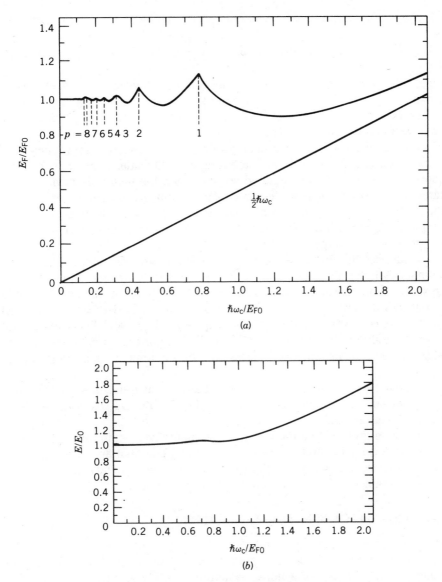

**FIGURE 12-7**  (a) The Fermi energy $E_F$ of a free electron system in a magnetic field as a function of $\hbar\omega_c/E_{F0}$, where $E_{F0}$ is the zero-field Fermi energy. The straight line is the energy of the lowest Landau level. (b) The total electron energy $E$ as a function of $\hbar\omega_c$. $E_0$ is the zero-field energy.

As implied by the above discussion, the Fermi energy is a maximum when the lowest level of a Landau cylinder crosses the Fermi level. Several maxima can be seen on the graph of Fig. 12-7a. For each the figure gives the value of $p$ for the highest cylinder with states below the Fermi level.

For extremely high fields, well above those attainable in the laboratory, only states on the $p = 0$ Landau cylinder are occupied. The sum in Eq. 12-10 then reduces to one term and the Fermi energy is given by

$$E_F = \tfrac{1}{2}\hbar\omega_c + \frac{\pi^4\hbar^4}{2m^{*3}\omega_c^2} n^2, \qquad (12\text{-}11)$$

where $n$ is the electron concentration. Just after the lowest $p = 1$ level crosses $E_F$, the second term of Eq. 12-11 dominates and $E_F$ decreases, but at higher fields the second term is much smaller than the first and the Fermi energy approaches the lowest Landau energy, $\tfrac{1}{2}\hbar\omega_c$.

Figure 12-7$b$ shows the total energy of the electron system as a function of $\hbar\omega_c$. It has a upward trend in an increasing field. Consider two concentric cylinders in reciprocal space, situated so that when the field **B** is turned on all states between them move to the same Landau cylinder. Some energies increase while others decrease, but the average remains the same. If these states are completely filled both before and after the field is turned on, the total electron energy does not change.

Now suppose the Fermi sphere cuts between the two cylinders, so states near the inner cylinder are filled while states near the outer cylinder are empty. All the states again coalesce to the same Landau cylinder when the field is turned on, but now the energies of more electrons increase than decrease so the total energy increases. Although the increase is quite small for weak fields and cannot be seen readily on the graph of Fig. 12-7$b$, it is important. At extremely strong fields all electrons are in $p = 0$ states and the total energy is proportional to $\hbar\omega_c$.

The total energy also oscillates about the general trend line. At least one oscillation can be seen on the graph of Fig. 12-7$b$. Later we will consider oscillations at weaker fields.

The results given above for $T = 0$ K are qualitatively valid for high temperatures. Because some states above the chemical potential are occupied while some below are not, oscillations of the chemical potential and total energy are not as pronounced as at the absolute zero of temperature.

## 12.2 FREE ELECTRON DIAMAGNETISM

*Magnetic Susceptibility.* Once the total energy is known as a function of the induction field the magnetic susceptibility can be computed. The magnetic contribution to the energy of a linear system is $-\mathbf{M} \cdot \mathbf{B} = -(\chi/\mu_0)B^2$, if $\chi \ll 1$. Thus $\chi = -\tfrac{1}{2}\mu_0 d^2 E_T/dB^2$. If the system is not linear we use a weak field approximation and evaluate the derivative for $B = 0$. The system is diamagnetic if its energy increases with increasing field and paramagnetic if its energy decreases. Here we consider contributions of electron motions to the susceptibility. Spins are treated in the next section.

A collection of circulating electrons has a net dipole moment, directed opposite to the applied field. Nevertheless, classical statistical mechanics predicts a net magnetization of zero. In thermodynamic equilibrium the probability that an electron has speed $v$ is proportional to $e^{-\beta v^2/2m}$, whether or not a field exists. As a result, the total energy of a free electron system is not affected by a field.

Collisions with phonons, defects, and sample boundaries bring about thermal equilibrium and, in the process, destroy the net dipole moment created by the field.

We turn to the quantum description. Since energies of electrons near the Fermi surface increase as a field is turned on, we conclude that the magnetization is opposite the field and that the system is diamagnetic. An algebraic expression for $\chi$ can be obtained for $k_B T \gg \hbar \omega_c$, but the derivation is complicated. We quote the result:

$$\chi = -\frac{\mu_0 n \mu_B^2 m^2}{2E_F m^{*2}}, \tag{12-12}$$

where $E_F$ is the zero-field Fermi energy. For free electrons the ratio $m/m^*$ is 1 and $\chi = -\mu_0 n \mu_B^2 / 2E_F$. An algebraic expression does not exist for nonparabolic bands. In the temperature range for which Eq. 12-12 is valid, the diamagnetic contribution of nearly free electrons to the susceptibility is nearly independent of the temperature.

**EXAMPLE 12-2**  Estimate the diamagnetic contribution of the free electrons to the susceptibility of potassium. Potassium is BCC with a cube edge of 5.225 Å. Take $m^*$ to be the free electron mass.

**SOLUTION**  Potassium has two free electrons per cubic unit cell, so the free electron concentration is $n = 2/(5.225 \times 10^{-10})^3 = 1.40 \times 10^{28}$ electrons/m$^3$. The Fermi energy is

$$E_F = \frac{\hbar^2}{2m}(3\pi^2 n)^{2/3} = \frac{(1.05 \times 10^{-34})^2}{2 \times 9.11 \times 10^{-19}}(3\pi^2 \times 1.40 \times 10^{28})^{2/3}$$

$$= 3.37 \times 10^{-19} \text{ J},$$

so

$$\chi = -\frac{\mu_0 n \mu_B^2}{2E_F} = -\frac{4\pi \times 10^{-7} \times 1.40 \times 10^{28} \times (9.27 \times 10^{-24})^2}{2 \times 3.37 \times 10^{-19}}$$

$$= -2.25 \times 10^{-6}. \qquad \blacksquare$$

*The de Haas-van Alphen Effect.*  At temperatures for which $k_B T$ is on the order of $\hbar \omega_c$ or less, the susceptibility of a free electron system is an oscillatory function of the field. This phenomenon, called the de Haas-van Alphen effect, is clearly shown by a plot of the susceptibility as a function of $1/B$, as in Fig. 12-8. The susceptibility is calculated by evaluating the second derivative of $E_T$ with respect to $B$, but not in the limit as $B$ becomes small.

The origin of the effect was mentioned earlier in connection with the Fermi energy of a free electron system in a magnetic field. We observed that the total energy is an oscillatory function of the induction field. To take a simple case,

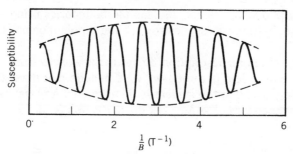

**FIGURE 12-8** The de Haas-van Alphen effect. A quantity proportional to the susceptibility is plotted as a function of the reciprocal of the magnetic field. Two periods, one long (dotted) and one short (solid), are present, indicating that the applied field is perpendicular to two extremal cross sections of the Fermi surface.

we now assume several thousand Landau cylinders pierce the Fermi surface. The Fermi level then remains essentially constant as the field increases.

Figure 12-9 shows a greatly expanded diagram of $k_z = 0$ energy levels in the vicinity of $E_F$. As the situation changes from that shown in ($a$) to that shown in ($b$), the total energy increases. The total number of electrons in Landau levels below the Fermi level remains constant but the separation of the levels increases. On the other hand, the energy decreases as the situation changes from that shown in ($b$) to that shown in ($c$). A Landau cylinder has crossed the Fermi surface and, as a result, is unoccupied. Electrons that previously occupied states

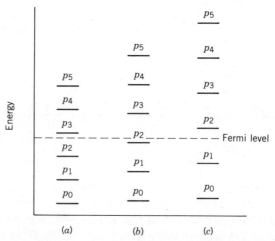

**FIGURE 12-9** The energy level structure of a free electron system in a magnetic field along the z axis. Only $k_z = 0$ energies are shown. The field strength increases from left to right. The total electron energy is greater for ($b$) than for ($a$) because the energy of each Landau level increases with the field. It is less for ($c$) than for ($b$) because the lowest level of a Landau cylinder has crossed the Fermi level and become unoccupied. A precise calculation takes into account variations of the Fermi energy, which are slight unless the field is extremely strong.

of that level are now at lower levels and the total energy is less. The increase in the degeneracy of each level assures there are enough states to hold the electrons. The process repeats as other cylinders cross the Fermi surface. An increase in energy means the sample is more diamagnetic than before, while a decrease means it is less.

We now find the period of oscillation. For simplicity assume the band and cyclotron effective masses have the same value. Then the cross sectional area of cylinder $p$ is given by $S = (2\pi m^* \omega_c / \hbar)(p + \frac{1}{2}) = (2\pi e B / \hbar)(p + \frac{1}{2})$ and

$$\frac{1}{B} = \frac{2\pi e}{\hbar S}(p + \frac{1}{2}). \tag{12-13}$$

Suppose the lowest $p$ level is initially at the Fermi level, then $(1/B)$ is decreased by $\Delta(1/B)$ so that the lowest $p - 1$ level is at the Fermi level. Afterwards cylinder $p - 1$ has the same cross section as cylinder $p$ had before. According to Eq. 12-13

$$\Delta\left(\frac{1}{B}\right) = \frac{2\pi e}{\hbar S}. \tag{12-14}$$

Each time $1/B$ decreases by $2\pi e/\hbar S$ another Landau cylinder crosses the Fermi surface. This quantity represents the period of oscillation of the susceptibility, considered as a function of $1/B$.

In Eq. 12-14, $S$ represents the cross section of a cylinder in reciprocal space as the cylinder surface crosses the Fermi sphere. It is therefore also the maximum cross section of the Fermi sphere. A measurement of the period of a de Haas-van Alphen oscillation for a metal with a parabolic conduction band gives the Fermi sphere radius.

The effect also occurs for a metal with nonparabolic energy bands. Landau cylinder cross sections are not circular and the Fermi surface is not spherical. As a result, a cylinder may cross different portions of the Fermi surface at different field strengths. For reasons we will not pursue, the susceptibility changes significantly only if the crossing occurs at a local maximum or minimum cross section of the Fermi surface. Some Fermi surfaces have two or more extremal cross sections for some directions of the applied field. If the magnetic field is along the [111] direction of copper, for example, extremal cross sections associated with both the central body and the necks contribute to the de Haas-van Alphen effect. A plot of $\chi$ as a function of $1/B$ is a linear combination of two or more oscillatory functions. Researchers may resort to Fourier analysis to sort out the trace.

De Haas-van Alphen experiments are important for the information they yield about the Fermi surfaces. Measurements are made for various orientations of the field relative to the sample and are used to map important features of the Fermi surface. Results are used to substantiate band theory calculations.

Other properties, such as electrical and thermal conductivity, show similar oscillations when plotted as functions of $1/B$. The occurrence of oscillations in the electrical conductivity is called the de Haas-Shubnikov effect. Optical properties are also affected by magnetic fields. By reorganizing energy levels, a magnetic field changes that part of the absorption spectrum that involves electron transitions to the conduction band from lower bands. These changes are known as magnetooptical effects. Electromagnetic radiation may also induce transitions from one Landau level to another. This phenomenon is called cyclotron resonance and will be discussed later.

The de Haas-van Alphen and related effects can be observed only at low temperatures. If $k_B T \gg \hbar \omega_c$, Landau levels above the Fermi level are occupied with high probability and changes in energy as a Landau cylinder crosses the Fermi surface are slight.

## 12.3 FREE ELECTRON PARAMAGNETISM

The intrinsic magnetic dipole moment of an electron is given by Eq. 11-8 with $S = \frac{1}{2}\hbar$. If the magnetic field is in the positive $z$ direction, a spin-up electron has energy $E_0 + (eB\hbar/2m)$ and a spin-down electron has energy $E_0 - (eB\hbar/2m)$. In each case the first term represents the electron energy in the absence of a field and the second arises from the interaction of the dipole with the field. In thermodynamic equilibrium more electrons have spin down than spin up and, since the dipole moments of spin-down electrons are in the same direction as the field, the spin system is paramagnetic.

To estimate the magnetization and susceptibility, we neglect diamagnetic effects and suppose the kinetic energy and dipole-field interaction energy are the only contributions to the total energy. In a field, the number of spin-up electrons with a given energy $E$ is the same as half the number with energy $E - e\hbar B/2m$ in the absence of a field. When the field is turned on, the energy of each spin-up electron is raised by $e\hbar B/2m$ to $E$. If $\rho(E)$ is the electron density of states and $f(E)$ is the Fermi-Dirac distribution function, then the number of spin-up electrons with energy between $E$ and $E + dE$ is $\frac{1}{2}f(E)\rho(E - e\hbar B/2m) \, dE$ and the number of spin-down electrons is $\frac{1}{2}f(E)\rho(E + e\hbar B/2m) \, dE$. The situation is diagrammed in Fig. 12-10.

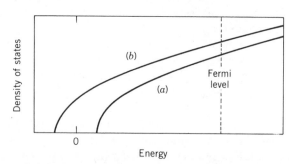

FIGURE 12-10 Density of states as a function of energy $E$ for (a) spin-up and (b) spin-down free electrons in a magnetic field. $E = 0$ is the minimum energy in the absence of a field. The spin-up distribution is shifted by $e\hbar B/2m$ toward higher energy, while the spin-down distribution is shifted by the same amount toward lower energy. At $T = 0$ K all states with $E < E_F$ are occupied and more electrons are spin down than spin up. The net spin magnetization is in the direction of the field.

The net magnetization is given by

$$M = \frac{e\hbar}{4m\tau_s} \int_{-\infty}^{+\infty} f(E)[\rho(E - e\hbar B/2m) - \rho(E + e\hbar B/2m)] \, dE, \quad (12\text{-}15)$$

where $\tau_s$ is the sample volume. Since the shift in energy is small $\rho(E \pm e\hbar B/2m)$ may be approximated by $\rho(E) \pm (e\hbar B/2m)[d\rho(E)/dE]$ and Eq. 12-15 becomes

$$M = -\frac{\mu_B^2 B}{\tau_s} \int_{-\infty}^{\infty} f(E) \frac{d\rho(E)}{dE} \, dE, \quad (12\text{-}16)$$

where $\mu_B$ has been substituted for $e\hbar/2m$.

To evaluate the integral for a metal, first carry out an integration by parts, then use the approximation $\int (df/dE)\rho \, dE = -\rho(E_F)$, valid for low temperatures. The result is $M = (\mu_B^2 B/\tau_s)\rho(E_F)$ and the spin contribution to the susceptibility is

$$\chi = \mu_0 \frac{\mu_B^2}{\tau_s} \rho(E_F). \quad (12\text{-}17)$$

This is a reasonable result. When a field is turned on, some electrons flip from spin up to spin down. At $T = 0$, only those with energy within $\mu_B B$ of the Fermi energy change their spin directions. Since this number is $\frac{1}{2}\rho(E_F)\mu_B B$ and each spin flip increases the magnetization by $2\mu_B/\tau_s$, $M = (\mu_B^2/\tau_s)\rho(E_F)B$ and $\chi$ is given by Eq. 12-17.

Higher order terms, which were omitted in evaluating the integral of Eq. 12-16, are temperature dependent. As Eq. D-7 (in Appendix D) indicates, the first omitted term is proportional to $T^2(d^2\rho/dE^2)$, evaluated for $E = E_F$. This term is usually quite small compared to the term retained, so the paramagnetic contribution to the susceptibility is nearly independent of the temperature.

For a free electron band containing $N$ electrons $\rho(E_F) = \frac{3}{2}N/E_F$ and $\chi = 3\mu_0 n \mu_B^2/2E_F$, where $n$ is the electron concentration. The paramagnetic contribution to $\chi$ is $3m^{*2}/m^2$ times the diamagnetic contribution, given by Eq. 12-12, so the net free electron susceptibility is

$$\chi = \frac{3\mu_0 n \mu_B^2}{2E_F} \left(1 - \frac{m^2}{3m^{*2}}\right), \quad (12\text{-}18)$$

where the first term is the spin contribution and the second is the orbital contribution.

For simple metals, such as the alkalis, $m^{*2} > \frac{1}{3}m^2$ and the conduction electron system is paramagnetic. Once core contributions are included, of course, the susceptibility of the sample may be negative. Silver and gold, for example, are diamagnetic although their conduction electrons are paramagnetic.

If $m^{*2} < \frac{1}{3}m^2$ the free electron system is diamagnetic. Although details must be modified to take the true shape of the band into account, Eq. 12-18 indicates that a metal with its Fermi surface in a flat band, with small effective mass, is diamagnetic. Some semimetals, such as bismuth, are examples.

Free electrons also contribute to the susceptibilities of transition and rare earth metals. For these materials, however, unfilled core contributions produce strong paramagnetism.

Analogous calculations can be carried out for electrons in the conduction band of a semiconductor. If the bottom of the band is far above the chemical potential compared to $k_B T$, the result is

$$\chi = \frac{\mu_0 n \mu_B^2}{k_B T} \left( 1 - \frac{m^2}{3m^{*2}} \right). \tag{12-19}$$

A similar expression is obtained for the valence band. Electron and hole contributions to $\chi$ are more strongly dependent on temperature than contributions of free electrons in metals. Although electron and hole contributions are typically positive, negative contributions of core states make the susceptibilities of most semiconductors negative. Both silicon and germanium, for example, are diamagnetic.

**EXAMPLE 12-3**   Estimate the contribution of free electron spins to the susceptibility of potassium, then find the total free electron susceptibility.

**SOLUTION**   If $m^* = m$, the spin contribution is three times the magnitude of the orbital contribution, which was found in Example 12-2 to be $2.25 \times 10^{-6}$. Thus the spin contribution is $6.75 \times 10^{-6}$ and $\chi = 6.75 \times 10^{-6} - 2.25 \times 10^{-6} = 4.50 \times 10^{-6}$.   ■

## 12.4   CHARGE TRANSPORT IN MAGNETIC FIELDS

In this section we consider several important phenomena that result from the motion of electrons in a magnetic field alone or in crossed electric and magnetic fields. We assume the fields are weak and ignore the reorganization into Landau levels.

*The Boltzmann Transport Equation.*   As in Section 9.2, let $f(\mathbf{k}, t)$ represent the number of electrons in a state with propagation constant $\mathbf{k}$. For a solid in thermal equilibrium, this function is the Fermi-Dirac distribution function $f_0$. At time $t + dt$ the number of electrons in a state with propagation constant $\mathbf{k}$ is the same as the number in a state with propagation constant $\mathbf{k} - (d\mathbf{k}/dt) \, dt$ at time $t$. That is, $f(\mathbf{k}, t + dt) = f[\mathbf{k} - (d\mathbf{k}/dt) \, dt, t]$ or $\partial f/\partial t = -\nabla_k f \cdot (d\mathbf{k}/dt)$. A relaxation term is added to the right side and $-(e/\hbar)(\mathcal{E} + \mathbf{v} \times \mathbf{B})$ is substituted for $d\mathbf{k}/dt$ to produce the Boltzmann transport equation for electrons in an electric field $\mathcal{E}$ and a magnetic induction field $\mathbf{B}$:

$$\frac{\partial f}{\partial t} = \frac{e}{\hbar} \nabla_k f \cdot (\mathcal{E} + \mathbf{v} \times \mathbf{B}) - \frac{f - f_0}{\bar{t}}, \tag{12-20}$$

where $\bar{t}$ is the relaxation time. The temperature is assumed to be uniform throughout the sample.

Since the fields are weak we may write $f = f_0 + f_1$, where $f_1$ is small, and take the product of $\nabla_k f_1 \cdot \mathcal{E}$ to be negligible. The same approximation cannot be made for the magnetic field term without losing the influence of the field. Note that $\nabla_k f_0 = (\partial f_0/\partial E) \nabla_k E = \hbar(\partial f_0/\partial E)\mathbf{v}$, a quantity that is perpendicular to $\mathbf{v} \times \mathbf{B}$. In the weak-field approximation the Boltzmann equation is

$$\frac{\partial f_1}{\partial t} = e\frac{\partial f}{\partial E}\mathbf{v} \cdot \mathcal{E} + \frac{e}{\hbar}\nabla_k f_1 \cdot (\mathbf{v} \times \mathbf{B}) - \frac{f_1}{t}. \qquad (12\text{-}21)$$

In the most general situation we consider, the electric field has the form $\mathcal{E} = \mathcal{E}_0 e^{i\omega t}$ and is perpendicular to a constant magnetic induction field $\mathbf{B}$. A static electric field can be considered by setting $\omega = 0$. If $\mathcal{E}$ is the electric component of an electromagnetic wave, we assume the wavelength is sufficiently long that we may neglect spatial variations of $\mathcal{E}$ and $f$.

Solutions to Eq. 12-21 have the form $f_1 = e\bar{t}(\partial f_0/\partial E)\mathbf{v} \cdot \mathbf{A}$, where $A$ is independent of $\mathbf{k}$ but is proportional to $e^{i\omega t}$. This expression is substituted into Eq. 12-21 to obtain

$$i\omega\bar{t}\mathbf{v} \cdot \mathbf{A} - \mathbf{v} \cdot \mathcal{E} - \frac{e\bar{t}}{\hbar}\nabla_k(\mathbf{v} \cdot \mathbf{A}) \cdot (\mathbf{v} \times \mathbf{B}) + \mathbf{v} \cdot \mathbf{A} = 0. \qquad (12\text{-}22)$$

To take the simplest example, we suppose the electrons are in a band with energy given by $E = \hbar^2 k^2/2m^*$, so $\mathbf{v} = \hbar\mathbf{k}/m^*$. Then $\nabla_k(\mathbf{v} \cdot \mathbf{A}) \cdot (\mathbf{v} \times \mathbf{B}) = (\hbar/m^*)\mathbf{A} \cdot (\mathbf{v} \times \mathbf{B}) = (\hbar/m^*)(\mathbf{B} \times \mathbf{A}) \cdot \mathbf{v}$ and Eq. 12-22 becomes

$$\mathbf{v} \cdot [i\omega\bar{t}\mathbf{A} - \mathcal{E} - \frac{e\bar{t}}{m^*}\mathbf{B} \times \mathbf{A} + \mathbf{A}] = 0. \qquad (12\text{-}23)$$

The vector in the brackets is not necessarily perpendicular to $\mathbf{v}$, so it vanishes. To find $\mathbf{A}$, note that $\mathcal{E}$, $\mathbf{B}$, and $\mathcal{E} \times \mathbf{B}$ are mutually orthogonal and write $\mathbf{A} = \alpha\mathcal{E} + \beta\mathbf{B} + \gamma\mathcal{E} \times \mathbf{B}$. Substitute this form into Eq. 12-23, use the vector identity $\mathbf{B} \times (\mathcal{E} \times \mathbf{B}) = B^2\mathcal{E} - \mathbf{B} \cdot \mathcal{E}\mathbf{E} = B^2\mathcal{E}$, and solve for the coefficients $\alpha$, $\beta$, and $\gamma$. The result for $\mathbf{A}$ is

$$\mathbf{A} = \frac{(1 + i\omega\bar{t})}{(1 + i\omega\bar{t})^2 + \omega_c^2\bar{t}^2}\mathcal{E} - \frac{e\bar{t}}{m^*}\frac{1}{(1 + i\omega\bar{t})^2 + \omega_c^2\bar{t}^2}\mathcal{E} \times \mathbf{B}, \qquad (12\text{-}24)$$

where $eB/m^*$ was replaced by $\omega_c$.

The discussion following Eq. 9-52 may be used as a guide to obtain an expression for the current density:

$$\mathbf{J} = -\frac{e}{\tau_s}\sum_{\text{states}}\mathbf{v}f_1(\mathbf{k}) = -\frac{e^2}{\tau_s}\sum_{\text{states}}\bar{t}\frac{\partial f_0}{\partial E}\mathbf{v}\mathbf{v} \cdot \mathbf{A} = \sigma_0\mathbf{A}, \qquad (12\text{-}25)$$

where the sums are over all states in the band and $\sigma_0$ is the dc conductivity in the absence of a magnetic field. In evaluating the sum over states we have assumed the band is isotropic. For a semiconductor $\bar{t}$ and $\omega_c$ represent appropriate averages over the band, while for a metal they have values corresponding to the Fermi energy. When Eq. 12-24 is used, Eq. 12-25 becomes

$$\mathbf{J} = \frac{\sigma_0(1 + i\omega\bar{t})}{(1 + i\omega\bar{t})^2 + \omega_c^2\bar{t}^2}\mathcal{E} - \frac{e\bar{t}}{m^*}\frac{\sigma_0}{(1 + i\omega\bar{t})^2 + \omega_c^2\bar{t}^2}\mathcal{E} \times \mathbf{B}. \qquad (12\text{-}26)$$

One term is along $\mathcal{E}$ and one is perpendicular to $\mathcal{E}$, so the conductivity of a sample in a magnetic field is a tensor. We now examine several special cases.

*Cyclotron Resonance.* Cyclotron resonance experiments are widely used to determine the effective masses of electrons and holes in semiconductors. The electron distribution for a partially filled band is displaced by the action of an electric field perpendicular to a static magnetic field, as illustrated in Fig. 12-11. The magnetic field then causes the distribution to rotate in reciprocal space and interactions with phonons and defects causes it to relax toward the thermal equilibrium distribution $f_0$. Resonance occurs when the electric field rotates with the propagation vectors, at the cyclotron frequency.

Take the static induction field to be $B_0\hat{z}$ and the electric field to have components $\mathcal{E}_x = \mathcal{E}_0 e^{i\omega t}$ and $\mathcal{E}_y = -i\mathcal{E}_0 e^{i\omega t}$. The real parts of the components form a vector that rotates in the counterclockwise direction when viewed from the positive z axis. The field might be associated with a right circularly polarized electromagnetic wave propagating in the direction of the static magnetic field. Following some algebraic manipulation, Eq. 12-26 yields

$$J_x = \frac{\sigma_0[1 - i(\omega - \omega_c)\bar{t}]}{1 + (\omega - \omega_c)^2\bar{t}^2}\mathcal{E}_0 e^{i\omega t} \tag{12-27}$$

and

$$J_y = -\frac{i\sigma_0[1 - i(\omega - \omega_c)\bar{t}]}{1 + (\omega - \omega_c)^2\bar{t}^2}\mathcal{E}_0 e^{i\omega t}. \tag{12-28}$$

$J_z$ is zero. $\mathbf{J}$ rotates about the z axis but it is not in phase with $\mathcal{E}$.

The power absorbed by the electron system is the average over a cycle of $P = \mathbf{J} \cdot \mathcal{E}$, where the real parts of the expressions for $\mathbf{J}$ and $\mathcal{E}$ must be used.

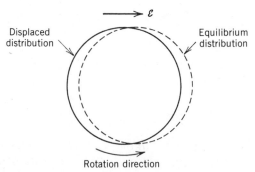

**FIGURE 12-11** Cyclotron resonance. An electric field to the right displaces electrons originally in states on the dotted circle to states on the solid circle. A magnetic field directed out of the page causes the electron distribution to rotate at the cyclotron frequency in the counterclockwise direction. Resonance occurs if the electric field rotates counterclockwise with the same frequency.

After evaluating $\mathbf{J} \cdot \mathcal{E}$, we replace $\sin^2\omega t$ and $\cos^2\omega t$ each by $\frac{1}{2}$ and $\sin \omega t \cos \omega t$ by 0. The result is

$$P_{\text{ave}} = \frac{\sigma_0 \mathcal{E}_0^2}{1 + (\omega - \omega_c)^2 \bar{t}^2}. \tag{12-29}$$

$P_{\text{ave}}$ is maximum for $\omega = \omega_c$, the resonance condition.

In a typical experiment a monochromatic circularly polarized wave, propagating along a static applied magnetic field, is incident on a sample. To find the absorption peak, power in the transmitted wave is monitored as the magnetic field strength is systematically varied. The resonance condition $eB/m^* = \omega$ is then used to find the cyclotron effective mass. Resonance data are often difficult to obtain experimentally. As Eq. 12-29 indicates, the peak is sharp if $\omega \bar{t} \gg 1$, but is washed out if $\omega \bar{t}$ is on the order of 1 or less. As a consequence, experiments are normally performed with high-frequency waves on extremely pure samples at low temperatures.

Cyclotron resonance experiments may also be carried out on holes. Expressions for the current density and absorbed power are the same as those given above for electrons except that $\omega_c$ is replaced by $-\omega_c$. Resonance occurs for left, not right, circularly polarized radiation propagating in the direction of **B**. For a typical semiconductor sample several peaks, at different field strengths, are obtained for each type polarization. These correspond to groups of electrons and holes with different cyclotron effective masses.

The skin depth is extremely small for high-frequency waves incident on a metal. Nevertheless, cyclotron resonances can be obtained by orienting the static magnetic field parallel to a sample surface and directing the wave along the field, so its electric field dips into the sample as it rotates. Electrons orbit around the direction of the magnetic field and once per orbit enter the narrow region near the surface where an electric field exists. A resonance occurs whenever the radiation frequency is a multiple of the cyclotron frequency. They are known as Azbel'-Kaner resonances.

Another phenomenon, called the Faraday effect, is closely related to cyclotron resonance. If a linearly polarized wave is transmitted through a sample in the direction of an applied magnetic field, the polarization direction is found to be rotated. A linearly polarized wave can be thought of as a combination of right and left circularly polarized waves. Since they travel at different speeds in the sample, the electric fields associated with them rotate through different angles during transmission. On exiting the sample, the combination is again linearly polarized but the direction of the resultant field is different from that of the incident field. Measured values of the rotation angle are used to calculate the cyclotron effective mass.

*The Hall Effect.* In a typical experimental arrangement, diagrammed in Fig. 12-12, a dc electric field is applied along the length of the sample, perpendicular to a uniform magnetic field. Charge carriers are deflected by the magnetic field to one side of the sample and produce a potential difference, called the Hall

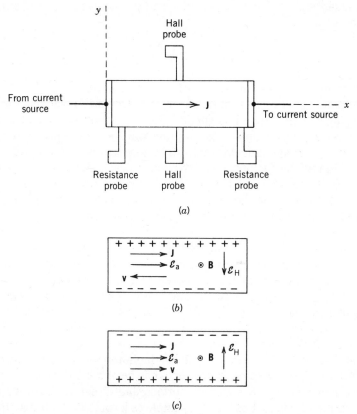

**FIGURE 12-12** (a) Geometry for a Hall effect experiment. The current is to the right and the magnetic field is out of the page. Carriers, whether positive or negative, are forced to the lower surface of the sample. They create a Hall electric field and a potential difference between the Hall probes. Resistance probes are used to determine the sample resistance. (b) The carriers are negative. Their velocity is to the left and the magnetic force on them is toward the lower surface. The Hall electric field points toward that surface and the upper surface is positive relative to the lower. (c) The carriers are positive. The magnetic force is again downward but now the upper surface is negative relative to the lower.

voltage, between points on opposite sides of the sample. Measured values of the Hall voltage, the magnetic field, and the current can be used to obtain information about the concentration of charge carriers and their mobilities.

In some cases the sign of the Hall voltage can be used to determine the sign of the charge carriers and, for example, can be used to distinguish $n$- from $p$-type semiconductors. Suppose the charge carriers are predominantly of one type, either electrons or holes. For the configuration shown in the diagram, the magnetic field deflects them to the lower side of the sample. That side is then positive relative to the opposite side if the carriers are holes and negative if they are electrons. The sign of the Hall voltage gives the sign of the carriers.

Charge collected on the sample sides produces an electric field, called the Hall field, and this field tends to repel other charges as they are forced to the side by the magnetic field. Steady state is reached when the transverse component of the current vanishes.

Take **B** to be in the positive $z$ direction, the applied electric field $\mathcal{E}_a$ to be in the positive $x$ direction, and the Hall field $\mathcal{E}_H$ to be along the $y$ axis. We suppose the charge carriers are electrons from a single band with effective mass $m^*$. Solutions to the Boltzmann transport equation again have the form $e\bar{t}(\partial f_0/\partial E)\mathbf{v} \cdot \mathbf{A}$ and $\mathbf{A}$ is given by Eq. 12-24, with $\omega = 0$. The current density is

$$\mathbf{J} = \sigma_0 \mathbf{A} = \sigma_0 \frac{\mathcal{E}_a - \omega_c \bar{t} \mathcal{E}_H}{1 + \omega_c^2 \bar{t}^2} \,\hat{\mathbf{x}} + \sigma_0 \frac{\mathcal{E}_H + \omega_c \bar{t} \mathcal{E}_a}{1 + \omega_c^2 \bar{t}^2} \,\hat{\mathbf{y}}. \tag{12-30}$$

In steady state $J_y = 0$, so

$$\mathcal{E}_H = -\omega_c \bar{t} \mathcal{E}_a. \tag{12-31}$$

The $x$ component of the current density is then $J_x = \sigma_0 \mathcal{E}_a$, just as if the magnetic field vanished. Substitute $\mathcal{E}_a = J_x/\sigma_0$ into Eq. 12-31 and rearrange to find

$$\frac{\mathcal{E}_H}{J_x B} = -\frac{\omega_c \bar{t}}{\sigma_0 B}. \tag{12-32}$$

The ratio defined by $R_H = \mathcal{E}_H/J_x B$ is called the Hall coefficient.

The current and Hall voltage are measured, not the current density and Hall field. The magnitude of the Hall field is given by $\mathcal{E}_H = V_H/w$, where $V_H$ is the Hall voltage and $w$ is the distance between the Hall voltage probes. We assume the positive terminal of the voltmeter is connected to the lower Hall probe in the figure, so $V_H$ is positive if the Hall field is in the positive $y$ direction. If the current density is uniform throughout the sample $J_x = I/A$, where $I$ is the current and $A$ is the sample cross section. Then, in terms of measured quantities,

$$R_H = \frac{A}{w} \frac{V_H}{BI}. \tag{12-33}$$

In a Hall effect experiment all quantities on the right side of Eq. 12-33 are measured, then $R_H$ is computed. It is negative if the carriers are electrons.

When $\sigma_0 = e^2 n \bar{t}/m^*$ and $\omega_c = eB/m^*$ are substituted into the right side of Eq. 12-32, that equation becomes $R_H = -1/en$, so

$$n = -\frac{1}{eR_H}. \tag{12-34}$$

We have canceled the cyclotron and band effective masses, a procedure that is not always valid. Equation 12-34 is used to find the carrier concentration from the experimentally determined value of $R_H$. Since $\sigma_0 = en\mu$, the electron mobility $\mu$ can also be found once $R_H$ and $\sigma_0$ are known. It is given by $\mu = -R_H \sigma_0$.

If charge is carried predominantly by holes, the sign of the Hall voltage is positive for the configuration described above. The Hall coefficient, still given by Eq. 12-33, is now related to the hole concentration $p$ by

$$p = +\frac{1}{eR_H} \tag{12-35}$$

for an isotropic parabolic band. The hole mobility is given by $\mu = +R_H\sigma_0$. $R_H$ is now positive.

Care must be taken in interpreting the measured transverse potential difference. If the two Hall probes are not directly opposite each other they are on different equipotential surfaces even when $\mathbf{B} = 0$. To compensate, a second measurement is taken with the magnetic field reversed and $V_H$ is evaluated as the average of the two measurements. The current in the sample and the magnitude of the magnetic field must be the same for the two measurements.

Equation 12-34 is valid for a simple metal, while Eqs. 12-34 and 12-35 are valid for many heavily doped semiconductors, $n$ and $p$ types, respectively.

**EXAMPLE 12-4** A Hall effect experiment is performed on a heavily doped $n$-type semiconductor sample with a length of 2.65 cm, a width of 1.70 cm, and a thickness of 0.0520 cm, in a magnetic induction field of 0.500 T. The current in the sample, along its length, is 200 μA. The potential difference along the length of the sample is 195 mV and the potential difference across its width is 21.4 mV. (a) What is the concentration of charge carriers? (b) What is their mobility?

**SOLUTION** (a) The Hall coefficient is

$$R_H = \frac{AV_H}{wBI} = -\frac{1.7 \times 10^{-2} \times 0.052 \times 10^{-2} \times 21.4 \times 10^{-3}}{1.7 \times 10^{-2} \times 0.5 \times 200 \times 10^{-6}} = -0.107 \text{ m}^3/\text{C}$$

so $n = -1/eR_H = 1/(1.6 \times 10^{-19} \times 0.107) = 5.84 \times 10^{19}$ electrons/m³. The resistance of the sample is $R = V/I = 195 \times 10^{-3}/200 \times 10^{-6} = 975\ \Omega$. Since $R = \ell/A\sigma = \ell/Aen\mu$, where $\mu$ is the mobility,

$$\mu = \frac{\ell}{AenR}$$

$$= \frac{2.65 \times 10^{-2}}{1.7 \times 10^{-2} \times 0.052 \times 10^{-2} \times 1.6 \times 10^{-19} \times 5.84 \times 10^{19} \times 975}$$

$$= 0.329 \text{ m}^2/\text{V} \cdot \text{s} .$$

We assumed the voltage probes are on the sample surface. In practice they usually are not. ∎

Both electrons and holes contribute to the Hall effect in a lightly doped or intrinsic semiconductor. Although both types of carriers are swept to the same side of the sample by a magnetic field, the Hall voltage does not necessarily vanish.

Suppose the current consists of two groups of carriers, perhaps with different effective masses, relaxation times, and concentrations. The current density is then given by $\mathbf{J} = \sigma_1 \mathbf{A}_1 + \sigma_2 \mathbf{A}_2$, where the subscripts label the groups. To derive an expression for the Hall coefficient, first set $J_y = 0$ and solve for $\mathcal{E}_H$ in terms of $\mathcal{E}_a$, then evaluate $R_H = \mathcal{E}_H / J_x B$. For simplicity we suppose $\omega_c \bar{t} \ll 1$ for each type particle. Then

$$\mathbf{J} = (\sigma_1 + \sigma_2)\mathcal{E}_a\hat{\mathbf{x}} + [(\sigma_1 + \sigma_2)\mathcal{E}_H + (\sigma_1\omega_{c1}\bar{t}_1 + \sigma_2\omega_{c2}\bar{t}_2)\mathcal{E}_a]\hat{\mathbf{y}}. \quad (12\text{-}36)$$

$J_y = 0$ leads to

$$\mathcal{E}_H = -\frac{\sigma_1\omega_{c1}\bar{t}_1 + \sigma_2\omega_{c2}\bar{t}_2}{\sigma_1 + \sigma_2}\mathcal{E}_a \quad (12\text{-}37)$$

and

$$R_H = \frac{\mathcal{E}_H}{J_x B} = -\frac{\sigma_1\omega_{c1}\bar{t}_1 + \sigma_2\omega_{c2}\bar{t}_2}{(\sigma_1 + \sigma_2)^2}. \quad (12\text{-}38)$$

If the particles labeled 1 are electrons and the particles labeled 2 are holes, we replace $\sigma_1$ by $en\mu_n$, $\omega_{c1}\bar{t}_1$ by $\mu_n B$, $\sigma_2$ by $ep\mu_p$, and $\omega_{c2}$ by $-\mu_p B$. Equation 12-38 then becomes

$$R_H = \frac{p\mu_p^2 - n\mu_n^2}{e(p\mu_p + n\mu_n)^2}. \quad (12\text{-}39)$$

For an intrinsic semiconductor, $n = p$ and

$$R_H = \frac{\mu_p - \mu_n}{en(\mu_p + \mu_n)}. \quad (12\text{-}40)$$

If more than one type carrier contributes to the Hall effect, values of the Hall coefficient and conductivity are not sufficient to obtain carrier concentrations and mobilities. Additional data are required. Sometimes independent values of the mobilities can be obtained from experiments on doped samples, for example. Then Eq. 12-39 and conductivity data can be used to find particle concentrations.

*The Quantum Hall Effect.* An interesting and important phenomenon occurs when a two-dimensional electron system is subjected to a strong magnetic field, perpendicular to the sample. In this context, a two-dimensional system is one for which all electron propagation vectors are parallel to the same plane. Both the Hall voltage and the sample resistance then show periodic behavior as functions of the Fermi level and, in fact, the resistance vanishes for certain values of $E_F$. Typically the applied field is about 15 to 20 T and the temperature is a few degrees above the absolute zero, although the effect has been observed for somewhat higher temperatures and weaker fields.

Two-dimensional electron systems exist near the oxide-semiconductor boundary in a metal-oxide-semiconductor (MOS) structure. Such a structure can be formed by coating one surface of the sample shown in Fig 12-12 with an oxide layer, then placing the layer in intimate contact with a metal plate. Current and voltage leads are attached to the ends of the semiconductor. In such a structure current exists only in a narrow channel near the interface. Furthermore, the Fermi level can be controlled by applying a potential difference, called a gate voltage, between the metal and semiconductor. For the sample of the diagram, the oxide and metal layers are parallel to the plane of the page and the gate voltage $V_G$ is appled perpendicularly to that plane. The gate voltage controls the number of electrons in the channel; as it varies, so does the Fermi energy. For the values of $V_G$ used, the Fermi energy is proportional to $V_G$.

Figure 12-13 shows data for a silicon MOS structure. $V_{pp}$ is the potential drop along the sample, parallel to the current and $V_H$ is the Hall voltage. Both are plotted as functions of $V_G$. $V_{pp}$ has a series of uniformly spaced zeros. The sample resistance vanishes at these points. $V_H$ has a plateau at each of the values of $V_G$ for which $V_{pp}$ vanishes.

Landau levels for a two-dimensional free electron system in a strong transverse magnetic field are given by $\hbar\omega_c(p + \frac{1}{2})$, where $\omega_c$ is the clyclotron angular frequency and $p$ is an integer. The spectrum consists of a series of separated levels. Overlap of levels associated with different values of $p$ does not occur, as

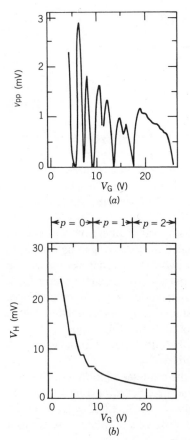

FIGURE 12-13 Integral quantum Hall effect in a silicon MOS structure. (a) Longitudinal voltage $V_{pp}$ as a function of gate voltage $V_G$. The Fermi energy is proportional to $V_G$. Zero resistance occurs when the Fermi level coincides with a Landau level. (b) Hall voltage as a function of gate voltage. Plateaus occur around values of $V_G$ for which the resitance is zero. The difference in the Hall voltage for two successive plateaus is exactly $e^2/2\pi\hbar$. (From K. von Klitzing, G. Dorda, and M. Pepper, *Phys. Rev. Lett.* **45**: 494, 1980. Used with permission.)

it does if electrons have propagation vectors with components along the magnetic field.

Resistance zeros occur because scattering of electrons by phonons, defects, and boundaries does not take place when the Fermi level coincides with a Landau level. Electrons cannot be scattered to other states with the same energy since these states are all occupied and, if the magnetic field is so strong the separation of levels is much larger than $k_BT$, they are scattered only rarely to other levels. As a result, the relaxation time is extremely long. Adjacent Landau levels differ in energy by $\hbar\omega_c$, so minima in $V_{pp}$ and plateaus in $V_H$ occur at intervals $\Delta E_F = \hbar\omega_c$. The effect can also be obtained by varying the magnetic field and thereby changing the separation of Landau levels until one of them coincides with the Fermi level.

We now calculate the values of the Hall voltage at the plateaus. The reciprocal lattice is two dimensional and in the absence of a magnetic field the number of **k** vectors per unit area is given by $S/4\pi^2$, where $S$ is the channel area. In a magnetic field, states are represented by circles. The radius $k$ of circle $p$ is given by $\hbar^2k^2/2m = \hbar\omega_c(p + \frac{1}{2})$, so its area is $\pi k^2 = (2\pi m\omega_c/\hbar)(p + \frac{1}{2}) = (2\pi eB/\hbar)(p + \frac{1}{2})$, once $\omega_c = eB/m$ is used. The area between the circles designated $p - 1$ and $p$ is $2\pi eB/\hbar$, so each circle represents $(2\pi eB/\hbar)(S/4\pi^2) = eBS/2\pi\hbar$ states. When the Fermi level is at a Landau level the number of states in an integer number of circles must equal the number of electrons in the channel. If states on $p$ circles are filled, $peB/2\pi\hbar = n_s$, where $n_s$ is the number of electrons per unit area of the channel.

The calculation does not ignore spin. In a strong magnetic field spin-up and spin-down electrons are well separated in energy and there are two sets of Landau circles, one for each value of the spin quantum number. Of the $p$ filled circles we are considering, somewhat more than half contain spin-down electrons and somewhat less than half contain spin-up electrons.

Assume the semiconductor is $n$ type. According to Eqs. 12-33 and 12-34, the Hall voltage for a three-dimensional sample is given by $V_H = wBI/Aen$, where $n$ is the electron concentration and $A$ is the sample cross section. We can adopt this expression to a two-dimensional sample by replacing $n$ with $n_s/t$, where $n_s$ is the number of electrons per unit area and $t$ is the channel thickness. Since $tw = A$, $V_H = BI/n_se$ and, since $n_s = peB/2\pi\hbar$,

$$\frac{V_H}{I} = \frac{2\pi\hbar}{e^2p}. \tag{12-41}$$

As $p$ takes on integer values the Hall voltage drops in a series of steps. The step height can be measured to better than one part in a million and the result leads to an extremely accurate determination of the constant $2\pi\hbar/e^2$. To five significant figures it is 25,813 $\Omega$. The quantum Hall effect may one day be used to define the ohm.

This model is greatly oversimplified. It does not account for the width of the plateaus in the plot of $V_H$, for example. Plateau width can be explained in terms of states localized at impurities, with energies between Landau levels. Theorists speculate that the plateaus do not appear in a perfectly pure ideal crystal. The argument is rather complicated and we will not pursue it here.

The simple model also does not account for the fractional quantum Hall effect, a phenomenon that occurs when the magnetic field is sufficiently strong that the lowest Landau level is only partially occupied. The fraction of states occupied in the lowest level seems to be a rational fraction: $\frac{1}{3}, \frac{2}{3}, \frac{2}{5}, \frac{3}{5}, \frac{4}{5}$, and $\frac{2}{7}$ have been observed. The effect is not clearly understood at this time.

## 12.5  REFERENCES

In addition to the references listed at the ends of Chapters 9 and 11, also see:

K. von Klitzing, "The Quantized Hall Effect" in *Rev. Mod. Phys.* **58**:519, 1986.
B. Lax and J. G. Mavroides, "Cyclotron Resonance" in *Solid State Physics;* (F. Seitz and D. Turnbull, Eds.), Vol. 11, p. 261 (New York: Academic, 1960).

## PROBLEMS

1.  The Fermi energy of sodium is $2.40 \times 10^{-19}$ J. Suppose a 0.25-T induction field is applied in the positive $z$ direction to a sodium sample. (a) What is the cyclotron frequency? (b) What is the radius of the classical orbit in direct space of an electron on the Fermi surface with $k_z = 0$? (c) What is the radius of the classical orbit of an electron on the Fermi surface with $k_z = \frac{1}{2}k_F$, where $k_F$ is the Fermi sphere radius? (d) Classically, how far in the direction of the field does the electron of part c travel in one cyclotron period?

2.  Free electron energies for a certain solid are given by

$$E = \frac{\hbar^2}{2} \left[ \frac{k_x^2}{m_x} + \frac{k_y^2}{m_y} + \frac{k_z^2}{m_z} \right],$$

where $m_x$, $m_y$, and $m_z$, are elements of the effective mass tensor. Show that the cyclotron angular frequency is given by $\omega_c = eB/\sqrt{m_x m_y}$ and that the cyclotron effective mass is $m^* = \sqrt{m_x m_y}$ for an induction field in the $z$ direction.

3.  The zero-field Fermi energy of sodium is $2.40 \times 10^{-19}$ J. Take the electron effective mass to be $9.11 \times 10^{-31}$ kg. (a) In a magnetic induction field of 0.50 T how many Landau cylinders pierce the zero-field Fermi surface? (b) How many pierce the Fermi surface if the field is 5.0 T? (c) For what values of the applied induction field does only one cylinder pierce the zero-field Fermi surface?

4.  Consider Landau cylinders for sodium in a magnetic induction field of 2.0 T. Take the zero-field Fermi energy to be $2.40 \times 10^{-19}$ J and the electron effective mass to be $9.11 \times 10^{31}$ kg. (a) What is the radius of the largest cylinder that pierces the zero-field Fermi surface? (b) What is the lowest

energy level associated with this cylinder? (c) What is the difference in radius of two adjacent cylinders that straddle the zero-field Fermi surface at its widest part? (d) What is the $z$ component of the velocity of an electron in a state on the smallest cylinder where it intersects the zero-field Fermi surface? (e) Assume the electron of part d obeys the laws of classical mechanics and describe its orbit in direct space.

5. Assume spin-up and spin-down free electrons share the same Landau levels. (a) Show that the contribution of Landau cylinder $p$ to the electron density of states is

$$\rho_p(E) = \frac{\tau_s}{4\pi^2} \left[\frac{2m}{\hbar^2}\right]^{3/2} \frac{\hbar\omega_c}{[E - (p + \frac{1}{2})\hbar\omega_c]^{1/2}},$$

if $E > (p + \frac{1}{2})\hbar\omega_c$. This must be summed over all cylinders with minimum energy less than $E$ to find the total density of states. (b) Use

$$N = \sum_p \int_{(p+\frac{1}{2})\hbar\omega}^{E_F} \rho_p(E) \, dE$$

to verify that the Fermi energy $E_F$ obeys Eq. 12-10. (d) Show that the total energy at 0 K is

$$E_T = \frac{\tau_s}{6\pi^2} \left[\frac{2m}{\hbar^2}\right]^{3/2} \hbar\omega_c \sum_p [E_F + (2p + 1)\hbar\omega_c][E_F - (p + \frac{1}{2})\hbar\omega_c]^{1/2},$$

where the sum is over all integers $p$ such that $E_F > (p + \frac{1}{2})\hbar\omega_c$.

6. Two metals have the same size and shape unit cells but the concentration of free electrons in metal A is twice the concentration in metal B. Take the effective masses to be the same and compare the free electron contributions to the susceptibilities of the two metals.

7. Calculate (a) the spin and (b) orbital contributions to the molar susceptibility of sodium. (c) Add the core contribution, given in Problem 3 of Chapter 11, and predict if sodium is diamagnetic or paramagnetic. The atomic concentration of sodium is $2.65 \times 10^{28}$ atoms/m$^3$ and each atom contributes one electron to the free electron band. Take the effective mass to be the free electron mass.

8. (a) Consider a free electron band, with energies given by $E = \hbar^2k^2/2m^*$. Neglect scattering. (i) Suppose all states are empty except for the one with $\mathbf{k} = k_0\hat{\mathbf{x}}$. What is the current density? (ii) A magnetic field $\mathbf{B} = B\hat{\mathbf{z}}$ is turned on for a time $t = \pi m^*/2eB$ (one-quarter of a cyclotron period). What is the current density after the field is turned off? (b) Answer the same questions if all states are filled except the one with $\mathbf{k} = k_0\hat{\mathbf{x}}$. (c) Answer the same questions if the energies are given by $E = E_0 - \hbar^2k^2/2m^*$ and only the state with $\mathbf{k} = k_0\hat{\mathbf{x}}$ is initially filled. (d) Answer the same questions for the band of part c with all states initially filled except the one with $\mathbf{k} = k_0\hat{\mathbf{x}}$.

9. Estimate the de Haas-van Alphen period for sodium, with a Fermi energy

of $2.4 \times 10^{-19}$ J. Starting with an induction field of exactly 0.5 T, what fractional change must be made in $B$ to observe 10 cycles of the susceptibility?

10. Consider two cylinders in reciprocal space, with radii $k_1$ and $k_2$. They extend from $k_z$ to $k_z + dk_z$. Show that the total energy of free electron states between the cylinders is

$$E_T = \frac{T_s}{8\pi^2} \frac{\hbar^2}{2m} [(k_2^4 - k_1^4) + 2(k_2^2 - k_1^2)] \, dk_z .$$

Take $k_1^2 = (2m^*/\hbar)\omega_c p$ and $k_2^2 = (2m^*/\hbar)\omega_c(p + 1)$, corresponding to two cylinders that straddle a Landau cylinder. Show that $E_T$ is the same as the total energy of states on the Landau cylinder with radius $k_1^2 = (2m^*/\hbar)\omega_c$ $(p + \frac{1}{2})$.

11. Use the equations derived in Problem 5 to demonstrate the de Haas-van Alphen effect: (a) Consider a free electron system with a Fermi energy of 2.000 eV when the applied magnetic field is such that $\hbar\omega_c = 0.6200$ eV. Find the electron concentration. (b) Now suppose the field is increased so $\hbar\omega_c = 0.8000$ eV. Verify that the Fermi energy is now 2.022 eV. (c) The field is increased again and $\hbar\omega_c$ becomes 0.8500 eV. Verify that the Fermi energy changes to 2.088 eV. (d) Calculate the average energy per electron for each of the fields above. Carry four significant figures. The cyclotron frequencies are much larger than those obtained with laboratory magnets and the Fermi energy varies much more than for realistic fields.

12. Electrons in the conduction band of silicon have an effective mass of about $1.1m_0$ and holes in the valence band have an effective mass of about $0.56m_0$, where $m_0$ is the free electron mass. (a) In a cyclotron resonance experiment, the sample is placed in a magnetic field and irradiated by a right circularly polarized wave with an angular frequency of $1.25 \times 10^{11}$ rad/s, traveling in the direction of the applied field. At what field strength do you expect to observe cyclotron resonance absorption? (b) At what field strength would resonance occur if the wave were left circularly polarized? (c) Suppose the power absorbed is greater than half its maximum value over a range of 0.015 T for the field found in part a. What is the carrier relaxation time? (b) Suppose now the temperature is increased and the relaxation time becomes shorter than that of part c by a factor of $10^2$. Over what range must the applied magnetic field be scanned to observe the half power points?

13. The magnetic field is doubled in a Hall effect experiment without changing the current. Assume all carriers are free electrons and tell what happens to (a) the Hall coefficient, (b) the Hall voltage, and (c) the Hall electric field.

14. At room temperature and above, what do you expect for the temperature dependence of the Hall coefficient of a semiconductor (a) if it is intrinsic? (b) if it is heavily doped $n$ type?

15. Use a two-level model to compute the Hall coefficient for silicon at 300 K. Take electron and hole mobilities to be 0.135 $m^2/V \cdot s$ and 0.0480 m²/V·s, respectively, and electron and hole effective masses to be $1.1 m_0$ and $0.56 m_0$, respectively. Here $m_0$ is the free electron mass. (a) Use Eqs. 8-54 and 8-60 to find the effective number of states per unit volume in each band and compute the Hall coefficient for intrinsic silicon. (b) Suppose the sample is doped with $2.5 \times 10^{16}$ acceptor atoms/m³. Assume each acceptor has accepted one electron and compute the Hall coefficient. (c) In each case identify the positive side of the sample in Fig. 12-12. (d) Why does the Hall voltage change sign? (e) For what acceptor concentration does the Hall coefficient vanish?

16. A Hall effect experiment is performed on a sample with two groups of carriers, of the same sign. Carriers in different groups have different effective masses, different relaxation times, and different concentrations. Show that the angle $\theta$ between the current density associated with group 1 and the applied electric field is given by

$$\tan e = \frac{n_2 \mu_2 (\mu_1 - \mu_2)}{n_1 \mu_1 + n_2 \mu_2} B .$$

# Chapter 13

## SUPERCONDUCTIVITY

Large electromagnet with superconducting coils, designed for the Colliding Beam
Accelerator at Brookhaven National Laboratory.

- **13.1 THE PHENOMENA OF SUPERCONDUCTIVITY**

- **13.2 THEORY OF SUPERCONDUCTORS**

- **13.3 ELECTRODYNAMICS OF SUPERCONDUCTORS**

- **13.4 JOSEPHSON EFFECTS**

The electrical resistivity of a normal metal in the form of a pure single crystal approaches zero only as the temperature approaches absolute zero. It does not vanish for any temperature above 0 K and, when defects are present, it has a nonzero limiting value. Resistivities of certain other metals, called superconductors, behave differently. The resistivity of a superconductor, even though noncrystalline and impure, is zero for a range of temperatures above absolute zero. Above the superconducting transition temperature, as the upper limit of the range is called, the metal has normal resistivity.

Twenty-seven of the chemical elements become superconducting at low temperatures. Metals in the left portions of transition series (Ti, V, Zr, Nb, Mo, Tc, Ru, Rh, La, Hf, Ta, W, Re, Os, and Ir) belong to this group, as do elements just to the right (Al, Zn, Ga, Cd, In, Sn, Hg, Tl, and Pb). Among the lanthanides and actinides, Lu, Th, and Pa are superconducting. Other elements, such as As, Ba, Ce, Ge, become superconducting when subjected to high pressure. In addition, a great many alloys and compounds, both stoichiometric and nonstoichiometric, are superconducting. On the other hand, alkali and noble metals are not, even at the lowest temperatures achieved thus far.

Superconductors are nearly perfect diamagnets: the magnetic induction field vanishes in the interior of a bulk sample. A superconductor does not expel fields of arbitrary strength, however. Fields with magnitudes greater than a certain critical value penetrate into the interior. For some superconductors, called type I, penetration coincides with the destruction of superconductivity. The metal becomes normal and has a nonvanishing resistivity. For others, called type II, the field strength must be increased to a second critical value before the resistivity ceases to vanish. In both cases a greater field strength is required at lower temperatures than at higher.

Superconductivity results from correlations between motions of electrons in a metal, induced by electron-phonon interactions. Its study has been important for our understanding of these interactions and has produced great insight into the physics of electrons in metals at low temperatures.

Superconductivity is also of technological importance. Zero resistivity has been exploited in the design of electromagnets and electric motors, the coils of which can carry large currents without the generation of heat and the associated loss of energy. Proposals have been made to use superconducting wires for long-distance lossless transmission of electric power.

Superconducting materials are the basis of extremely sensitive magnetic field and electromagnetic radiation detectors and, as such, have found widespread use in many scientific research fields and in medicine. With the recent discovery of high-temperature superconductors, the number of technological applications is expected to increase enormously.

## 13.1 THE PHENOMENA OF SUPERCONDUCTIVITY

This section contains a summary of some of the important properties of superconductors. Theory will be discussed in the next section.

*Persistent Currents.* Figure 13-1 shows the resistivity of mercury as a function of temperature. Above a temperature of about 4.2 K it roughly follows the $T^5$ law expected of normal metals. Below 4.2 K the metal is superconducting. The transition occurs in a temperature range of less than $10^{-5}$ K.

Transition temperatures for other superconductors are given in Tables 13-1 and 13-2. They range from 0.012 K for tungsten to about 90 K for a Y-Ba-Cu-O compound. Transitions for defect-free elements are sharp, like that of mercury. For strained alloys, on the other hand, the transition may take place over a temperature range as wide as 0.1 K. Obtaining temperatures below 20 K or so involves the use of liquid helium and is expensive. A great deal of effort is directed toward finding superconductors with transition temperatures above 77 K so liquid nitrogen can be used as a coolant. The discovery of materials that are superconducting at room temperature is an important goal of many scientists working in the field.

One way to test for zero resistivity is to measure the current in a loop of superconducting wire in the absence of an electric field. If the resistivity is truly zero the current does not diminish with time. Current can be induced by directing a static magnetic field through the loop when it is in the normal state, then cooling it to below the transition temperature and reducing the field to zero. At later times the current is determined by measuring its magnetic field. Currents

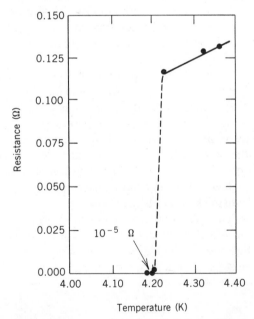

**FIGURE 13-1** The resistance of a mercury sample as a function of temperature. As the temperature is lowered past the transition temperature, the resistance of the sample drops precipitously. Unlike normal metals, superconductors have zero resistivity for a range of temperatures above the absolute zero. (From K. H. Onnes, *Commun. Kamerlingh Onnes Lab. Univ. Leiden* Suppl. 346, 1913).

**TABLE 13-1** Transition Temperature $T_c$ and Critical Field $H_c$ at $T = 0$ K for Selected Type I Superconductors

| Solid | $T_c$ (K) | $H_c$ (A/m) |
|---|---|---|
| **ELEMENTS** | | |
| Aluminum | 1.196 | $7.88 \times 10^3$ |
| Cadmium | 0.56 | $2.36 \times 10^3$ |
| Indium | 3.4035 | $2.25 \times 10^4$ |
| Lead | 7.193 | $6.39 \times 10^4$ |
| Mercury | 4.154 | $3.02 \times 10^4$ |
| Molybdenum | 0.917 | $7.80 \times 10^3$ |
| Niobium | 9.26 | $1.58 \times 10^5$ |
| Osmium | 0.655 | $5.17 \times 10^3$ |
| Tellurium | 2.39 | $1.36 \times 10^4$ |
| Tin | 3.722 | $2.43 \times 10^4$ |
| Tungsten | 0.012 | $8.51 \times 10^1$ |
| Vanadium | 5.30 | $8.12 \times 10^4$ |
| Zinc | 0.852 | $4.22 \times 10^3$ |
| Zirconium | 0.546 | $3.74 \times 10^3$ |
| **COMPOUNDS** | | |
| $BaBi_3$ | 5.69 | $5.89 \times 10^4$ |
| $Bi_2Pt$ | 0.155 | $7.96 \times 10^2$ |
| $Cr_{0.1}Ti_{0.3}V_{0.6}$ | 5.6 | $1.08 \times 10^5$ |
| $In_{0.8}Tl_{0.2}$ | 3.223 | $2.01 \times 10^4$ |
| $Mg_{0.47}Tl_{0.53}$ | 2.75 | $1.75 \times 10^4$ |
| $NbSn_2$ | 2.60 | $4.93 \times 10^4$ |
| $PbTl_{0.27}$ | 6.43 | $6.02 \times 10^4$ |

*Source: American Institute of Physics Handbook* (D. W. Gray, Ed.) (New York: McGraw-Hill, 1963).

in superconducting loops have been observed to remain undiminished over periods longer than a year. Errors inherent in the instruments used to measure the magnetic field set a lower limit on the persistence time. The limit is several tens of thousands of years in some cases. As we will see, theoretical arguments lead us to believe the resistivity is exactly zero.

*Thermodynamic Properties.* Specific heat data provide some clues to the fundamental processes that give rise to superconductivity. Figure 13-2 shows the electron contribution to the specific heat of aluminum as a function of temperature. The most obvious feature is a discontinuity at the transition temperature. Experimentally the magnitude of the discontinuity is found to be proportional to the transition temperature $T_c$ and, in particular, to be given approximately by $1.43\gamma T_c$, where $\gamma$ is the parameter that appears in Eq. 8-80 for the electron contribution to the specific heat of a normal metal. Since a similar discontinuity does not occur in the phonon contribution, we conclude that the supercon-

**TABLE 13-2.** Transition Temperatures $T_c$ and Critical Fields $H_{c1}$ and $H_{c2}$ at $T = 0$ K for Selected Type II Superconductors

| Solid | $T_c$ (K) | $H_{c1}$ (A/m) | $H_{c2}$ (A/m) |
|---|---|---|---|
| $Al_2CMo_3$ | 9.8 –10.2 | $7.24 \times 10^3$ | $1.24 \times 10^7$ |
| $C_{0.44}Mo_{0.56}$ | 12.5 –13.5 | $6.92 \times 10^3$ | $7.84 \times 10^6$ |
| $Cr_{0.10}Ti_{0.30}V_{0.60}$ | 5.6 | $5.65 \times 10^3$ | $6.72 \times 10^6$ |
| $In_{0.96}Pb_{0.04}$ | 3.68 | $7.96 \times 10^3$ | $9.55 \times 10^3$ |
| $Mo_{0.16}Ti_{0.84}$ | 4.18 | $2.23 \times 10^3$ | $7.85 \times 10^7$ |
| $Nb_3Sn$ | 18.05 | — | — |
| $Nb_3Ge$ | 23.2 | — | — |
| $Nb_3Al$ | 17.5 | — | — |
| $O_2SrTi$ | 0.43 | $3.90 \times 10^2$ | $4.01 \times 10^4$ |
| $SiV_3$ | 17.0 | $4.38 \times 10^4$ | $1.24 \times 10^7$ |
| $Ti_{0.775}V_{0.225}$ | 4.7 | $1.91 \times 10^3$ | $1.37 \times 10^7$ |
| $Ti_{075}V_{0.25}$ | 5.3 | $2.31 \times 10^3$ | $1.58 \times 10^7$ |
| $Ti_{0.615}V_{0.385}$ | 7.07 | $3.98 \times 10^3$ | $2.7 \times 10^6$ |
| $Ti_{0.516}V_{0.484}$ | 7.20 | $4.93 \times 10^3$ | $2.23 \times 10^6$ |
| $Ti_{0.415}V_{0.585}$ | 7.49 | $6.21 \times 10^3$ | $1.99 \times 10^6$ |

**HIGH-TEMPERATURE SUPERCONDUCTORS**

| | | | |
|---|---|---|---|
| $La_{1.8}Sr_{0.2}CuO_4$ | 36.2 | — | — |
| $(Y_{0.6}Ba_{0.4})_2CuO_4$ | $\approx 90$ | — | $\approx 1.3 \times 10^8$ |

*Sources: American Institute of Physics Handbook;* (D. W. Gray, Ed.) (New York: McGraw-Hill, 1963): C. Kittel, *Introduction to Solid State Physics* (New York: Wiley, 1986); R. J. Cava, R. B. van Dover, B. Batlogg, and F. A. Rietman, *Phys. Rev. Lett.* **58:**408, 1987; M. K. Wu, J. R. Ashburn, C. J. Torng, P. H. Hor, R. L. Gao, Z. J. Huang, Y. Q. Wang, and C. W. Chu, *Phys. Rev. Lett.* **58:**908, 1987.

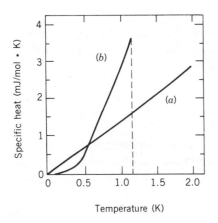

**FIGURE 13-2** The electron contribution to the specific heat of (a) normal and (b) superconducting aluminum as a function of temperature. Data for the normal sample below the transition temperature were obtained by applying a magnetic field greater than the critical field. The specific heat is discontinuous at the transition temperature, marked by a dotted line. (From N. E. Phillips, *Phys. Rev.* 114: 676, 1959. Used with permission.)

ducting transition is associated with a change in the electron system, not the phonon system.

Although superconductivity is associated with electron states, specific heat and resistance data indicate that phonons do play a role. Specifically, the transition temperature depends on the mass of the atoms in the solid. The dependence can be confirmed by measuring transition temperatures for superconductors composed of different isotopes of the same chemical element, which differ only in the masses of their atoms. In particular, the transition temperature is found to be proportional to $M^{-\alpha}$, where $\alpha$ has a value of about 0.5, although it is slightly different for different superconductors. Historically, the isotope effect pointed the way to a workable theory of superconductivity.

The specific heat is a measure of the change in entropy with temperature. Specific heat data indicate that, below the superconducting transition temperature, the entropy of a superconductor falls more rapidly than that of similar normal metal. We conclude that a superconducting electron system is, in some sense, more ordered than a normal electron system.

Specific heat data can be fitted to a function of the form $Ae^{-2\beta\Delta}$, where $A$ and $\Delta$ are parameters that may be different for different superconductors and may also be temperature dependent. At low temperature, however, both become independent of temperature and the specific heat becomes an exponential function of $1/k_B T$. Such behavior is indicative of a gap on the order of $2\Delta$ in the electron energy spectrum.

Experiments show that $\Delta$ is proportional to the transition temperature $T_c$ and, in fact, can be approximated by $\Delta \approx 1.8 k_B T_c$ at low temperatures. A superconductor with a transition temperature of 10 K has a gap of about $3 \times 10^{-3}$ eV. Evidently, most known superconductors have low transition temperatures because their energy gaps are small. We speculate that high-temperature superconductors have large gaps, although the gaps have not been measured directly yet.

The existence of superconducting gaps can be confirmed by measuring the absorption of electromagnetic radiation. A gap of $10^{-3}$ eV corresponds to an angular frequency of $1.5 \times 10^{12}$ rad/s or a wavelength of $1 \times 10^{-3}$ m, in the far infrared. Although a metal is highly reflecting at microwave frequencies, transmission through an extremely thin film can be detected. It increases as the film is cooled through the superconducting transition point. Microwave photons have energies that are less than the superconducting gap and so are not absorbed by electrons. In the normal state the gap does not exist and they are absorbed.

In special circumstances, some materials are superconducting although their energy spectra do not have gaps. One instance occurs, for example, when a small concentration of magnetic impurities is introduced into a nonsuperconducting sample. We will not discuss gapless superconductivity here.

*Flux Exclusion and the Meissner Effect.* Magnetic flux is excluded from the interior of a superconducting sample in an applied magnetic field. Figure 13-3

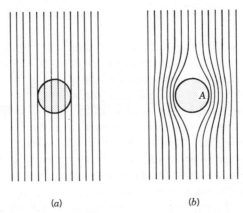

(a)                    (b)

**FIGURE 13-3** Lines of magnetic induction in the vicinity of (a) a normal metal cylinder and (b) a superconducting cylinder. Far from each cylinder the fields are the applied fields. The induction field is excluded from the superconducting sample and, for most geometries, is greater than the applied field just outside some parts of the surface. The field at $A$, for example, is twice the applied field.

shows induction field lines near normal and superconducting samples in identical applied magnetic fields. For the geometry shown, field lines that penetrate the normal sample skirt the edges of the superconducting sample, making the induction field higher at points near the surface than at analogous points near the normal sample. Flux exclusion can be confirmed by measuring the magnetic field in the vicinity of the sample.

Flux exclusion is not complete near a sample surface. Rather, the induction field decays exponentially with distance into the sample. The characteristic penetration depth is typically a few hundred angstroms.

Current exists at the surface of a superconductor in a magnetic field, but not in its interior. According to the steady state Maxwell equation $\nabla \times \mathbf{B} = \mu_0 \mathbf{J}$, so $\mathbf{J} = 0$ in the interior of a field free region. Surface currents, on the other hand, create an induction field that cancels the applied field in the interior and augments it in the region just outside the surface.

Superconductors are often described as perfect diamagnets, with magnetic susceptibilities of $-1$. To discuss flux exclusion in terms of magnetization and susceptibility, think of supercurrents as magnetization currents. The magnetization is given by $\mathbf{M} = \chi \mathbf{H}$, so $\mathbf{B} = \mu_0(\mathbf{H} + \mathbf{M}) = \mu_0(1 + \chi)\mathbf{H}$ and, since $\mathbf{B} = 0$ in the interior, $\chi = -1$.

**EXAMPLE 13-1** A magnetic field $H_a$ is applied parallel to the axis of a long cylindrical superconducting sample. What is the surface density of super-current?

**SOLUTION** Supercurrent flows around the circumference of the cylinder and produces a magnetic induction field in the interior. If $K$ is the supercurrent per unit length of cylinder, the field it produces is given by $B_s = \mu_0 K$ (see Example 11-1). $B_s$ exactly cancels the applied induction field $\mu_0 H_a$, so $K = H_a$. When directions are taken into account the result can be written $\mathbf{K} = \hat{\mathbf{n}} \times \mathbf{H_a}$, where $\hat{\mathbf{n}}$ is the unit outward normal to the cylinder.    ■

Flux exclusion is plausible if a magnetic field is turned on while a sample is in a superconducting state. An emf and current are induced by the changing field. When the field reaches its final strength the emf vanishes but a current remains in a narrow boundary region near the sample surface and its field cancels the applied field in the interior.

What happens if a normal sample is placed in a magnetic field, then cooled to below the transition temperature? We might expect the induction field to be the same before and after cooling. We base our conclusion on Faraday's law $\partial \mathbf{B}/\partial t = -\nabla \times \mathcal{E}$, which becomes $\partial \mathbf{B}/\partial t = 0$ in the absence of an electric field.

The conclusion is *not* true. In fact, supercurrents are generated and flux is excluded from the interior as the temperature passes the transition point. For type I superconductors the expulsion of flux is sudden and, except for a narrow boundary region, complete. The expulsion of flux that occurs as a type I sample becomes superconducting is called the Meissner effect. We will see later how theory explains this surprising result.

*Critical Magnetic Fields.*   If the magnetic field applied to a superconductor is sufficiently strong it destroys superconductivity and the sample becomes normal. Here we discuss type I superconductors. A weak induction field applied parallel to the axis of a cylindrical sample is completely excluded from the sample, except for a narrow boundary region. As the applied field strength is increased, penetration suddenly takes place when the critical field is reached. Magnetization as a function of applied field is diagrammed in Fig. 13-4.

Critical fields for type I superconductors are temperature dependent. Except near $T_c$, the critical field is nearly quadratic in the temperature, as Fig. 13-5 shows for a typical type I superconductor. Table 13-1 lists values of $H_c(0)$, the critical field at $T = 0$ K.

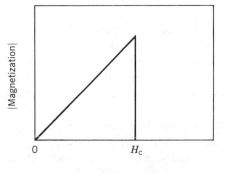

Applied magnetic field

**FIGURE 13-4**  The magnitude of the magnetization **M** as a function of applied field $H_a$ for a type I superconductor. **M** is given by $B/\mu_0$, where **B** is the induction field produced by supercurrent on the sample surface. $\mathbf{M} = -\mathbf{H_a}$ for $H_a$ less than the critical field $H_c$ and $\mathbf{M} = 0$ for $H_a$ greater than $H_c$.

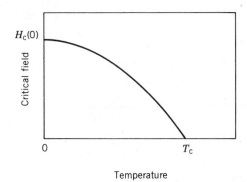

$H_c(0)$

Critical field

0                            $T_c$

Temperature

**FIGURE 13-5** The critical field $H_c$ as a function of temperature for a typical type I superconductor. It is a maximum for $T = 0$ and is zero for $T > T_c$.

We can use the critical field to estimate the change in energy that occurs at a superconducting transition. Consider a superconductor at $T = 0$ K and calculate the magnetic energy that must be supplied to convert it to the normal state by increasing the external magnetic field. The energy per unit volume required to change the magnetization by $d\mathbf{M}$ is $dE = -\mu_0\mathbf{H} \cdot d\mathbf{M}$ or, since $\mathbf{M} = -\mathbf{H}$ in a superconductor, $dE = \mu_0\mathbf{H} \cdot d\mathbf{H}$. Integrate this from $\mathbf{H} = 0$ to $\mathbf{H} = H_c(0)$ and find $\Delta E = \frac{1}{2}\mu_0 H_c^2(0)$. This must be the same as the change in energy per unit volume that occurs when the sample passes through the transition point in the absence of an applied field. For a critical field of $5 \times 10^4$ A/m, $\Delta E = 1.6 \times 10^3$ J/m³.

Once the critical field is reached and the sample turns normal, flux penetration occurs. If the sample is a long cylinder parallel to the applied field, the field strength is the same at all points on the surface and flux penetration takes place uniformly. For other geometries, penetration at one place on the surface may require a different applied field strength than penetration at another. For example, if the sample is a cylinder with its axis perpendicular to the applied field, as in Fig. 13-3, the field at the point $A$ is twice the applied field. The sample becomes normal in the neighborhood of $A$ when the applied field reaches $\frac{1}{2}H_c$. For applied fields greater than $\frac{1}{2}H_c$ but less than $H_c$, the sample is laced with strips of normal metal, as depicted in Fig. 13-6. Field lines pass through these regions and are excluded from regions between. Such a condition is known as an intermediate state. As the applied field strength is increased the normal regions grow in size and number until the whole sample is normal.

The resistance of a sample in an intermediate state is zero. Although some parts are normal, others are superconducting. The sample acts like a collection of zero-resistance wires in parallel. Superconducting paths short-circuit normal paths and the net resistance vanishes.

The existence of a critical field limits the current that can be carried by a superconducting sample. The sample becomes normal if the field created by the current exceeds the critical field. For a long straight wire of circular cross section carrying current $I$, the field in the exterior region is given by $H = I/2\pi r$, where $r$ is the radial distance from the center of the wire. The field at the surface is

Lines of magnetic induction

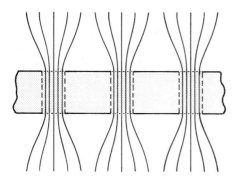

**FIGURE 13-6** A schematic diagram of the cross section of a type I superconducting sample in an intermediate state. The sample is laced with laminar normal regions through which magnetic flux passes. Flux is excluded from superconducting regions between.

$I/2\pi R$, where $R$ is the radius of the wire and, if this exceeds $H_c$, the outer region of the wire becomes normal. Current still exists but only on the surface of the superconducting portion, a cylinder with a radius smaller than $R$. The field there is greater than the critical field so these portions also become normal. The argument can be continued to show that the whole sample becomes normal.

**EXAMPLE 13-2** Calculate the maximum current that can be carried by a niobium wire with a radius of 0.050 mm, at the absolute zero of temperature.

**SOLUTION** The maximum current is given by $I_{max} = 2\pi R H_c$, where $R$ is the wire radius and $H_c$ is the critical field at $T = 0$ K. Since $H_c = 1.58 \times 10^5$ A/m (see Table 13-1),

$$I_{max} = 2\pi \times 0.05 \times 10^{-3} \times 1.58 \times 10^5 = 50 \text{ A} .$$

∎

*Type II Superconductors.* Two critical fields are important for type II superconductors. At the lower, denoted by $H_{c1}$, an applied magnetic field penetrates the sample while at the higher, denoted by $H_{c2}$, the resistance becomes nonzero. For fields between these two values, fine strands of normal metal are interspersed with superconducting metal. Magnetic flux passes through the sample in the normal portions but is excluded from the superconducting portions. The magnetization is shown as a function of applied field in Fig. 13-7.

The situation is somewhat similar to an intermediate state of a type I superconductor. The two situations should not be confused, however. For a cylindrical type I sample, with its axis along the applied field, flux is completely excluded for fields below $H_c$ and penetrates everywhere for fields above $H_c$. For a similar type II sample in the same orientation, flux is completely excluded for fields below $H_{c1}$ but penetrates in fine threads for fields above. Table 13-2 lists critical fields for some type II superconductors.

Regions where flux penetrates the surface can be observed by means of an electron microscope after the surface is coated with a ferromagnetic powder.

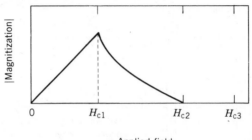

FIGURE 13-7 The magnitude of the magnetization **M** as a function of applied field $H_a$ for a type II superconductor. Below $H_{c1}$ flux is excluded and $\mathbf{M} = -\mathbf{H_a}$. Between $H_{c1}$ and $H_{c2}$ flux penetrates the sample in filaments. Between $H_{c2}$ and $H_{c3}$ the bulk of the sample is normal but a narrow region at the surface is superconducting. Above $H_{c3}$ the entire sample is normal.

Neutron scattering can also be used to determine the distribution of normal regions in a type II superconductor.

The threads of normal material that run through a sample when the applied field is above $H_{c1}$ are called vortex regions. As the applied field increases in strength, the number of such regions increases until, at the upper critical field, the whole sample becomes normal. The resistance of the sample is zero for fields below $H_{c2}$. As for an intermediate state of a type I superconductor, zero-resistance current paths short-circuit normal paths.

The structure of a vortex region is complicated. Its core is normal and contains a thin filament of magnetic flux. The induction field decays exponentially with distance away from the core, like the field at the surface of a type I sample in an external field. Supercurrents in the boundary region screen the field so it vanishes in the superconducting material between vortex regions.

Narrow regions near the surfaces of some type II samples remain superconducting even when the applied field is above $H_{c2}$. A third critical field, $H_{c3}$, is associated with these samples. For fields above $H_{c3}$, even the boundary region is normal.

Note that $H_{c2}$ is relatively high for some type II superconductors listed in Table 13-2. Although mechanical properties, such as brittleness, limit the usefulness of some, many are used to build superconducting motors and magnets. A high critical field, of course, means that a sample can carry a high current and remain superconducting.

## 13.2 THEORY OF SUPERCONDUCTORS

A theory, based on fundamental principles of quantum mechanics, was developed during the 1950s by J. Bardeen, L. N. Cooper, and J. R. Schrieffer. Known as the BCS theory, after the last initials of its authors, this theory and refinements made later explain all the phenomena of superconductivity discussed above and allow detailed calculations of many properties of superconductors. Unfortunately, the theory is mathematically quite complex and we will not be able to give details.

*Cooper Pairs.* According to BCS theory, superconductivity depends on the existence of an attractive force acting between electrons, strong enough to overcome their mutual electrostatic repulsion. This is not a stringent requirement, however, because the electric field of each electron in a metal is effectively screened by other electrons. Nevertheless, the existence of an attractive force must be demonstrated and such a force must be shown to lead to superconducting properties. These demonstrations are major triumphs of the theory.

An attractive force arises from electron-phonon interactions. Figure 13-8 illustrates a classical model. One electron pulls ions in its vicinity from their equilibrium sites and, as a result, the region around the electron contains more positive charge than otherwise. The region remains positive for a time after the electron leaves and a second electron is attracted toward it. In a quantum mechanical description, the first electron changes the number of phonons in one or more vibrational modes and they, in turn, influence the wave function of a second electron.

The attractive force is a maximum when the two electrons have propagation vectors that are equal in magnitude and opposite in direction and also spins that are in opposite directions. The spin condition arises because the energy is lowered if the spatial part of the wave function for the pair is symmetric under interchange of particle coordinates. Then the spin part is antisymmetric and the spins are oppositely directed. Pairs of electrons that meet these conditions, called Cooper pairs, are the fundamental building blocks of BCS theory. Organization of the electron system into pairs produces the ordering suggested by specific heat measurements.

A Cooper pair has zero net momentum or, in classical language, its center of mass is at rest. We may consider a Cooper pair to be single particle with spin zero and charge $-2e$. Such a particle obeys Bose-Einstein statistics and, as a result, any number may occupy the lowest energy state. In fact, the ground state of the electron system may be described by saying each electron is paired and all pairs are in the lowest energy state, the one with zero crystal momentum.

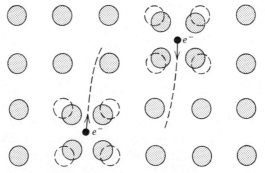

**FIGURE 13-8** A classical model of an attractive electron-electron interaction, mediated by the ions of the sample. Each electron pulls surrounding ions from their sites and leaves a wake of positive charge, which attracts another electron. Dotted circles represent the equilibrium positions of ions which have been displaced.

Pairing is not unique to superconductivity. We may, if we wish, consider electrons in the ground state of a normal metal to be paired in the same way. What is important for superconductivity is the attractive force that exists between electrons of a pair.

An analysis similar to the one used in Section 7.4 to obtain the energy spectrum of nearly free electrons can be used to show how a net attractive interaction, no matter how weak, might lead to reduced energy for an electron pair. Consider a free electron system at $T = 0$ K. Electron wave functions have the form $e^{i\mathbf{k}\cdot\mathbf{r}}$ and all states within the Fermi sphere are occupied. In the absence of an attractive force, we may take the wave function for a Cooper pair to be the product of two free electron functions, with oppositely directed propagation vectors. That is, $\psi = Ae^{i\mathbf{k}\cdot\mathbf{r}_1}e^{-i\mathbf{k}\cdot\mathbf{r}_2} = e^{i\mathbf{k}\cdot\mathbf{r}}$, where $\mathbf{r} = \mathbf{r}_1 - \mathbf{r}_2$. When squared, the magnitude of this function gives the probability per unit volume that the electrons are separated by a distance $r$. To obtain the complete wave function, $\psi$ must be multiplied by a function associated with the center of mass of the pair. Since the total momentum is zero this is just a constant. Later, when we discuss supercurrents, we will revise the wave function.

To simplify the calculation we suppose the electrons of the pair are in states at the Fermi surface, so the magnitude of each propagation vector is $k_F$ and the energy of each electron, in the absence of an attractive force, is $E_F$.

The attractive force scatters pairs from one pair state to another, so the wave function becomes a linear combination of plane waves. That is,

$$\psi(\mathbf{r}) = \sum_k A_k e^{i\mathbf{k}\cdot\mathbf{r}}. \tag{13-1}$$

where $A_k$ does not depend on $r$. We do not have an explicit form for the attractive force but we suppose the potential energy $U$ associated with it is a function of the distance $r$ between electrons. The Schrödinger equation for the pair is then

$$-\frac{\hbar^2}{2m}[\nabla_1^2 + \nabla_2^2]\psi + U(\mathbf{r})\psi = (2E_F + \delta E)\psi, \tag{13-2}$$

where $\nabla_1^2$ operates on the coordinates of electron 1, $\nabla_2^2$ operates on the coordinates of electron 2, and $\delta E$ is the change in energy due to the attractive force.

Equation 13-1 is substituted into Eq. 13-2, $\nabla_1^2\psi = -k^2\psi$ and $\nabla_2^2\psi = -k^2\psi$ are used, and the terms are rearranged slightly to obtain

$$\sum_k \left[\frac{\hbar^2 k^2}{m} - 2E_F - \delta E\right] A_k e^{i\mathbf{k}\cdot\mathbf{r}} = -\sum_k A_k U(r)e^{i\mathbf{k}\cdot\mathbf{r}}. \tag{13-3}$$

For ease in writing let $\xi_k = (\hbar^2 k^2/2m) - E_F$. It is the free electron energy of one of the states in the sum, relative to the Fermi energy. Equation 13-3 then becomes

$$\sum_k (2\xi_k - \delta E)A_k e^{i\mathbf{k}\cdot\mathbf{r}} = -\sum_k A_k U(r)e^{i\mathbf{k}\cdot\mathbf{r}}. \tag{13-4}$$

Multiply by $e^{-i\mathbf{k}'\cdot\mathbf{r}}$, then integrate over the volume of the sample. The integral of $e^{i(\mathbf{k}\,-\,\mathbf{k}')\cdot\mathbf{r}}$ vanishes unless $\mathbf{k} = \mathbf{k}'$ and it equals the sample volume $\tau_s$ if $\mathbf{k} = \mathbf{k}'$. Thus

$$A_{k'} = -\frac{\sum A_k U_{kk'}}{2\xi_{k'} - \delta E},\qquad(13\text{-}5)$$

where

$$U_{kk'} = \frac{1}{\tau_s}\int U(\mathbf{r})e^{i(\mathbf{k}-\mathbf{k}')\cdot\mathbf{r}}\,d\tau\qquad(13\text{-}6)$$

and the sum in the numerator is over $\mathbf{k}$.

We now make use of the special characteristics of the system that lead to a lowering of the energy and therefore to superconductivity. First, because all states within the Fermi sphere are occupied the attractive force cannot scatter electrons to states with $k < k_F$. We take $A_k$ to be zero 0 for $k < k_F$. Second, because the attractive force is a result of the electron-phonon interaction it cannot scatter electrons to states with energy greater than $E_F + \hbar\omega$, where $\omega$ is the maximum angular frequency in the phonon spectrum. We approximate $\omega$ by the Debye angular frequency $\omega_D$. The sum in Eq. 13-5 does not contain any terms for which either $\mathbf{k}$ or $\mathbf{k}'$ refer to states outside an energy shell of width $\hbar\omega_D$, extending upward from the Fermi energy. In lieu of an accurate expression for $U_{kk'}$ we take it to be the constant $-U$ if both $k$ and $\mathbf{k}'$ refer to states within the shell. $U$ is positive for an attractive force. Equation 13-5 becomes

$$A_{k'} = U\frac{\sum A_k}{2\xi_{k'} - \delta E}.\qquad(13\text{-}7)$$

To find $\delta E$, use $\mathbf{k}'$ as a summation index and sum Eq. 13-7 over states in the shell. Since the sum over $\mathbf{k}'$ on the left and the sum over $\mathbf{k}$ on the right are identical,

$$1 = U\sum_{k'}\frac{1}{2\xi_{k'} - \delta E}.\qquad(13\text{-}8)$$

is obtained. The density of states is used to convert the sum to an integral. Since the spins of electrons in a Cooper pair are in opposite directions, we use the density of states appropriate for one type spin, either up or down. If $\rho(E)$ is half the usual density of states, including spin, then Eq. 13-8 becomes

$$1 = U\int_0^{\hbar\omega_D}\frac{\rho}{2\xi - \delta E}\,d\xi.\qquad(13\text{-}9)$$

Since $\hbar\omega_D$ is quite small compared to the range over which $\rho$ varies significantly, we may take $\rho$ to be a constant, the one spin density of states at the Fermi level. The integral can then be evaluated in closed form: Eq. 13-9 becomes $1 = \frac{1}{2}U\rho\,\ln[(\delta E - 2\hbar\omega_D)/\delta E]$ and

$$\delta E = -\frac{2\hbar\omega_D}{e^{2/U\rho} - 1} \approx -2\hbar\omega_D e^{-2/U\rho},\qquad(13\text{-}10)$$

where the last expression is valid for $U\rho \ll 1$. For all known superconductors $U\rho < 0.5$ and for most it is considerably smaller.

The change in energy is negative. The electrons in a Cooper pair are said to be quasi-bound because their energy is less than that of two free electrons. They are not truly bound since their total energy is positive relative to the bottom of the conduction band.

Quasi-binding occurs only because electrons of a pair cannot be scattered to states inside the Fermi sphere. Quantum mechanics tells us that no state exists for a three-dimensional potential well unless the depth of the well is greater than some minimum value. The electron-phonon interaction is weak and does not produce quasi-binding if the wave function mixes many states with energy inside as well as outside the Fermi sphere.

Because a pair wave function is a sum of plane waves, not a single plane wave, the average distance between electrons is finite. We use the Heisenberg uncertainty principle to estimate the average separation of two electrons in a pair. The uncertainty $\delta k$ in the magnitude of the propagation vector is related to $\delta E$ by $\delta E = \delta(\hbar^2 k^2/2m) = \hbar^2 k \delta k/m = \hbar^2 k_F \delta k/m$, so $\delta k = m\delta E/\hbar^2 k_F$. We replaced $k$ by $k_F$ because the magnitude of the propagation constant for each plane wave in the sum is nearly $k_F$. The extent of the wave function is $\delta x = 1/\delta k = \hbar^2 k_F/m\delta E = (2/k_F)(E_F/\delta E)$, where $E_F = \hbar^2 k_F^2/2m$. Now $E_F/\delta E$ is typically $10^3$ or more and $k_F$ is typically $10^{10}$ m$^{-1}$, so $\delta x$ is on the order of 100 nm. At any time the centers of many millions of pairs are between the electrons of any one pair.

*The Superconducting Ground State.* The analysis described above deals with a single Cooper pair at the Fermi surface and must be modified to discuss the superconducting ground state. Because the mathematics is quite complex, only a qualitative description will be given, along with some results.

Consider first a Cooper pair with energy $E_F - \delta E$, where $\delta E$ is given by Eq. 13-10. Since its wave function contains functions for free electron states above the Fermi level, the electrons occupy free electron states at the Fermi level only part of the time. The rest of the time these states are free for occupation by electrons with less energy. Thus electrons below the Fermi level may also form Cooper pairs. Superconductivity is a cooperative phenomenon: the formation of Cooper pairs near the Fermi surface facilitates the formation of other pairs inside.

But the process cannot go very far. If free electron states below the Fermi level are unoccupied, pairs may be scattered to them from above as well as from below, thus reducing the tendency toward quasi-binding. The distribution that minimizes the total energy is shown in Fig. 13-9. Although the curve is for $T = 0$ K, it somewhat resembles the Fermi-Dirac distribution function for a higher temperature. Some states with energy above the Fermi level are occupied while some states with energy below are not. Well below the Fermi level the occupation number is 1.

The energy range of partially occupied states is designated $2\Delta_0$ and, although the derivation is more complicated than the derivation of Eq. 13-10 for the energy

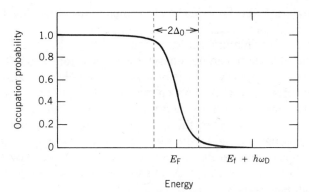

FIGURE 13-9 Occupation probability for one electron states of a superconductor in its ground state. The function is $\frac{1}{2}[1 - \xi(\Delta_0^2 + \xi^2)^{-1/2}]$, where $\xi = E - E_F$ and $\Delta_0$ is the gap parameter. The range of energy for which states are occupied only part of the time is about $2\Delta_0$.

of a Cooper pair near Fermi surface, $\Delta_0$ is given by a similar expression:

$$\Delta_0 = 2\hbar\omega_D e^{-1/U\rho} \tag{13-11}$$

in the limit of small $U\rho$. $\Delta_0$ is often interpreted as the energy per electron required to break a Cooper pair, but we must remember it results from the collective motions of all the electrons and cannot be attributed to a single Cooper pair.

Equation 13-11 is not valid for all superconductors. For what are known as strong coupling superconductors, the potential energy of the attractive force is more complicated than the simple function we used and somewhat different results are obtained for $\Delta_0$. Lead and mercury are examples.

The total energy of the electron system can be computed and compared with the total energy of a normal system at the absolute zero of temperature. As might be guessed from the electron distribution function, the result depends on $\Delta_0$. BCS theory gives

$$E_n - E_s = \frac{1}{2}\rho(E_F)\Delta_0^2. \tag{13-12}$$

For a system of $N$ free electrons $\rho(E_F) = 3N/4E_F$, so

$$E_n - E_s = \frac{3}{8}N\Delta_0 \left(\frac{\Delta_0}{E_F}\right). \tag{13-13}$$

Equation 13-13 is consistent with a model in which a small fraction $\frac{3}{4}\Delta_0/E_F$ of the electrons, associated with states near the Fermi level, form quasi-bound pairs. We may use this interpretation provided we do not forget the collective nature of superconductivity.

*Superconducting Energy Gaps.* The electron-phonon interaction also changes the energy spectrum for excited states. If the energy of a state outside the Fermi surface of a normal metal is $E_n(\mathbf{k})$ then, according to BCS theory, the corresponding state for a superconductor is $E_s(\mathbf{k}) = E_F + [(E_n - E_F)^2 + \Delta_0^2]^{1/2}$, where

$\Delta_0$ is given by Eq. 13-11. Energies of states below the Fermi level are given by $E_s = E_F - [(E_n - E_F)^2 + \Delta_0^2]^{1/2}$.

At $T = 0$ K the minimum energy for an electron above the Fermi level is $E_F + \Delta_0$. Similarly, the maximum energy for an electron below the Fermi level is $E_F - \Delta_0$. Thus an energy of at least $2\Delta_0$ is required to lift an electron from below the Fermi level to above. An energy gap of magnitude $2\Delta_0$ straddles the Fermi level; the electron system cannot have energy in the gap.

Electrons with energy above the gap are unpaired. They are called normal electrons to distinguish them from electrons below the gap, called superconducting electrons. At temperatures above absolute zero and below the transition temperature the system is a mixture of superconducting and normal electrons. At $T = 0$ K all electrons are superconducting.

Since superconductivity is mediated by electron-phonon interactions we might erroneously expect gaps to be on the order of $\hbar\omega_D$ and transition temperatures to be on the order of Debye temperatures ($\hbar/k_B)\omega_D$. Thus, we might expect many materials to be superconducting to well above room temperature. Observed values of transition temperatures are less. BCS theory gives

$$k_B T_c = 0.565\Delta_0 = 1.13\hbar\omega_D e^{-1/U\rho}, \tag{13-14}$$

for the relationship between the transition temperature and the gap parameter. The last equality is valid for small $U\rho$. The cooperative nature of the superconducting ground state introduces the exponential into Eqs. 13-11 nad 13-14 and reduces the gap and transition temperature.

Electron systems for which the product $U\rho$ is large have relatively large superconducting gaps and relatively high transition temperatures. Thus materials for which the electron-phonon interaction is strong are superconducting at higher temperatures than those for which it is weak. Both experimental and theoretical effort is currently directed toward electron-phonon interactions in high-temperature superconductors, such as the Y-Ba-Cu-O compound listed in Table 13-2. Scientists hope that detailed knowledge of the interactions will enable them to find superconductors with still higher transition temperatures.

Electron-phonon interactions are also responsible for the resistivity of a metal in its normal state and a strong interaction leads to high resistivity. We conclude that metals that are relatively poor electrical conductors in their normal states make the best superconductors, with relatively large superconducting gaps and relatively high transition temperatures. This conclusion is borne out by observation.

Equation 13-11 indicates that a solid becomes superconducting at some temperature if an attractive force, no matter how small, exists between electrons. Materials for which transitions have not been observed may, in fact, be superconductors below currently obtainable temperatures.

The superconducting gap varies with temperature. BCS theory gives an integral equation for its value. For weak coupling superconductors the ratio $\Delta(T)/\Delta_0$ depends only on the transition temperature and the curve plotted in Fig. 13-10 is valid for all weak coupling superconductors. At low temperatures

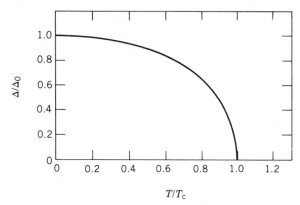

**FIGURE 13-10** The superconducting gap parameter $\Delta$ as a function of temperature. The gap has its greatest value at $T = 0$ K and is zero for $T > T_c$.

$\Delta(T)$ is nearly constant, then when $T$ is about $\frac{1}{2}T_c$ it begins a precipitous decline toward 0.

The temperature dependence of $\Delta(T)$ results from the collective nature of the superconducting state. As the temperature increases electrons are excited across the gap, leaving some states within the Fermi surface unoccupied, at least part of the time. As a result, the gap is reduced and the energy required to excite other electrons is decreased. The curve becomes steep when a significant number of electrons have been excited and its slope is infinite at the transition temperature. At $T_c$ and above the gap vanishes. All electrons are then normal and the sample is no longer superconducting.

*Predictions of the Theory.* Here we indicate how BCS theory explains some of the phenomena of superconductivity. Although the discussion is highly qualitative, the results quoted are obtained from the theory by detailed calculations. We defer discussion of the Meissner effect until the next section.

The *isotope effect* is predicted by Eqs. 13-11 and 13-14. The Debye frequency is proportional to $M^{-1/2}$, where $M$ is the atomic mass, and the proportionality extends to the gap and transition temperature. A smaller atomic mass leads to a higher Debye frequency, so the sum in Eq. 13-1 includes free electron states with energies higher above the Fermi level than otherwise.

*Persistent currents* are also easily explained. A uniform constant electric field displaces the electron distribution in reciprocal space according to $d\delta\mathbf{k}/dt = -(e/\hbar)\mathcal{E}$ so, after time $t$, the displacement is $\delta\mathbf{k} = -(e/\hbar)\mathcal{E}t$. A single particle state at $\mathbf{k} + \delta\mathbf{k}$ is now paired with the single particle state at $-\mathbf{k} + \delta\mathbf{k}$ and the superconducting gap is displaced along with the distribution.

The equation of motion for $\delta\mathbf{k}$ does not contain a relaxation term. Relaxation occurs in a normal metal because electrons are scattered by phonons and defects to states left empty by displacement of the distribution. Scattering involves energy exchanges between the electron system and the phonon or defect systems and can always occur in a normal metal because energy differences between initial and final states are extremely small. In a superconductor, on the other hand, each scattering event must involve an energy of at least $2\Delta$. If the energy required is not available, scattering does not occur.

We may think of a Cooper pair as a particle with mass $2m$, charge $-2e$, and momentum $2\hbar\,\delta\mathbf{k}$. Its velocity is $2\hbar\,\delta\mathbf{k}/2m = (\hbar/m)\,\delta\mathbf{k}$. If the concentration of superconducting electrons is $n_s$, the current density is given by

$$J = -2e(n_s/2)v = -\frac{e\hbar n_s}{m}\,\delta k. \tag{13-15}$$

If the current is initially zero when an electric field is turned on, $\delta\mathbf{k} = -(e/\hbar)\mathscr{E}t$ and

$$\mathbf{J} = \frac{n_s e^2}{m}\,\mathscr{E}t. \tag{13-16}$$

Since no mechanism exists to scatter electrons back to the thermal equilibrium distribution, characterized by $\delta\mathbf{k} = 0$, the current density continues to increase as long as an electric field exists and is constant after the field is turned off.

BCS theory also relates the *critical magnetic field* and the gap parameter of a type I superconductor. Equate the difference in normal and superconducting energies, given by Eq. 13-12, to $\tfrac{1}{2}\mu_0 H_c^2(0)$ to obtain

$$H_c^2(0) = \frac{\rho(E_F)}{\mu_0\tau_s}\,\Delta_0^2. \tag{13-17}$$

Since $\Delta_0$ is proportional to $T_c$, so is $H_c$. Observations of weak coupling superconductors with nearly the same density of states support Eq. 13-17 to within a few percent. Discrepancies occur for strong coupling superconductors, such as lead and mercury.

BCS theory predicts the critical field obeys

$$H_c(T) = H_c(0)\left[1 - 1.06\left(\frac{T}{T_c}\right)^2\right] \tag{13-18}$$

for $T \ll T_c$ and

$$H_c(T) = 1.74H_c(0)\left[1 - \frac{T}{T_c}\right] \tag{13-19}$$

for $T$ near $T_c$. For most superconductors computed and experimental values of $H_c(T)$ agree within a few percent. Equation 13-19 is often used to estimate $H_c(0)$ from measurements of $H_c$ near the transition temperature. This technique is especially important if $H_c(0)$ is well above fields that can be produced in the laboratory.

**EXAMPLE 13-3** Use the free electron model to estimate the gap parameter $\Delta_0$ for a superconducting sample with $5.0 \times 10^{28}$ electrons/$m^3$ and a critical field of $7.0 \times 10^3$ A/m.

**SOLUTION**   The Fermi energy is

$$E_F = \frac{\hbar^2}{2m} (3\pi^2 n)^{2/3} = \frac{(1.05 \times 10^{-34})^2}{2 \times 9.11 \times 10^{-31}} (3\pi^2 \times 5.0 \times 10^{28})^{2/3}$$
$$= 7.86 \times 10^{-19} \text{ J} .$$

Use $\rho = 3n\tau_s/4E_F$ and Eq. 13-17 to find

$$\Delta_0 = \left[ \frac{4\mu_0 E_F}{3n} \right]^{1/2} H_c = \left[ \frac{4 \times 4\pi \times 10^{-7} \times 7.86 \times 10^{-19}}{3 \times 5.0 \times 10^{28}} \right]^{1/2} \times 7.0 \times 10^3$$
$$= 3.59 \times 10^{-23} \text{ J} = 2.25 \times 10^{-4} \text{ eV} .$$

∎

BCS theory also predicts a discontinuity in the *specific heat* at $T = T_c$. The fractional change $(C_n - C_s)/C_n$ is calculated to be 1.43. For low temperatures theory gives

$$C_s = 1.34\gamma T_c \left[ \frac{\Delta_0}{k_B T} \right]^{3/2} e^{-\Delta_0/k_B T} . \tag{13-20}$$

As mentioned previously, the exponential behavior at low temperatures is indicative of a system with an energy gap. The gap and the exponential behavior of the specific heat are natural outcomes of the theory.

## 13.3  ELECTRODYNAMICS OF SUPERCONDUCTORS

*The London Equation.*   Equation 13-16 gives the current density for a system of superconducting electrons; its derivative with respect to time is

$$\frac{d\mathbf{J}}{dt} = \frac{n_s e^2}{m} \boldsymbol{\mathcal{E}} . \tag{13-21}$$

As Eq. 13-21 indicates, a steady state supercurrent exists only if the electric field vanishes. In a wire carrying a steady supercurrent, the potential difference between any two points is zero. This behavior is a direct consequence of a vanishing resistivity. By way of contrast, a steady current density in a normal metal is proportional to the field.

We use Eq. 13-21 to discuss the Meissner effect. Take the curl of both sides and use Faraday's law, $\nabla \times \boldsymbol{\mathcal{E}} = -\partial \mathbf{B}/\partial t$ to substitute for $\nabla \times \boldsymbol{\mathcal{E}}$. Then Eq. 13-21 becomes

$$\frac{\partial}{\partial t} \left[ \frac{n_s e^2}{m} \mathbf{B} + \nabla \times \mathbf{J} \right] = 0 . \tag{13-22}$$

Equation 13-22 is valid for any material with zero resistivity. It does not by itself lead to the Meissner effect. However, the Meissner effect is predicted if the quantity in brackets vanishes. That is,

$$\nabla \times \mathbf{J} = -\frac{n_s e^2}{m} \mathbf{B} \tag{13-23}$$

in a superconductor. Equation 13-23 is called the London equation after F. London and H. London, who postulated its validity for supercurrents.

To see how the London equation leads to the Meissner effect, consider a sample above its transition temperature, in a weak magnetic field. Both $\mathbf{J}$ and $n_s$ vanish, so Eq. 13-23 is satisfied even though an induction field penetrates the sample. According to Eq. 13-22, $(n_s e^2/m)\mathbf{B} + \nabla \times \mathbf{J}$ remains zero as the sample is cooled to below its transition point. Since an electric field does not exist, $\mathbf{J}$ remains zero in the interior. Because $n_s$ no longer vanishes, Eq. 13-23 requires that $\mathbf{B} = 0$ in the interior. Actually neither $\mathbf{J}$ nor $\mathbf{B}$ vanish in a narrow boundary region near the surface, but Eq. 13-23 is still valid. We will discuss the boundary layer shortly.

To understand why the London equation is obeyed by superconductors and to examine the limits of its validity we must know more about wave functions for Cooper pairs.

*The Macroscopic Wave Function.* The wave function given by Eq. 13-1 and used to find the energy of a Cooper pair is a function of the relative coordinates of the two electrons in a pair. It does not take into account motion of the center of mass. Since center of mass motion is important when a Cooper pair is subjected to electric and magnetic fields, we now consider it.

As we learned in the last section, the propagation vector for each electron changes by $\delta\mathbf{k} = -(e/\hbar)\mathcal{E}t$ in a constant uniform electric field. The wave function for a pair becomes

$$\psi = \sum_k A_k e^{i(\mathbf{k}+\delta\mathbf{k})\cdot\mathbf{r}_1} e^{i(-\mathbf{k}+\delta\mathbf{k})\cdot\mathbf{r}_2} = e^{2i\delta\mathbf{k}\cdot\mathbf{R}} \sum_k A_k e^{i\mathbf{k}\cdot\mathbf{r}}, \qquad (13\text{-}24)$$

where $\mathbf{R} = \frac{1}{2}(\mathbf{r}_1 + \mathbf{r}_2)$ is the position of the center of mass and $\mathbf{r} = \mathbf{r}_1 - \mathbf{r}_2$ is the relative displacement of the electrons, as before. The first exponential factor in Eq. 13-24 is the center of mass wave function for a Cooper pair with total momentum $2\hbar\delta\mathbf{k}$. We now give a more general definition.

Let $\Psi(\mathbf{R})$ be the wave function for the center of mass of a Cooper pair. We could define it so that $|\Psi(\mathbf{R})|^2 d\tau$ gives the probability for finding the center of mass of the pair in the infinitesimal volume element $d\tau$ at $\mathbf{R}$. However, $\Psi(\mathbf{R})$ is the same for every Cooper pair. In the absence of an electric field it is a constant and when a field is present every pair reacts in the same way. We define $\Psi(\mathbf{R})$ so that $|\Psi(\mathbf{R})|^2$ gives the concentration of Cooper pairs $n_p$ at $\mathbf{R}$. In general, $\Psi$ is complex so it can be written in the form

$$\Psi(\mathbf{R}) = n_p^{1/2} e^{i\phi(\mathbf{R})}, \qquad (13\text{-}25)$$

where $\phi$ is a phase angle that might vary from place to place in the sample.

The spatial dependence of the phase angle is related to the supercurrent density. In quantum mechanics the momentum operator is $-i\hbar\nabla$ and, for particles of charge $q$ and mass $M$, the current density is

$$\mathbf{J} = -i\frac{q\hbar}{2M}[\Psi^*\nabla\Psi - \Psi\nabla\Psi^*]. \qquad (13\text{-}26)$$

To find the supercurrent density, substitute Eq. 13-25 into Eq. 13-26, then take $q = -2e$ and $M = 2m$. If the pair concentration is uniform,

$$\mathbf{J} = -\frac{e\hbar}{m}\, n_{\mathrm{p}}\nabla\phi\,. \tag{13-27}$$

That is, the current density is proportional to the gradient of the phase angle. When no current exists, $\nabla\phi = 0$ and the center of mass wave function may be taken to be the constant $n_{\mathrm{p}}^{1/2}$. On the other hand, when current is present $\phi$ is a function of position in the sample. If, for example, $\Psi$ is the plane wave $e^{i2\delta\mathbf{k}\cdot\mathbf{R}}$ then $\nabla\phi = 2\delta\mathbf{k}$ and $\mathbf{J} = -2e\hbar\,\delta\mathbf{k}\,n_{\mathrm{p}}/m$, in agreement with Eq. 13-16 when $\frac{1}{2}n_{\mathrm{s}}$ is substituted for $n_{\mathrm{p}}$ and $-(e/\hbar)\mathcal{E}t$ is substituted for $\delta\mathbf{k}$.

$\Psi$ is called a macroscopic wave function because its magnitude and phase can be interpreted in terms of macroscopically measurable quantities, the pair concentration and the current density. For a normal metal we calculate the contribution of an electron to the current density by substituting a single particle wave function into Eq. 13-26. The contributions are then summed. This procedure is correct because no correlation exists between the phases of the single particle functions. For a superconductor we essentially sum the wave functions of all Cooper pairs, then substitute the sum into Eq. 13-26. This procedure is correct for these materials because all the wave functions have the same magnitude and phase.

If a magnetic field is present the expression for the current density must be modified. The quantum mechanical operator $-i\hbar\nabla$ is no longer associated with particle momentum alone but rather with a sum of particle and field momenta. Specifically for a charge $q$, $-i\hbar\nabla$ is associated with $\mathbf{p} + q\mathbf{A}$, where $\mathbf{p}$ is the particle momentum and $\mathbf{A}$ is the vector potential of the field, defined so $\mathbf{B} = \nabla\times\mathbf{A}$. To find the current density we use the operator $-i\hbar\nabla - q\mathbf{A}$, which is associated with the particle momentum. $\mathbf{J}$ is given by

$$\mathbf{J} = -i\frac{\hbar q}{2M}(\Psi^{*}\nabla\Psi - \Psi\nabla\Psi^{*}) - \frac{q^{2}}{M}\mathbf{A}\Psi^{*}\Psi\,. \tag{13-28}$$

For Cooper pairs replace $q$ by $-2e$, $M$ by $2m$, and $\Psi$ by $n_{\mathrm{p}}^{1/2}e^{i\phi}$ to find

$$\mathbf{J} = -\left[\frac{\hbar e}{m}\nabla\phi + \frac{2e^{2}}{m}\mathbf{A}\right]n_{\mathrm{p}}\,, \tag{13-29}$$

provided the concentration $n_{\mathrm{p}}$ is uniform. The current density has two terms, one proportional to the gradient of the phase angle and one proportional to the vector potential.

Equation 13-29 leads immediately to the London equation. Take the curl of both sides and make use of the identity $\nabla\times\nabla\phi = 0$ to find

$$\nabla\times\mathbf{J} = -\frac{2n_{\mathrm{p}}e^{2}}{m}\mathbf{B} = -\frac{n_{\mathrm{s}}e^{2}}{m}\mathbf{B}\,, \tag{13-30}$$

in agreement with Eq. 13-23. The London equation follows naturally from the quantum mechanics of Cooper pairs.

*Penetration Depth.* For a superconductor in a magnetic field, the induction field does not change discontinuously from its value outside the sample to zero inside. Rather it decreases in magnitude with distance from the boundary. The London equation, in conjunction with Maxwell's equations, gives an expression for the steady state field just inside the boundary.

Take the curl of both sides of the Maxwell equation $\nabla \times \mathbf{B} = \mu_0 \mathbf{J}$, then substitute for $\nabla \times \mathbf{J}$ from Eq. 13-30 to obtain

$$\nabla \times \nabla \times \mathbf{B} = -\mu_0 \frac{n_s e^2}{m} \mathbf{B}. \tag{13-31}$$

For any vector $\mathbf{B}$, $\nabla \times \nabla \times \mathbf{B} = \nabla(\nabla \cdot \mathbf{B}) - \nabla^2\mathbf{B}$, but since $\mathbf{B}$ is a magnetic induction field, $\nabla \cdot \mathbf{B} = 0$ and Eq. 13-31 becomes

$$\nabla^2\mathbf{B} = \mu_0 \frac{n_s e^2}{m} \mathbf{B}. \tag{13-32}$$

This differential equation is to be solved for $\mathbf{B}$.

Consider the situation depicted in Fig. 13-11. $\mathbf{B}$ is in the $z$ direction, parallel to $xz$ face of the sample, and is a function of $y$. Equation 13-32 becomes

$$\frac{d^2B_z}{dy^2} = \mu_0 \frac{n_s e^2}{m} B_z \tag{13-33}$$

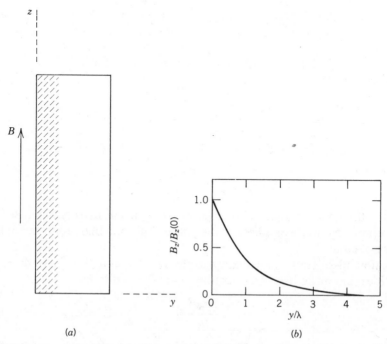

(a)                                                                (b)

**FIGURE 13-11** (a) A slab of superconducting material in an external field. The induction field penetrates the sample in a narrow region near the $xz$ face. (b) Inside the sample it is proportional to $e^{-y/\lambda}$, where $\lambda$ is the penetration depth.

and has the solution

$$B_z(y) = B_z(0)e^{-y/\lambda}, \tag{13-34}$$

where $B_z(0)$ is the field at the boundary and

$$\lambda^2 = \frac{m}{\mu_0 n_s e^2}. \tag{13-35}$$

The parameter $\lambda$, called the London penetration depth, is a measure of the distance the induction field penetrates into the sample. For $n_s = 10^{28}$ electrons/m$^2$, $\lambda$ is about 50 nm. As the temperature increases and the concentration of superconducting electrons decreases the penetration depth increases until finally, when $n_s = 0$, the field penetrates throughout the sample. Some London penetration depths at $T = 0$ K are listed in Table 13-3.

$\mathbf{J} = \nabla \times \mathbf{B}/\mu_0$ gives the current density in the penetration region. For the situation depicted, $\nabla \times \mathbf{B} = (dB_z/dy)\hat{\mathbf{x}} = -[B_z(0)/\lambda]e^{-y/\lambda}\hat{\mathbf{x}}$, so

$$\mathbf{J} = -\frac{B_z(0)}{\mu_0\lambda} e^{-y/\lambda}\hat{\mathbf{x}}. \tag{13-36}$$

The current density also decays exponentially with distance into the sample.

Because an induction field may exist within the penetration region near the surface of a sample, the Meissner effect is not complete for thin films of superconducting material. Figure 13-12 shows results for a thin superconducting film in a field parallel to its surface.

*Flux Quantization.* The magnetic flux through any closed loop that lies entirely in superconducting material is a multiple of $\Phi_0 = \pi\hbar/e$. Consider the ring of type I superconducting wire shown in Fig. 13-13 and suppose lines of a magnetic induction field $\mathbf{B}$ pass through the central region. The loop drawn as a dotted line lies entirely within the wire and is sufficiently far from the surface that both $\mathbf{B}$ and $\mathbf{J}$ are zero at all points on it.

The magnetic flux through the loop is given by the integral $\Phi = \int \mathbf{B} \cdot d\mathbf{S}$ over the surface enclosed by the loop. Since $\mathbf{B} = \nabla \times \mathbf{A}$, this is $\Phi = \int \nabla \times \mathbf{A} \cdot d\mathbf{S} = \oint \mathbf{A} \cdot d\boldsymbol{\ell}$, where $d\boldsymbol{\ell}$ is an infinitesimal line element tangent to the loop and the last integral is a line integral around the loop. Stokes theorem was used to convert the surface integral to a line integral.

**TABLE 13-3** London Penetration Depth $\lambda$ and Intrinsic Coherence Length $\xi_0$ of Selected Superconductors at $T = 0$ K

| Solid | $\lambda$ $(10^{-8}$ m) | $\xi_0$ $(10^{-8}$ m) |
|---|---|---|
| Aluminum | 1.6 | 160 |
| Cadmium | 11.0 | 76 |
| Lead | 3.7 | 8.3 |
| Niobium | 3.9 | 3.8 |
| Tin | 3.4 | 23 |

*Source:* C. Kittel, *Introduction to Solid State Physics* (New York: Wiley, 1986).

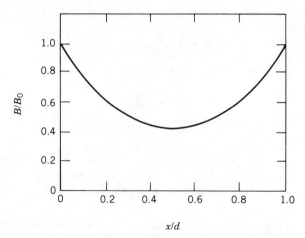

FIGURE 13-12 The magnitude of the magnetic induction field **B** as a function of distance $x$ into a thin superconducting film. $B_0$ is the field at both the left and right surfaces. The plot is for $\lambda = \ell/3$, where $\ell$ is the thickness of the film.

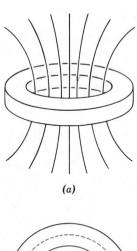

(a)

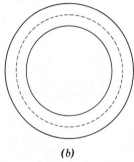

(b)

FIGURE 13-13 (a) A ring of superconducting wire with magnetic flux through the central region. (b) A cross section of the ring. The current density and field are both zero along the dotted line. The flux through the ring is a multiple of $\pi\hbar/e$.

Use Eq. 13-29 to find the vector potential at points on the loop in terms of the gradient of the wave function phase angle. Since $\mathbf{J} = 0$, $\mathbf{A} = -(\hbar/2e)\nabla\phi$ and $\Phi = (\hbar/2e)\oint\nabla\phi \cdot d\boldsymbol{\ell} = (\hbar/2e)\Delta\phi$, where $\Delta\phi$ is the change in the phase around the complete loop. Since $\Psi$ is single valued, $\Delta\phi$ is a multiple of $2\pi$ and

$$\Phi = s\frac{\pi\hbar}{e}, \tag{13-37}$$

where $s$ is an integer. The quantum of flux $\Phi_0 = \pi\hbar/e = 2.06 \times 10^{-15}$ Wb is called a fluxoid.

The flux through the loop is produced by external sources and by supercurrents on the ring surface. As the applied flux changes, the supercurrent and the flux it produces adjust so Eq. 13-37 is valid. The adjustment, of course, takes place via the macroscopic wave function, which remains single valued as the external flux changes. The phase change around the loop is always a multiple of $2\pi$.

Equation 13-37 also holds for vortices in type II superconductors. In fact, each vortex has the minimum flux possible, $\pi\hbar/e$. We can use the value of $\Phi_0$ to estimate the critical field $H_{c1}$ required for the onset of the vortex state, in terms of the penetration depth $\lambda$. The induction field around a vortex is given by $B_0 e^{-\lambda/r}$, where $B_0$ is the field at the center, so the total flux is

$$\Phi = \int B \, dS = 2\pi B_0 \int_0^\infty e^{-r/\lambda} r \, dr = 2\pi B_0 \lambda^2. \tag{13-38}$$

The total flux must be $\pi\hbar/e$ so $B_0 = \hbar/2e\lambda^2$ and

$$H_{c1} = \frac{B_0}{\mu_0} = \frac{\hbar}{2\mu_0 e\lambda^2}. \tag{13-39}$$

Small penetration depths lead to high critical fields.

The quantization of magnetic flux and the persistence of supercurrents are linked. If current in a superconducting ring decays the flux through the ring changes and, since the flux must change by a multiple of $\pi\hbar/e$, current decay can occur only in quantum jumps. Decay does not occur because the phonon and defect systems cannot bring about sufficiently large changes in the current.

A superconducting ring traps flux. Suppose a magnetic field is turned on while the ring is in its normal state, then the temperature is reduced to below the transition point. Flux is excluded from the ring itself but passes through the hole.

Now suppose the applied field is turned off. Since magnetic field lines are closed, they must cross the superconductor if the flux through the hole is to decrease. Since this is not possible, we conclude that the flux through the hole does not change as the applied field changes. Instead, supercurrents generated around the ring maintain the flux through the hole.

*Coherence Length.* The intrinsic or Pippard coherence length, denoted by $\xi_0$, is an important parameter used to characterize a superconductor. It is a measure of the average separation of electrons in a Cooper pair, shown in Section 13.2

to be $\delta x = (2/k_F)(E_F/\delta E)$. Use $E_F = \hbar^2 k_F^2/2m$, $v_F = \hbar k_F/m$, and $\delta E = \Delta_0$ to write $\delta x = \hbar v_F/\Delta_0$. The intrinsic coherence length is taken to be

$$\xi_0 = \frac{\hbar v_F}{\pi \Delta_0}. \qquad (13\text{-}40)$$

The factor $1/\pi$ is included for convenience in writing other expressions in which $\xi_0$ appears. Some intrinsic coherence lengths are listed in Table 13-3.

Consider the penetration region at the surface of a superconductor in an applied magnetic field. If $\xi_0$ is much greater than the penetration depth, the magnetic induction field varies significantly over distances that are short compared to the average separation of electrons in a Cooper pair. The region in which the field exists does not then coincide with the region in which there is a supercurrent.

If $\xi_0 \gg \lambda$, the second term of Eq. 13-29 is no longer proportional to $\mathbf{A}(\mathbf{r})$ but it can be written instead as $-\int \Gamma(\mathbf{r} - \mathbf{r}')\mathbf{A}(\mathbf{r})\, d\tau'$, where primed coordinates are used as variables of integration. The supercurrent density at any point is influenced by the vector potential at other points. $\Gamma(\mathbf{r} - \mathbf{r}')$ can be evaluated by solving a Schrödinger-like equation for $\Psi$. Although an algebraic form cannot be obtained, numerical values can be computed. Typically the function is as plotted in Fig. 13-14.

The London equation is obtained if $\Gamma(\mathbf{r} - \mathbf{r}')$ is zero for every value of $\mathbf{r}'$ except $\mathbf{r}' = \mathbf{r}$. As Fig. 13-14 indicates, $\Gamma$ is a maximum for $\mathbf{r}' = \mathbf{r}$ but it does not vanish for other values. The distance over which it is significantly different from 0 is about $\xi_0$. If $\lambda \gg \xi_0$, $\xi_0$ can be approximated by 0 and the London equation is valid.

Relative values of the penetration depth and coherence length determine whether the sample is type I or type II. For pure type I superconductors the ratio $\lambda/\xi_0$ is less than $1/\sqrt{2}$, while for pure type II superconductors $\lambda/\xi_0$ is

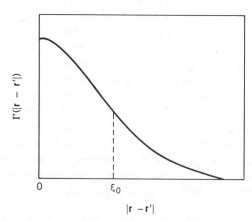

**FIGURE 13-14** $\Gamma(\mathbf{r} - \mathbf{r}')$ plotted as a function of $|\mathbf{r} - \mathbf{r}'|$. The intrinsic coherence length $\xi_0$ is marked by a dotted line. The ratio $\lambda/\xi$ of the penetration depth to the coherence length determines whether a sample is type I or type II.

greater than $1/\sqrt{2}$. The condition comes from energy considerations. A contribution to the total energy is associated with the field and supercurrents in the boundary region between normal and superconducting material. For type I superconductors the contribution is positive and the energy of the system is reduced by nearly complete expulsion of flux, provided the applied field strength is less than the critical field. For type II superconductors the boundary contribution is negative and energy is reduced by the formation of vortices.

As the external field strength increases, vortex formation in a type II superconductor continues until the distance between vortex cores is about equal to the coherence length. Just before the whole sample becomes normal the flux associated with each vortex, $\pi\hbar/e$, passes through a tube with cross section $\pi\xi_0^2$. A calculation similar to the one used to obtain Eq. 13-39 can be used to estimate the critical field $H_{c2}$. The result is

$$H_{c2} = \frac{\hbar}{2\mu_0 e \xi_0^2}. \tag{13-41}$$

The coherence length is modified by electron-defect interactions. For a sample containing a high defect concentration, $\xi_0$ is replaced by $\xi = (\xi_0 \ell)^{1/2}$, where $\ell$ is the mean free path for normal electrons at the temperature of the sample. For low impurity concentrations $\xi$ is nearly $\xi_0$.

The penetration depth is also altered by impurities and, for high impurity content, is given by $\lambda = \lambda_0(\xi_0/\ell)^{1/2}$, where $\lambda_0$ is the London penetration depth. Note that the addition of impurities causes the coherence length to decrease and the penetration depth to increase. Because the coherence length and penetration depth depend on impurity concentration in this manner, the addition of impurities to a type I superconductor may convert it to type II.

## 13.4 JOSEPHSON EFFECTS

*Normal (Giaver) Tunneling.* Consider the situation depicted in Fig. 13-15a. One strip is superconducting while the other is normal. At the junction the strips are separated from each other by a film of insulating material, perhaps an oxide layer 5 nm or so thick. It is thick enough to inhibit the flow of electrons but not so thick that the flow is stopped altogether. The insulator presents a potential energy barrier to the electrons but, being quantum mechanical objects, they may tunnel through.

The three materials are in thermodynamic equilibrium and at $T = 0$ K have a common Fermi level, at the midpoint of the superconducting gap as illustrated in Fig. 13-15b. First suppose no potential difference is applied to the junction. No electrons flow across the junction in either direction because no empty states have energies less than those of filled states. If, however, a potential difference $V$ is applied across the junction, with the superconductor more positive than the normal metal, energy levels on the left are lowered by $eV$ relative to those on the right, as in Fig. 13-15c. Electrons flow when the potential difference is greater than $\Delta_0/e$. Figure 13-15d shows the current $I$ as a function of $V$ for a typical junction. If $T > 0$ K the gap is narrower than at $T = 0$ K and, in addition,

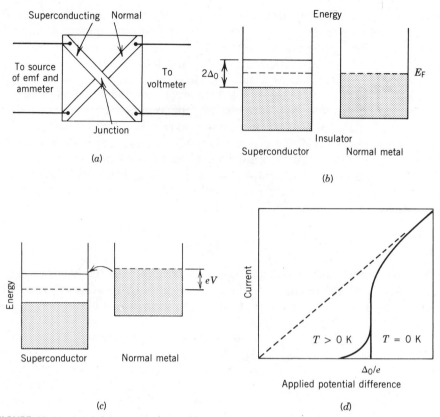

**FIGURE 13-15** (a) A junction is formed by evaporating normal and superconducting metal strips on a glass plate, separated from each other by a thin insulating film. (b) Energy levels in the superconductor and normal metal when no potential difference is applied. The current vanishes. (c) Energy levels after a potential difference $V$ is applied. Electrons flow from the normal metal to the superconductor. (d) Current as a function of applied potential difference. At $T = 0$ K the current is zero for $eV < \Delta_0$. For $T > 0$ K there is a slight current for $eV < \Delta_0$. The Ohm's law prediction is shown by a dotted line.

states above the Fermi level in the normal metal are occupied. Electrons flow at lower potential differences than for $T = 0$ K.

The process is called normal tunneling since electrons to the right of the junction are normal. Measurements of the current as a function of applied voltage are used to determine the superconducting gap.

*The dc Josephson Effect.* The experimental setup is the same as before but now both strips are superconducting and, to maintain a current, the voltage source is replaced by a constant current source. Macroscopic wave functions are associated with Cooper pairs in each superconductor and the functions are linked through the barrier, as illustrated in Fig. 13-16. Suppose the two superconducting strips are made of the same material, so they have the same pair concentration. Then the wave amplitude is the same on the two sides of the junction but, as

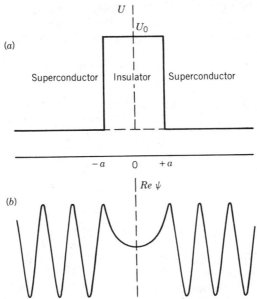

(a)

U |
|U_0

Superconductor | Insulator | Superconductor

−a   0   +a

(b)

| Re ψ

**FIGURE 13-16** (a) The potential energy barrier $U$ and (b) the real part of the macroscopic wave function $\Psi$ in a Josephson junction. The barrier produces a phase shift in the wave function.

the figure shows, the barrier introduces a change in phase across the junction. The phase difference is intimately associated with the supercurrent through the junction.

Suppose the barrier extends from $x = -a$ to $x = +a$ and has height $U_0$, as shown in the diagram. In the insulator the macroscopic wave function has the form

$$\Psi = Ae^{\alpha x} + Be^{-\alpha x}, \qquad (13\text{-}42)$$

where $A$ and $B$ are constants and $\alpha = \sqrt{2mU_0/\hbar^2}$. If the phase is $\phi_1$ at the left boundary and $\phi_2$ at the right boundary, then continuity of the wave function dictates that

$$Ae^{-\alpha a} + Be^{-\alpha a} = n_p^{1/2}e^{i\phi_1} \qquad (13\text{-}43)$$

and

$$Ae^{\alpha a} + Be^{-\alpha a} = n_p^{1/2}e^{i\phi_2}. \qquad (13\text{-}44)$$

Equations 13-43 and 13-44 are solved for $A$ and $B$:

$$A = \frac{e^{i\phi_2}e^{\alpha a} - e^{i\phi_1}e^{-\alpha a}}{e^{2\alpha a} - e^{-2\alpha a}} n_p^{1/2} \qquad (13\text{-}45)$$

and

$$B = \frac{e^{i\phi_1}e^{\alpha a} - e^{i\phi_2}e^{-\alpha a}}{e^{2\alpha a} - e^{-2\alpha a}} n_p^{1/2}. \qquad (13\text{-}46)$$

The current in the insulator can be computed using Eq. 13-26, with $q$ replaced by $-2e$ and $M$ replaced by $2m$. Once Eq. 13-42 is used to substitute for $\Psi$ and Eqs. 13-45 and 13-46 are used to substitute for $A$ and $B$, the result is

$$J = i\frac{e\hbar\alpha}{m}[AB^* - A^*B]$$

$$= \frac{4e\hbar\alpha}{m}n_s\frac{\sin(\phi_1 - \phi_2)}{e^{2\alpha a} - e^{-2\alpha a}}. \tag{13-47}$$

This is usually written

$$J = J_0\sin(\phi_1 - \phi_2), \tag{13-48}$$

where

$$J_0 = \frac{4e\hbar\alpha}{m}\frac{n_s}{e^{2\alpha a} - e^{-2\alpha a}}. \tag{13-49}$$

$J_0$ represents the maximum supercurrent density that can be supported by the junction with zero potential difference across it. For $J < J_0$, the current is proportional to the sine of the phase difference from one side of the junction to the other.

For given superconductors, $J_0$ is determined chiefly by properties of the insulating layer. If the potential energy barrier is wide and strong, $\alpha a \gg 1$ and $J_0$ is small. The current density can be approximated by $J_0 = (4e\hbar\alpha/m)n_se^{-2\alpha a}$. For a narrow and weak barrier, $J_0$ is approximately $(e\hbar/ma)n_s$.

The junction may contain a normal current as well as a supercurrent. For current densities below $J_0$ the supercurrent dominates but if the current density is above $J_0$ the current is entirely normal. Consequently, a Josephson junction can be used as a switching device. Suppose the current is adjusted upward from zero. As long as $J < J_0$ the potential difference across the junction is zero and the current follows the line $AB$ in Fig. 13-17. To obtain $J > J_0$ the current source creates a potential difference across the junction. The device is then being

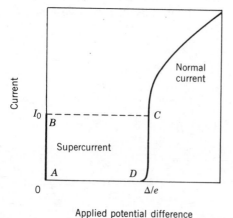

FIGURE 13-17   dc current as a function of bias potential across a Josephson junction. With zero bias the supercurrent can have any value less than $I_0$. To attain larger currents a bias of at least $\Delta/e$ is required.

operated at point $C$, on the normal curve. To increase the current from $C$, the potential difference across the junction must be increased.

A Josephson junction displays a hysteresis effect. If the current is reduced from point $C$ or above, it does not follow the curve through $B$ to $A$. Instead, it follows the normal curve to $D$. There is a supercurrent but, with a potential difference across the junction, it oscillates rapidly. Only the dc portion of the current, which is normal, is plotted in Fig. 13-17.

*The ac Josephson Effect.* If a potential difference is applied across the junction and $J$ is maintained at a value less than $J_0$, the supercurrent oscillates. If the potential at the left side of the insulator is $V$ above the potential at the right side, then the energy of a Cooper pair on the left is lower than the energy of one on the right by 2 eV. If the center of mass energy of a Cooper pair is $E$, a factor of the form $e^{-iEt/\hbar}$ should be included in the wave function. For a dc experiment the factor is the same for both sides of a junction and has no influence on the current. Now we must take it into account.

We take the wave function to be $n_p^{1/2}e^{i(\phi_1 + 2eVt/\hbar)}$ at the left edge of the insulator and $n_p^{1/2}e^{i\phi_2}$ at the right edge. Then Eqs. 13-45 and 13-46 become

$$Ae^{-\alpha a} + Be^{\alpha a} = n_p^{1/2}e^{i(\phi_1 + 2eVt/\hbar)} \tag{13-50}$$

and

$$Ae^{\alpha a} + Be^{-\alpha a} = n_p^{1/2}e^{i\phi_2}, \tag{13-51}$$

respectively. Equations 13-50 and 13-51 are solved for $A$ and $B$, then Eq. 13-26 is used to find an expression for the current density. The result is

$$J = \frac{4e\hbar\alpha}{m} n_s \frac{\sin(\phi_1 - \phi_2 - 2eVt/\hbar)}{e^{2\alpha a} - e^{-2\alpha a}} \tag{13-52}$$

or

$$J = J_0\sin(\phi_1 - \phi_2 - 2eVt/\hbar), \tag{13-53}$$

where $J_0$ is again given by Eq. 13-49. The supercurrent is now sinusoidal with an angular frequency $\omega = 2eV/\hbar$. For $V = 1\ \mu V$, $\omega \approx 3 \times 10^9$ rad/s, corresponding to a frequency of about 500 MHz.

A particularly interesting phenomenon occurs if electromagnetic radiation is incident on a biased junction. The potential difference across the insulator is then given by $V_0 + V_1\sin(\omega t)$, where $V_0$ is the dc potential difference, $V_1$ is the potential amplitude associated with the incident radiation, and $\omega$ is the radiation frequency. According to Eq. 13-53 the supercurrent density is

$$J = J_0\sin[\Delta\phi - 2eV_0t/\hbar + (2eV_1/\hbar)\sin(\omega t)], \tag{13-54}$$

where $\Delta\phi$ is the phase difference at $t = 0$. The expression for $J$ can be written as the sum of an infinite number of sinusoidal terms, each with a different frequency. All harmonics of the radiation frequency occur.

To obtain an explicit form, use $\sin(A + B) = \sin A \cos B + \cos A \sin B$ to write

$$J = J_0\{\sin(\Delta\phi - 2eV_0 t/\hbar)\cos[(2eV_1/\hbar)\sin(\omega t)]$$
$$+ \cos(\Delta\phi - 2eV_0 t/\hbar)\sin[(2eV_1/\hbar)\sin(\omega t)]\}. \qquad (13\text{-}55)$$

Both $\cos[(2eV_1/\hbar)\sin(\omega t)]$ and $\sin[(2eV_1/\hbar)\sin(\omega t)]$ can be expanded as Fourier series:

$$\cos[(2eV_1/\hbar)\sin(\omega t)] = \sum_{s=0}^{\infty} A_s\cos(s\omega t) \qquad (13\text{-}56)$$

and

$$\sin[(2eV_1/\hbar)\sin(\omega t)] = \sum_{s=0}^{\infty} A_s\sin(s\omega t), \qquad (13\text{-}57)$$

respectively. The explicit form of the coefficients $A_s$ does not concern us here. When Eqs. 13-56 and 13-57 are substituted into Eq. 13-55 the result is

$$J = J_0 \sum_{s=0}^{\infty} A_s\sin\left[\Delta\phi_0 - \left(\frac{2eV_0}{\hbar} - s\omega\right)t\right]. \qquad (13\text{-}58)$$

As the externally applied potential is increased, a dc current occurs whenever

$$V_0 = s\frac{\hbar\omega}{2e}, \qquad (13\text{-}59)$$

where $s$ is an integer. The dc current density for a Josephson junction in a radiation field is shown in Fig. 13-18. Steps occur whenever the potential difference across the insulator is a multiple of $\hbar\omega/2e$.

Since both frequency and voltage can be measured with high accuracy, Eq. 13-59 is the basis of an excellent technique for evaluating the ratio $\hbar/e$. Accuracy

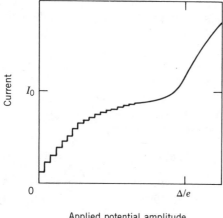

Current
$I_0$

0

$\Delta/e$

Applied potential amplitude

**FIGURE 13-18** dc current as a function of bias potential $V$ for a Josephson junction in an electromagnetic radiation field. The steps have been exaggerated to show the effect. At megahertz frequencies, the range between $V = 0$ and the onset of normal tunneling typically contains hundreds of steps.

extends to eight or more significant figures. The National Bureau of Standards uses Eq. 13-59 and frequency measurements to define the volt. Josephson junctions are also used extensively as electromagnetic radiation detectors.

A sinusoidal potential difference is sometimes applied to a junction to test for superconductivity. The analysis described above is valid and a plot of dc current as a function of applied potential amplitude displays steps if the junction is made of superconductors.

*Superconducting Quantum Interference.* The phase difference across a junction is altered in the presence of a magnetic field. Equation 13-29 gives the relationship between the current density, phase gradient, and magnetic vector potential. We use it to analyze the circuit shown in Fig. 13-19, consisting of two Josephson junctions in parallel. For simplicity we take the two insulating films to be identical and to be linked by two identical superconductors.

Evaluate the line integral of Eq. 13-29 along each of the paths shown. Along path 1, from $a$ to $b$, the result is

$$\int_a^b \mathbf{J} \cdot d\boldsymbol{\ell} = -n_p \left[ \frac{\hbar e}{m} \int_a^b \nabla\phi \cdot d\boldsymbol{\ell} + \frac{2e^2}{m} \int_a^b \mathbf{A} \cdot d\boldsymbol{\ell} \right]$$

$$= -n_p \left[ \frac{\hbar e}{m} (\Delta\phi)_1 + \frac{2e^2}{m} \int_a^b \mathbf{A} \cdot d\boldsymbol{\ell} \right], \qquad (13\text{-}60)$$

where $(\Delta\phi)_1$ is the change in phase across junction 1. Except for the narrow insulating portion, the path of integration lies in the interior of a superconductor, so the left side of Eq. 13-60 is nearly zero. We write

$$(\Delta\phi)_1 = -\frac{2e}{\hbar} \int_a^b \mathbf{A} \cdot d\boldsymbol{\ell}. \qquad (13\text{-}61)$$

Similarly, for path 2,

$$(\Delta\phi)_2 = -\frac{2e}{\hbar} \int_a^b \mathbf{A} \cdot d\boldsymbol{\ell}. \qquad (13\text{-}62)$$

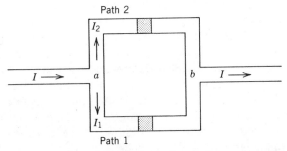

**FIGURE 13-19** Schematic diagram of a SQUID circuit, consisting of two Josephson junctions in parallel. Magnetic flux passes through the central region and the circuit is operated below the superconducting transition temperature.

Subtract Eq. 13-61 from Eq. 13-62. The difference of the two integrals is the line integral of **A** around the complete loop and so equals the magnetic flux through the loop. Thus

$$(\Delta\phi)_2 - (\Delta\phi)_1 = \frac{2e}{\hbar}\,\Phi\,, \qquad (13\text{-}63)$$

where the flux is taken to be positive if it is into the page. Note that the phase change is the same across the two junctions only if the flux through the loop vanishes.

In light of Eq. 13-63, we may write $(\Delta\phi)_1 = \phi_0 - (e/\hbar)\Phi$ and $(\Delta\phi)_2 = \phi_0 + (e/\hbar)\Phi$, where $\phi_0$ is a constant. The current densities along the two paths are then $J_1 = J_0\sin[\phi_0 - (e/\hbar)\Phi]$ and $J_2 = J_0\sin[\phi_0 + (e/\hbar)\Phi]$, respectively. The total current into point $a$ is the sum of the current through the two junctions:

$$I = I_0\left[\sin\left(\phi_0 - \frac{e}{\hbar}\,\Phi\right) + \sin\left(\phi_0 + \frac{e}{\hbar}\,\Phi\right)\right]$$

$$= 2I_0\sin(\phi_0)\cos\left(\frac{e}{\hbar}\,\Phi\right), \qquad (13\text{-}64)$$

where $I_0$ is the maximum supercurrent supported by either junction. The magnetic field also affects the maximum current through an individual junction so $I_0$ cannot be calculated precisely by multiplying $J_0$, given by Eq. 13-49, by the junction area.

According to Eq. 13-64 the current is an oscillating function of the magnetic flux through the loop. Its magnitude is a maximum whenever $e\Phi/\hbar$ is a multiple of $\pi$ or, what is the same, $\Phi$ is a multiple of $\pi\hbar/e$, the quantum of flux. The oscillations come from interference of the two currents. Variation of the magnetic field changes their relative phases and hence their sum.

Devices based on two junction loops are called SQUIDs, short for *superconducting quantum interference devices*. One such device is a magnetometer, used to measure magnetic fields. An applied field is measured by monitoring the current as the flux increases from zero and counting the number of maxima. Flux can be measured with a precision much smaller than $\pi\hbar/e$, so a SQUID is sensitive to extremely small fields. If the loop area is 1 cm$^2$, for example, induction fields less than $10^{-11}$ T can be measured easily.

SQUIDs are used in laboratory and commercial instruments when magnetic fields or field gradients must be measured accurately. Medical uses include the detection of magnetic fields produced by currents of the nervous system and by iron in the digestive system. SQUIDs are also used in geological studies to detect magnetic structures, often well below the earth's surface.

SQUIDs are also used as sensitive ammeters and voltmeters. The magnetic induction field produced by an unknown current is measured using a SQUID and then is compared to the field of a known current. To use a SQUID as a voltmeter, the unknown potential difference is applied across a resistor with a known resistance and the current is measured.

## 13.5   REFERENCES

### General

P. B. Allen and B. Mitrović, "Theory of Superconducting $T_c$" in *Solid State Physics* (H. Ehrenreich, F. Seitz, and D. Turnbull, Eds.), Vol. 37, p. 1 (New York: Academic, 1982).
J. M. Blatt, *Theory of Superconductivity* (New York: Academic, 1964).
R. Dalven, *Introduction to Applied Solid State Physics* (New York: Plenum, 1980).
P. G. de Gennes, *Superconductivity of Metals and Alloys* (New York: Benjamin, 1966).
R. D. Parks (Ed.), *Superconductivity* (New York: Dekker, 1969).
M. Tinkham, *Introduction to Superconductivity* (New York: McGraw-Hill, 1975).

### Applications

A. Barone and G. Paterno, *Physics and Applications of the Josephson Effect* (New York: Wiley, 1982).
J. Clarke, "SQUIDs, Brains and Gravity Waves" in *Phys. Today* **39**(3):36, March 1986.
H. Hayakawa, "Josephson Computer Technology" in *Phys. Today* **39**(3):46, March 1986.
D. Larbalestier, G. Fisk, B. Montgomery, and D. Hawksworth, "High-Field Super-conductivity" in *Phys. Today* **39**(3):24, March 1986.
P. L. Richards, "Analog Superconducting Electronics" in *Phys. Today* **39**(3):54, March 1986.
L. Solymar, *Superconductive Tunnelling and Applications* (New York: Wiley, 1972).

## PROBLEMS

1. For the type I superconductors listed in Table 13-1 plot the critical field as a function of transition temperature. Most points fall on or near a straight line.

2. The heat capacity can be calculated from $C = -T(\partial^2 E/\partial T^2)$, where $E$ is the free energy. (a) Derive an expression for the discontinuity in the heat capacity at $T = T_c$ in terms of the critical field. In particular, show that

$$C_s - C_n = \mu_0 \tau_s T_c \left[ \frac{\partial H_c}{\partial T} \right]^2$$

   at $T = T_c$. *Hint:* $H_c$ is linear in $T$ for temperatures near $T_c$. (b) Use Eq. 13-19 and data from Table 13-1 to estimate the discontinuity per unit volume for aluminum and for zinc.

3. A long straight wire with circular cross section and radius $R$ carries current $I$, uniformly distributed over its cross section. Suppose the wire is made of a linear magnetic material with magnetic susceptibility $\chi$. Magnetic field lines are circles with centers on the axis of the wire. (a) Show that inside the wire a distance $r$ from its axis the magnitude of the magnetic field is given by $Ir/2\pi R^2$, the magnitude of the induction field is given by $(1 + \chi)Ir/2\pi R^2$, and the magnitude of the magnetization is given by

$\chi Ir/2\pi R^2$. In what directions are these fields? Also show that the total current, including magnetization current, in the interior of the wire is $(1 + \chi)I$ and that, in addition, there is a surface current given by $-\chi I$. (b) Find the limits of these expressions as $\chi \to -1$.

4. A superconducting sphere of radius $R$ is in a uniform induction field $B_0\hat{z}$. (a) Show that

$$\mathbf{B} = \hat{a}_r \left[ A - \frac{2C}{r^3} \right] \cos \theta - \hat{a}_\theta \left[ A + \frac{C}{r^3} \right] \sin \theta$$

satisfies $\nabla \cdot \mathbf{B} = 0$ and $\nabla \times \mathbf{B} = 0$, valid outside the sphere. Here $A$ and $C$ are constants, $\hat{a}_r$ is the unit vector $\hat{x} \sin \theta \cos \phi + \hat{y} \sin \theta \sin \phi + \hat{z} \cos \theta$, $\hat{a}_\theta$ is the unit vector $\hat{x} \cos \theta \cos \phi + \hat{y} \cos \theta \sin \phi - \hat{z} \sin \theta$, and $r$, $\theta$, and $\phi$ are the usual spherical coordinates. The first unit vector is radially outward while the second is tangent to a circle of constant $\phi$ and points in the direction of increasing $\theta$. (b) Find expressions for the constants $A$ and $C$ so that $\mathbf{B}$ satisfies the following boundary conditions: it becomes $B_0\hat{z}$ far from the sphere and its radial component $[A - 2C/r^3] \cos \theta$ vanishes at the sphere surface. (c) Find the maximum induction field strength on the sphere surface. At what points on the surface is the field a maximum? (d) Suppose the sphere is a type I superconductor and that the critical field is $H_c$. What is the maximum applied field strength for which flux is excluded from the bulk material?

5. For the superconducting sphere of Problem 4 in a weak field, show that supercurrent flows along circles of constant $\theta$ on the sphere surface and that the surface current density is $(3/2\mu_0)/B_0\sin \theta$. Ampere's law in the form $\oint \mathbf{B} \cdot d\boldsymbol{\ell} = \mu_0 I$ can be used. The integration contour is formed by two lines of constant $\theta$ and $\phi$ with length $dr$ and two lines of constant $r$ and $\phi$ with length $d\ell = R\, d\theta$, one just inside the sphere and one just outside.

6. Assume Eq. 13-18 is valid for the temperatures given below and use data from Table 13-1 to estimate the critical field for lead at 2, 4, and 6 K. For each of these temperatures, what is the maximum supercurrent that can be carried by a lead wire with a radius of 0.10 mm?

7. Use the free electron model to estimate the gap parameter $\Delta_0$ and the transition temperature $T_c$ for (a) aluminum ($1.81 \times 10^{29}$ electrons/m$^3$) and (b) zinc ($1.31 \times 10^{29}$ electrons/m$^3$). Critical fields can be found in Table 13-1. (c) For each of these superconductors estimate the product $\rho U$ of the density of states for one spin and the electron-phonon interaction parameter. The Debye temperatures are 428 K (Al) and 327 K (Zn).

8. According to the BCS theory the temperature dependence of the gap parameter $\Delta(T)$ for a weak coupling superconductor can be found by solving

$$\frac{\Delta}{\Delta_0} = \tanh \left( \frac{T_c\Delta}{T\Delta_0} \right).$$

Let $\alpha = T_c\Delta/T\Delta_0$ and plot $\tanh\alpha$ as a function of $\alpha$. Use the plot to show that: (a) $\Delta \to \Delta_0$ as $T \to 0$ K and (b) $\Delta = 0$ for $T \geq T_c$. (c) Use an expansion of $\tanh\alpha$ for $\alpha \ll 1$ to show that

$$\frac{\Delta}{\Delta_0} \approx \sqrt{3}\left[1 - \frac{T}{T_c}\right]^{1/2}\left[\frac{T}{T_c}\right]$$

for $T$ near $T_c$. (d) Use the full expression to evaluate $\Delta/\Delta_0$ for $T/T_c = 0.25$, 0.50, and 0.75.

9. Consider an infinite superconducting plate of thickness $d$, parallel to the $xy$ plane, and suppose the magnetic field at each face is $H_0\hat{\mathbf{y}}$. Find an expression for the magnetic field everywhere in the plate in terms of $H_0$ and the penetration depth $\lambda$.

10. For the plate of Problem 9 show that the magnetic energy density in the sample is given by

$$U_M = \frac{B_0^2}{2\mu_0}\left[1 - \frac{\cosh(x/\lambda)}{\cosh(d/2\lambda)}\right]$$

and that its average over the plate is given by

$$U_{M\,ave} = \frac{B_0^2}{2\mu_0}\left[1 - \frac{2\lambda}{d}\tanh(d/2\lambda)\right].$$

The plate becomes normal when $H_{M\,ave} = \frac{1}{2}H_c^2$, where $H_c$ is the critical field for the bulk material. Show that in the limit $d/\lambda \gg 1$ the plate is normal for

$$H_0^2 > \frac{H_c^2}{1 - (2\lambda/d)}$$

and that in the limit $d/\lambda \ll 1$ it is normal for

$$H_0^2 > 12\frac{\lambda^2}{d^2}H_c^2.$$

A thin plate is superconducting at higher applied fields than the bulk material.

11. Use data from Table 13-2 to estimate penetration depths for the five titanium-vanadium alloys listed. What do you think is responsible for the differences in penetration depth?

12. Use the free electron model to estimate the intrinsic coherence length $\xi_0$ for aluminum ($1.81 \times 10^{29}$ electrons/m$^3$). The gap parameter $\Delta_0$ was calculated in Problem 7. Compare your answer with the value given in Table 13-3.

13. A Josephson junction with an area of $8.9 \times 10^{-6}$ m$^2$ is made of two identical superconductors, each with a concentration of $1.8 \times 10^{29}$ electrons/m$^3$.

The insulator that separates them has a resistivity of 31.1 $\Omega \cdot$ m and presents a potential energy barrier of $1.25 \times 10^{-4}$ eV. (a) If the maximum supercurrent supported by the junction is 2.22 mA, what is the thickness of the insulating film? (b) To obtain a current of 4.44 mA what potential difference must be applied across the junction? (c) When a 5.0-$\mu$V potential difference is applied what is the amplitude and frequency of the ac supercurrent?

# Chapter 14

## PHYSICS OF SEMICONDUCTOR DEVICES

The inventors of the transistor; Walter H. Brattain (left), William Shockley (center), and John Bardeen (right). Inset: the first transistor.

In this chapter we discuss two processes that are important for semiconductor device technology and apply the results to several devices. The first process is recombination, in which an electron from the conduction band fills a hole in the valence band. The second is diffusion, in which an originally nonuniform carrier distribution, consisting of either electrons or holes, tends to become more uniform as the carriers spread through a sample.

Sufficiently high-frequency electromagnetic radiation induces electron transitions from the valence to the conduction bands. The promotion of electrons across the gap and their subsequent recombination play important roles in photoconductivity, the increase in electrical conductivity that occurs when a sample is illuminated by electromagnetic radiation. Photoconductivity is exploited in video cameras, light detectors, and light intensity meters. Semiconductors that emit light on recombination are used in light-emitting diodes and solid state lasers.

A *p-n* junction consists of *p* and *n* semiconducting meterials in intimate contact. Such junctions are the basic building blocks of many solid state devices, including solar cells and integrated circuit elements. The third section of the chapter is devoted to the characteristics of *p-n* junctions; some junction devices are discussed in the fourth section.

## 14.1 EXCESS CARRIERS AND PHOTOCONDUCTIVITY

*Direct and Indirect Semiconductors.* The most likely sequence of events that occurs when light is absorbed by a semiconductor is illustrated in Fig. 14-1a. An electron in the valence band absorbs a photon and makes a transition across the gap to the conduction band. As a result of interactions with the phonon system, it then makes transitions within the conduction band until it reaches a state near the minimum of the band. When an electron jumps the gap it leaves a hole in the valence band and subsequent electron-phonon interactions alter the electron distribution until the hole reaches a state near the maximum of the band.

Following the absorption of radiation, the sample is not in thermal equilibrium: both conduction and valence bands contain excess carriers. When the light source is turned off equilibrium is achieved by recombination processes in which an appropriate number of electrons from the conduction band fill holes in the valence band.

In some cases recombination occurs by simple inversion of the absorption process, called direct recombination. A downward transition across the gap is accompanied by the emission of a photon and the angular frequency of the radiation is given by $\hbar\omega = E_c(\mathbf{k}) - E_v(\mathbf{k})$, where $E_c$ and $E_v$ are the initial and final electron energies, respectively. Total crystal momentum is conserved in such a transition. Since the photon momentum is much less than the electron crystal momentum, the initial and final electron crystal momenta are essentially the same. A radiative transition is represented by a vertical line on a band diagram.

The final state must be empty before the transition. Most electrons are in

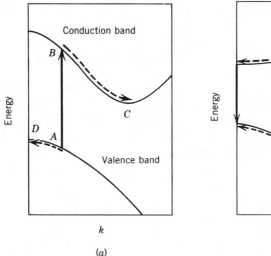
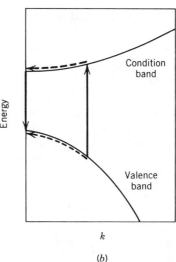

**FIGURE 14-1** (a) Band structure diagram depicting the absorption of a photon by an electron. An electron originally at $A$ absorbs a photon and makes a transition to $B$. As a result of phonon interactions it then makes transitions to $C$, in the vicinity of the conduction band minimum. Phonon interactions cause the hole at $A$ to make transitions to $D$, near the valence band maximum. (b) Bands in a direct semiconductor. The minimum of the conduction band and maximum of the valence band are associated with the same crystal momentum, in this case $k = 0$. The electron drops across the gap and recombines with a hole.

states at the bottom of the conduction band, while most holes are in states at the top of the valence band. Conservation of crystal momentum precludes radiative transitions unless the minimum of the conduction band and the maximum of the valence band correspond to the same crystal momentum, as illustrated in Fig. 14-1$b$. If they do, the material is called a direct semiconductor. Emitted photons have energies equal to $E_g$ and the radiation is called gap light.

Semiconductors for which the conduction band minimum and valence band maximum correspond to different crystal momenta are said to be indirect. The band diagram of Fig. 14-1$a$, for example, is for an indirect semiconductor. Silicon, germanium, and gallium phosphide are indirect, while gallium arsenide, indium phosphide, and cadmium sulfide are direct.

Recombination also takes place in indirect semiconductors. In the dominant process, intermediate transitions are made by both an electron and a hole to a state localized at an impurity and they recombine there. Radiation does not usually accompany indirect transitions. Instead, energy is transferred to the phonon system.

*Direct Recombination.* Consider a direct semiconductor with excess electron and hole concentrations, possibly produced by the absorption of light. We are interested in recombination after the light is turned off. The electron concentration $n$ depends on the time and we write $n(t) = n_0 + \delta n(t)$, where $n_0$ is the equilibrium concentration and $\delta n(t)$ is the excess. The hole concentration also depends on the time and we write $p(t) = p_0 + \delta p(t)$. Since electron-hole generation necessarily means a hole is created for every electron excited across the gap and, since every electron that drops across the gap fills an empty state, $\delta n(t) = \delta p(t)$ for all values of $t$. If the sample is doped, $n_0$ and $p_0$ do not have the same values. Nevertheless, doped or not, $dn/dt = dp/dt$.

The rate at which electrons and holes recombine is very nearly proportional to the product $np$. It is proportional to $n$ because every electron near the conduction band minimum has nearly the same probability of making a transition across the gap and it is proportional to $p$ because every hole near the valence band maximum has nearly the same probability of being filled. If only recombination events are considered $dn/dt = -\alpha_R np$, where $\alpha_R$ is a proportionality constant, called the recombination coefficient.

The recombination coefficient depends primarily on the extent to which probability densities for states near the bottom of the conduction band overlap probability densities for states near the top of the valence band. A transition is likely if an electron has a high probability of being at the same place as a hole.

We must also take into account the thermal excitation of electrons from the valence to the conduction band. If the electron system is in thermal equilibrium, $dn/dt = 0$. Then the rate of thermal excitation matches the rate of recombination, so the former must be $+\alpha_R n_0 p_0$. When both processes are taken into account

$$\frac{dn}{dt} = -\alpha_R np + \alpha_R n_0 p_0 \tag{14-1}$$

or, when $n = n_0 + \delta n$ and $p = p_0 + \delta n$ are used,

$$\frac{d\,\delta n}{dt} = -\alpha_R(n_0 + p_0)\,\delta n - \alpha_R(\delta n)^2. \tag{14-2}$$

Equation 14-2 is a nonlinear differential equation for the excess electron concentration as a function of time. If $\delta n$ is large initially, algebraic solutions do not exist but the equation can be solved numerically.

An algebraic solution can be obtained, however, if $\delta n \ll n_0 + p_0$. The last term of Eq. 14-2 can then be neglected and

$$\frac{d\,\delta n}{dt} = -\alpha_R(n_0 + p_0)\,\delta n. \tag{14-3}$$

Equation 14-3 has the solution

$$\delta n(t) = \delta n(0)e^{-t/\tau}, \tag{14-4}$$

where $\delta n(0)$ is the value of $\delta n$ at $t = 0$, when the light is turned off, and

$$\tau = \frac{1}{\alpha_R(n_0 + p_0)}. \tag{14-5}$$

According to Eq. 14-4, the electron concentration approaches its equilibrium value exponentially. $\tau$ is the average time spent by an electron in the conduction

band before recombining and is called the low-level recombination lifetime.*
Equation 14-3 is often written

$$\frac{d\,\delta n}{dt} = -\frac{\delta n}{\tau}.$$ (14-6)

Recombination influences the electrical conductivity $\sigma$ of a sample. According
to Eq. 9-62 $\sigma = e(n\mu_n + p\mu_p)$ for a sample with scalar mobilities: $\mu_n$ for electrons and $\mu_p$ for holes. Equation 14-4 can be used to write

$$\sigma = \sigma_0 + e(\mu_n + \mu_p)\,\delta n(0)e^{-t/\tau},$$ (14-7)

where $\sigma_0 = e(n_0\mu_n + p_0\mu_p)$ is the conductivity when the system is in thermal
equilibrium. The conductivity has its greatest value at the time the light is turned
off and excess carrier concentrations are the greatest. It then decays exponentially toward its thermal equilibrium value.

The conductivity can be measured as indicated in Fig. 14-2. The sample is
illuminated by a strobe light, which flashes periodically. Current is supplied by
a constant current source and the potential difference across the sample is
displayed as a function of time on an oscilloscope. In practice only the time-
varying portion of the potential difference is displayed and the dc portion is
read from a voltmeter. For the geometry shown, the conductivity is calculated
using $\sigma(t) = \ell I/AV(t)$, where $\ell$ is the distance between potential probes, $A$ is

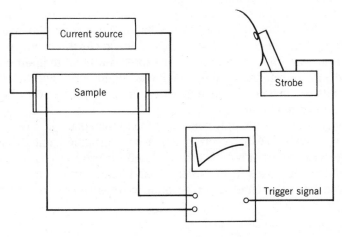

**FIGURE 14-2** Experimental arrangement for measuring photoconductivity. When the strobe
flashes, the conductivity of the sample increases. When the light goes off, the conductivity
decays to its dark value as excess carriers recombine. The oscilloscope is triggered by the light
flash and shows the time-dependent part of the potential difference across the sample.

---

*In this chapter $\tau$ is used primarily to denote a recombination lifetime rather than a volume. In the
few instances when it is used as a volume its meaning is clearly stated.

the cross section of the sample, and $I$ is the current. $V(t)$ is the total potential difference, the sum of the ac and dc portions. According to Eq. 14-7, the natural logarithm of the time dependent part of $\sigma$, plotted as a function of time, is a straight line with slope $-1/\tau$. Typically $\tau$ is on the order of $10^{-7}$ s for direct semiconductors but recombination lifetimes that are several orders of magnitude shorter have been observed.

**EXAMPLE 14-1**   Estimate the recombination coefficient for a direct intrinsic semiconductor with $1.7 \times 10^{19}$ electrons/m³ in the conduction band at thermal equilibrium. The recombination lifetime is measured to be $5.0 \times 10^{-6}$ s.

**SOLUTION**   According to Eq. 14-5

$$\alpha_R = \frac{1}{\tau(n_0 + p_0)} = \frac{1}{5.0 \times 10^{-7} \times 2 \times 1.7 \times 10^{19}} = 5.9 \times 10^{-4} \text{ m}^3/\text{s} ,$$

a typical value for direct semiconductors. ∎

*Indirect Recombination.*   Two fundamental recombination processes involving impurity states are diagrammed in Fig. 14-3. In $(a)$ the impurity is an acceptor and is neutral when the state is empty. It captures an electron from the conduction band, then a hole from the valence band. The second step is tantamount to the release of the electron to the valence band. In $(b)$ the impurity is a donor and is neutral when the state is filled. It captures a hole, then an electron. In either case an electron and hole recombine at the impurity and the impurity is returned to its original condition. Impurities that act in either of these ways are called recombination centers.

Recombination is not the only event that might take place at a recombination center. For process $(a)$ the electron might be thermally excited back to the conduction band before a hole is captured and for process $(b)$ the hole might be thermally excited back to the valence band before an electron is captured. We must take thermal excitation into account, as we did for direct recombination.

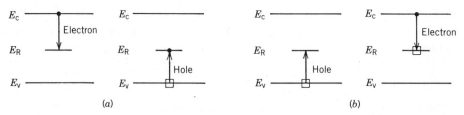

FIGURE 14-3  Two types of recombination centers. In $(a)$ an electron from the conduction band drops to an impurity level where it recombines with a hole from the valence band. In $(b)$ a hole is captured first, then an electron recombines with it.

Consider the process depicted in Fig. 14-3a. Suppose the concentration of recombination centers is $n_R$ and a state with energy $E_R$ in the gap is associated with each center. The rate at which electrons make transitions from the conduction band to the impurity level is proportional to the concentration of electrons in the conduction band, to the concentration of centers, and to the probability that a center is unoccupied. Write $\alpha_n n_R n[1 - f(E_R)]$ for the transition rate. Here $f(E_R)$ is the probability a center is occupied and $\alpha_n$ is a constant of proportionality. The rate at which electrons make transitions back to the conduction band is proportional to the concentration of centers and to the probability a center is occupied, so we may write $C n_R f(E_R)$ for this rate. Here $C$ is a constant of proportionality. At thermal equilibrium the two rates are the same so $C = \alpha_n n_0[1 - f_0(E_R)]/f_0(E_R)$, where $f_0$ is the Fermi-Dirac distribution function. For convenience in writing, let $n_1 = n_0[1 - f_0(E_R)]/f_0(E_R)$ or, more explicitly, $n_1 = n_0 e^{\beta(E_R - \eta)}$, where $\eta$ is the chemical potential. Then $C = \alpha_n n_1$.

The net rate at which electrons fill recombination centers is the difference between the rate at which they enter from the conduction band and the rate at which they are thermally excited, so

$$\frac{dn}{dt} = -\alpha_n n_R n[1 - f(E_R) + C n_R f(E_R)]$$

$$= -\alpha_n n_R[n - n f(E_R) - n_1 f(E_R)] . \qquad (14\text{-}8)$$

A similar analysis can be carried out for holes, with the result

$$\frac{dp}{dt} = -\alpha_p n_R[p f(E_R) - p_1 + p_1 f(E_R)] , \qquad (14\text{-}9)$$

where $p_1 = p_0 f_0(E_R)/[1 - f_0(E_R)] = p_0 e^{-\beta(E_R - \eta)}$.

Equate the right sides of Eqs. 14-8 and 14-9 to each other and solve for $f(E_R)$:

$$f(E_R) = \frac{\alpha_n n + \alpha_p p_1}{\alpha_n(n + n_1) + \alpha_p(p + p_1)} . \qquad (14\text{-}10)$$

Substitute the result into either Eq. 14-8 or 14-9 to obtain

$$\frac{dn}{dt} = \frac{dp}{dt} = -\frac{n_R \alpha_n \alpha_p (np - n_i^2)}{\alpha_n(n + n_1) + \alpha_p(p + p_1)} , \qquad (14\text{-}11)$$

where $n_i$ is the intrinsic electron concentration and $n_1 p_1 = n_0 p_0 = n_i^2$ was used.

For low-level excitation $\delta n$ and $\delta p$ are small compared to $n_0 + p_0$ and Eq. 14-11 becomes

$$\frac{d\,\delta n}{dt} = -\frac{n_R \alpha_n \alpha_p (n_0 + p_0)}{\alpha_n(n_0 + n_1) + \alpha_p(p_0 + p_1)} \delta n . \qquad (14\text{-}12)$$

Equation 14-12 has the same form as Eq. 14-6, with the recombination lifetime given by

$$\tau = \frac{\alpha_n(n_0 + n_1) + \alpha_p(p_0 + p_1)}{n_R \alpha_n \alpha_p (n_0 + p_0)} . \qquad (14\text{-}13)$$

In addition to the concentration of recombination centers and the recombination coefficients $\alpha_n$ and $\alpha_p$, the recombination lifetime depends on the locations of the recombination level and chemical potential within the gap. Figure 14-4 illustrates the variation with chemical potential for a recombination level near the center of the gap. If the sample is heavily doped $n$ type, $n_0$ is large and $p_0$ is small. The chemical potential is near the bottom of the conduction band, well above the recombination level, so $n_1$ is small. The parameter $p_1$ is larger than $p_0$ but it is still several orders of magnitude or more smaller than $n_0$. In Eq. 14-13 we neglect $n_1$, $p_0$, and $p_1$ compared to $n_0$ and write

$$\tau = \frac{1}{n_R \alpha_p} \tag{14-14}$$

for the recombination lifetime. This is the limiting value on the right side of the diagram. For a heavily doped $n$-type sample the recombination lifetime is determined primarily by the rate at which recombination centers are emptied into the valence band.

If the sample is heavily doped $p$ type, with the chemical potential near the top of the valence band,

$$\tau = \frac{1}{n_R \alpha_n} . \tag{14-15}$$

This is the limiting value on the left side of the diagram. It is determined primarily by the rate at which electrons fall to the impurity level. The maximum of $\tau$ occurs for values of the chemical potential near the center of the gap. Its exact position depends on the values of $\alpha_n$, $\alpha_p$, $E_R$, and $\eta$, but if $\alpha_n = \alpha_p$ it occurs when the chemical potential has its intrinsic value.

A comparison of either Eq. 14-14 or 14-15 with Eq. 14-5 shows that indirect

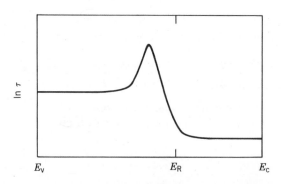

FIGURE 14-4 Natural logarithm of the recombination lifetime as a function of the chemical potential for an indirect semiconductor containing recombination centers. $E_v$ is the top of the valence band, $E_c$ is the bottom of the conduction band, and $E_R$ is the recombination level. In practice, the chemical potential is changed by changing dopant concentrations.

lifetimes are longer than direct lifetimes because $n_R$ is usually much less than $n_0 + p_0$. Typical values range from several hundred microseconds to several hundred milliseconds. Indirect recombination can occur in direct semiconductors but, since indirect lifetimes are much longer than direct lifetimes, nearly all recombination is direct. On the other hand, direct processes are rare in indirect semiconductors so recombination occurs primarily via recombination centers.

Direct and indirect recombination lifetimes behave quite differently as functions of temperature. Both $n_0$ and $p_0$ increase with increasing temperature so, according to Eq. 14-5, a direct lifetime decreases. For an indirect semiconductor, on the other hand, an increase in temperature means an increase in the probability that the first carrier captured is excited back to its original band before a second carrier is captured. Thus an indirect lifetime increases with increasing temperature.

For some types of recombination centers a captured carrier has a higher probability of being reexcited than of recombining, no matter what the temperature. Nickel and zinc in silicon or germanium are examples of such centers, known as traps. A zinc atom in silicon, for example, is a double acceptor. Each atom accepts an electron in a level about 0.31 eV above the valence band and a second electron in a level about 0.55 eV above the valence band, near the center of the gap. If the sample is heavily doped $n$ type, impurity states have a high probability of being filled and zinc acts as a hole trap. When a zinc atom captures a hole in the upper level, the atom remains negatively charged, making it unlikely that an electron from the conduction band will fill the state before the hole drops back to the valence band. Similarly, sulfur is a double donor in silicon and germanium and acts as a trap for electrons in heavily doped $p$-type samples.

If traps are present, their influence can be seen in photoconductivity data. Traps decrease the conductivity immediately after the light is turned off, when they are being filled, and increase it at times that are long compared to the recombination lifetime. Carriers are then released from traps and recombine via recombination centers.

*Optical Generation of Electron-Hole Pairs.* The creation of excess electron-hole pairs by light is described by giving the number of pairs generated per unit volume per unit time, a quantity called the optical generation rate and denoted by $g_{op}$. It depends on the number of photons incident on the sample per unit time and on the probability an electron in the valence band absorbs an incident photon.

Consider an illuminated semiconductor with recombination lifetime $\tau$. If the excess carrier concentrations are small, Eq. 14-6 is valid provided we add $g_{op}$ to the right side:

$$\frac{d\,\delta n}{dt} = -\frac{\delta n}{\tau} + g_{op}. \tag{14-16}$$

A similar equation holds for holes. If $\delta n$ is zero at time $t = 0$, when the light is turned on, then

$$\delta n(t) = \tau g_{op}[1 - e^{-t/\tau}] \qquad (14\text{-}17)$$

is the appropriate solution. In the limit as $t$ becomes much longer than $\tau$, $\delta n$ reaches the steady state value $\tau g_{op}$. As the following example illustrates, Hall effect and conductivity data can be used to calculate $g_{op}$ for a given light source and sample.

**EXAMPLE 14-2**  Hall effect experiments performed on a certain semiconductor yield $\mu_n = 0.850 \text{ m}^2/\text{V} \cdot \text{s}$ and $\mu_p = 0.0400 \text{ m}^2/\text{V} \cdot \text{s}$ for the electron and hole mobilities, respectively. The sample is illuminated for a long time, then the light source is turned off and the conductivity is monitored. The time-dependent part is $2.95 \ \Omega^{-1} \cdot \text{m}^{-1}$ when the light is turned off and is half that value $3.80 \times 10^{-6}$ s later. What is the recombination lifetime for electrons in the sample and what is the optical generation rate for the light and sample?

**SOLUTION**  Suppose the light is turned off at $t = 0$ and the excess carrier concentrations at that time are both $\tau g_{op}$. Then the time-dependent part of the conductivity is given by

$$\delta \sigma(t) = e\tau g_{op}(\mu_n + \mu_p)e^{-t/\tau} .$$

At $t = 3.80 \times 10^{-6}$ s, $\delta\sigma = \frac{1}{2}e\tau g_{op}(\mu_n + \mu_p)$ so $\frac{1}{2} = e^{-t/\tau}$, or

$$\tau = \frac{t}{\ln 2} = \frac{3.80 \times 10^{-6}}{\ln 2} = 5.48 \times 10^{-6} \text{ s} .$$

At $t = 0$, $\delta\sigma = e\tau g_{op}(\mu_n + \mu_p)$ so

$$g_{op} = \frac{\delta\sigma(0)}{e\tau(\mu_n + \mu_p)} = \frac{2.95}{1.60 \times 10^{-19} \times 5.48 \times 10^{-6}(0.850 + 0.040)}$$

$$= 3.78 \times 10^{24} \text{ pairs/m}^3 \cdot \text{s} . \qquad \blacksquare$$

## 14.2  DIFFUSION OF CARRIERS

In Section 9.4 we considered the diffusion of electrons and holes arising from a concentration gradient in a sample with a nonuniform temperature. Here we consider a concentration gradient in a sample with a uniform temperature. If, for example, only part of a sample is illuminated by light, excess electrons and holes are created in the illuminated region and diffuse into the dark region beyond. Electron current is directed from the dark region toward the illuminated region, while hole current is in the opposite direction.

*The Diffusion Process.*  An expression for the diffusion current can be derived from the Boltzmann equation, Eq. 9-73. The distribution function $f(\mathbf{r}, \mathbf{k})$ gives the probability an electron in the neighborhood of **r** occupies a state with propagation vector **k**. For the steady state in the absence of an electric field,

$$f(\mathbf{r}, \mathbf{k}) = f_0(\mathbf{k}) - \bar{t}\mathbf{v} \cdot \nabla f(\mathbf{r}, \mathbf{k}),\tag{14-18}$$

where $f_0$ is the Fermi-Dirac distribution function, **v** is the velocity of an electron with propagation constant **k**, and $\bar{t}$ is the relaxation time.

When the conduction band contains excess electrons the system is not in thermal equilibrium. Nevertheless, we can choose a parameter $\eta_n(\mathbf{r})$, called the electron quasi chemical potential,* so that $f = [e^{\beta(E-\eta_n)} + 1]^{-1}$ leads to the correct electron concentration in the neighborhood of **r**. If $\eta_n$ is in the gap, far from the band edges, the electron concentration is proportional to $e^{\beta\eta_n}$. On the other hand, in the intrinsic material, with chemical potential $\eta_i$, it is proportional to $e^{\beta\eta_i}$. Thus

$$n(\mathbf{r}) = n_i e^{\beta(\eta_n - \eta_i)},\tag{14-19}$$

where $n_i$ is the equilibrium concentration.

In a similar manner the hole concentration is given by

$$p(\mathbf{r}) = p_i e^{-\beta(\eta_p - \eta_i)},\tag{14-20}$$

where $\eta_p(\mathbf{r})$ is the hole quasi chemical potential. For Eq. 14-20 to be valid, $\eta_p$ must be in the gap and far from the band edges. When the sample is in thermal equilibrium the two quasi chemical potentials have the same value, the true chemical potential for the sample. Equations 14-19 and 14-20 together imply $np = n_i^2 e^{\beta(\eta_n - \eta_p)}$ and, if $\eta_n = \eta_p$, then $pn = n_i^2$. When excess carriers are present, however, $\eta_n \neq \eta_p$ and $np \neq n_i^2$.

Since $f = [e^{\beta(E-\eta_n)} + 1]^{-1}$, $\nabla f = (\partial f/\partial \eta_n) \nabla \eta_n = -(\partial f/\partial E) \nabla \eta_n$, where the second equality is valid because $\eta_n$ enters $f$ only in the combination $E - \eta_n$. According to Eq. 14-19, $\nabla n = \beta n \nabla \eta_n$, so $\nabla \eta_n = (1/\beta n) \nabla n$ and $\nabla f = -(\partial f/\partial E)(1/\beta n) \nabla n$. If the concentration gradient is small we replace $\partial f/\partial E$ with $\partial f_0/\partial E$ and Eq. 14-18 becomes

$$f(\mathbf{r}, \mathbf{k}) = f_0(\mathbf{k}) - \frac{\bar{t}}{\beta n} \frac{\partial f_0}{\partial E} \mathbf{v} \cdot \nabla n.\tag{14-21}$$

To find an expression for the electron current density $\mathbf{J}(\mathbf{r})$, sum the contributions of individual electrons. The term arising from $f_0$ vanishes and

$$\mathbf{J}(\mathbf{r}) = -\frac{e}{\beta\tau_s n} \sum_{\text{states}} \bar{t}\mathbf{v}\mathbf{v} \cdot \nabla n \frac{\partial f_0}{\partial E}.\tag{14-22}$$

Assume the band and relaxation time are isotropic in **k** and take $\nabla n$ to be in the $z$ direction. Then

$$J_z = -\frac{ek_B T}{3\tau_s n} \frac{dn}{dz} \sum_{\text{states}} \bar{t}v^2 \frac{\partial f_0}{\partial E},\tag{14-23}$$

---

*In semiconductor work this quantity is called the electron quasi Fermi energy.

where $v_z^2$ was replaced by $\frac{1}{3}v^2$ and $\beta$ was replaced by $1/k_BT$. Since the electron mobility is given by

$$\mu_n = -\frac{e}{3\tau_s n} \sum_{\text{states}} \bar{t}v^2 \frac{\partial f_0}{\partial E}, \tag{14-24}$$

the diffusion current is

$$J_z = k_BT\mu_n \frac{dn}{dz}. \tag{14-25}$$

For a concentration gradient in any direction

$$\mathbf{J} = k_BT\mu_n \, \nabla n. \tag{14-26}$$

Electrons flow from regions of high concentration toward regions of low concentration, opposite the gradient and, as Eq. 14-26 indicates, the current density is in the direction of the concentration gradient.

Equation 14-26 is usually written

$$\mathbf{J} = eD_n \, \nabla n(\mathbf{r}), \tag{14-27}$$

where

$$D_n = \frac{k_BT}{e} \mu_n. \tag{14-28}$$

$D_n$ is called the electron diffusion constant and Eq. 14-28 is known as the Einstein relation between the mobility and diffusion constant. A similar derivation can be carried out for holes, with the result

$$\mathbf{J}(\mathbf{r}) = -eD_p \, \nabla p(\mathbf{r}), \tag{14-29}$$

where $D_p$ is the hole diffusion constant, given by

$$D_p = \frac{k_BT}{e} \mu_p. \tag{14-30}$$

The hole current is opposite the concentration gradient.

At room temperature the diffusion constant for electrons in silicon is about $3.5 \times 10^{-3}$ m²/s while that for holes is about $1.25 \times 10^{-3}$ m²/s. Diffusión constants for electrons and holes in germanium are about $1.0 \times 10^{-2}$ m²/s and $5.0 \times 10^{-3}$ m²/s, respectively, at room temperature.

When an electric field exists in the sample, the total electron current density $\mathbf{J}_n$ is the sum of drift and diffusion contributions. That is,

$$\mathbf{J}_n(\mathbf{r}) = e\mu_n n(\mathbf{r})\mathcal{E} + eD_n \, \nabla n(\mathbf{r}). \tag{14-31}$$

Similarly, the total hole current density is

$$\mathbf{J}_p(\mathbf{r}) = e\mu_p p(\mathbf{r})\mathcal{E} - eD_p \, \nabla p(\mathbf{r}). \tag{14-32}$$

*Continuity and Diffusion-Drift Equations.* If charge is conserved, the current density $\mathbf{J}$ and charge density $\rho$ obey the continuity equation

$$\nabla \cdot \mathbf{J} + \frac{\partial \rho}{\partial t} = 0 . \qquad (14\text{-}33)$$

Integrate over any volume. The first term becomes $\int \nabla \cdot \mathbf{J} \, d\tau = \oint \mathbf{J} \cdot \hat{n} \, dS = I$, where $\hat{n}$ is the unit outward normal to the surface bounding the integration volume and $I$ is the current through the surface. Gauss's theorem was used to convert the volume integral to a surface integral. The second term becomes $-dQ/dt$, where $Q$ is the charge in the volume. Thus Eq. 14-33 is equivalent to $I = -dQ/dt$. The charge in a region changes only because current flows through its bounding surface.

Equation 14-33 is valid for the total charge and current density in any region of a semiconductor but, since recombination takes place, it is not valid for electrons and holes separately. Consider the hole current in the portion of a sample depicted in Fig. 14-5. Take it to be in the positive $z$ direction, and assume the current and charge densities are functions of $z$ only. The charge in the region from $z$ to $z + dz$ changes with time if the current at the right boundary is different from the current at the left or if holes recombine within the volume.

The rate at which charge enters the left boundary is $AJ_p(z)$ and the rate at which it leaves the right boundary is $AJ_p(z + dz)$, where $A$ is the sample cross section. The net outflow of charge per unit time is $A[J_p(z + dz) - J_p(z)]$ or, in the limit as the width of the region becomes small, $A[\partial J_p/\partial z] \, dz$. If the excess hole concentration is small compared to $n_0 + p_0$, the rate at which charge disappears from the region by recombination is given by $eA[\delta p(z)/\tau] \, dz$, where $\tau$ is the recombination lifetime.

The net rate at which charge leaves the region, whether by moving across the boundary or by recombination, is $A[\partial J_p/\partial z] \, dz + eA[\delta p/\tau] \, dz$. This must equal $-eA[\partial p/\partial t] \, dz$, so

$$\frac{\partial p}{\partial t} = -\frac{1}{e} \frac{\partial J_p}{\partial z} - \frac{\delta p}{\tau} . \qquad (14\text{-}34)$$

In three dimensions $\partial J_p/\partial z$ is replaced by the divergence of $J_p$ and

$$\frac{\partial p}{\partial t} = -\frac{1}{e} \nabla \cdot \mathbf{J}_p - \frac{\delta p}{\tau} . \qquad (14\text{-}35)$$

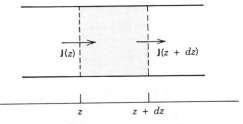

**FIGURE 14-5** A portion of a long sample-carrying current. The carrier concentration changes in the region from $z$ to $z + dz$ because the current is different at the two boundaries or because carriers recombine within the region.

A similar equation,

$$\frac{\partial n}{\partial t} = +\frac{1}{e} \nabla \cdot \mathbf{J}_n - \frac{\delta n}{\tau},$$ (14-36)

holds for electrons. Equations 14-35 and 14-36 are the continuity equations for holes and electrons, respectively.

The diffusion-drift equation for holes is derived by substituting Eq. 14-32 into Eq. 14-35. Assume the electric field is uniform and the sample is homogeneous, so $\mathcal{E}, p_0, \mu_p,$ and $D_p$ are independent of position in the sample. Since $p_0$ is constant and uniform, $\partial p / \partial t = \partial\, \delta p / \partial t$ and $\nabla p = \nabla\, \delta p$. Thus

$$\frac{\partial\, \delta p}{\partial t} = D_p \nabla^2\, \delta p - \mu_p \nabla\, \delta p \cdot \mathcal{E} - \frac{\delta p}{\tau}.$$ (14-37)

Similarly, for electrons

$$\frac{\partial\, \delta n}{\partial t} = D_n \nabla^2\, \delta n + \mu_n \nabla\, \delta n \cdot \mathcal{E} - \frac{\delta n}{\tau}.$$ (14-38)

The diffusion-drift equations are to be solved for $\delta p$ and $\delta n$. We now discuss an important example.

*Steady State Carrier Injection.* Consider a long homogeneous sample that extends along the positive $z$ axis, with one end at $z = 0$. Suppose holes are steadily created at $z = 0$, perhaps by illuminating the end. They simultaneously diffuse into the sample and recombine with electrons. We wish to find an expression for the steady state hole distribution.

In the steady state $\partial\, \delta p / \partial t = 0$ everywhere and Eq. 14-37, with $\mathcal{E} = 0$, reduces to

$$D_p \frac{d^2\, \delta p}{dz^2} - \frac{\delta p}{\tau} = 0,$$ (14-39)

a differential equation that has the general solution

$$\delta p(z) = A e^{-z/L_p} + B e^{+z/L_p}.$$ (14-40)

$A$ and $B$ are constants determined by boundary conditions and

$$L_p^2 = D_p \tau.$$ (14-41)

If the sample is sufficiently long, $\delta p$ vanishes as $z$ becomes large, so $B = 0$. $A$ then represents the excess concentration $\delta p(0)$ at $z = 0$, the excess created by the light. Equation 14-40 becomes

$$\delta p(z) = \delta p(0) e^{-z/L_p}.$$ (14-42)

The excess hole concentration decreases exponentially with distance into the sample. The distance $L_p$, called the hole diffusion length, is the average distance holes diffuse into the sample before recombining.

**EXAMPLE 14-3**  Find an expression for the current density in a long homogeneous sample if the hole concentration $\delta p(0)$ is held constant at one end by steady injection. What current must be supplied to maintain the steady state? Evaluate the current if $\delta p(0) = 3.90 \times 10^{15}$ holes/m³. Assume the sample is circular with radius 5.00 mm, the hole diffusion constant is $8.50 \times 10^{-3}$ m²/s, and the recombination lifetime is $4.80 \times 10^{-6}$ s.

**SOLUTION**  The current density $J_p(z)$ is given by $-eD_p dp/dz$ and, since $dp/dz = d\,\delta p/dz = -[\delta p(0)/L_p]e^{-z/L_p}$,

$$J_p(z) = \frac{eD}{L_p}\,\delta p(0)e^{-z/L_p}\,.$$

The current through the cross section at $z$ is $I = J_p(z)A$, where $A$ is the cross section. At $z = 0$.

$$I = J_p(0)A = \frac{eD_p}{L_p}A\,\delta p(0)\,.$$

For the given sample $L_p = \sqrt{D_p \tau} = \sqrt{8.50 \times 10^{-3} \times 4.80 \times 10^{-6}} = 2.02 \times 10^{-4}$ m and

$$I = \frac{1.60 \times 10^{-19} \times 8.50 \times 10^{-3}}{2.02 \times 10^{-4}}\,\pi(5.00 \times 10^{-3})^2 \times 3.9 \times 10^{15}$$

$$= 2.06 \times 10^{-6}\,\text{A}\,. \qquad\blacksquare$$

As shown in the example, the current that must be supplied to maintain the steady state is given by

$$I = \frac{eD}{L_p}A\,\delta p(0)\,. \qquad (14\text{-}43)$$

This is exactly the current required to replenish all excess charge in the sample in a time equal to the recombination lifetime. The total excess charge $Q$ is

$$Q = eA \int_0^\infty \delta p(z)\,dz = e\,\delta p(0)A \int_0^\infty e^{-z/L_p}\,dz = e\,\delta p(0)AL_p\,. \qquad (14\text{-}44)$$

If the charge is replenished steadily in time $\tau$ the current is given by $I = Q/\tau = e\,\delta p(0)AL_p/\tau$ or, since $\tau = L_p^2/D_p$, by $I = eD_pA\,\delta p(0)/L_p$, in agreement with Eq. 14-43.

## 14.3  *p-n* JUNCTIONS

A *p-n* junction consists of an *n*-type and a *p*-type semiconductor joined together as diagrammed in Fig. 14-6. In actual practice, the host material is usually the same throughout and a procedure known as counter doping is used. The entire

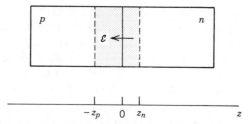

**FIGURE 14-6** An unbiased $p$-$n$ junction in thermal equilibrium. The metallurgical junction is at $z = 0$ and the transition region extends from $z = -z_p$ to $z = +z_n$. An electric field exists in the region, created chiefly by positive donor ions on the $n$ side and negative acceptor ions on the $p$ side.

sample may start as $n$ type, for example, then enough acceptors are added to a selected region to form $p$-type material.

Doping may be carried out in several ways. For example, impurity atoms may be diffused into the sample at elevated temperature from an atmosphere containing them. To locate a junction the sample surface is covered by a metal mask with holes at points where impurities are to go. Hundreds of junctions may be placed in a region with dimensions less than a millimeter. Alloyed junctions are fabricated by placing a small amount of a metal on the surface. When heated, metal atoms diffuse into the semiconductor and replace host atoms. Ion implantation is also used, especially when the diffusion rate for the impurity is small. Impurity ions are fired at high speeds into the sample.

These methods all result in impurity concentration gradients and produce junctions that are difficult to analyze mathematically. For simplicity, we consider an abrupt junction, in which impurity concentrations are uniform on each side.

*The Contact Potential.* For purposes of discussion, suppose the junction is made by placing originally separate $n$- and $p$-type semiconductors in intimate contact. When contact is first made electrons diffuse from the $n$ to the $p$ side, where they recombine with holes. Similarly, holes diffuse from the $p$ to the $n$ side and recombine with electrons. Electrons leave behind positively charged donors on the $n$ side, holes leave behind negatively charged acceptors on the $p$ side, and the uncovered impurities create an electric field directed from the $n$ toward the $p$ side. A potential difference exists across the junction, with the $n$ side more positive than the $p$ side. The field is called the contact field and the potential difference is called the contact potential.

The force of the contact field on electrons is directed toward the $n$ side while its force on holes is directed toward the $p$ side. It stops the diffusion of carriers: when equilibrium is reached both hole and electron currents vanish. Carrier concentrations are depleted in a narrow region, called the transition or depletion region, which straddles the metallurgical junction. Outside the transition region, the electric field vanishes and carrier concentrations are the same as in the separated semiconductors. Since the electric field in the transition region is precisely the field that causes the current to vanish, the contact potential cannot be measured directly with a voltmeter.

The condition that the current vanish can be used to find an expression for the contact potential. Suppose the sample lies along the $z$ axis, as in Fig. 14-6, with the junction at $z = 0$ and the transition region extending from $z = -z_p$ on the $p$ side to $z = +z_n$ on the $n$ side. Consider first the hole current density, given by Eq. 14-32. Set $\mathbf{J}_p$ equal to zero, replace $\mathcal{E}$ with $-dV/dz$, and use the Einstein relationship to replace $D_p$ with $(k_B T/e)\mu_p$. Then

$$\frac{e}{k_B T}\frac{dV}{dz} = -\frac{1}{p}\frac{dp}{dz}. \tag{14-45}$$

Integrate from $z = -z_p$ to $z = +z_n$. Let $V_p$ be the potential and $p_p$ the hole concentration at the left boundary of the transition region and let $V_n$ be the potential and $p_n$ the hole concentration at the right boundary. Then $(e/k_B T)(V_n - V_p) = \ln(p_n/p_p)$. $V_n - V_p$ is the contact potential $V_0$, so

$$V_0 = \frac{k_B T}{e} \ln \frac{p_p}{p_n}. \tag{14-46}$$

Since $p_p > p_n$, $V_0$ is positive.

An important special case occurs if the host material is the same on the two sides, both sides are heavily doped, and the junction is in the extrinsic temperature region. Then $p_p = N_a$ and $p_n = n_i^2/N_d$, where $N_a$ is the concentration of acceptors on the $p$ side, $N_d$ is the concentration of donors on the $n$ side, and $n_i$ is the intrinsic electron concentration. The contact potential is given by

$$V_0 = \frac{k_B T}{e} \ln \frac{N_a N_d}{n_i^2}. \tag{14-47}$$

Equation 14-47 emphasizes the role played by dopants in determining $V_0$.

A similar derivation carried out using the electron current yields

$$V_0 = \frac{k_B T}{e} \ln \frac{n_n}{n_p}, \tag{14-48}$$

where $n_n$ is the electron concentration to the right of the transition region and $n_p$ is the electron concentration to the left. Because $n_n p_n$ and $n_p p_p$ both equal $n_i^2$, Eqs. 14-46 and 14-48 produce identical values for $V_0$.

The chemical potential has the same value on the two sides. Suppose it is $\eta_n$ on the $n$ side and $\eta_p$ on the $p$ side. If we measure energies from the top of the valence band on the $p$ side, then

$$p_p = \frac{1}{\tau_s}\sum_{\text{states}} e^{\beta(E-\eta_p)} \tag{14-49}$$

and

$$p_n = \frac{1}{\tau_s}\sum_{\text{states}} e^{\beta(E-eV_0-\eta_n)}, \tag{14-50}$$

SO

$$\frac{p_p}{p_n} = e^{\beta e V_0} e^{\beta(\eta_n - \eta_p)}. \tag{14-51}$$

Comparison with Eq. 14-46 reveals that $\eta_n = \eta_p$. The proof depends on $\eta$ being far from the band edges. Even if it is not, however, it must be the same on the two sides since the sample is in thermal equilibrium and hence has a single chemical potential.

Fig. 14-7a shows energy levels for the n- and p-type materials before the junction is formed. On the n side the chemical potential is near the bottom of the conduction band, while on the p side it is near the top of the valence band. Figure 14-7b shows the levels after the junction is formed. Levels on the n side are lower than corresponding levels on the p side by $eV_0$. Since the chemical potential is the same on the two sides, $eV_0$ is the difference in the chemical potentials of the two separated materials.

*The Transition Region.* The contact field is similar to the field in a parallel plate capacitor, although the charge that produces it is distributed throughout the transition region rather than just on its surfaces. Except for fringing, the contact field vanishes beyond the transition region.

Maxwell's equation for the electric field is $d\mathcal{E}/dz = \rho/\epsilon_0$, where $\rho$ is the charge density. In the transition region on the n side $\rho$ is nearly $en_d$ since carrier

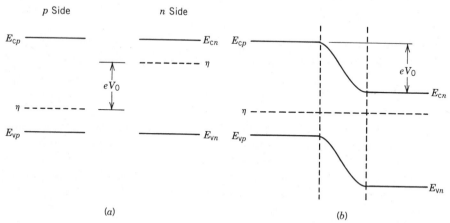

(a)                                                        (b)

**FIGURE 14-7** Energy levels in an unbiased p-n junction. $E_{cp}$ and $E_{cn}$ are the conduction band minima for the p and n materials, respectively, while $E_{vp}$ and $E_{vn}$ are the valence band maxima. (a) Separated materials. Doping causes the chemical potential to be near the valence band in the p material and near the conduction band in the n material. (b) The materials are joined to form a junction. The vertical dotted lines represent boundaries of the transition region. Levels on the n side are shifted down relative to levels on the p side by $eV_0$, where $V_0$ is the contact potential, and the chemical potential is the same on the two sides. The difference in the chemical potentials of the separated materials becomes $eV_0$ when the junction is formed.

concentrations are quite small. Thus

$$\mathcal{E} = \frac{en_d}{\epsilon_0}(z - z_n) \tag{14-52}$$

there. The constant of integration was chosen so $\mathcal{E} = 0$ at $z = z_n$. In the transition region on the $p$ side $\rho$ is nearly $-en_a$ and

$$\mathcal{E} = -\frac{en_a}{\epsilon_0}(z + z_p), \tag{14-53}$$

where the integration constant was chosen so $\mathcal{E} = 0$ at $z = -z_p$. Since the field is continuous at the junction, Eqs. 14-52 and 14-53 must give the same result for $z = 0$ and

$$n_d z_n = n_a z_p. \tag{14-54}$$

According to Eq. 14-54, boundaries of the transition region are positioned so the region contains as much positive charge on the $n$ side as negative charge on the $p$ side. The transition region extends farther into the side with lighter doping than into the side with heavier doping.

Values of $z_n$ and $z_p$ can be obtained in terms of the potential difference across the junction. On the $p$ side $-dV/dz = -en_a(z + z_p)$. Integrate from $z = -z_p$ to $z = 0$ to find

$$V_p = V_j - \frac{en_a}{2\epsilon_0} z_p^2, \tag{14-55}$$

where $V_j$ is the potential at the junction. To obtain the potential on the $n$ side, integrate $-dV/dz = en_d(z - z_n)$ from $z = 0$ to $z = z_n$ to find

$$V_n = V_j + \frac{en_d}{2\epsilon_0} z_n^2. \tag{14-56}$$

The potential difference $V = V_n - V_p$ is

$$V = \frac{e}{2\epsilon_0}(n_d z_n^2 + n_a z_p^2). \tag{14-57}$$

Equations 14-54 and 14-57 are solved simultaneously to obtain

$$z_n = \left[ \frac{2V\epsilon_0}{e} \frac{n_a}{n_d} \frac{1}{n_d + n_a} \right]^{1/2} \tag{14-58}$$

and

$$z_p = \left[ \frac{2V\epsilon_0}{e} \frac{n_d}{n_a} \frac{1}{n_d + n_a} \right]^{1/2}. \tag{14-59}$$

The total width $W$ of the transition region is given by

$$W = z_n + z_p = \left[ \frac{2V\epsilon_0}{e} \frac{n_d + n_a}{n_d n_a} \right]^{1/2}. \tag{14-60}$$

For a junction in thermal equilibrium, $V$ is the contact potential. However, Eqs. 14-58, 14-59, and 14-60 are valid even when an external potential difference is applied. As we will see, the dependence of the transition region width on the potential difference is important for the operation of some semiconductor devices. The following example indicates the magnitudes of some quantities for a typical $p$-$n$ junction.

**EXAMPLE 14-4** A $p$-$n$ junction is formed in a silicon sample at $T = 300$ K. One side has $7.80 \times 10^{19}$ donors/m³ while the other has $4.40 \times 10^{20}$ acceptors/m³. Suppose each donor is singly ionized and each acceptor has captured one electron. Find the contact potential, the extent of the transition region on each side, the total width of the transition region, and the electric field at the junction. The intrinsic electron concentration is $1.50 \times 10^{16}$ electrons/m³.

**SOLUTION** The equilibrium hole concentrations are $p_p = n_a = 4.40 \times 10^{20}$ holes/m³ on the $p$ side and $p_n = n_i^2/n_d = (1.50 \times 10^{16})^2/7.80 \times 10^{19} = 2.88 \times 10^{12}$ holes/m³ on the $n$ side. The contact potential is

$$V_0 = \frac{k_B T}{e} \ln \frac{p_p}{p_n} = \frac{1.38 \times 10^{-23} \times 300}{1.60 \times 10^{-19}} \ln \frac{4.40 \times 10^{20}}{2.88 \times 10^{12}} = 0.488 \text{ V} .$$

According to Eq. 14-58,

$$z_n = \left[ \frac{2V_0 \epsilon_0}{e} \frac{n_a}{n_d} \frac{1}{n_d + n_a} \right]^{1/2}$$

$$= \left[ \frac{2 \times 0.488 \times 8.85 \times 10^{-12}}{1.60 \times 10^{-19}} \frac{4.40 \times 10^{20}}{7.80 \times 10^{19}} \frac{1}{7.80 \times 10^{19} + 4.40 \times 10^{20}} \right]^{1/2}$$

$$= 7.67 \times 10^{-7} \text{ m}$$

and, according to Eq. 14-54,

$$z_p = \frac{n_d}{n_a} z_n = \frac{7.80 \times 10^{19}}{4.40 \times 10^{20}} \times 7.67 \times 10^{-7} = 1.36 \times 10^{-7} \text{ m} .$$

The field at the junction is given by Eq. 14-52, with $z_n = 0$:

$$\mathcal{E} = -\frac{en_d}{\epsilon_0} z_n = -\frac{1.60 \times 10^{-19} \times 7.80 \times 10^{19}}{8.85 \times 10^{-12}} \times 7.67 \times 10^{-7}$$

$$= -1.08 \times 10^6 \text{ V/m} .$$

By laboratory standards, a large field exists in the transition region. ∎

*A Biased Junction.* A junction is biased by applying a potential difference, as shown in Figs. 14-8a and b. If the positive terminal of the source is applied to the $p$ side, the junction is said to be forward biased. It is reverse biased if the negative terminal is applied to the $p$ side. The current $I$ through a biased junction

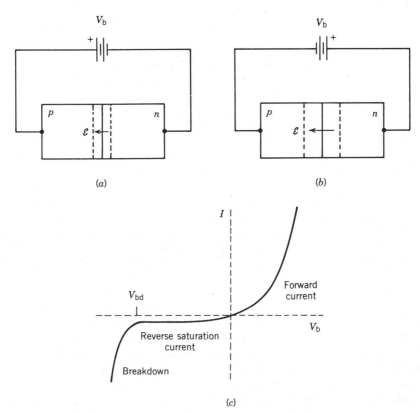

**FIGURE 14-8** (*a*) A forward and (*b*) a reverse biased *p-n* junction. In (*a*) the positive terminal of the battery is connected to the *p* side of the junction while in (*b*) it is connected to the *n* side. Forward bias weakens the electric field and narrows the transition region. Reverse bias strengthens the field and widens the transition region. (*c*) Current *I* through a *p-n* junction as a function of bias $V_b$. The current increases dramatically with bias in the forward direction while for reverse bias it is nearly independent of $V_b$ until the breakdown bias $V_{bd}$ is reached.

is shown in Fig. 14-8*c* as a function of the bias potential $V_b$. $V_b$ is taken to be positive for forward bias and negative for reverse bias; current from the *p* to the *n* side is positive while current in the opposite direction is negative.

The current vanishes if $V_b$ is zero and increases sharply as $V_b$ increases in the forward direction. On the other hand, a small reverse bias produces only an extremely small current. It is essentially constant over a broad range of bias potentials and is known as the reverse saturation current. When the reverse bias is increased beyond a certain value, called the breakdown bias, the current increases dramatically. We now investigate the three bias regimes.

*Forward Bias.* Figure 14-9 shows electron energy levels on the two sides of a forward biased junction. Since the applied potential opposes the contact potential, the total potential across the junction is smaller than $V_0$. The electric field in the transition region is weakened, so both electron and hole drift currents

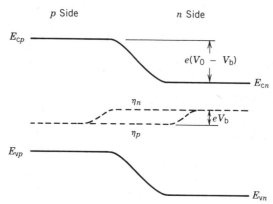

**FIGURE 14-9** Energy levels for a forward biased $p$-$n$ junction. Levels on the $n$ side are $e(V_0 - V_b)$ lower than corresponding levels on the $p$ side. The quasi chemical potentials, $\eta_p$ for holes and $\eta_n$ for electrons, are shown as dotted lines. They are characteristic of an excess hole concentration on the $n$ side and an excess electron concentration on the $p$ side.

decrease. More importantly, levels on the $n$ side are raised by $eV_b$ relative to the levels shown in Fig. 14-7b. The bottom of the conduction band on the $n$ side, for example, is now $e(V_0 - V_b)$ below the bottom of the conduction band on the $p$ side. As a result, the energy barrier between the two sides is lowered and both the number of electrons injected from the $n$ to the $p$ side and the number of holes injected from the $p$ to the $n$ side increase. Consequently, both electron and hole diffusion currents increase. Drift and diffusion contributions to the current do not cancel and a net current flows from the $p$ to the $n$ side, against the electric field.

Since the resistance of the transition region is much greater than that of the end regions, most of the applied potential difference $V_b$ appears across the transition region and the electric field is extremely small in the end regions. The diagram does not show its influence. The levels actually slope upward slightly from left to right.

Although the current is the same through every cross section of the sample, its nature is different at different places. Figure 14-10 shows the various contributions. Far from the transition region on the $p$ side the current is due almost entirely to hole drift. The electron concentration is small and uniform, so the electron current is small. The hole concentration is large and uniform, so the hole drift component is large while the hole diffusion component is small. Similarly, far from the transition region on the $n$ side the current is due almost entirely to electron drift.

Near the transition region on the $p$ side electrons injected from the $n$ side diffuse to the left and recombine. Similarly, on the $n$ side holes injected from the $p$ side diffuse to the right and recombine. On both sides the carrier concentrations vary with position and both electron and hole currents have significant diffusion components.

A simple model can be used to estimate the total current as a function of bias

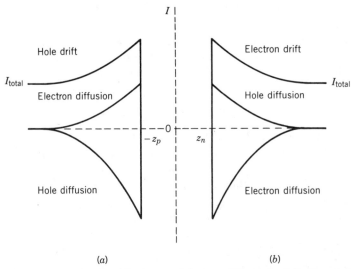

$(a)$ $(b)$

**FIGURE 14-10** (a) Hole drift, hole diffusion, and electron diffusion contributions to the current on the $p$ side of a $p - n$ junction. (b) The same contributions to the current on the $n$ side.

potential. First consider hole diffusion outside the transition on the $n$ side. With a forward bias applied, the sample is not in thermal equilibrium and a quasi chemical potential $\eta_p(z)$ must be used to describe the hole concentration. If $E_{vp}$ is the energy at the top of the valence band on the $p$ side then the hole concentration at $-z_p$ is

$$p(-z_p) = N_v e^{\beta[E_{vp} - \eta_p(-z_p)]} , \tag{14-61}$$

where $N_v$ is the effective density of states per unit volume for the valence band. The energy at the top of the valence band on the $n$ side is $E_{vp} - e(V_0 - V_b)$, so the hole concentration at $z = z_n$ is given by

$$p(z_n) = N_v e^{\beta[E_{vp} - eV_0 + eV_b - \eta_p(z_n)]} . \tag{14-62}$$

The ratio is

$$\frac{p(z_n)}{p(-z_p)} = e^{-\beta e(V_0 - V_b)} e^{\beta[\eta_p(-z_p) - \eta_p(z_n)]} . \tag{14-63}$$

We must now find a value for $\eta_p(-z_p) - \eta_p(z_n)$. Accurate values for $p(z)$, and hence for $\eta_p(z)$, can be obtained by solving the diffusion-drift differential equation simultaneously with Poisson's equation. Closed form solutions do not exist but numerical solutions can be obtained. Rather than use them, we make the simplifying assumption that the quasi chemical potential is uniform across the transition region, as drawn in Fig. 14-9. Then Eq. 14-63 becomes

$$\frac{p(z_n)}{p(-z_p)} = e^{-\beta e(V_0 - V_b)} . \tag{14-64}$$

Equation 14-64 can be used to find the excess hole concentration at $z_n$ in terms of the equilibrium concentration $p_n$ on the $n$ side. Only an extremely small fraction of the holes on the $p$ side recombine with electrons, so we may approximate $p(-z_p)$ by $p_p$ in Eq. 14-64 and write $p(z_n) = p_p e^{-\beta e(V_0 - V_b)}$. According to Eq. 14-46, $p_p = p_n e^{\beta e V_0}$, so $p(z_n) = p_n e^{\beta e V_b}$ and the excess hole concentration at $z_n$ is

$$\delta p(z_n) = p(z_n) - p_n = p_n[e^{\beta e V_b} - 1]. \tag{14-65}$$

After crossing the transition region, holes diffuse into the $n$ side. The expression for hole diffusion current density obtained in Example 14-3 can be used once it is modified slightly. The origin must be changed from $z = 0$ to $z = z_n$ and $\delta p(0)$ must be replaced by $\delta p(z_n)$. The result is

$$J_p = \frac{eD_p}{L_p} p_n[e^{\beta e V_b} - 1]e^{-(z - z_n)/L_p} \tag{14-66}$$

for $z > z_n$. On the $n$ side of the junction the hole concentration is extremely small and, as a result, the drift component of the hole current can be neglected in comparison to the diffusion component. For practical purposes, Eq. 14-66 gives the total hole current density on the $n$ side.

A similar calculation can be carried out for electrons. To the left of the transition region $(z < -z_p)$ the electron diffusion current is given by

$$J_n = \frac{eD_n}{L_n} n_p[e^{\beta e V_b} - 1]e^{(z + z_p)/L_n}, \tag{14-67}$$

where the solution to the diffusion-drift equation that meets the boundary condition $\delta n \to 0$ as $z \to -\infty$ was used. On the $p$ side the electron drift current is extremely small, so Eq. 14-67 gives the total electron current density there.

We are now in a position to find the total current in the junction. Pair generation and recombination events are extremely rare in the transition region, so nearly all holes that enter the region at the left leave it at the right. Thus the total current at the left edge of the transition region is given very closely by the sum of the electron current at the left edge and the hole current at the right edge. Set $z = z_n$ in Eq. 14-66 and $z = -z_p$ in Eq. 14-67, then sum the two equations to obtain the total current density at the left edge of the transition region:

$$J = e\left[\frac{D_n n_p}{L_n} + \frac{D_p p_n}{L_p}\right]\left[e^{eV_b/k_B T} - 1\right]. \tag{14-68}$$

If the current density is uniform, the current is given by the product of $J$ and the cross section $A$ of the junction: $I = JA$. Although we have found the current only at the left edge of the transition region it is the same for all cross sections.

Forward current density in an actual junction is given roughly by Eq. 14-68 provided the bias is low. At high forward bias, however, the current density in a typical junction becomes much larger than predicted by the equation. In fact, potential differences greater than a critical value, called the forward bias limit, cannot be maintained across the junction.

The limiting value of the forward bias is nearly the contact potential. When $V_b > V_0$ the conduction band on the $n$ side is above the conduction band on the $p$ side and electrons are dumped from the $n$ to the $p$ side. Similarly, holes are dumped from the $p$ to the $n$ side. In addition, the electric field in the transition region is reversed so it points from the $p$ toward the $n$ side, thereby facilitating carrier flow. The junction then has an extremely small electrical resistance.

*Reverse Bias and Breakdown.* The same analysis can be carried out for a reverse biased junction. Now the electric field in the transition region is greater than the equilibrium field and energy levels on the $n$ side are lower than analogous levels on the $p$ side by $e(V_0 + |V_b|)$. The drift current in the transition region is greater than the diffusion current. The mathematics is the same for forward and reverse bias and Eq. 14-68 is again obtained. For reverse bias, however, $V_b$ is negative.

If $e|V_b| \gg k_B T$, then Eq. 14-68 becomes

$$ J = -e \left[ \frac{D_n n_p}{L_n} + \frac{D_p p_n}{L_p} \right], \tag{14-69} $$

indicating a current density that is independent of bias. $JA$ is the reverse saturation current.

In a reverse biased junction holes flow from the $n$ to the $p$ side while electrons flow from the $p$ to the $n$ side, both in the direction of the electric force on them. The reverse current is limited by the number of available carriers. When $V_b$ is negative and much larger than $k_B T$, Eq. 14-65 predicts $\delta p(z_n) = -p_n$, so $p(z_n) = 0$. Every hole near the transition region on the $n$ side diffuses to the transition region where it is swept to the $p$ side. An analogous situation occurs for electrons.

But the current does increase if the reverse bias is large enough. One of two possible mechanisms operate, called Zener and avalanche breakdown, respectively.

Zener breakdown occurs when the bias potential lowers the conduction band minimum on the $n$ side to below the valence band maximum on the $p$ side, as illustrated in Fig. 14-11. Since electrons from the $p$ side valence band can then cross to the $n$ side conduction band, the number of electrons available for the current increases enormously. Zener breakdown occurs when $e(V_0 + |V_b|) > E_g$, where $E_g$ is the band gap.

In avalanche breakdown additional carriers are generated by collisions with atoms in the transition region. The electric field accelerates carriers as they cross the region and if the energy of a carrier becomes great enough it may liberate another carrier in a collision. Both old and new carriers now accelerate and may generate additional carriers. The number of collisions suffered by a single carrier is roughly equal to the ratio of the transition region width to the mean free path. As this may be several thousand or more, the number of collision

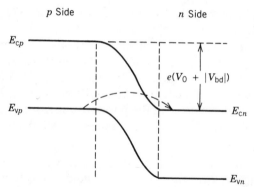

$p$ Side    $n$ Side

**FIGURE 14-11** Energy levels in a $p$-$n$ junction at the onset of Zener breakdown. Reverse bias lowers the bottom of the conduction band on the $n$ side to the top of the valence band on the $p$ side. Electrons are dumped from the $p$ to the $n$ side.

generated carriers is quite large. For most junctions avalanche breakdown occurs at a smaller bias than Zener breakdown.

As a final remark we mention that the term breakdown does not imply an irreversible change in the junction. Once a reverse bias is reduced to below the breakdown bias the current becomes the reverse saturation current.

## 14.4 SEMICONDUCTOR DEVICES

*Junction Diodes.* A semiconductor junction used as a circuit element is called a diode and is represented by the symbol →|. A diode has low resistance for current in the direction of the arrow and high resistance for current in the opposite direction. An ideal diode has zero resistance for forward bias and infinite resistance for reverse bias. Diodes have a variety of uses and, for each use, a specific set of design considerations. We start with what is perhaps the earliest use, rectification of a time-dependent signal.

If an ac signal is applied as shown in Fig. 14-12a, the potential difference across the resistor is as shown in b. When the upper terminal of the source is positive the diode conducts and the potential difference across the resistor mimics the input signal. On the other hand, when the upper terminal is negative the current is essentially zero and the potential difference across the resistor vanishes. The time average of the input signal vanishes but the time average of the output signal does not: the potential difference across the resistor has both dc and ac components. The ac component can be effectively removed by placing a capacitor in parallel with the resistor.

A diode with a small reverse saturation current, a large reverse breakdown bias, and a small forward bias limit approximates an ideal diode. That is, it has large resistance for reverse bias and nearly zero resistance for forward bias. These characteristics cannot be obtained simultaneously and compromises must be made in the design. For example, the reverse saturation current can be made small by heavily doping the two sides of the junction, thus making $n_p$ and $p_n$

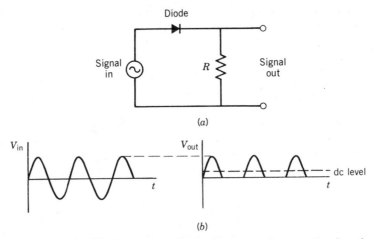

(a)

(b)

**FIGURE 14-12** (a) A simple diode rectifying circuit. (b) Input and output signals as functions of time for an ideal diode. The signal is passed only when the upper terminal of the generator is positive. The output signal has a dc component, the time average of the signal shown.

small. But heavy doping also gives rise to a large contact potential and hence to a large forward bias limit.

For rectifying diodes a small reverse saturation current is usually more important than a small forward bias limit so heavily doped wide gap materials are used. Silicon is generally preferred over germanium, for example.

Reverse breakdown is exploited in a special class of devices known as Zener diodes. Consider the circuit shown in Fig. 14-12a and suppose the input signal consists of a 12-V dc component, applied as a reverse bias, and a 1-V peak-to-peak ac ripple, as shown in Fig. 14-13. The ripple might be unwanted noise, for example. If the reverse breakdown bias of the diode is less than 11 V, the potential difference across the resistor remains at the breakdown value, a constant.

Zener diodes are used to stabilize potentials in circuits and as voltage regu-

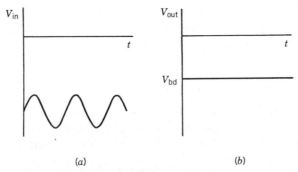

(a)                                              (b)

**FIGURE 14-13** The operation of a Zener diode. The input signal is a reverse bias that remains greater in magnitude than the breakdown bias $V_{bd}$. The output is nearly a constant potential difference, equal to $V_{bd}$.

lators. They are often used to provide fixed reference voltages. The electric field in the transition region, the width of the transition region, and the carrier mean free path are important design parameters.

*Photodiodes.* Photodiodes are used as electromagnetic radiation detectors and as solar cells. Before we discuss them, we investigate steady injection into a uniformly illuminated homogeneous sample.

In Section 14.2 we studied the steady injection of holes into a long sample along the positive $z$ axis with one end at $z = 0$. Now suppose the sample is uniformly illuminated by light with an optical generation rate $g_{op}$. Equation 14-39 becomes

$$D_p \frac{d^2 \, \delta p}{dz^2} = \frac{\delta p}{\tau} - g_{op} \, . \tag{14-70}$$

For steady state injection at $z = 0$ the appropriate solution is

$$\delta p(z) = [\delta p(0) - \tau g_{op}]e^{-z/L_p} + \tau g_{op} \tag{14-71}$$

and the hole current density is

$$J_p(z) = -eD_p \frac{d \, \delta p}{dz} = \frac{eD_p}{L_p} [\delta p(0) - \tau g_{op}]e^{-z/L_p} \, . \tag{14-72}$$

Equation 14-72 can be applied to the $p$-$n$ junction of Fig. 14-8. To find the hole current density at the right side of the transition region, replace $\delta p(0)$ by $p_n(e^{\beta eV_b} - 1)$ and $z$ by 0 in Eq. 14-72. The result is

$$J_p = \frac{eL_p}{\tau_p} [e^{\beta eV_b} - 1] - eL_p g_{op} \tag{14-73}$$

once $D_p = L_p^2 \tau_p$ is used. Here $\tau_p$ is the recombination lifetime for holes on the $n$ side of the junction. Similarly, the electron current density at the left side of the transition region is

$$J_n = \frac{eL_n}{\tau_n} [e^{\beta eV_b} - 1] - eL_n g_{op} \, . \tag{14-74}$$

If generation and recombination in the transition region can be neglected the total current density in the junction is

$$J = \left[ \frac{eL_n}{\tau_n} n_p + \frac{eL_p}{\tau_p} p_n \right] [e^{\beta eV_b} - 1] - e(L_n + L_p)g_{op} \, . \tag{14-75}$$

According to Eq. 14-75, the curve of Fig. 14-8c is displaced downward by $eA(L_n + L_p)g_{op}$, where $A$ is the cross section of the junction. Figure 14-14 shows the result.

At zero bias, with the junction short-circuited, the current is

$$I_{sc} = -eA(L_n + L_p)g_{op} \, , \tag{14-76}$$

where the minus sign indicates a reverse current, from the $n$ to the $p$ side. $AL_p g_{op}$ gives the number of holes created by the light per unit time within a diffusion length of the right edge of the transition region. Excess holes diffuse to the

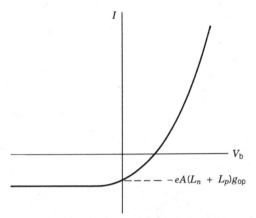

**FIGURE 14-14** Current $I$ through an illuminated diode as a function of bias $V_b$. The curve of Fig. 14-8c is displaced downward by $eA(L_n + L_p)g_{op}$, where $A$ is the junction cross section, $L_n$ is the diffusion length for electrons, $L_p$ is the diffusion length for holes, and $g_{op}$ is the optical generation rate.

transition region where they are swept by the electric field to the other side. Similarly, electrons created within a diffusion length of the left edge are swept across the transition region and contribute to the current. Carriers created inside the transition region have been ignored, although they are important for some junctions.

If the current vanishes, as it does when the circuit is not completed, a potential difference $V_b$ appears across the junction. Its value can be found by solving Eq. 14-75 after setting $J = 0$. $V_b$ is the sum of an external bias, if one is applied, and a potential created in addition to $V_0$ by charges in the transition region. The appearance of a potential difference across an illuminated unbiased junction is called the photovoltaic effect.

To operate the junction as a solar cell, the circuit of Fig. 14-15a is used. The junction acts like a battery, with the $p$ side at a higher potential than the $n$ side.

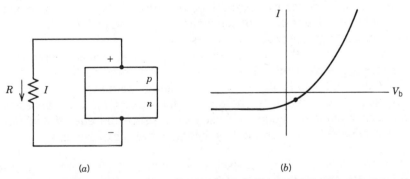

(a)                                              (b)

**FIGURE 14-15** A photodiode operating as a solar cell. The circuit is shown in (a) and the current as a function of bias is shown in (b). The dot marks a typical operating point. $V_b$ is produced internally.

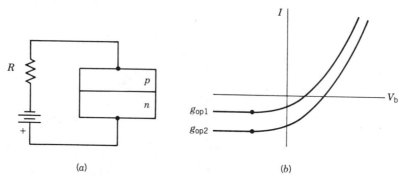

**FIGURE 14-16** A photodiode operating as a light detector. (a) The circuit and (b) the current as a function of bias for two light intensities, with $g_{op2} > g_{op1}$. Dots mark typical operating points. The current through the resistor is proportional to the optical generation rate. Dots mark typical operating points.

Inside, current is from the $n$ to the $p$ side, so power is supplied by the junction to the external circuit. Solar cells are designed to maximize the power delivered, given by $IV_b$.

They are also designed to intercept as much light as possible. Usually they are flat plates several centimeters on a side with junctions near the surface. A thin layer of $n$-type material, say, is placed on a $p$-type substrate. The thickness of the surface layer must be less than a diffusion length so carriers created in the layer diffuse to the transition region before recombining. Normally a large number of such junctions are formed into an array, with an area of perhaps several square meters or more.

Reverse biased photodiodes are used to measure light intensity. Current in the circuit of Fig. 14-16a is nearly independent of bias over a wide range but it is proportional to the optical generation rate, as indicated in (b). Since current in the junction is from the $n$ side to the $p$ side, power is delivered from the external battery to the junction.

Often the response time of a detector is an important design consideration. Some applications require the current to follow microsecond pulses of light, for example, and the time for excess carriers to diffuse to the transition region is unacceptably long. Light detector junctions are designed with wide transition regions, so most light-absorbing processes occur there and excess carriers are rapidly swept across by the electric field. Such devices are called depletion layer photodiodes. Since absorption is weak for light with an angular frequency less than $E_g/\hbar$, the band gap is also an important design parameter.

Energetic particles other than photons also induce electron transitions across the gap. Semiconductor diodes are used to measure the energy and flux of particles produced in nuclear and high energy physics experiments.

*Light-Emitting Diodes (LEDs).* Doped direct semiconductors are used to fabricate LEDs. Recombination takes place via transitions across the gap and gap light is emitted. Such diodes are in common use as displays on meters, calculators, and other instruments. They also used in semiconductor lasers.

Since the angular frequency of the emitted light is $E_g/\hbar$, the band gap is clearly an important design parameter. It can be controlled by varying the composition of the material. For example, solid mixtures of phosphorous in gallium arsenide have gaps that increase with phosphorous content. The composite, represented by $GaAs_{1-x}P_x$, is direct for $x$ between 0 and about 0.44, then becomes indirect. Gap light for pure GaAs is in the infrared but the wavelength decreases as $x$ increases and for $x \approx 0.40$ the familiar red light of many displays is produced. Radiative transitions also occur for $x > 0.44$ if the material is doped with nitrogen. Transitions from nitrogen levels are accompanied by light emission in yellow-green portion of the spectrum.

*Bipolar Junction Transistors.* Current through a reverse biased *p-n* junction is limited by the rate at which excess carriers are generated within a diffusion length of the transition region. In the last section we saw that the generation rate could be increased by illuminating the junction. Excess carrier concentrations can also be increased by injecting carriers from a forward biased junction.

A *p-n-p* transistor, diagrammed in Fig. 14-17*a*, consists of a forward biased junction on the left and a reverse biased junction on the right. The left region is called the emitter, the central region the base, and the right region the collector. We suppose the emitter region is much more heavily doped than either the base or collector regions, so holes are the primary carriers. They are injected across the emitter junction into the base region, then form part of the reverse current through the collector junction. The flow of electrons and holes is indicated in (*b*).

The width of the base region is usually much less than a hole diffusion length. To increase the diffusion length the region is fabricated of indirect material with a low concentration of recombination centers. Then nearly all holes that enter the base region reach the collector.

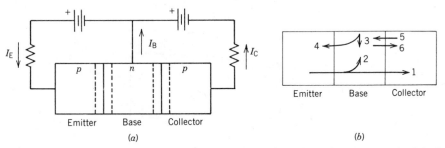

(a)                                         (b)

**FIGURE 14-17** (a) A *p-n-p* bipolar junction transistor. The emitter junction, on the left, is forward biased and injects holes into the base region. The collector junction, on the right, is reverse biased. The current through it depends on the injection rate across the emitter junction and on the rate of recombination in the base region. (b) Schematic representation of particle flow in a *p-n-p* junction transistor. Holes (1) are injected across the emitter junction and become part of the reverse current across the collector junction. Some (2) recombine with electrons (3) in the base region. Electrons (4) cross the emitter junction and recombine. Carriers (5 and 6) are thermally generated within a diffusion length of the collector junction and cross that junction.

For most applications, electron flow into the base region is used to control the hole current from emitter to collector. Some electrons recombine with holes in the base region while others cross the emitter junction to recombine in the emitter region. Electrons are resupplied by the base current $I_B$.

A change in base current changes the number of electrons available for recombination and leads to a change in the collector current. Since a small change in the base current leads to a large change in the collector current, bipolar transistors are widely used as amplifiers. The base current is the input signal while the collector current is the output signal.

*Field Effect Transistors (FETs).*    In an FET the transition region width controls the current and is, in turn, determined by a bias potential. The idea is illustrated in Fig. 14-18a, which shows an *n*-type sample with a *p*-type region imbedded

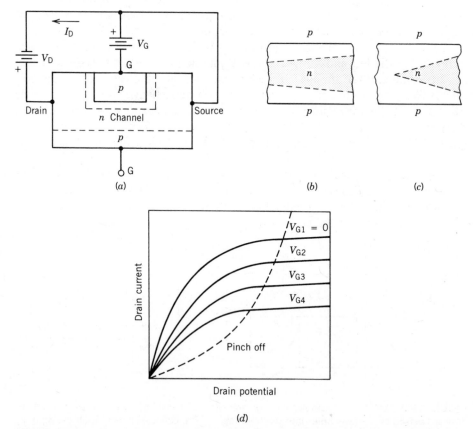

(a)                    (b)                    (c)

(d)

**FIGURE 14-18** (a) An *n*-channel FET. The *p* material is heavily doped so the transition region is wide. The width and resistance of the channel are controlled by the negative gate bias $V_G$. (b) The channel shape when the drain is at a higher potential than the source. (c) The channel at pinch off. (d) Characteristic curves for an *n*-channel FET. Drain current $I_D$ is plotted as a function of drain potential $V_D$ for various values of gate bias $V_G$. $V_G$ increases in magnitude from top to bottom.

into each side. The $p$ regions are much more heavily doped than the $n$ region so the transition regions extend far into the $n$ material. Current flows in the $n$ material, through the channel formed between the high-resistivity transition regions. For the situation illustrated, electron flow is from the right side, called the electron source, to left side, called the electron drain.

The $p$ regions that form the channel sides are called gates and the two gate terminals, marked G on the diagram, are electrically connected to each other and are biased negatively relative to the source. As the gate potential $V_G$ becomes more negative the transition regions widen and the resistance of the channel increases. Equation 14-60 can be used to calculate the channel width.

The transistor illustrated is an $n$-channel FET. FETs with $p$ channels operate in a similar manner but their gates, made of $n$-type material, are biased positively relative to the source.

To understand how an $n$-channel FET works, we examine the drain current $I_D$ as a function of drain potential $V_D$, as shown in Fig. 14-18$d$ for various values of gate bias. Consider first the curve for $V_G = 0$. Along the channel the potential $V$ increases from source to drain, so the bias is greater at the left than at the right and, as a consequence, the transition regions are wider at the left than at the right. Figure 14-18$b$ shows the geometry. As $V_D$ increases the channel narrows and its resistance increases, so the curve of Fig. 14-18$d$ bends. At some value of $V_D$ the transition regions come together, as illustrated in Fig. 14-18$c$, and a condition known as pinch off occurs. For $V_D$ above pinch off, the current is nearly independent of $V_D$.

A negative gate bias reduces the value of $V_D$ at which pinch off occurs and also reduces the value of the drain current above pinch off. This explains the family of curves shown in the plot. Since a small change in $V_G$ produces a large change in drain current, an FET can be used for amplification.

*MOSFETs.*    The geometry for one type metal-oxide-semiconductor field effect transistor (MOSFET) is shown in Fig. 14-19$a$. Two heavily doped $n$ regions are imbedded in a $p$-type substrate with an oxide layer covering its surface. Metal contacts for the drain and source penetrate the oxide layer but the gate-substrate separation is typically on the order of 0.1 $\mu$m. When the gate is unbiased a drain-source potential difference produces only negligible current. The two $p$-$n$ junctions are in series with one forward biased and the other reverse biased, so the combination offers high resistance. When the gate is biased positive relative to the substrate, a low-resistance channel forms between the source and drain, as illustrated in Fig. 14-19$b$. Its width varies because the electric potential varies along the channel. Current is determined by the gate potential as well as by the drain potential.

MOSFETs are widely used in digital electronic circuits. Normally the gate potential must exceed some minimum value before a channel is formed. The device can be put in its low-resistance state, labeled "on," by applying a gate bias above the minimum for channel formation and put in its high-resistance or "off" state by removing the bias. The state of conduction is determined by testing the current through the device for a given drain-source potential difference.

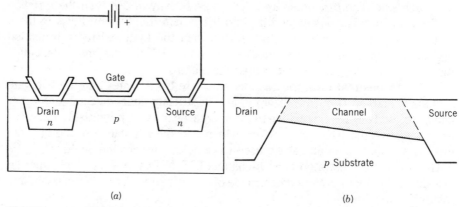

(a)            (b)

**FIGURE 14-19** (a) The geometry of a MOSFET. When the gate is biased positive, a highly conducting channel is formed in the p substrate between source and drain. Current can be switched on and off by regulating the gate bias. (b) The geometry of the channel.

A MOSFET that operates as described above is said to run in the enhancement mode. MOSFETs can also be designed so channels exist when the gate bias is zero. Such a MOSFET is normally "on" and a gate bias must be applied to turn it "off." These devices are said to run in the depletion mode.

Logic gates are constructed by appropriately connecting several MOSFETs. Current flows through an "AND" gate, for example, only if two MOSFETS are "on." It flows through an "OR" gate when either or both of two MOSFETs are "on" and it flows through a "XOR" (exclusive "OR") gate when either of two MOSFETs are "on," but not both. Gates such as these are widely used in instrument control and computer circuits.

## 14.5 REFERENCES

R. Dalven, *Introduction to Applied Solid State Physics* (New York: Plenum, 1980).

A. J. Diefenderfer, *Principles of Electronic Instrumentation* (Philadelphia: Saunders, 1972).

A. S. Grove, *Physics and Technology of Semiconductor Devices* (New York: Wiley, 1967).

J. P. McKelvey, *Solid-State and Semiconductor Physics* (New York: Harper & Row, 1966).

D. H. Navon, *Semiconductor Microdevices and Materials* (New York: Holt, Rinehart & Winston, 1986).

K. Seeger, *Semiconductor Physics* (Berlin: Springer-Verlag, 1973).

B. G. Streetman, *Solid State Electronic Devices* (Englewood Cliffs, NJ: Prentice-Hall, 1980).

S. M. Sze, *Physics of Semiconductor Devices* (New York: Wiley, 1981).

A. van der Ziel, *Solid State Physical Electronics* (Englewood Cliffs, NJ: Prentice-Hall, 1976).

For details of solid state devices presented at an introductory level see the *Modular Series on Solid State Devices* edited by R. F. Pierret and G. W. Neudeck and published by Addison-Wesley (Reading, MA). Titles in the series are

R. F. Pierret, *Field Effect Devices* (1983).
R. F. Pierret, *Semiconductor Fundamentals* (1983).
G. W. Neudeck, *The Bipolar Junction Transistor* (1983).
G. W. Neudeck, *The PN Junction Diode* (1983).

## PROBLEMS

1. A direct semiconductor is uniformly illuminated, then the light is turned off at time $t = 0$. The excess electron concentration is given by Eq. 14-4. (a) Show that the number of electrons per unit volume that recombine in an infinitesimal time interval $dt$ is $[\delta n(0)/\tau]e^{-t/\tau}dt$ and that the average time an electron spends in the conduction band is given by

$$\langle t \rangle = \frac{1}{\tau} \int_0^\infty te^{-t/\tau}dt.$$

   Carry out the integration to show that $\langle t \rangle = \tau$. (b) What fraction of the original excess electrons have recombined by time $t = \tau$?

2. Suppose the optical generation rate for a uniformly illuminated semiconductor is $g_{op}$. (a) Show that the steady state excess electron concentration is given by

$$\delta n = \tfrac{1}{2}[(n_0 + p_0)^2 + 4\tau(n_0 + p_0)g_{op}]^{1/2} - \tfrac{1}{2}(n_0 + p_0),$$

   where $n_0$ and $p_0$ are the equilibrium electron and hole concentrations, respectively, and $\tau$ is the recombination lifetime. Do not assume low-level excitation. (b) Show that $\delta n = \tau g_{op}$ if $\tau g_{op} \ll (n_0 + p_0)$. (c) Suppose $n_0 = 5.50 \times 10^{19}$ electrons/m³, $p_0 = 2.10 \times 10^{20}$ holes/m³, $\tau = 6.30 \times 10^{-7}$ s, and $g_{op} = 1.10 \times 10^{26}$ pairs/m³·s. Calculate the rate at which pairs recombine and the rate at which they are thermally generated in the steady state.

3. The recombination lifetime for carriers in a sample can be calculated from photoconductivity data. Suppose steady uniform current $I$ flows in a sample of length $\ell$ and cross section $A$. A light flash at time $t = 0$ uniformly illuminates the sample and, after the flash, the potential difference across the sample is $V_0 + \delta V(t)$, where $V_0$ is the potential difference when the sample has been dark for a long time. (a) Show that the deviation of the conductivity from its dark value is given by

$$\delta\sigma(t) = -\frac{\ell I}{A} \frac{\delta V}{V_0(V_0 + \delta V)}.$$

   (b) $\delta V(t)$ is read from an oscilloscope and used to calculate $\delta\sigma$ for various times after the flash. $V_0 = 0.535$ V and the following table gives $\delta V$ as a function of time:

| $t$ ($\mu$s) | $\delta V$ ($\mu V$) |
|---|---|
| 0 | $-101$ |
| 2 | $-97.0$ |
| 4 | $-93.4$ |
| 6 | $-90.0$ |
| 8 | $-86.6$ |
| 10 | $-83.4$ |
| 12 | $-80.4$ |
| 14 | $-77.4$ |
| 16 | $-74.5$ |
| 18 | $-71.8$ |
| 20 | $-69.1$ |

Calculate the recombination lifetime. Are measurements of $\ell$, $I$, and $A$ important?

4. Infrared radiation with a frequency of $3.50 \times 10^{14}$ Hz uniformly illuminates a direct semiconductor with a band gap of 1.20 eV and energy is absorbed at the rate of 1.50 mW. (a) What is the rate at which photons are absorbed? (b) What is the rate at which thermal energy is generated? (c) What is the frequency of the radiation emitted by recombining pairs? (d) What is the power emitted?

5. A recombination level is near the center of the band gap of an indirect semiconductor. Use a two-energy-level model: assume $N_v$ states per unit volume in the valence band, all with energy $E_v$, and $N_c$ states per unit volume in the conduction band, all with energy $E_c$. Also assume $E_c - E_v \gg k_B T$. (a) Show the recombination lifetime is given approximately by Eq. 14-14 if the chemical potential equals the conduction band energy and (b) is given approximately by Eq. 14-15 if the chemical potential equals the valence band energy. (c) Find an expression for the recombination lifetime if the chemical potential is at the impurity level, then specialize to the case for which the electron and hole transition rates $\alpha_n$ and $\alpha_p$ are equal.

6. (a) A direct semiconductor has a band gap of 1.10 eV. Use a two-level model with $5.30 \times 10^{25}$ states/m$^3$ in the conduction band and an identical number in the valence band. Take the recombination coefficient $\alpha$ to be $2.20 \times 10^{-11}$ pairs/m$^3 \cdot$s and compute the recombination lifetime for a temperature of 300 K and a chemical potential at the gap center. (b) An indirect semiconductor has the same band gap, the same number of valence band states, and the same number of conduction band states as the direct semiconductor. It has $4.7 \times 10^{15}$ recombination centers/m$^3$, each with an energy level 0.460 eV above the valence band. Take the electron and hole transition coefficients to be $\alpha_n = \alpha_p = 2.20 \times 10^{-11}$ pairs/m$^3 \cdot$s and compute the recombination lifetime for a temperature of 300 K and a chemical potential at the gap center.

7. Germanium at 300 K has an intrinsic electron concentration of $2.50 \times 10^{19}$ electrons/m$^3$. A sample is doped with $6.5 \times 10^{19}$ donors/m$^3$. (a) What are the electron and hole quasi chemical potentials, relative to the intrinsic chemical potential? (b) The sample is uniformly illuminated by light and the optical generation rate is $3.5 \times 10^{25}$ pairs/m$^3 \cdot$ s. Assume a recombination lifetime of $7.3 \times 10^{-7}$ s and calculate the steady state electron and hole quasi chemical potentials, relative to the intrinsic chemical potential.

8. Consider a homogeneous sample with $N$ electron states per unit volume, all with the same energy $E$. (a) Assume $E - \eta \gg k_B T$, where $\eta$ is the chemical potential. If an electric field is turned on in the positive $z$ direction, the electron concentration at $z$ is given by

$$n(z) = N e^{-\beta(E - eV - \eta)},$$

where $V$ is the electric potential at $z$. Suppose the circuit is not completed, so the drift and diffusion currents cancel in the steady state. Assume Eq. 14-31 is valid and use the vanishing of the current to prove the Einstein relation for the electron diffusion constant and mobility. (b) Assume $eV \ll k_B T$ and find the relation between the diffusion constant and mobility if Fermi-Dirac statistics must be used.

9. (a) At 300 K the relaxation time for electrons in a certain semiconductor is $3.1 \times 10^{-12}$ s. Assume the conduction band is parabolic and the effective mass is $1.1 m_0$, where $m_0$ is the free electron mass. Compute the electron mobility and diffusion constant. (b) Assume the mean free path is independent of temperature and that the mean thermal energy is proportional to $k_B T$. Estimate the electron mobility and diffusion constant at 150 K.

10. Consider the steady state injection of holes into one end of a long homogeneous sample, along the positive $z$ axis. $\delta N_p$ holes start at $z = 0$. Show that $(\delta N_p / L_p) e^{-z/L_p}\, dz$ holes recombine in the infinitesimal interval from $z$ to $z + dz$ and that $L_p$ is the average distance traveled by the holes before they recombine.

11. The sample of Problem 10 has a cross section of $2.2 \times 10^{-5}$ m$^2$ and the excess hole concentration at $z = 0$ is maintained at $1.8 \times 10^{12}$ holes/m$^3$. The diffusion constant for holes is $5.5 \times 10^{-3}$ m$^2$/s and their recombination lifetime is $2.4 \times 10^{-7}$ s. For the steady state find the number of holes that pass $z = L_p$ per unit time, the number that pass $z = 2L_p$ per unit time, and the number that recombine per unit time in the region from $z = L_p$ to $z = 2L_p$.

12. (a) Start with Eq. 14-29 and show that the hole diffusion current is given by $\mathbf{J}_p = p\mu_p \nabla \eta_p$. (b) For steady state injection of holes into the sample of Problem 10, show that the hole quasi chemical potential is given by

$$\eta_p = \eta_i - k_B T \ln \frac{p_0}{p_i} - k_B T \ln \left[ 1 + \frac{\delta p(0)}{p_0} e^{-z/L_p} \right],$$

where $\eta_i$ is the intrinsic chemical potential, $p_0$ is the equilibrium hole concentration, $p_i$ is the intrinsic hole concentration, and $L_p$ is the hole diffusion length. (c) Use the results of parts a and b to show $J = (eD_p/L_p)\,\delta p(0)e^{-z/L_p}$, in agreement with an expression derived in Example 14-3.

13. A symmetric $p$-$n$ junction, with a uniform cross section of $5.0 \times 10^{-7}$ m$^2$, has minority carrier concentrations $n_p = p_n = 7.1 \times 10^{13}$ carriers/m$^3$, with diffusion constants $D_n = D_p = 4.3 \times 10^{-3}$ m$^3$/s, and diffusion lengths $L_n = L_p = 2.30 \times 10^{-5}$ m. For a forward bias of 0.50 V find (a) the total current, (b) the hole current at the transition region boundary on the $n$ side, and (c) the electron current at the transition region boundary on the $p$ side.

14. A 0.50-V reverse bias is applied to the $p$-$n$ junction of Problem 13. Find (a) the total current, (b) the hole current at the transition region boundary on the $n$ side, and (c) the electron current at the transition region boundary on the $p$ side.

15. A $p$-$n$ junction has a contact potential of 0.40 V at 300 K and its transition region extends three times as far into the $p$ side as into the $n$ side. The intrinsic electron concentration is $2.5 \times 10^{19}$ electrons/m$^3$ on each side. Assume both sides are heavily doped but the chemical potential is in the gap, far from both the conduction and valence bands. Calculate the donor concentration on the $n$ side and the acceptor concentration on the $p$ side. How far does the transition region extend into each side? What is the magnitude of the electric field at the metallurgical junction?

16. A $p$-$n$ junction, with a cross section of $3.6 \times 10^{-7}$ m$^2$, is fabricated of silicon, with an intrinsic electron concentration of $1.50 \times 10^{16}$ electrons/m$^3$ at 300 K. The $n$ side has $2.40 \times 10^{17}$ donors/m$^3$, while the $p$ side has $3.40 \times 10^{16}$ acceptors/m$^3$. Assume each donor is singly ionized and each acceptor has captured one electron. What is: (a) the contact potential, (b) the width of the transition region, (c) the magnitude of the electric field at the metallurgical junction, and (d) the total impurity charge in the transition region on the $n$ side of the metallurgical junction?

17. For the junction of Problem 16 use Eq. 14-68 to calculate the current for (a) a forward bias of 0.050 V, (b) a forward bias of 0.20 V, (c) a reverse bias of 0.050 V, and (d) a reverse bias of 0.20 V. Assume an electron diffusion constant of $2.5 \times 10^{-3}$ m$^3$/s, a hole diffusion constant of $1.25 \times 10^{-3}$ m$^2$/s, and a recombination lifetime of $4.70 \times 10^{-7}$ s.

18. For the $p$-$n$ junction of Problem 16 use Eq. 14-65 to calculate the excess hole concentration at the transition region boundary on the $n$ side for (a) a forward bias of 0.050 V, (b) a forward bias of 0.20 V, (c) a reverse bias of 0.050 V, and (d) a reverse bias of 0.20 V. In each case compare the excess hole concentration to the equilibrium concentration $p_n$ on the $n$ side and to the equilibrium concentration $p_p$ on the $p$ side.

19. For a uniformly illuminated $p$-$n$ junction in an open circuit, show that the

potential difference $V_b$ across the junction is given by

$$V_b = \frac{k_B T}{e} \ln \left[ 1 + \frac{I_{sc}}{I_R} \right],$$

where $I_{sc}$ is the magnitude of the short-circuit current and $I_R$ is the magnitude of the reverse saturation current. (b) Estimate $V_b$ at 300 K for an optical generation rate of $6.20 \times 10^{24}$ pairs/m$^3 \cdot$s. The hole concentration on the $n$ side is $7.5 \times 10^{18}$ holes/m$^3$ and the hole diffusion length is $5.5 \times 10^{-5}$ m. The electron concentration on the $p$ side is $4.2 \times 10^{18}$ electrons/m$^3$ and the electron diffusion length is $7.4 \times 10^{-5}$ m. The recombination lifetime is $9.30 \times 10^{-7}$ s on both sides and the cross section of the sample is $6.4 \times 10^{-4}$ m$^2$.

20. The circuit of Fig. 14-15 is used to operate a diode as a solar cell. (a) Show that the power output is a maximum if the potential difference $V_b$ across the cell obeys

$$\left[ 1 + \frac{eV_b}{k_B T} \right] e^{\beta e V_b} = 1 + \frac{I_{sc}}{I_R},$$

where $I_{sc}$ is the magnitude of the short-circuit current and $I_R$ is the magnitude of the reverse saturation current. (b) Suppose the diode and light described in Problem 19 are used and estimate the potential difference when the output power is a maximum. Use a systematic trial and error method or a computer programmed to find roots. (c) What is the current through the junction when output power is a maximum? (d) What should the total resistance of the circuit be to achieve maximum power? (e) What is the maximum power attainable?

# Appendix A

## EQUATIONS OF ELECTROMAGNETISM

| *In SI Units* | *In Gaussian Units* |
|---|---|
| $\nabla \cdot \mathcal{E} = \dfrac{\rho}{\epsilon_0}$ | $\nabla \cdot \mathcal{E} = 4\pi\rho$ |
| $\nabla \cdot \mathbf{B} = 0$ | $\nabla \cdot \mathbf{B} = 0$ |
| $\nabla \times \mathcal{E} = -\dfrac{\partial \mathbf{B}}{\partial t}$ | $\nabla \times \mathcal{E} = -\dfrac{1}{c}\dfrac{\partial \mathbf{B}}{\partial t}$ |
| $\nabla \times \mathbf{B} = \mu_0 \mathbf{J} + \mu_0\epsilon_0 \dfrac{\partial \mathcal{E}}{\partial t}$ | $\nabla \times \mathbf{B} = \dfrac{4\pi}{c} \mathbf{J} + \dfrac{1}{c}\dfrac{\partial \mathcal{E}}{\partial t}$ |
| $\oint \mathcal{E} \cdot d\mathbf{S} = \dfrac{q}{\epsilon_0}$ | $\oint \mathcal{E} \cdot d\mathbf{S} = 4\pi q$ |
| $\oint \mathbf{B} \cdot d\mathbf{S} = 0$ | $\oint \mathbf{B} \cdot d\mathbf{S} = 0$ |
| $\oint \mathcal{E} \cdot d\ell = -\dfrac{d\Phi_B}{dt}$ | $\oint \mathcal{E} \cdot d\ell = -\dfrac{1}{c}\dfrac{d\Phi_B}{dt}$ |
| $\oint \mathbf{B} \cdot d\ell = \mu_0 I + \mu_0\epsilon_0 \dfrac{d\Phi_E}{dt}$ | $\oint \mathbf{B} \cdot d\ell = \dfrac{4\pi}{c} I + \dfrac{1}{c}\dfrac{d\Phi_E}{dt}$ |
| $\mathbf{D} = \epsilon_0\mathcal{E} + \mathbf{P}$ | $\mathbf{D} = \mathcal{E} + 4\pi\mathbf{P}$ |
| $\mathbf{H} = \dfrac{1}{\mu_0} \mathbf{B} - \mathbf{M}$ | $\mathbf{H} = \mathbf{B} - 4\pi\mathbf{M}$ |
| $\mathbf{F} = q\mathcal{E} + q\mathbf{v} \times \mathbf{B}$ | $\mathbf{F} = q\mathcal{E} + \dfrac{q}{c}\mathbf{v} \times \mathbf{B}$ |

$\Phi_\mathcal{E} = \int \mathcal{E} \cdot d\mathbf{S}$ and $\Phi_B = \int \mathbf{B} \cdot d\mathbf{S}$ in both systems of units.

# Appendix B

## FOURIER SERIES

A Fourier series is a series of sine and cosine functions, or else exponential functions, used to represent a function of interest. For any of the functions we consider in our applications, the appropriate series converges to the function. Such series are used extensively in solid state physics to represent electron wave functions and atomic displacement. In this appendix, we give some general properties.

Let $f(x)$ be a periodic function of one variable with period $a$: $f(x) = f(x + na)$ for all values of $x$ and for all integers $n$. If $f(x)$ is single valued, if it has a finite number of maxima and minima in any range of width $a$, if it is continuous or has a finite number of finite discontinuities in any range of width $a$, and if the integral $\int |f(x)|^2 \, dx$ over an interval of width $a$ is finite, then $f(x)$ can be represented by a convergent Fourier series:

$$f(x) = \tfrac{1}{2}A_0 + \sum_{\ell=0}^{\infty} A_\ell \cos\left(\frac{2\pi \ell x}{a}\right) + \sum_{\ell=0}^{\infty} B_\ell \sin\left(\frac{2\pi \ell x}{a}\right), \qquad \text{(B-1)}$$

where the sums are over all positive integers. The Fourier coefficients are given by

$$A_\ell = \frac{2}{a} \int_{-a/2}^{+a/2} f(x) \cos\left(\frac{2\pi \ell x}{a}\right) dx \qquad \text{(B-2)}$$

and

$$B_\ell = \frac{2}{a} \int_{-a/2}^{+a/2} f(x) \sin\left(\frac{2\pi \ell x}{a}\right) dx. \qquad \text{(B-3)}$$

If the function is known, Eqs. B-2 and B-3 can be used to find the coefficients. In most practical applications the function is not known, but we nevertheless assume it can be expanded as a Fourier series and, for example, substitute the series into an appropriate differential equation. The differential equation is then used to determine the Fourier coefficients. In some cases, only a few terms in the series are needed to reproduce the function with sufficient accuracy. In many solid state applications, however, several thousand terms might be required and a high-speed computer is used to evaluate the sums.

If $f(x)$ is continuous at some point $x_0$, then the series converges to $f(x_0)$. If $f(x)$ is discontinuous at $x_0$, then the limiting value of $f(x)$ as $x$ approaches $x_0$ from above is different from the limiting value as it approaches from below. For $x = x_0$, the series converges to the average of the two limiting values.

Notice that the series consists of sinusoidal functions with periods that are submultiples of the period of $f(x)$. No sinusoidal function with a period greater than that of $f(x)$ enters the sum. Short period sinusoidal functions are needed to reproduce details of $f(x)$, all of which must, of necessity, occur in ranges smaller than $a$. Qualitatively, the more rapidly $f(x)$ varies, the larger the coefficients associated with high $\hbar$ terms.

Since $\cos(2\pi nx/a) = \frac{1}{2}[e^{i2\pi\hbar x/a} + e^{-i2\pi\hbar x/a}]$ and $\sin(2\pi\hbar x/a) = \frac{1}{2}i[e^{i2\pi\hbar x/a} - e^{-i2\pi\hbar x/a}]$, a Fourier series can also be written in terms of exponential functions:

$$f(x) = \sum_{\hbar=-\infty}^{+\infty} C_\hbar e^{i2\pi\hbar x/a}, \tag{B-4}$$

where

$$C_\hbar = \frac{1}{a} \int_{-a/2}^{+a/2} f(x)e^{-i2\pi\hbar x/a}\, dx. \tag{B-5}$$

Both positive and negative exponents must be included, so $\hbar$ takes on both positive and negative integer values. A series of exponentials is often more convenient than a series of sine and cosine functions.

The expression for $C_\hbar$ clearly follows from Eq. B-4. We first note that

$$\int_{-a/2}^{+a/2} e^{i2\pi(\hbar-\hbar')/a}\, dx = a\delta_{\hbar\hbar'}, \tag{B-6}$$

where $\delta_{\hbar\hbar'}$ is the Kronecker delta, a quantity that is 1 if $\hbar = \hbar'$ and 0 if $\hbar \neq \hbar'$. Multiply Eq. B-4 by $e^{-i2\pi\hbar'x/a}$ and integrate the result from $-a/2$ to $+a/2$ to obtain

$$\int_{-a/2}^{+a/2} f(x)e^{-i2\pi\hbar'x/a}\, dx = \sum_\hbar c_\hbar a\delta_{\hbar\hbar'}. \tag{B-7}$$

Every term in the sum vanishes except the $\hbar = \hbar'$ term, so the sum is equal to $aC_{\hbar'}$. Equation B-5 follows immediately when $\hbar'$ is replaced by $\hbar$.

The statements made above are easily generalized to three dimensions. Suppose $f(\mathbf{r})$ is periodic and its periodicity is described by a lattice with primitive vectors $\mathbf{a}$, $\mathbf{b}$, and $\mathbf{c}$. That is, $f(\mathbf{r} + n_1\mathbf{a} + n_2\mathbf{b} + n_3\mathbf{c}) = f(\mathbf{r})$ for any $\mathbf{r}$ and any three integers $n_1$, $n_2$, and $n_3$.

Write $\mathbf{r} = u\mathbf{a} + v\mathbf{b} + w\mathbf{c}$ and consider $u$, $v$, and $w$ to be the independent variables instead of the components of $\mathbf{r}$. The function $f(\mathbf{r})$ is periodic in each of the variables and, for each, its period is 1. First consider $f(\mathbf{r})$ to be a function of $u$ and expand it as a one-dimensional Fourier series. The coefficients depend on $v$ and $w$. Each coefficient may now be expanded as a series in $v$, with coefficients that depend on $w$. Finally each of these coefficients may be written as a series in $w$. The end result is

$$f(\mathbf{r}) = \sum_{\hbar=-\infty}^{+\infty} \sum_{\ell=-\infty}^{+\infty} \sum_{\ell=-\infty}^{+\infty} C_{\hbar\ell\ell}e^{i2\pi(\hbar u + \ell v + \ell w)}, \tag{B-8}$$

where

$$C_{hkl} = \int_{-1/2}^{+1/2} \int_{-1/2}^{+1/2} \int_{-1/2}^{+1/2} f(\mathbf{r})e^{-i2\pi(hu+kv+lw)} \, du \, dv \, dw \,. \qquad \text{(B-9)}$$

Now $2\pi(hu + kv + lw) = \mathbf{G} \cdot \mathbf{r}$, where $\mathbf{G}$ is the reciprocal lattice vector $h\mathbf{A} + k\mathbf{B} + l\mathbf{C}$. The volume of a unit cell is $\tau = |\mathbf{a} \cdot \mathbf{b} \times \mathbf{c}|$ and an infinitesimal volume element is $d\tau = |\mathbf{a} \cdot \mathbf{b} \times \mathbf{c}| \, du \, dv \, dw = \tau \, du \, dv \, dw$. Equations B-8 and B-9 can therefore be written

$$f(\mathbf{r}) = \sum_{\mathbf{G}} C(\mathbf{G})e^{i\mathbf{G} \cdot \mathbf{r}} \qquad \text{(B-10)}$$

and

$$C(\mathbf{G}) = \frac{1}{\tau} \int_{\substack{\text{unit} \\ \text{cell}}} f(\mathbf{r})e^{-i\mathbf{G} \cdot \mathbf{r}} \, d\tau \,, \qquad \text{(B-11)}$$

respectively. We have written $C_{hkl}$ as $C(\mathbf{G})$. The sum in Eq. B-10 is over all reciprocal lattice vectors.

# *Appendix C*

---

# EVALUATION OF
# MADELUNG CONSTANTS

The Madelung constant $\alpha$ for an ionic crystal with two atoms in its primitive unit cell is given by

$$\alpha = R_0 \left[ \sum_{\mathbf{R}} \frac{1}{|\mathbf{R} + \mathbf{P}|} - \sum_{\mathbf{R}}' \frac{1}{|\mathbf{R}|} \right], \tag{C-1}$$

where $\mathbf{R}$ is a lattice vector, $\mathbf{p}$ is a basis vector, and $R_0$ is the nearest-neighbor distance. The prime on the second summation symbol indicates that the $\mathbf{R} = 0$ term omitted. Each sum in Eq. C-1 has the form

$$S = \sum_{\mathbf{R}} \frac{1}{|\mathbf{R} + \mathbf{r}|}, \tag{C-2}$$

where $\mathbf{R}$ runs over all lattice vectors. Here $\mathbf{r}$ is either 0 or $\mathbf{p}$. The sum diverges but the combination that appears in Eq. C-1 is finite. We will be able to isolate divergent terms and cancel them when the two sums are combined.

Make use of the mathematical identity*

$$\frac{1}{x} = \frac{2}{\sqrt{\pi}} \int_0^\infty e^{-x^2\rho^2} \, d\rho \tag{C-3}$$

and divide the range of integration into two parts, one from 0 to an arbitrary number $\eta$ and the other from $\eta$ to $\infty$. Then

$$S = \frac{2}{\sqrt{\pi}} \int_0^\eta \left[ \sum_{\mathbf{R}} e^{-|\mathbf{R}+\mathbf{r}|^2\rho^2} \right] d\rho + \frac{2}{\sqrt{\pi}} \int_\eta^\infty \left[ \sum_{\mathbf{R}} e^{-|\mathbf{R}+\mathbf{r}|^2\rho^2} \right] d\rho. \tag{C-4}$$

Each integrand consists of a series of Gaussian functions. In the first, $\rho$ is less than $\eta$ and each of the functions in the sum has a long range. We will represent this integral by its Fourier series and expect only long wavelength terms to be significant. If $\eta$ is sufficiently small, the Fourier series converges rapidly. In the second integrand, $\rho$ is greater than $\eta$ and the Gaussian functions have short ranges. If $\eta$ is large the sum over $\mathbf{R}$ converges rapidly. We hope to select $\eta$ so $S$ can be computed with only a few terms in the Fourier series and a few in the sum over $\mathbf{R}$.

As a function of $\mathbf{r}$, the sum of the Gaussian functions in the first integral of Eq. C-4 is periodic with the periodicity of the lattice. It can be written as the

---

*See, for example, *Standard Mathematical Tables* (Boca Raton, FL: CRC Press).

Fourier series

$$\sum_{\mathbf{R}} e^{-|\mathbf{R}+\mathbf{r}|^2\rho^2} = \sum_{\mathbf{G}} F(\mathbf{G})e^{i\mathbf{G}\cdot\mathbf{r}}, \qquad (C-5)$$

where

$$F(\mathbf{G}) = \frac{1}{\tau_s} \sum_{\mathbf{R}} \int e^{-|\mathbf{R}+\mathbf{r}|^2\rho^2} e^{-i\mathbf{G}\cdot\mathbf{r}} \, d\tau. \qquad (C-6)$$

The integral is over the volume of $\tau_s$ of the sample. If we insert the factor $e^{-i\mathbf{G}\cdot\mathbf{R}}$ ($= 1$) into the integrand of Eq. C-6, then use the components of $\mathbf{r} + \mathbf{R}$ as variables of integration, we see that all terms in the sum have the same value. We evaluate the $\mathbf{R} = 0$ integral and multiply it by $N$, the number of unit cells in the sample. Place the $z$ axis along $\mathbf{r}$ and use spherical coordinates:

$$\int\int\int e^{-r^2\rho^2} e^{-iGr\cos\theta} r^2 \, dr \sin\theta \, d\theta \, d\phi$$

$$= 2\pi \int_0^\infty \left[ \frac{e^{-r^2\rho^2} e^{-iGr\cos\theta}}{-iGr} \right]_{\cos\theta=-1}^{+1} r^2 \, dr$$

$$= 4\pi \int_0^\infty \frac{e^{-r^2\rho^2} \sin(Gr)}{G} r \, dr = \pi\sqrt{\pi} \, \frac{e^{-G^2/4\rho^2}}{\rho^3}. \qquad (C-7)$$

Thus

$$F(\mathbf{G}) = \frac{N}{\tau_s} \pi\sqrt{\pi} \, \frac{e^{-G^2/4\rho^2}}{\rho^3} \qquad (C-8)$$

and the first term of Eq. C-4 becomes

$$S_1 = \frac{2\pi N}{\tau_s} \sum_{\mathbf{G}} \int_0^\eta \frac{e^{-G^2/4\rho^2}}{\rho^3} e^{i\mathbf{G}\cdot\mathbf{r}} \, d\rho. \qquad (C-9)$$

For $\mathbf{G} \neq 0$, the integration over $\rho$ can be carried out easily. If $u = G/2\rho$ is the variable of integration, then

$$\int_0^\eta \frac{e^{-G^2/4\rho^2}}{\rho^3} \, d\rho = -\frac{4}{G^2} \int_\infty^{G/2\eta} u e^{-u^2} \, du = \frac{2}{G^2} e^{-G^2/4\eta^2}. \qquad (C-10)$$

So

$$S_1 = \frac{4\pi N}{\tau_s} \sum_{\mathbf{G}}' \frac{e^{-G^2/4\eta^2}}{G^2} e^{i\mathbf{G}\cdot\mathbf{r}} + \frac{2\pi N}{\tau_s} \int_0^\eta \frac{d\rho}{\rho^3}. \qquad (C-11)$$

The prime on the summation symbol indicates that the $\mathbf{G} = 0$ term is omitted; it is written separately as the last term. The integral diverges but it has exactly the same form for $\mathbf{r} = \mathbf{p}$ as for $\mathbf{r} = 0$ and the two divergent terms cancel when we evaluate Eq. C-1.

The integral that appears in the second term of Eq. C-4 can be written

$$\frac{2}{\sqrt{\pi}} \int_{\eta}^{\infty} e^{-|R+r|^2\rho^2} \, d\rho = \frac{\text{erfc}(\eta|R+r|)}{|R+r|}, \tag{C-12}$$

where $\text{erfc}(u)$ is the complimentary error function, tabulated in many handbooks.*

We are now ready to sum the two contributions that appear in Eq. C-4. For $r = 0$ we must remember to omit the $R = 0$ term. This is easily done in summing over the complimentary error functions. We must, however, also subtract the $R = 0$ contribution to $S_1$ since all values of $R$ were used in deriving Eq. C-11. This contribution is

$$\frac{2}{\sqrt{\pi}} \int_{0}^{\eta} d\rho = \frac{2\eta}{\sqrt{\pi}}. \tag{C-13}$$

The result for the Madelung constant $\alpha$ is

$$\alpha = R_0 \left[ \frac{\pi N}{\tau_s \eta^2} \sum_{G}{}' \frac{e^{-G^2/4\eta^2}}{G^2/4\eta^2} (e^{i G \cdot P} - 1) + \eta \sum_{R} \frac{\text{erfc}(\eta|R+p|)}{\eta|R+p|} \right.$$
$$\left. - \eta \sum_{R}{}' \frac{\text{erfc}(\eta R)}{\eta R} + \frac{2\eta}{\sqrt{\pi}} \right]. \tag{C-14}$$

This expression looks complicated but it is not. In the sum over $G$ we arrange the terms in order of increasing values of $|G|$. If $\eta$ is small only the first terms, corresponding to the shortest reciprocal lattice vectors, contribute significantly. Likewise, we arrange terms in the other sums in order of increasing values of $|R|$ or $|R + p|$, as appropriate. If $\eta$ is large only the first terms are important.

We hope to pick $\eta$ so only a few terms need be considered in any of the sums. An appropriate value is roughly the reciprocal of a primitive unit cell dimension. The last term of Eq. C-14 then dominates.

As an example, consider the CsCl structure. It is simple cubic so lattice vectors are given by $R = n_1 a \hat{x} + n_2 a \hat{y} + n_3 a \hat{z}$ (where a is the cube edge) and reciprocal lattice vectors are given by $G = (2\pi k/a)\hat{x} + (2\pi k/a)\hat{y} + (2\pi \ell/a)\hat{z}$. The basis vector $p$ is $\frac{1}{2}a(\hat{x} + \hat{y} + \hat{z})$.

Take $\eta = 2/a$ and use $\tau_s = Na^3$. Look first at the sum over reciprocal lattice vectors. Aside from $G = 0$, the shortest vectors are $\pm(2\pi/a)\hat{x}$, $\pm(2\pi/a)\hat{y}$, and $\pm(2\pi/a)\hat{z}$. For each of them, $e^{-G^2/4\eta^2}/(G^2/4\eta^2) = 3.4370 \times 10^{-2}$ and $e^{i G \cdot P} - 1 = -2$. The total contribution of the six terms is $6 \times (\pi/4) \times (R_0/a) \times 3.4370 \times 10^{-2} \times (-2) = -0.32393 R_0/a$.

The next longest reciprocal lattice vectors are $\pm(2\pi/a)\hat{x} \pm (2\pi/a)\hat{y}$, $\pm(2\pi/a)\hat{x} \pm (2\pi/a)\hat{z}$, and $\pm(2\pi/a)\hat{y} \pm (2\pi/a)\hat{z}$. For each of them, $e^{i G \cdot P} - 1 = 0$ so they make no contribution to $\alpha$. The next longest are $\pm(2\pi/a)\hat{x} \pm (2\pi/a)\hat{y} \pm (2\pi/a)\hat{z}$. For each of them, $e^{-G^2/4\eta^2}/(G^2/4\eta^2) = 8.2397 \times 10^{-5}$ and $e^{i G \cdot P} - 1 = -2$. Their

---

*See, for example, *Standard Mathematical Tables* (Boca Raton, FL: CRC Press).

total contribution is $8 \times (\pi/4) \times (R_0/a) \times 8.2397 \times 10^{-5} \times (-2) = -1.0354 \times 10^{-3}R_0/a$. The next set contributes 0 and the next contributes on the order of $10^{-5}R_0/a$. We neglect these and succeeding terms.

Next consider the sum of terms containing $|\mathbf{R} + \mathbf{p}|$. Eight terms have $|\mathbf{R} + \mathbf{p}| = \sqrt{0.75}a$. For each of them, $\mathrm{erfc}(|\mathbf{R} + \mathbf{p}|\eta)/|\mathbf{R} + \mathbf{p}|\eta = 8.2593 \times 10^{-3}$ so their total contribution is $8 \times 8.2593 \times 10^{-3} \times 2 \times R_0/a = 0.13215R_0/a$. Other terms contribute on the order of $10^{-5}R_0/a$ or less and we neglect them.

Six lattice vectors have length $a$ and, for each of them, $\mathrm{erfc}(R\eta)/R\eta = 2.3387 \times 10^{-3}$. Together they contribute $2.8064 \times 10^{-2}R_0/a$ to $\alpha$. Twelve vectors have length $\sqrt{2}a$. For each of them, $\mathrm{erfc}(R\eta)/R\eta = 2.2260 \times 10^{-5}$ and their total contribution is $5.3424 \times 10^{-4}R_0/a$.

Finally $2R_0\eta/\sqrt{\pi} = 2.2568R_0/a$. When all the contributions are summed, the result is $\alpha = 2.0353R_0/a$. Now $R_0 = (\sqrt{3}/2)a$ so $\alpha = 1.7626$, correct to 5 significant figures. Greater precision can be obtained if more terms are retained in the series.

# *Appendix D*

## INTEGRALS CONTAINING THE FERMI-DIRAC DISTRIBUTION FUNCTION

We wish to evaluate integrals of the form

$$I = \int_{E_0}^{\infty} h'(E)f(E)\, dE, \tag{D-1}$$

where $h'(E)$ is the derivative with respect to $E$ of a function $h(E)$ that vanishes at the lower limit $E_0$ and $f(E)$ is the Fermi-Dirac distribution function, given by

$$f(E) = \frac{1}{e^{\beta(E-\eta)} + 1}. \tag{D-2}$$

Here $\beta = 1/k_B T$. We assume $E_0$ is well below the chemical potential $\eta$, so $f(E_0) = 1$.

Integration of Eq. D-1 by parts gives

$$I = h(E)f(E) \Big|_{E_0}^{\infty} - \int_{E_0}^{\infty} h(E)f'(E)\, dE = -\int_{E_0}^{\infty} h(E)f'(E)\, dE, \tag{D-3}$$

since $h(E_0) = 0$ and $f(\infty) = 0$. The Fermi-Dirac function is nearly constant everywhere except in the vicinity of $E = \eta$, so $f'(E)$ is nearly zero except near $E = \eta$. We expand $h(E)$ in a Taylor series:

$$h(E) = h(\eta) + (E - \eta)h'(\eta) + \tfrac{1}{2}(E - \eta)^2 h''(\eta) + \cdots, \tag{D-4}$$

where primes indicate derivatives with respect to $E$. Then

$$I = -h(\eta) \int_{E_0}^{\infty} f'(E)\, dE - h'(\eta) \int_{E_0}^{\infty} (E - \eta)f'(E)\, dE$$

$$- \tfrac{1}{2}h''(\eta) \int_{E_0}^{\infty} (E - \eta)^2 f'(E)\, dE + \cdots. \tag{D-5}$$

Since $f(E_0) = 1$ and $f(\infty) = 0$, the first term yields $+h(\eta)$. We may replace the lower limit by $-\infty$ in the other terms. The second term is 0 because the integrand is an odd function of $E - \eta$. The integral in the third term can be evaluated by letting $\beta(E - \eta) = x$. Then $(E - \eta)^2 = x^2/\beta^2$ and $dE = dx/\beta$. The derivative of $f$ is given by

$$f'(E) = \frac{d}{dE} \frac{1}{e^{\beta(E-\eta)} + 1} = -\frac{\beta e^{\beta(E-\eta)}}{[e^{\beta(E-\eta)} + 1]^2}$$

$$= -\frac{\beta e^x}{[e^x + 1]^2} = -\frac{\beta}{e^x + 2 + e^{-x}} = -\frac{\beta}{2} \frac{1}{1 + \cosh x}. \tag{D-6}$$

The integrand of the third term is $-(1/2\beta^2)x^2/(1 + \cosh x)$ and the integral has the value $-\pi^2/3\beta^2$, so

$$I = h(\eta) + \frac{\pi^2}{6} (k_B T)^2 h''(\eta) + \cdots. \tag{D-7}$$

Equation D-7 gives $I$ to lowest order in $T^2$. The next term is proportional to $(k_B T)^4$ and to the fourth derivative of $h$. In most circimstances we can neglect it.

# Appendix E

## ELECTRON TRANSITIONS IN A UNIFORM ELECTRIC FIELD

We start with the time-dependent Schrödinger equation for an electron in a uniform electric field and show its crystal momentum changes according to

$$\frac{d\hbar\mathbf{k}}{dt} = -e\mathcal{E}. \tag{E-1}$$

For simplicity, consider a one-dimensional crystal along the $x$ axis and assume the field is in the positive $x$ direction. The potential energy of the electron-field interaction is then $e\mathcal{E}x$, so the Schrödinger equation is

$$-\frac{\hbar^2}{2m}\frac{\partial^2\Psi}{\partial x^2} + U\Psi + e\mathcal{E}x\Psi = i\hbar\frac{\partial\Psi}{\partial t}, \tag{E-2}$$

where $U(x)$ is the potential energy function in the absence of the field.

Write $\Psi$ as a linear combination of wave functions associated with electron states in the absence of the field:

$$\Psi(x, t) = \sum_{k'} A(k', t)\psi(k', x), \tag{E-3}$$

where the sum is over all propagation constants in the Brillouin zone. We assume the only functions that enter are those associated with the original band of the electron. The band index is suppressed. The coefficients $A(k', t)$ are the quantities of interest here. $|A(k', t)|^2$ gives the probability that the state with wave constant $k'$ is occupied at time $t$.

In the course of the derivation, we will need the result

$$\int_{\text{sample}} \psi^*(k, x)\psi(k', x)\, dx = \delta_{kk'}. \tag{E-4}$$

This is a special case of a general property of wave functions that satisfy the same time-independent Schrödinger equation.* If $k' \neq k$ the integral vanishes. If $k' = k$, the integral is 1, indicating that the wave function is normalized.

We will also need the following theorem: if $k$ and $k'$ are two propagation constants in the Brillouin zone and $f(k', x)$ is any function with the translational symmetry of the direct lattice, then

$$\int_{\text{sample}} f(k', x)e^{i(k'-k)x}\, dx = N\delta_{kk'}\int_{\text{cell}} f(k, x)\, dx, \tag{E-5}$$

where $N$ is the number of unit cells in the sample. To prove Eq. E-5, the integral

---

*Details can be found in most quantum mechanics texts. See, for example, D. Park, *Introduction to the Quantum Theory* (New York: McGraw-Hill, 1964); or L. I. Schiff, *Quantum Mechanics* (New York: McGraw-Hill, 1968).

on the left is evaluated one cell at a time. Since $f(k', x)$ is periodic, the result is

$$\int_{\text{sample}} f(k', x)e^{i(k'-k)x}\,dx = \left[\sum_n e^{i(k'-k)na}\right]\int_{\text{cell}} f(k', x)e^{i(k'-k)x}\,dx, \quad \text{(E-6)}$$

where $a$ is the cell width. Since $k - k'$ has the form $2\pi \hslash/Na$, where $\hslash$ is an integer, the sum vanishes unless $\hslash$ is a multiple of $N$. Thus $k' - k$ is a reciprocal lattice vector, and, since both $k$ and $k'$ are in the Brillouin zone, the only possibility is $k' - k = 0$. Thus the sum yields $N$ and Eq. E-5 follows immediately.

To find the time dependence of $A(k, t)$ substitute Eq. E-3 into Eq. E-2 and use the Schrödinger equation obeyed by $\psi(k', x)$ to replace $-(\hslash^2/2m)(\partial^2\psi(k',x)/\partial x^2) + U(x)\psi(k', x)$ with $E(k')\psi(k', x)$. The result is

$$\sum_{k'} E(k')A(k', t)\psi(k', x) + \sum_{k'} e\mathcal{E}xA(k', t)\psi(k', x)$$

$$= \sum_{k'} i\hslash \frac{\partial A(k', t)}{\partial t}\psi(k', x). \quad \text{(E-7)}$$

Multiply by $\psi^*(k, x)$, then integrate over the sample to obtain

$$A(k, t)E(k) + e\mathcal{E}\sum_{k'} A(k', t)\int \psi^*(k, x)x\psi(k', x)\,dx = i\hslash \frac{\partial A(k, t)}{\partial t}. \quad \text{(E-8)}$$

Equation E-4 was used to evaluate some of the integrals.

To evaluate the sum in Eq. E-8, replace $\psi(k', x)$ by $e^{ik'x}u(k', x)$ and $\psi^*(k, x)$ by $e^{-ikx}u^*(k, x)$, then observe that $xe^{-ikx} = i\,\partial e^{-ikx}/\partial k$. In detail:

$$\sum_{k'} A(k', t)\int \psi^*(k, x)x\psi(k', x)\,dx$$

$$= i\sum_{k'} A(k', t)\int \left[\frac{\partial}{\partial k}e^{-ikx}\right]u^*(k, x)e^{ik'x}u(k', x)\,dx$$

$$= i\frac{\partial}{\partial k}\left[\sum_{k'} A(k', t)\int u^*(k, x)u(k', x)e^{i(k'-k)x}\,dx\right]$$

$$-i\sum_{k'} A(k', t)\int \frac{\partial u^*(k, x)}{\partial k}u(k', x)e^{i(k'-k)x}\,dx, \quad \text{(E-9)}$$

where the product rule for differentiation was used to write the second equality. The two integrals can be evaluated using Eqs. E-4 and E-5. The first is just $\int\psi^*(k, x)\psi(k', x)\,dx = \delta_{kk'}$. The second is $N\delta_{kk'}\int[\partial u^*(k, x)/\partial k]u(k', x)\,dx$, where the integral is over a unit cell. Equation E-8 becomes

$$\frac{\partial A(k, t)}{\partial t} = -\frac{i}{\hslash}A(k, t)E(k) + \frac{e\mathcal{E}}{\hslash}\frac{\partial A(k, t)}{\partial k}$$

$$- N\frac{e\mathcal{E}}{\hslash}A(k, t)\int_{\text{cell}} \frac{\partial u(k, x)}{\partial k}u(k, x)\,dx. \quad \text{(E-10)}$$

To obtain $\partial |A|^2/\partial t = (\partial A^*/\partial t)A + A^*(\partial A/\partial t)$, multiply Eq. E-10 by $A^*$, then multiply the complex conjugate of Eq. E-10 by $A$ and sum the two results. The terms containing $E(k)$ cancel as do the terms containing integrals. This can be seen by writing

$$\int_{cell} \frac{\partial u^*}{\partial k} u \, dx + \int_{cell} u^* \frac{\partial u}{\partial k} \, dx = \frac{\partial}{\partial k} \int_{cell} |u|^2 \, dx \qquad \text{(E-11)}$$

and noting that the integral on the right side is $1/N$ for every value of $k$. Thus

$$\frac{\partial}{\partial t} |A(k, t)|^2 = +\frac{e\mathscr{E}}{\hbar} \frac{\partial}{\partial k} |A(k, t)|^2 . \qquad \text{(E-12)}$$

Equation E-12 has the solution $|A|^2 = f(k + e\mathscr{E}t/\hbar)$, where $f$ is an arbitrary function. If the electron is initially in a state with propagation constant $k_0$, then $f(k)$ is a function that is 1 for $k = k_0$ and 0 for all other values of $k$. At time $t$, $|A|^2 = 0$ for all values of $k$ except the one for which $k + e\mathscr{E}t/\hbar = k_0$. That is, the electron is in the state with propagation constant $k = k_0 - e\mathscr{E}t/\hbar$, in agreement with Eq. E-1.

## PHOTO CREDITS

# INDEX

# Vector Relationships

**A, B, C**, and **D** are vectors, $u$ and $v$ are scalars. Subscripts $x$, $y$, and $z$ denote Cartesian components. $\hat{\mathbf{x}}$, $\hat{\mathbf{y}}$, and $\hat{\mathbf{z}}$ are Cartesian unit vectors.

## CARTESIAN FORMS

$$\mathbf{A} = A_x\hat{\mathbf{x}} + A_y\hat{\mathbf{y}} + A_z\hat{\mathbf{z}}$$

$$u\mathbf{A} = uA_x\hat{\mathbf{x}} + uA_y\hat{\mathbf{y}} + uA_z\hat{\mathbf{z}}$$

$$\frac{\partial \mathbf{A}}{\partial x} = \frac{\partial A_x}{\partial x}\hat{\mathbf{x}} + \frac{\partial A_y}{\partial y}\hat{\mathbf{y}} + \frac{\partial A_z}{\partial z}\hat{\mathbf{z}}$$

$$\mathbf{A} \cdot \mathbf{B} = A_xB_x + A_yB_y + A_zB_z$$

$$\mathbf{A} \times \mathbf{B} = (A_yB_z - A_zB_y)\hat{\mathbf{x}} + (A_zB_x - A_xB_z)\hat{\mathbf{y}} + (A_xB_y - A_yB_x)\hat{\mathbf{z}}$$

$$\nabla u = \frac{\partial u}{\partial x}\hat{\mathbf{x}} + \frac{\partial u}{\partial y}\hat{\mathbf{y}} + \frac{\partial u}{\partial z}\hat{\mathbf{z}}$$

$$\nabla \cdot \mathbf{A} = \frac{\partial A_x}{\partial x} + \frac{\partial A_y}{\partial y} + \frac{\partial A_z}{\partial z}$$

$$\nabla \times \mathbf{A} = \left[\frac{\partial A_z}{\partial y} - \frac{\partial A_y}{\partial z}\right]\hat{\mathbf{x}} + \left[\frac{\partial A_x}{\partial z} - \frac{\partial A_z}{\partial x}\right]\hat{\mathbf{y}} + \left[\frac{\partial A_y}{\partial x} - \frac{\partial A_x}{\partial y}\right]\hat{\mathbf{z}}$$

$$\nabla^2 u = \frac{\partial^2 u}{\partial x^2} + \frac{\partial^2 u}{\partial y^2} + \frac{\partial^2 u}{\partial z^2}$$

$$\nabla^2 \mathbf{A} = \frac{\partial^2 \mathbf{A}}{\partial x^2} + \frac{\partial^2 \mathbf{A}}{\partial y^2} + \frac{\partial^2 \mathbf{A}}{\partial z^2}$$

$$(\mathbf{A} \cdot \nabla)\mathbf{B} = A_x\frac{\partial \mathbf{B}}{\partial x} + A_y\frac{\partial \mathbf{B}}{\partial y} + A_z\frac{\partial \mathbf{B}}{\partial z}$$

## Date Due

| | | | |
|---|---|---|---|
| | | | |
| | | | |
| | | | |
| | | | |
| | | | |
| | | | |
| | | | |